Aviation Fuel Problems and Solutions: From Refinery to Wingtip

Moshe Rabaev · Joanna M Bauldreay ·
Frederick Passman · Raj Shah · Liel Mazar
Editors

Aviation Fuel Problems and Solutions: From Refinery to Wingtip

Springer

Editors
Moshe Rabaev
Moshe Rabaev Consulting
Be'er Sheva, Israel

Ben-Gurion University of the Negev
Be'er Sheva, Israel

Frederick Passman
Biodeterioration Control Associates, Inc.
Princeton, NJ, USA

Liel Mazar
Independent Consultant
Jerusalem, Israel

Joanna M Bauldreay
Management
Bauldreay Jet Fuel Consulting Limited
Chester, UK

Raj Shah
Koehler Instrument Company
Holtsville, NJ, USA

ISBN 978-3-032-05513-2 ISBN 978-3-032-05514-9 (eBook)
https://doi.org/10.1007/978-3-032-05514-9

This Springer imprint is published by the registered company Springer Nature Switzerland AG
The registered company address is: Gewerbestrasse 11, 6330 Cham, Switzerland

To my beloved wife Irina, whose unwavering support and love have been my anchor; to my daughters Bat Sheva, Neomi, and Eden, who fill my life with joy, inspiration, and purpose; and to my parents and parents-in-law, whose love and wisdom have laid the foundation for everything I have become

—Moshe Rabaev

For Henry and Eleanor, who I hope will grow up to be scientifically savvy

—Joanna M Bauldreay

To my wife, partner, and lodestar Willa, who keeps me on course throughout life's journey

—Frederick Passman

To Kian, who always inspires and taught me patience and gratitude

—Raj Shah

To my parents, whose love, support, and encouragement provided the foundation for this achievement, and with special thanks to

Moshe Rabaev and Yalfal Siyum, for their invaluable mentorship in the field of aviation fuels

—Liel Mazar

The editors also express their sincere appreciation to all the contributing authors and staff for the dedication and patience in providing their vital assistance in making this extensive work become a reality.

Foreword

Aviation represents a unique and important part of the global transportation system, carrying 4.5 billion passengers in 2023 to link the countries of the world across oceans and continents. The industry is built upon scientific understanding, with attention to detail for aircraft and engine design to deliver high levels of reliability, efficiency, and performance. Integral to this engineering achievement is a third critical element, the fuel, a single component on the aircraft which must be manufactured and distributed across a world-wide infrastructure to arrive fit for use. Since the first aviation fuel specification of 1917 the industry has sought to apply its combined knowledge to achieve this goal. Throughout this time fuel specifications have been revised and updated, distribution systems improved, and product quality monitored in an on-going quest to provide fit-for-flight product reliably in every region of the globe. The path has not always been easy, with technical challenges and complications along the way, some unique, some more fundamental.

In this book the reader is invited to enter the world of aviation fuel manufacture and supply, to learn about specifications, production, distribution, and coming environmental challenges. More importantly, the text takes the reader beyond standard day-to-day operations into the more complex field of fuel quality and operational problems, illustrating how these have been identified and resolved through the application of scientific and engineering knowledge. For those new to aviation, the chapters offer an excellent insight into aviation fuel. For those already members of the aviation community, the text offers an invaluable reference with numerous examples of problems which can impact fuel supply and quality together with successful solutions. Such knowledge represents decades of combined expertise and experience, the authors of each chapter bringing their own style and skill to generate an interesting and informative account and open the door to the fascinating world of fuels for flight.

January 2026

Alisdair Q. Clark
Puttenham, UK

Foreword

Preface

Aviation fuel properties and characteristics are fundamental in modern aviation's safety, efficiency, and sustainability. There are excellent existing comprehensive publications such as "*Aviation Fuels*" by Maxwell Smith, and "*Aviation Fuel Properties Handbook*" by the CRC, "*Aviation Fuels Technical Review*" by Chevron, ASTM materials, EI guides, and JIG publications. Many of these have traditionally focused on fuel specifications, quality, and manufacturing processes for fuels derived from crude oil. However, many of these classic references were published nearly a half-century ago. The rapid pace of technological and regulatory change over the past two decades calls for an updated perspective.

This book, *Aviation Fuel Problems and Solutions: From Refinery to Wingtip* aims to address this need by bringing together the collective knowledge of today's industry experts in an engaging and technically helpful format. Compiling such a comprehensive body of knowledge is a significant task, but through international collaboration, we have assembled a team of experienced professionals, each contributing their expertise. This collaborative approach ensures that the content is broad and deep, reflecting the latest developments and challenges in the field.

The book is organized into chapters authored by individuals or teams of experts, each focusing on specific aspects of aviation fuel. The chapters are arranged to carry the reader through an understanding of product and infrastructure to a deeper understanding of factors that affect aviation fuels as they move from refinery tank farms through the distribution system and into aircraft. Chapter 1 provides an overview of fuel specifications. Chapter 2 describes how conventional aviation fuels are produced at refineries. Chapter 3 surveys the current status of sustainable aviation fuels (SAFs), including production and acceptability validation. Chapter 4 reviews the properties and quality considerations associated with aviation gasoline. Chapter 5 provides an overview of aviation fuel additives. It discusses additive chemistries and their use in specific aviation fuels. Chapter 6 reviews the interactions between aviation fuel and the various polymeric and other non-metallic materials they come into contact with. Chapter 7 discusses storage and thermal stability, including the chemical reactions facilitating degradation. Chapter 8 provides an overview of supply chain risk management. There are financial risks associated with fuel distribution from refinery to aircraft. To a considerable extent, maintaining aviation fuel in fit-for-use conditions depends on appropriate storage, handling and filtration at each stage of the distribution system. Chapter 9 then describes the non-filtration fuel handling equipment comprising fuel distribution systems. Assessing fuel conditions at critical distribution stages depends on sample collection and testing. Chapter 10 describes the types of filters used to remove particulate contamination and water that can accumulate in fuel as it proceeds on its path from the refinery to the wingtip. Chapter 11 examines the critical considerations related to sample collection. Importantly, it relates sample collection options to the intended use of the sample fuel. For example, samples used for specification testing will differ from those used to assess the presence of contaminants. Chapter 12 provides an overview of aviation fuel and fuel system microbiology. Microbiological contamination can be difficult to detect, but if it is not detected and controlled adequately, it can cause damage to fuel system components and degrade fuel quality. Chapter 13 provides an overview of the factors contributing to fuel corrosivity, the chemical processes associated with fuel system corrosion, and practices that can mitigate corrosion risks. Chapter 14 reviews laboratory research methods for testing contaminants in aviation fuels. Fuel specification test methods typically provide high-level information that lacks sufficient granularity to facilitate root cause analysis (RCA) when fuel quality issues arise. Finally, Chap. 15 discusses aviation fuel system failures that could contribute to accidents and safety incidents.

This book aims to address a knowledge gap for those who provide solutions and manage crises in the aviation fuel industry. It serves as a new resource that builds on and expands previous texts by including contemporary examples of fuel handling, quality control, problem-solving, remediation methods, and operational insights. While some authors have included background information that may be found elsewhere, these details are retained to explain how and why problems may arise. They emphasize that if processes are executed effectively, problems are less likely to occur—"prevention is better than cure." The "well-to-wing" approach ensures a comprehensive understanding of the entire fuel supply chain and its impact on aviation operations.

This first edition of *Aviation Fuel Problems and Solutions* is not intended to replace existing handbooks but to be an additional resource for the aviation fuel specialist community. By offering shared knowledge and insights from a global perspective, we hope to foster further research and innovation within the field. We refer readers to the *EI 1550: Handbook on Equipment Used for the Maintenance and Delivery of Clean Aviation Fuel* for a more in-depth understanding of filtration challenges. Similarly, electrostatic issues are comprehensively covered in CRC, API, and other publications.

Our goal is to provide an informative and practical resource, helping fuel engineers, researchers, and aviation professionals navigate the complex landscape of aviation fuels. We hope this book will inspire further exploration and contribute to safer, more efficient, and more sustainable aviation practices.

Be'er Sheva, Israel — Moshe Rabaev
Chester, UK — Joanna M Bauldreay
Princeton, USA — Frederick Passman
Holtsville, USA — Raj Shah
Jerusalem, Israel — Liel Mazar
January 2026

Acknowledgments

The editors gratefully acknowledge the invaluable contributions of numerous individuals and organizations that made this book possible. Members of the general aviation industry, engine and aircraft manufacturers, fuel producers, distributors, pilots' associations and related groups have shared their insights into piston engine fuels. This knowledge, exchanged over many years at ASTM International, the CRC Aviation Committee, the Aviation Fuels Committee, and the Energy Institute, as well as through the International Association for Stability, Handling, and Use of Liquid Fuels, Inc. (IASH), has greatly benefited the industry. The wealth of historical data, field experience, and laboratory test work provided by these groups has been invaluable in preparing this book.

Additionally, the IASH symposium platform served as a key venue to promote ideas, recruit authors, and involved reviewers.

We extend special gratitude to Alisdair Clark for his leadership, deep involvement, and partnership from the inception of this idea through to its publication.

Our heartfelt thanks go to the reviewers whose expertise ensured the high quality of this work: David Abdallah, Dave Atwood, Elizabeth M. Bowman, Naphtali Brodsky, John Buffin, Jack Burgazli, Myrsini Chronopoulou, Jim Gammon, Tina Gleaves, Daniel E. Kadlecek, Jason Lee, Chris Lewis, Brenda Little, Madi Mohtadi, Micki Overberg, Paul Rawson, Francois Ropital, Joel Schmitigal, Stan Seto, and Ron Wilkinson.

We are grateful to the companies, organizations, and authors who provided copyright permissions for the use of their photos and images, as well as Inna Kaprovsky, Faran Levi, Maor Saban, and Yair Cohen for their photographic contributions, and Johnatan Moler for his molecular structure illustrations that enhance the visual quality of this work.

We thank all contributors, colleagues, and supporters who played a role in making this book a reality. Your efforts have helped create a valuable resource for the aviation fuels industry.

And finally, our families, who supported and inspired us throughout this journey. Your patience, encouragement, and understanding made this project possible.

Contents

About the Editors

Dr. Moshe Rabaev is an active researcher and independent consultant with more than 25 years of experience in aviation fuels, advanced materials, and aviation-related manufacturing and maintenance processes. A retired officer of the Israeli Air Force, he previously led its Science and Materials Division. His work focuses on strategic research and technology development in areas such as sustainable aviation fuel (SAF), fuels and lubricants, failure analysis and root cause analysis (RCA), composite materials, corrosion, polymers, analytical chemistry, and additive manufacturing. He chairs the Central Committee for Chemical Standards at the Standards Institution of Israel and serves on multiple specialized standards committees. He is a board member of the International Association for Stability, Handling, and Use of Liquid Fuels, Inc. (IASH) and an active member of ASTM International, Subcommittee D02.J for Aviation Fuels. Dr. Rabaev also holds a senior researcher position at the Blechner Center for Industrial Catalysis and Process Development at Ben-Gurion University of the Negev. He has authored numerous scientific publications, books, and conference papers. Throughout his career, he has received several professional awards, including the Merit of Excellence from the Israeli Minister of Defense and the Award for Technological Innovation from the Chief of the General Staff of the Israel Defense Forces.

Dr. Joanna M Bauldreay is an independent consultant (2019+), with over 30 years' industrial R&D experience, including championing new jet fuels from laboratories to commercialization. Her consulting work, focused on jet fuel technical matters, includes helping customers through processes and technical requirements to produce "Fit for Purpose Fuels for Future Flights." Throughout she has been involved with key technical aviation stakeholder groups and research consortia, including: Coordinating Research Council (CRC) Aviation Fuels Committee (past Chair, Steering Committee, Chair of Emerging Fuels Group), UK MoD Aviation Fuels Committee, International Association for the Stability, Handling and Use of Liquid Fuels (IASH), Commercial Aviation Alternative Fuels Initiative (CAAFI), UK Clearing House Advisory Group, UK Jet Zero Council subgroups, NATO AVT-309, UK SUPERGEN Bioenergy Hub Industrial Advisory Board, and ASTM International, D02 Subcommittee J. Awards received include: IASH Lifetime Achievement, CAAFI "Most valuable chemist" CRC Recognition for leadership, and ASTM Award of Appreciation.

Dr. Frederick Passman is an ASTM Fellow, STLE Fellow and Certified Metalworking Fluids Specialist with more than 50 years' experience in environmental-industrial microbiology. Since 1973, Dr. Passman has conducted research and consulted the government and private industry on diverse topics related to microbial ecology and biodeterioration. Dr. Passman is a member of numerous professional and trade organizations including ASTM, the International Biodegradation and Biodeterioration Society (IBBS) and STLE. Dr. Passman has twice received STLE's Wilber Deutsch Memorial Award for writing excellence, is a recipient of STLE's P. M. Ku Award and Raymond L. Thibault Excellence in Education Award. He has also received ASTM's Award of Merit, and ASTM Committee D02's George V. Dyroff Award and Eagle Award. Dr. Passman also received the International Association on the Stability, Use, and Handling of Liquid Fuels' Lifetime Achievement Award. He has more than 70 publications to his name.

Dr. Raj Shah is Director at Koehler Instrument Company in New York and a globally recognized expert in petroleum testing with over 30 years of experience. A Ph.D. graduate from The Pennsylvania State University, Dr. Shah has authored hundreds of publications and is an elected fellow by his peers at The Energy Institute, STLE, NLGI, IChemE, The Chartered management Institute, Royal Society of Chemistry, Institute of Physics and few others. With decades of leadership in ASTM, he has chaired key subcommittees and co-created the ASTM hands-on courses in fuel technology. Dr. Shah's exceptional contributions have been honored with multiple awards, such as the ASTM Eagle Award and three ASTM awards of Excellence. Dr. Shah is also an adjunct professor of chemical engineering at State University of New York, Stony brook, NY. As a co-editor of seminal works like both the editions of ASTM Fuels and Lubricants Handbook, Dr. Shah continues to advance scientific knowledge as a chartered petroleum engineer while serving on several advisory boards and contributing to professional societies globally.

Mr. Liel Mazar is a fuels and petroleum products engineer with over seven years of experience specializing in aviation fuels quality assurance, infrastructure, laboratory analysis, and incident management. He has extensive expertise in aviation fuel systems, infrastructure, laboratory testing and interpretation, sustainable aviation fuels (SAF), and international regulation. Liel is well-versed in JIG standards and a range of global fuel-related protocols, contributing to the development of Israeli Standards 6294, 5563, and 7566. Liel has led key initiatives within the Israeli Air Force, including developing microbial fuel testing laboratories, advancing SAF integration, and adopting new fuel testing technologies. His roles have spanned leading teams, engineering policy and quality assurance in fuel handling, and resolving complex anomalies in aviation fuel systems.

Acronyms

A4A	Airlines for America
AAIB	Air Accidents Investigation Branch
AD	Airworthiness Directive
AddCo	Additive Company
AES	Atomic Emission Spectrometer
AFC	Aviation Fuel Committee
AFM	Airplane Flight Manual
AFQRJOS	The Aviation Fuel Quality Requirements for Jointly Operated Systems
AFRL	Air Force Research Laboratory
AIAI	Aviation Incidents and Accidents Investigation
AMK	Anti Misting Kerosene
AMM	Aircraft Maintenance Manual
ANAC	Agência Nacional de Aviação Civil
ANSV	Agênzia Nazionale per la Sicurezza del Volo
AO	Antioxidant
APB	Acid Producing Bacteria
API	American Petroleum Institute
APU	Auxiliary Power Unit
ARAIC	Aircraft and Railway Accidents Investigation Commission
ARP	Office of Airports
ARs	Accident Reports
ASCENT	Aviation Sustainability Center
AST	Above Ground Storage Tanks
ATA	Air Transport Association
ATJ	Alcohol to Jet
ATP	Adenosine Triphosphate
ATR	Attenuated Total Reflectance
ATSB	Australian Transport Safety Bureau
AutoGas	Automotive Gasoline
AvGas	Aviation Gasoline

BCs	Intermediate Bulk Containers
BDL	Below Detection Levels
BOCLE	Ball on Cylinder Lubricity Evaluator
BPR	Biocidal Products Regulation
CAAC	Civil Aviation Administration of China
CAAFI	Commercial Aviation Alternative Fuels Initiative
CAROL	Case Analysis and Reporting Online
CCB	Clean, Clear, and Bright
CCR	Continuous Catalytic Reformers
CDF	Clean, Dry, on Specification Fuel
CEPS	Central European Pipeline System
CFU	Colony Forming Units
CGSB	Canadian General Standards Board
CI	Compression Ignition
CI	Corrosion Inhibitor
CI/LI	Corrosion Inhibitor/Lubricity Improver
CID	Controlled Impact Demonstration
CIS	Commonwealth of Independent States
COA	Certificate of Analysis
COC	Chain of Custody
COQ	Certificate of Quality
CORSIA	Carbon Offset and Reduction Scheme for International Aviation
CP	Cathodic Protection
CRC	Collision/Reaction Cell
CRC	Coordinating Research Council
CSI	Controlled Service Introduction
CTR	Controlled Traffic Region
CWT	Center Wing Fuel Tank
DARPA	Defense Advanced Research Projects Agency
DBB	Double Block and Bleed
DCA	Deposit Control Additives
DD	Dirt Defense
DEF	Diesel Exhaust Fluid
Def Stan 91-091	Defence Standard 91-091
dia	Diameter
DIB	Double Isolation and Bleed
DiEGME	Di-Ethylene Glycol Monomethyl Ether
DNA	Deoxyribonucleic Acid
DPF	Diesel Particulate Filter
DRA	Drag Reducer Additive
DSC	Differential Scanning Calorimeter
EAA	Experimental Aircraft Association
EASA	European Aviation Safety Agency
ECAM	Electronic Centralized Aircraft Monitor

ECCAIRS	European Coordination Center for Accident and Incident Reporting Systems
ECU	Electronic Control Unit
EDB	Ethylene Dibromide
EDS	Energy Dispersive Spectroscopy
EDTO	Extended Diversion Time Operation
EFE	External Factor Evaluation
EGME	Ethylene Glycol Monomethyl Ether
EI	Energy Institute
ELISAs	Enzyme-Linked Immunoassays
EPA	Environmental Protection Agency
EPDM	Ethylene Propylene Diene Terpolymer
EPR	Engine Pressure Ratio
EPS	Extracellular Polymeric Substance
ESF	Explosion Suppressant Foams
ETFE	Ethylene Tetrafluoroethylene
ETR	Ellipsometric Tube Rater
EU	European Union
EWS	Electronic Water Sensor
FAA	Federal Aviation Administration
FADEC	Full Authority Digital Electronic Control System
FAME	Fatty Acid Methyl Ester
FCC	Fluidized Catalytic Cracker
FCU	Fuel Control Unit
FFP	Fuel Flow Proportioner
FIMS	Field Ionization Mass Spectrometric
FMU	Fuel Metering Unit
FOD	Foreign Object Damage
FOHE	Fuel-Oil Heat Exchanger
FPC	Floating Piston Cylinder
FQIS	Fuel Quantity Indicator System
FSII	Fuel System Icing Inhibitor
FT	Fischer-Tropsch
FTIR	Fourier Transform Infrared
FWS	Filter/Water Separator
GC	Gas Chromatography
GFRP	Glass-Fiber Reinforced Plastic
gpm	Gallons Per Minute
GRP	Glass-Reinforced Plastic
HCCG	Heavy Catalytically Cracked Gasoline
HDPE	High Density Polyethylene
HDS	Hydrodesulfurization
HEFA	Hydroprocessed Esters and Fatty Acids
HEI	High-Explosive Incendiary
HiReTS	High Reynolds Number Thermal Stability

HMU	Hydro-Mechanical Unit
HP	High Pressure
HPU	Hydrogen Producing Unit
HVGO	High-Vacuum Gasoil
IAF	Israeli Air Force
IASH	International Association for the Stability and Handling of Liquid Fuels
IATA	International Air Transport Association
IBC	Intermediate Bulk Containers
ICAO	International Civil Aviation Organization
ICE	Internal Combustion Engines
ICP	Inductively Coupled Plasma
IDL	Instrument Detection Limits
IFE	Internal Factor Evaluation
IFOs	Intermediate/Residual Fuel Oils
IFQP	IATA Fuel Quality Pool
IG	Inert Gas
IGS	Inert Gas Systems
IJFSLG	International Jet Fuel Specification Liaison Group
ILS	Interlaboratory Study
IRB	Iron Related Bacteria
ISLG	International Specification Liaison Group
iSTARS	Integrated Safety Trend Analysis and Reporting System
ITR	Interferometric Tube Rater
ITS	Internal Transcribed Spacer
JCAB	Japan's Civil Aviation Bureau
JFTOT	Jet Fuel Thermal Oxidation Tester
JIG	Joint Inspection Group
JP	Jet Propellant
JPTS	Jet Propellant Thermally Stable
KED	Kinetic Energy Discrimination
KHT	Kerosene Hydrotreater
LFDs	Lateral Flow Devices
LHSV	Liquid Hourly Space Velocity
LIA	Lubricity Improver Additive
LIF	Laser Induced Fluorescence
LLCO	Light Light Cycle Oil
LMOA	Low Molecular Weight Organic Acids
LOD	Limit of Detection
LOQ	Limit of Quantification
LP	Low Pressure
MDA	Metal Deactivator Additive
MIC	Microbiologically Influenced Corrosion
MMM	Molecular Microbiology Methods
MOC	Management of Change

MOD	Ministry of Defence
MoGAS	Automotive Gasoline
MON	Motor Octane Number
MPPL	Multiproduct Pipelines
mRNA	Messenger RNA
MS	Mass Spectrometer
MSEP	Micro Separometer Analysis
MTBE	Methyl Tertiary Butyl Ether
NATA	National Air Transportation Association
NBR	Nitrile Butadiene Rubber
NIFTR	Near-Isothermal Flowing Test Rig
NIR	Near-Infra Red
NJFCP	National Jet Fuel Combustion Program
NPA	Servicing Tank Vehicle
NPS	Nominal Pore Size
NTSB	National Transportation Safety Board
OBIGGS	Onboard Inert Gas Generation System
OEM	Original Equipment Manufacturer
OES	Optical Emission Spectroscopy
OOS	Out of Specification
OSHA	Occupational Safety and Health Administration
OTU	Operational Taxonomic Unit
PAC	Polyacrylamide Gels
PHMG-SS	Poly (Hexamethylene Guanidine) Hydrochloride-Sodium Stearate
PIC	Pilot-in-Command
PL	Particulate Load
PLA	Position Lever Angle
POS	Point of Sale
PP	Polypropylene
PRV	Pressure Regulating Valve
PTFE	Poly Tetra Fluoro Ethylene
PTO	Power Take-off
QA	Quality Assurance
QC	Quality Control
QCM	Quartz Crystal Microbalance
qPCR	Quantitative Polymerase Chain Reaction
QPL	Qualified Product List
RC	Release Certificate
RCA	Root Cause Analysis
RCs	Release Certificates
RDE	Rotating Disc Electrode
RFID	Radio-Frequency Identification
RLU	Relative Light Units
RNA	Ribonucleic Acid

RON	Research Octane Number
rRNA	Ribosomal RNA
RT	Recertification Test
RTV	Room Temperature Vulcanizations
SAF	Sustainable Aviation Fuels
SAFO	Safety Alert for Operators
SAIB	Special Airworthiness Information Bulletin
SAP	Superabsorbent Polymer
SARPs	Standards and Recommended Practices
SATF	Synthetic Aviation Turbine Fuels
SBC	Synthetic Blending Components
SCC	Stress Corrosion Cracking
SCR	Selective Catalytic Reduction
SCRM	Supply Chain Risk Management
SDA	Static Dissipator Additive
SDI	Static Discharge Ignition
SDOs	Standards Development Organizations
SDS	Safety Data Sheet
SEM	Scanning Electron Microscopy
SG	Specific Gravity
SIB	Safety Information Bulletin
SMR	Steam Methane Reformer
SOPs	Standard Operating Procedures
SOS	Source of Supply
SPE	Solid Phase Extraction
SPKs	Synthetic Paraffinic Kerosenes
SRB	Sulfate Reducing Bacteria
STC	Supplemental Type Certificates
SVA	Stator Vane Actuator
TAN	Total Acid Number
TCP	Tricresyl Phosphate
TDR	Tube Deposit Rater
TDS	Total Dissolved Solids
TEL	Tetraethyl Lead
TGA	Thermal Gravimetric Analyzer
TLC	Thin-Layer Chromatography
TOFA	Tall Oil Fatty Acids
TPD	Thermally Programmed Desorption
tRNA	Transfer RNA
TSO	Technical Standard Orders
TSS	Total Suspended Solids
TWA	Trans World Airlines
UAV	Unmanned Air Vehicles
UCL	Upper Control Limit
UDRI	University of Dayton Research Institute

UERF	Uncontained Engine Rotor Failure
ULSD	Ultra Low Sulfur Diesel
UOP	Universal Oil Products
USAF	United States Air Force
UST	Underground Storage Tanks
VGO	Vacuum Gas Oil
VTR	Visual Tube Rating
WBF	Water Barrier Filters
WISM	Water Separation Index Modified
WTI	West Texas Intermediate
XRF	X-Ray Fluorescence

Chapter 1
An Introduction to Aviation Fuels and Specifications

Joanna M. Bauldreay

Abstract Flight safety has been the primary driver for the development of aviation fuels and their governing specifications since first piston-engined aircraft and later turbine engine-aircraft were developed. For over 70 years in the case of turbine fuels, problems and issues with fuel storage, distribution and usage have led to the specifications being modified as much as the aircraft and engines have changed. Producing fuels to match current specifications reduces the risk of fuel-related problems not associated with handling challenges. This chapter provides an overview of fuel and specification changes, particularly in relation to civil jet fuels, and explains the main performance requirements embodied in the fuel specifications. These are living documents and nowadays allow for the use of synthetic fuel components; additional feedstocks and specification changes to accommodate them can be expected in response to environmental pressures.

1 An Introduction to Aviation Fuels and Specifications

Prevention is better than the cure, a phrase which is applicable to aviation fuel specifications which have been developed over many years to give the level of flight reliability the world enjoys today transporting over 4 billion passengers each year across the globe [1]. Aviation fuel specifications are the cornerstone in the prevention of problems developing anywhere in the production, distribution, and storage system. This chapter will allude to some aspects of these developments and what forced changes, but it does not set out to provide an exhaustive, property by property historical perspective. Anyone wanting to understand how the industry has developed and reached today's jet fuel and aviation gasoline specifications may wish to read excellent documents already available [2–6].

Note: Readers must not use this book as a basis for their quality assurance activities related to aviation fuels. They should always use the current version of the specification relevant to their operations.

J. M. Bauldreay (✉)
Bauldreay Jet Fuel Consulting Limited, Chester, United Kingdom
e-mail: Joanna@Bauldreay.com

M. Rabaev et al. (eds.), *Aviation Fuel Problems and Solutions: From Refinery to Wingtip*, https://doi.org/10.1007/978-3-032-05514-9_1

1.1 Scope of This Chapter

This chapter provides a high-level look at the specifications used for civil and military jet fuels including a limited discussion of why some specification requirements have been set.

The following chapters provide a more detailed examination of specific jet fuel issues and challenges, starting with the basics of manufacture (Chap. 2), new synthetic jet fuels (Chap. 3), and additives (Chap. 5). Chapter 4 deals specifically with aviation gasoline issues. Later chapters discuss the challenges downstream of fuel production.

Both this and subsequent chapters may mention specialist military fuels, such as rocket and missile fuels. There are very few manufacturers of such products, which are supplied directly to the limited pool of end users, and any operational issues are unlikely to be relevant to the broader aviation fuel community. However, where a relevant item is recorded, efforts have been made to include some details to assist the reader.

1.2 Overview of Aviation Fuels and Their Applications

Turbine and piston-engined aircraft are designed to use fuels that have appropriate combustion properties and that also fulfil all the activities and roles needed before they are combusted, namely in the distribution and aircraft fuel system. Turbine-engined aircraft have mainly been developed to use kerosene fuels originating from the first jet fuel specification, issued in 1943 by the UK Ministry of Defence's Research Directorate of Engineering, RDE/F/KER/210, though there have been wide-cut fuels spanning the gasoline to kerosene range originating from United States Air Force (USAF) military jet fuel specification, AN-F-32a, where availability and/or more extreme conditions have had to be accommodated.

Piston-engined aircraft mainly developed using spark ignition gasoline engines featuring high octane fuels. However, historically, compression ignition (diesel) engines have also been used. This alternative approach, operating on a kerosene fuel, is regaining interest and development today.

Fuels must meet the needs of aircraft operated by civil and military users, covering private pilots to large-scale commercial applications. Ground vehicle operations, which tend to be very country or region specific, can accommodate many fuel grades or classes[1] within a single specification to manage seasonal variations or marketing offers. In the aviation world, many operations are international, and airport fuel supply and storage are on a very different scale to road fuel supply. A choice of grades

[1] The European automotive diesel specification, EN 590 [7], lists eleven classes of fuel, covering summer and winter operations. It allows northern Sweden to southern Greece to be supplied with EN 590 fuels that cover their climatic differences.

of fuel at one location is not a realistic option, and having grades made to internationally recognized specifications is essential to avoid operational and purchasing confusion for airlines and suppliers.

1.3 *Brief Overview of Civil Jet Fuel Categories*

1.3.1 Main Grades

Jet A and Jet A-1 are the two most common grades of jet fuel in civil applications. The grades are covered by more than one specification, with subtle differences in the requirements, but internationally the main specifications are from the UK and ASTM International, currently Defence Standard 91-091 (Def Stan 91-091) [8] and ASTM D1655 [9], respectively. Where other Jet A and Jet A-1 specifications exist, they usually follow these lead specifications, translated into relevant national languages; see Table 1 for a listing of the jet fuel grades in main international specifications.

Oil companies have produced some helpful documents that help to compare assorted specifications [14–16]. There is also ongoing activity by the International Specification Liaison Group (ISLG) of International Air Transport Association's (IATA) technical fuels group to seek specification alignment wherever possible, with their Guidance Material comparing specifications [17].

These main specifications now allow for the use of synthetic components. These are either blended in with conventional jet fuel, using named very specific routes (Sasol Fischer–Tropsch (FT) blends) or more commonly via ASTM D7566 [18], or are listed as components that can be co-processed with conventional refineries in hydroprocessing or more upstream process units. Chapter 3 discusses the development of the ASTM D7566 specification, which describes "drop-in" fuels that once batched can enter the delivery system as Def Stan 91-091 or ASTM D1655 fuels.

Table 1 Jet fuel grades in main international specifications

Grade	Specification			
	Def Stan 91-091	ASTM D1655	CGSB 3.23 [10] (Canada)	DSEA 134/B [11] (France)
Jet A	No	Yes	Yes	No
Jet A-1	Yes	Yes	Yes	Yes
Jet B	No	See ASTM D6615 [12]	See CGSB 3.22 [13]	No

1.3.2 AFQRJOS "Check List Jet A-1"

Most airlines operate internationally, and to reduce the challenges to both suppliers and users of working with a plethora of specifications, members of the Joint Inspection Group (JIG) introduced a unifying document to use for fuel purchases. The Aviation Fuel Quality Requirements for Jointly Operated Systems (AFQRJOS) represents the most stringent elements for Jet A-1 set out in Def Stan 91-091, ASTM D1655 and the IATA Fuel Guidelines for Jet A-1. A fuel meeting AFQRJOS, referred to as "Jet A-1 to Check List" or "Check List Jet A-1", allows fuel to be supplied to either Def Stan or ASTM specifications for Jet A-1. Check List is not a specification, but it is frequently referred to in documents used for fuel release from many refineries and airport storage locations and also in purchase agreements.

1.3.3 Other Specifications

The People's Republic of China makes several grades of jet fuel. Most relevant to international users, specification GB 6537-2018 [19] describes one grade, Jet Fuel 3, that is like Def Stan 91-091 Jet A-1, though not necessarily matching all aspects of the latter's latest versions. The specification currently allows for two synthetic hydrocarbon kerosene fractions to be blended with conventional crude-derived kerosenes. Currently, additive usage is slightly different, as discussed in Chap. 5.

Russia, Commonwealth of Independent States (CIS) and former CIS countries have used a specification (GOST 10227, [20]) covering two main grades of lower flash point, low freezing point jet fuel, TS-1 and RT, more suited to the extreme temperatures they must encounter. A specification to cover manufacture and supply of a jet fuel grade like Jet A-1, using a combination of international and Russian test methods, has also been developed (GOST R 52050 [21]).

All the above relate to specifications that are used in production and final airline purchasing of jet fuel. Many pipelines, particularly in North America, also have jet fuel specifications which apply for fuels entering and leaving the systems. These may simply match standard specifications or feature some additional margin to ensure quality at point of delivery. For example, thermal stability requirements, particularly for the fungible systems in North America, are often more stringent than those in the primary specifications. Thermal stability testing at 275 °C, not 260 °C, has been used, to minimise occasional issues which might cause this parameter to degrade during transportation over hundreds of miles of pipeline and through intermediate terminals.

1.4 Brief Overview of Military Jet Fuel Categories

1.4.1 Main Fuels

Miliary fuels and their specifications have developed in parallel with those in civil applications. In the UK, military fuel started as a kerosene product while in the USA, earlier military fuels extended into the gasoline range. Indeed, Jet Propellant 4 (JP-4) was the main USAF fuel until the early 1990s when they moved over to JP-8. More recently the main USAF fuel in the USA has become F-24, an additivated version of Jet A, allowing access to civil fuel supply.

Military and civil jet fuels refer to different specifications, but in practice refineries generally make civil jet fuel and the military upgrades it or has it delivered upgraded at a convenient agreed point before usage. A notable exception is NATO F-44/JP-5, where the minimum flash point of +60 °C necessitates more restrictive distillation cut points. JP-8 is effectively based on Jet A-1 meeting MIL-DTL-83133 [22] and Def Stan 91-091 requirements, while F-24 is based on Jet A meeting ASTM D1655. Upgrading involves using additives to meet the additional requirements of military operations, for example long-term storage, thermal management in supersonic aircraft and afterburner operation. The military additives are covered in more detail in Chap. 5.

There have been periods when military specifications have led the way, such as in the generic approvals of Fischer–Tropsch (FT) and then hydroprocessed esters and fatty acids (HEFA) synthetic components. Nowadays, it is more common for civil specifications, and the related approval process put in place by the industry, to drive development in parallel with the military.

1.4.2 Specialist Applications

The bulk of military operations using jet fuels involve air forces and most of their needs are adequately covered by using specially additivated civil fuels such as JP-8 (F-34 or Avtur/FSII), or F-24 (in the US), with standard F-35 (Def Stan 91-091) widely used in the UK and locations outside US. There are, however, applications that warrant different fuels.

Naval aircraft carrier operations use a high flash point fuel such as JP-5 or Avcat/FSII F-44 for additional safety. The base fuel can be made in some refineries (see Chap. 2) without compromising civil fuel needs, but generally it is not available from many refineries.

Wide-cut fuels are used to ensure adequate volatility or cope with lower ground temperatures. They also offer scope to increase the amount of usable fuel that can be supplied if supplies are limited, such as in times of conflict. Examples are JP-4 and Avtag/FSII F-40, which are like the rarely made civil Jet B with additional, appropriate additives.

Very high-altitude operations may need specialist fuels, such as jet propellant thermally stable (JPTS). This is kerosene based but made to order with a thermal stability additive package. Missile fuels tend to be very specialist chemical blends. JPTS, missile fuels, and the relevant specifications covering them are out of scope for this book.

1.4.3 Interoperability

Just as civil operators wish to use the same 1 or 2 grades of jet fuel internationally, military operators also want security of supply of known quality fuels. NATO fuel committee members have defined fuels that allow interoperability [23], see Table 2. Nowadays these are based on kerosene fuels, and most member nations follow UK or US specifications; some, like France, Germany, and Canada, have their own specifications but these align well with those of the UK and USA.

Separately, some military operators need their jet fuels to cover operations in other applications than aircraft. There are "Single Fuel" concepts covering both land and sea operations. Supplying jet fuel, diesel fuel, gasoline, and marine fuel when supplies are limited and over long, difficult distribution chains is a challenge. The US Army no longer operates gasoline vehicles; their diesel-engined ground equipment (static and moving) is expected to be able to operate on their jet fuel, JP-8. Likewise, US Navy ships are expected to be able to operate on their primary jet fuel, JP-5. Other countries have similar challenges, not just in times of conflict, particularly when local ground fuel quality is very poor; using jet fuel, which must meet very stringent conditions, is an acceptable compromise and equipment suppliers can be requested to produce engines, etc., that allow for this occasional usage of jet fuel. This may require diesel exhaust systems to be bypassed. The Single Fuel concept is an area of continuing interest to the military. In particular, the implications of jet

Table 2 NATO jet fuel grades (latest issues unless indicated otherwise)

NATO code	UK joint service designation	USA	Canada
F-24 (Jet A)	No equivalent	NATO AFLP-3747 [24]	No equivalent
F-34	Avtur/FSII (Def Stan 91-087) [25]	JP-8 (MIL-DTL-83133)	CAN/CGSB 3.24 [26]
F-35 (Jet A-1)	Avtur	F-35 (MIL-DTL-83133)	CAN/CGSB 3.23 [27]
F-37	No equivalent	JP-8 + 100 (MIL-DTL-83133)	CAN/CGSB 3.24
F-40	Avtag/FSII (Def Stan 91-088) [28]	JP-4 (MIL-DTL-5624) [29]	No equivalent
F-44	Avcat/FSII (Def Stan 91-086) [30]	JP-5 (MIL-DTL-5624)	CAN/CGSB 3.24

fuels made with synthetic (mainly sustainable) components are being addressed by several NATO and related military research groups.

2 Specifications

2.1 *What is a Specification?*

According to the Collins English Dictionary[2] a specification is "a detailed description of the criteria for the constituents, construction, appearance, performance, etc., of a material, apparatus, etc., or of the standard of workmanship required in its manufacture". For aviation fuel specifications, which cover all these listed elements, there is obviously a chicken and egg situation, with both the fuel and the systems for which it is designed taking some time and iterations to bed in with one other. Ultimately, however, an aviation fuel specification is a set of criteria that fuel purchasers or fuel users request to ensure that the fuel they have is fit for purpose for their aircraft and operations. It is also what the fuel manufacturer uses as a production requirement to ensure fuel quality.

Importantly, specifications apply to fuels, and aircraft/platforms (engines and airframes) are only certified to use fuels that meet the named specifications. Fuels and aircraft/platforms both evolve and change.

Specifications do not provide a recipe for making a fuel—they are not compositional specifications but performance specifications. They list requirements and restrictions, some of which are compositional, that, if followed, should provide fit for purpose fuels to the end users. A fuel shall meet all the criteria of the specification, not just fulfilling the laboratory requirements in "Table 1", to be considered "on spec".

Specifications have been expected to cover all stages from manufacture to final usage, but in practice no single operator can provide all the necessary controls to ensure this. Each performs their part, be it in manufacture, distribution, delivery, or use of the product, following appropriate regulations to give a reliable, fit-for-purpose system. By way of example, current specifications are starting to drop controls that are unenforceable or irrelevant at the point of manufacture and instead are referring to documents, particularly the EI/JIG 1530 Standard on Jet product quality [31], to ensure that fuel is made "on spec" and can arrive at the airport "on spec". (EI/JIG 1530 is a voluntary standard but is increasingly appearing in purchasing contracts and being updated to reflect users' comments). At the airport, fuel handling procedures such as those in JIG and Air Transport Association (ATA) [32] documents set out how to maintain the fuel quality right up to delivery in the aircraft fuel tanks.

---

[2] English Dictionary, Complete and Unabridged, Sixth Edition 2003.

2.2 *Typical Range of Requirements*

Contents and layouts of specifications differ between issuing authorities, but they generally include the items to control the production, testing and release of finished fuels, and to avoid previously encountered problems linked to fuel quality. They usually contain a record of the most recent revisions, plus management and technical elements. Def Stan 91-091 and ASTM D1655 contents and layouts are summarized in Tables 3 and 4, respectively, though omitting elements that focus on legal, health and safety elements (mainly in Section 1).

Table 3 Defence Standard 91-091 content and layout

Category	Content
Section 1	
Revision note	Summary of changes since the last version
Historical record	Lists standards that the current version supersedes
Standard clauses	Information about who publishes the standard, for whom, how the standard has been reached, and how to deal with potential contractual or tendering activities. Legal obligations. Usage and liability exclusions
Section 2	
Scope	Standard sets out requirements for one grade of kerosine type aviation turbine fuel, to be satisfactory for use in specified aircraft or engines
Normative references	Points to documents and publications referenced in the standard are being listed in Section 3
Materials	Lists conventional feedstocks, refers to synthetic components that may be used, and additives (with more details in Annexes). Toxicological evidence may be required
Quality assurance	Batches and traceability are required. Certificate of Quality (CoQ) shall be issued and contain specific information, for fuel samples taken as per IP 475 [33] and ASTM D4057 [34], and management of change processes shall be used to limit the scope for incidental materials to cause operation problems
Testing	Table 1—Quantitative data to be supplied, using permitted listed methods, in a CoQ. (Alternative methods are listed in Annex E). There are notes to accompany the Table, with limitations on usage or reference to method similarity, etc. Where more than one method is listed per test requirement, a referee method is often indicated

(continued)

Table 3 (continued)

Category	Content
Containers and marking of containers	Flags need for appropriate containers, with necessary coatings and paint finishes, and markings, for supply of fuel samples. Exact details may be in contracts
Annexes	
Annex A—List of qualified additives	Provides general information about hydrocarbon diluents that may be needed, then lists the categories of additives that shall or may be used to obtain satisfactory storage, performance, and handling. See below, and Chap. 5
Annex B—Additional requirements applicable to fuels from non-conventional sources	Background and current approach to accepting synthetic kerosine components in the making of Jet A-1, and additional reporting requirements in the Certificate of Quality (CoQ). Refers to historical one-off, step-by-step approvals of coal-based materials from Sasol and the current drop-in fuels made according to ASTM D7566. Also describes synthetic streams that can be introduced into jet production by co-processing routes. Documentation and handling requirements of components and blends are outlined, and the point of manufacture (which shall be upstream of an airport except for very small deliveries) is defined. Points to EI 1533 [35] standard for more information on handling of such fuels and components This is a rapidly evolving area. See more in Chap. 3
Annex C—Product integrity management	More generally, manufacturing and distribution locations are required to have adequate product quality and management of change processes to maintain product quality that may be compromised by incidental materials. Formal risk assessments shall be involved ahead of any production changes. Also sets limits on materials that may enter fuel downstream of production (in particular Fatty Acid Methyl Ester, FAME, and Pipeline Drag Reducer additive, DRA) in multiproduct pipes and systems, and points to JIG bulletins that advise on managing FAME carryover
Annex D—Product certification and traceability	Sets out 4 types of key documentation needed from point of manufacture onwards, with reference to EI JIG 1530. Lists the minimum information to be supplied with each type of document It also explains the importance of batch traceability back to the original point of manufacture and lists the information to be supplied to enable this Accepts that it may not be possible to ensure traceability in fungible systems but indicates the requirements that shall still apply Covers how to handle and certify Jet A-1 created by blending multiple Jet A-1 batches in the distribution system

(continued)

Table 3 (continued)

Category	Content
Annex E—Alternative test methods for use with Table 1 test requirements	"Table 5" lists acceptable test methods to use instead of those listed in "Table 1"; these are not referee methods which are to be used in case of disputes
Annex F—Additional information on fuel properties	Provides useful information on: sampling and particulate contamination; electrical conductivity; lubricity; colour (as an indicator of fuel quality); guidance on biobased content
Annex G—Technically equivalent ISO methods for Tables 1 and 5 test methods	Lists a small number of ISO methods that are deemed technically equivalent to the IP and ASTM methods listed in the "Table 1" and "Table 5" requirements
Section 3	
Normative references	List of publications referred to in the text. Provides link to access Defence Standards Warning that the current issue will be assumed to apply
Definitions	Where relevant, reads across to ISO/EC definitions guidance
Abbreviations	Abbreviations (including acronyms) and descriptions
Changes since last issue	Detailed list of changes, listed by clause, page and with reason for the change, which can be editorial or technical or for harmonization purposes

Table 4 ASTM D1655 layout and content

Category	Content
1 Scope	To be used to set purchasing specifications, for Jet A and Jet A-1 suitable for civil and military applications. Can cover from production to aircraft though points to other documents to provide quality assurance needed in distribution. States it does not cover all safety concerns associated with its use
2 Referenced documents	Lists ASTM, EI, API, JIG, ANSI, and other standards referred to in the standard
3 Terminology	References main ASTM D4175 [36] document for general terminology and defines terminology specific to the standard. COQ to contain results of all requirements of Table 1 and requirements of Table 2 for additive usage. Additional requirements for fuels containing synthesized components are set out in Annex A1
4 General	Required properties of the fuel at the time and place of delivery
5 Classification	High level description of two grades of aviation turbine fuel, Jet A and Jet A-1

(continued)

Table 4 (continued)

Category	Content
6 Materials and manufacture	Describes general composition, the nature of the standard being performance not composition based, and the predominant sources of the refined hydrocarbons. Also lists additives, discusses dilution and reporting, and identified incidental materials (usually encountered post-manufacture)
7 Detailed requirements	Shall conform to requirements in Table 1; precision discussed
8 Workmanship, finish, and appearance	Covers appearance (visually free of undissolved water, sediment, and suspended matter—not measured as part of Table 1) and odour (shall not be nauseating or irritating)
9 Sampling	Recommends proper sampling procedure and recommends sample containers best used to prevent trace contamination impacting thermal stability, water separation, electrical conductivity, and other properties
10 Report	Stresses need to agree reporting between seller and purchaser. Lists the core contents of any Certificate of Quality or Certificate of Analysis forms, covering methods used, Table 1 test and Table 2 additive information
11 Test methods	Lists allowed methods by performance area and refers to referee methods used in case of disputes
12 Keywords	Lists main words that relate to this specification
Annexes	
A1 Fuels from non-conventional sources	Provides overview of specifically approved fuels (Sasol blends), drop-in fuels meeting ASTM D7566 and co-processed fuels. Lists additional test results that will need to be reported for the specific fuels and discusses potential issues with bias for each property
Appendices	
X1 Performance characteristics of aviation turbine fuels	Overview of key performance areas, with reference to more detailed ASTM Manual 1 [37] and ASTM MNL 37 [38], and the CRC handbook of aviation fuel properties [39]. Specific sections on: significance and use; thermal stability; combustion; fuel metering and aircraft range; fuel atomization; fluidity at low temperatures; compatibility with elastomer and the metals in the fuel system and turbin; fuel storage stability; fuel lubricity; fuel handling; fuel cleanliness and contamination; surfactants; cleanliness at time of fuel custody transfer at airport; miscellaneous (additives, sample containers, colour, biobased carbon content); identified incidental materials. Each subsection lists tests relevant to the performance area. Some items, like colour and lubricity, are not part of ASTM D1655's Table 1, but this section explains their importance

(continued)

Table 4 (continued)

Category	Content
X2 Control of upstream and process additives, distribution and other uses	Discusses sources of additives, from crude to distribution, and extent to which producers should control them, where known, using, say, management of change processes. Also flags usage of D1655 fuels for compression-ignition aircraft engines, and the potential to estimate properties relevant to them (e.g., cetane) that are not included in D1655
X3 Forms for reporting inspection data on aviation turbine fuels	Suggested form for reporting inspection data on Certificate of Quality or Certificate of Analyses. Non-mandatory but they incorporate the requirements of D1655
References	Lists specific reports referred to in the appendices
Summary of changes	Highlights content change since previous issue

Military specifications contain similar categories of information but have been omitted for brevity.

2.2.1 Materials

Both specifications list the most common feedstocks for the refined hydrocarbons making up jet fuel to include crude oil, natural gas liquid condensates, heavy oil, shale oil, and oil sands. Components from other sources are only permitted in accordance with additional requirements as detailed in the specification Annexes.

The aviation industry has decades of experience with conventional fuels, produced as described in Chap. 2. Fuels meeting given specifications will have a range of other inherent properties that can be assumed because of the compositional features the specification brings with it. The CRC Handbook of Aviation Fuel Properties [38, 39] is an excellent source of information about these related properties, most relevant to engineers and designers working with original equipment manufacturers (OEMs). Such properties do not need to be measured routinely if the fuel is made in the traditional manner from traditional materials but have been a key focus area in the test programmes leading to the approval of synthetic jet fuel components (Chap. 3).

2.2.2 Additives

These may be used to improve fuel performance characteristics, or to improve handling properties between production and end usage. Table 5 provides an indication of the additives' purposes and the most common point of introduction into the fuel (this can depend on the customer and/or the main specification being used). The table does not aim to address whether usage is mandatory or optional, as this depends on the specification. Military users generally list more mandatory additives, reflecting their more extreme performance requirements such as afterburners that increase the

Table 5 Overview of additive types, purpose, and place of usage (This split into Fuel Performance versus Fuel Handling does not match that given in ASTM D1655; the author sees lubricity as a performance issue and Def Stan 91-091 is more explicit about having to measure it)

Additive class	Purpose	Most common place of addition			
		Manufacture	Downstream terminal	Airfield	Aircraft
Improve fuel performance					
Antioxidant (AO)	Prevent peroxidation and gum formation during storage. Also, may improve thermal stability	x			
Metal deactivator additive (MDA)	Improve thermal stability by chelating metals in fuel	x	x		
Lubricity improver additive (LIA)	Provide boundary lubrication to "hard" fuels	(x)	x	x	
Fuel system icing inhibitor (FSII)	Prevents free water freezing and blocking aircraft filters		(x) For military	x	x
+100[a]	Increases fuel thermal stability			x	
For fuel handling and maintenance					
Static dissipator additive (SDA)	Help to discharge electrostatic charges if fuel relaxation times are short	x	x	(x)	
Leak detection additive	Maintenance only. To detect leaks when there are no non-additive options			x	
Biocide			(x)	x	x
Incidental additives from downstream activities—to be managed below set limits					
FAME	Component of biodiesel blends		From (bio)diesel		
Drag reducing additive (DRA)	Increases volumes of fuel that can be pumped and delivered		From pipeline operations		

[a]Military usage only

thermal stressing of fuels and usage of fuels after long periods of storage. Chapter 5 provides more details on all aspects of additive usage.

2.3 "Table 1" Requirements of Specifications

"Table 1" is the most visible part of specifications, because it is reflected in Certificates of Quality (CoQ). Table 6 provides an example of a Def Stan 91-091 Table 1; notes to clarify limits, applicability, and referee and alternative methods are provided in the specification but are not being shown here. When used to report test results in a CoQ, the actual methods used shall be indicated, not the range of permitted methods. The CoQ shall also contain information about the additives used, covering qualification reference and amount used and ideally also reporting "nil" levels. See more in Chap. 5.

Table 6 Example of Def Stan 91-091 Table 1

Test		Property	Units	Limits		Method
1	Appearance					
	1.1	Visual appearance	–	Clear, bright, and visually free from solid matter and undissolved water at ambient temperature		Visual
	1.2	Colour	–	Report		ASTM D156 [40] or ASTM D6045 [41]
	1.3	Particulate at point of manufacture (shall meet 1 of the following)				
	1.3.1	Particulate contamination	mg/L	1.0 max		ASTM D5452 [42]
	1.3.2	Particulate, cumulative channel counts:	Individual channel counts and ISO code	Channel counts	ISO code	IP 565 [43] or IP 577 [44]
		≥4 micron(c)	–	Report	Max 19	
		≥6 micron(c)	–	Report	Max 17	
		≥14 micron(c)	–	Report	Max 14	
		≥21 micron(c)	–	Report	Report	
		≥25 micron(c)	–	Report	Report	

(continued)

Table 6 (continued)

Test		Property	Units	Limits		Method
		≥30 micron(c)	–	Report	Max 13	
2	Composition					
	2.1	Total acidity	mg KOH/g	Max 0.015		IP 354 [45]/ ASTM D3242 [46]
	2.2	Aromatic hydrocarbon types (1 of following shall be met)				
	2.2.1 or	Aromatics	% v/v	Max 25.0		IP 156 [47]/ ASTM D1319 [48]
	2.2.2	Total aromatics	%v/v	Max 26.5		IP 436 [49]/ ASTM D6379 [50]
	2.3	Sulphur, total	% m/m	Max 0.30		IP 336 [51]
	2.4	Mercaptans (1 of following shall be met)				
	2.4.1 or	Sulphur, Mercaptan	% m/m	Max 0.0030		ASTM D3227 [52]
	2.4.2	Doctor test	–	Doctor Negative		IP 30 [53]
	2.5	Refining components, at point of manufacture				
	2.5.1	Non hydroprocessed components	% v/v	Report		
	2.5.2	Severely hydroprocessed components	% v/v	Report		
	2.5.3	Synthetic components	% Volume	Report (limits in Annex B)		See Annex B
3	Volatility					
	3.1	Distillation				IP 123 [54]/ ASTM D86 [55]
	3.1.1	Initial boiling point	°C	Report		
	3.1.2	10% recovery	°C	Max 205.0		
	3.1.3	50% recovery	°C	Report		
	3.1.4	90% recovery	°C	Report		
	3.1.5	End point	°C	Max 300.0		
	3.1.6	Residue	% v/v	Max 1.5		
	3.1.7	Loss	% v/v	Max 1.5		
	3.2	Flash point	°C	Max 38.0		IP 170 [56]
	3.3	Density at 15 °C	kg/m^3	Min 775.0	Max 840.0	IP 365 [57]/ ASTM D4052 [58]

(continued)

Table 6 (continued)

Test		Property	Units	Limits	Method
4	Fluidity				
	4.1	Freezing point	°C	Max minus 47.0	IP 16 [59]/ ASTM D2386 [60]
	4.2	Viscosity at − 20 °C	cSt	Max 8.000	IP 71 [61]/ ASTM D445 [62]
5	Combustion				
	5.1	Smoke point (1 of following shall be met)			
	5.1.1 or	Smoke point	mm	Max 25.0	IP 598 [63]/ ASTM D1322 [64]
	5.1.2	Smoke point	mm	Min 18.0	IP 598/ASTM D1322
		Naphthalenes	% v/v	Max 3.00	ASTM D1840 [65]
	5.2	Specific energy	MJ/kg	Min 42.80	Detail in notes
6	Corrosion				
	6.1	Copper strip (2 h at 100 °C)	Class	Max 1	IP 154 [66]/ ASTM D130 [67]
7	Thermal stability				IP 323 [68]/ ASTM D3241 [69]
	7.1	Test temperature	°C	Min 260	
	7.2	Tube rating (1 of following shall be met)			
	7.2.1	Annex B VTR or	–	Less than 3. No peacock (P) or abnormal (A)	
	7.2.2	Annex C ITR or Annex D ETR, average over area of 2.5 mm^2	nm	Max 85	
	7.3	Differential pressure	mm Hg	Max 25	
8	Contaminants				
	8.1	Existent gum	mg/100 ml	Max 7	IP 540 [70]
9	Water separation characteristics				
	9.1	Microseparometer at point of manufacture (1 of following shall be met)			ASTM D3948 [71]
	9.1.1	MSEP without SDA	Rating	Min 85	
	9.1.2	MSEP with SDA	Rating	Min 70	

(continued)

Table 6 (continued)

Test		Property	Units	Limits		Method
10	Conductivity					
	10.1	Electrical conductivity	pS/m	Min 50	Max 600	IP 274 [72]/ ASTM D2624 [73]
11	Lubricity					
	11.1	Wear scar diameter	mm	Max 0.85		ASTM D5001 [74]

Similar information, other than appearance, is required by ASTM D1655. Table 7 highlights this, comparing the tests that these two specifications list that cover specific performance features for a conventionally made jet fuel (Additional criteria, not listed here, are required for co-processed fuels and semi-synthetic blend).

Table 7 Main performance areas explicitly covered in civil specifications

Property	Def Stan 91-091	ASTM D7566
Appearance	Visual appearance Colour Particulates at point of manufacture	Visual appearance
Composition	Total acidity Total aromatics Total sulphur Mercaptan sulphur Refining composition, including synthetic content	Total acidity Total aromatics Total sulphur Mercaptan sulphur
Volatility	Distillation Flash point Density at 15 °C	Distillation Flash point Density at 15 °C
Fluidity	Freezing point Viscosity at −20 °C	Freezing point Viscosity at −20 °C
Combustion	Smoke point or Smoke point + naphthalenes Specific energy	Smoke point or Smoke point + naphthalenes Net heat of combustion (Specific Energy)
Corrosion	Copper strip	Copper strip

(continued)

Table 7 (continued)

Property	Def Stan 91-091	ASTM D7566
Thermal Stability	Thermal stability at 260 °C	Thermal stability at 260 °C
Contaminants	Existent gum	Existent gum Microseparometer
Water separation characteristics	Microseparometer rating at point of manufacture	
Conductivity	Electrical conductivity	Electrical conductivity, if required
Lubricity	Wear scar diameter (if composition meets certain criteria)	

3 Evolution of Specifications

Lamp/illuminating kerosene was selected for Whittle's jet engine because of fuels research by the Gas Turbine Panel and led to the first aviation turbine fuel specification RDE/F/KER/210. Today's main jet fuel specifications have evolved over 70 years; for example, Table 8 lists the UK Jet fuel specification development. As shown in Table 7 the specifications contain requirements for more than fuel combustion, which was the initial minimum requirement for illuminating kerosine. The fuel must retain its quality from point of manufacture to the point of use, so it must be clean and stable. It must flow properly in the aircraft tanks, not degrading as temperatures increase (impacting fuel cooled oil heat exchangers, injectors, etc.) and not cause filters to block or pumps to seize. It must have low enough viscosity to be injected as a suitably sized spray to ignite in the combustor and burn cleanly with minimal radiated heat which might degrade turbine blade life.

Specifications are living documents; many properties, limits and methods listed in them have changed over the decades. The aviation industry aspires to respond to aviation quality issues seen at any point from manufacturing to usage, in a timely manner. The principal concern is with flight reliability and safety which must not be compromised. This means that the specification must be able to change rapidly to prevent problems or to include new or modified analytical methods or manufacturing routes to benefit quality control and supply.

Table 8 UK Jet fuel specification development

Name	Issue	Date of issue (Implementation date latterly becoming 3 and now 2 months after issue)
RDE/F/KER/210	1	1943
DERD 2482	1	1947
DERD 2494	1	1957
	2	1961
	3	1964
	4	1966
	5	1968
	6	1971
	7	1971
	8	1978
	9	1983
	10	30 June 1988
Def Stan 91-091	1	1 September 1994
	2	8 May 1996
	3	12 November 1999
	4	14 June 2002
	4 amendment 1	30 January 2004
	5	8 February 2005
	5 amendment 1	31 March 2006
	5 amendment 2	9 March 2007
	6	8 April 2008
	6 amendment 1	25 August 2008
	7	18 February 2011
	7 amendment 1	16 December 2011
	7 amendment 2	1 December 2012
	7 amendment 3	2 February 2015

(continued)

Table 8 (continued)

Name	Issue	Date of issue (Implementation date latterly becoming 3 and now 2 months after issue)
	8	26 August 2016 (not implemented)
	9	3 October 2016
	10	28 September 2018
	11	28 October 2019
	12	14 September 2020
	13	28 January 2022
	14	4 March 2022
	15	10 November 2023 (not implemented)
	16	14 December 2023
	17	28 June 2024
	18	28 December 2024

3.1 Reasons for Specification Changes

Specifications can or may change for several reasons. Briefly these are:

- Performance shortcomings with "on spec" fuel, suggesting that the fuel is not fit for purpose. This is rare but the most important reason for a specification to change and is discussed in the next section.
- Changing user requirements, e.g., +100 additives were sought to improve fuel thermal stability for use in future aircraft with greater thermal stresses.
- Other special operational needs (mainly for the military).
- To achieve better alignment between international specifications. For example, laboratory methods, limits, etc.
- Alignment with other fuels under same authority (military with civil, as with the Defence Standard series of specifications).
- Redundancy of 1 or more property limits (e.g., removal of an olefins limit).
- Commercial reality, such as an additive no longer being commercially available so listing it in a specification is redundant.
- Availability of new additives to fulfil existing or additional performance or handling needs.
- The development of new test methods with better accuracy and precision for measuring existing properties. This is perhaps the most common reason for a specification change. One example is the introduction of metrology methods to provide a quantitative measure of deposit thickness in thermal oxidation tests, rather than referring to a comparative visual colour standard.
- The development of new methods for measuring properties recently identified as being relevant.

- Development of new fuels, covering semisynthetic fuels made according to ASTM D7566 and fuels made by co-processing synthetic materials with conventional refinery streams.
- To correct grammar and formatting, and to reword sections for greater clarity.
- The specification can also respond at times when security of supply is affected; for instance, there were changes to freeze point limits to accommodate fuel shortages in the early seventies. [Aside: Jet fuel, based on kerosene, is not the main product from refineries but is the cut taken between gasoline and diesel. Meeting gasoline and diesel demands in times of crude shortages may affect the distillation range available for kerosine production].

A few examples of such changes from 1970 onwards are listed in Table 9.

There are other factors that may drive fuel changes, affecting specifications if they enable other technologies or desirable outcomes. These include the changing crude slates that can be processed, changing (expanding) the refinery technologies that are used for kerosine production, security of energy supply (the original reason for approving synthetic fuel components), the drive to reduce environmental issues such as noise, NO_x, SO_x emissions, changed flight patterns. In addition, there is the need to reduce CO_2 footprint through use of aviation fuels manufactured from sustainable feedstocks. Specification changes can be proposed but ultimately, they need to be safe and affordable, without impacting fuel supply and still meeting the performance requirements of the aviation industry.

3.2 Performance Shortcomings

Most fuel-related problems are because a fuel has arrived "off-specification" due to deterioration and/or being contaminated somewhere in the distribution system. This is usually attributable to shortcomings with the fuel handling and is not a specification problem; in addition, there is often a remediation option to regain product quality which is the focus of the present book.

The real challenge is when a fuel that was made and delivered "on specification" causes operational problems. Such instances are very rare but require detailed investigation, the most extreme cases being overseen by the relevant aviation (safety) authority, so the root cause can be established, and the specification updated to prevent recurrence.

In recent years, four examples of such incidents have occurred necessitating specification changes, see Table 10. In all these cases, for geographical or airline scheduling reasons, the impacted aircraft were exposed to a limited slate of fuels. Anyone undertaking management of change risk evaluations for new fuel production should establish whether that fuel could be a sole source fuel for some operators and not assume that an aircraft's average fuel diet will prevent any problems occurring.

Table 9 Examples of specification changes for reasons other than performance issues

Issue	Time	Spec change
Availability of kerosine range fuels	1970s, change implemented 1980	Jet A-1 freeze point relaxed to −47 °C from −50 °C
Metal deactivator additive became available (to manage thermal stability issues linked to trace metal contamination)	1983	Included as an optional additive
Copper content of copper sweetened fuels deemed redundant because of thermal stability test	1988	Dropped 150 ppb max copper content requirement
ASA-3 static dissipator production stopped	1995	Removed from list of approved SDAs
Olefins limit deemed redundant—thermal stability test was more sensitive	1999	Removed 5% v/v limit on olefins
Silver strip test, only being required by one military air force	1999	Removed. Helped production flexibility particularly with hydrocrackers. Option for the customer to add Ag (Silver) strip if needed
Security of supply of jet fuel in South Africa	1999	First synthetic blend approval, with specific Fischer–Tropsch stream from Sasol's Secunda plant included in Def Stan 91-091
Semi-synthetic fuels	2009	First introduction of fuels made to ASTM D7566
Thermal stability rating challenges by visual rating method	2014	Introduction of first metrology method for measuring deposit thickness
Smoke point limit of 19 mm	2016	Reduced to 18 mm in Def Stan 91-091, to align with ASTM D1655
Co-processed fuels	2019	First introduction of co-processing clause
Kathon™ FP 1.5 Biocide	2023	Removed from approved biocides

However, some fuel specification changes that have been proposed from time to time have eventually been rejected due to the impact on global fuel availability. This was the case after an inflight explosion due to an aircraft electrical fault which might have been prevented with higher flash point product. Design requirements and use of deoxygenation (inerting) systems for a fuel tanks, both for new aircraft and as retrofits, offered a better, more direct solution to address the issue.

Table 10 Performance shortcoming issues that have led to specification changes

Issue	Time	Specification change
Issues with Main Engine Control units on flights to and from Brazil, and shortened maintenance intervals. Initially established that passing ASTM D1660 [75] Coker test fuels did not pass D1655 Thermal Oxidation test at 260 °C. Later issues indicated high mercaptan sulphur levels were impacting thermal stability. There was also a contributing influence of trace lead from the refinery sweetening process. Note: Airline was operating dedicated aircraft to and from Brazil, not altering their routes or fuel source (single supply), so this increased their exposure to the fuel	Mid 1980s	Increased thermal stability requirement in Brazilian spec Further controls on mercaptan sulphur levels in Brazilian spec QAV-1
Engine pumps failed on take-off, with flights aborted in New Zealand. Involved aircraft, many with older pumps, that had a constant diet of fuel with poor lubricity characteristics. Note: Long haul aircraft, seeing a mixed diet of fuels, did not have problems	1996, with changes implemented 1999	Def Stan 91-091 added lubricity requirement for fuels with high levels of severely hydroprocessed fuels
Aviation gasoline contaminated with excessive refinery processing additive (ethylene diamine) in Australia causing carburettor failure, see more in Chap. 4	Late 1999-early 2000; changes implemented 2002	Def Stan 91-091 added Management of change requirement
Pipeline Drag Reducing Additive has become popular in multi-product pipelines, to increase efficiency during movement of fuels (particularly diesel) through pipelines. There have been several reported incidents of aircraft engine issues, which led to an incidental contamination limit requirement in jet fuel specifications. Due to the nature of this high molecular weight polymer, it is not approved to be present in jet fuel. The limit in the specifications is a functional definition of "Nil" as it represents the limit of detection	Changes implemented 2018	Def Stan 91-091 added Drag Reducing Additive to the incidental materials table Annex C

3.3 *Making Changes*

Specifications change because of inputs from interested parties in the aviation fuel supply chain. All new problems need to be shared with the community and improvements to the specifications suggested. This is particularly important for anything that falls under "performance shortcomings", where "on specification" fuel causes problems. Is the specification not well enough written or is it not capturing the most recent events, design or operational changes that cause a fuel to be unfit for purpose? Are there problems with methodology and/or its usage that are being missed?

There is no all-knowing, all-seeing body that will make these things happen automatically. Historically such changes were driven by large aviation original equipment manufacturers (OEMs), major oil companies, additive companies, airlines, military, and test equipment manufacturers. However, this demographic is changing with new players entering the market, either through novel manufacturing technology, distribution system involvement, airport facilities management, or other aspect of aviation fuel supply. It is not sufficient to assume that the traditional industry will cover the costs and manpower to support future specification development; all involved need to play an active role.

4 Future Trends

The specifications will continue to change for administrative, commercial, methodology, and fuel change reasons. Climate change and the security of supply-driven moves to produce sustainable fuels will probably account for the biggest changes, but conventional petroleum-derived fuels are unlikely to stay static. The bigger changes anticipated for the main specifications will relate to an increasing range of co-processed fuels. More routes (and feedstocks) will be approved in the ASTM D7566 specification for semi-synthetic drop-in fuels, which is also likely to expand to cover fully synthetic drop-in fuels. Another specification for non-drop-in synthetic fuels is also expected. In addition, a further challenge may be the manufacture of fuels closer to the point of use, with blending locations needing to build experience similar to traditional refineries for product control, sampling, and laboratory analysis.

While the focus of the foregoing has been on aviation turbine fuel specifications, aviation gasoline specifications, such as ASTM D910 [76], D7547 [77], and Defence Standard 91-090 [78], have followed a similar path. Indeed, today, perhaps one of the most significant challenges awaits the industry, the transition from grades containing tetraethyl lead octane enhancement additive to unleaded alternatives. Further details can be found in Chap. 4.

Acknowledgements The author would like to thank Tina Gleaves and Alisdair Clark for their helpful comments and additional insights during the preparation of this chapter.

References

1. Aviation: benefits beyond borders. Air Transport Action Group (2020)
2. Goodger EM (ed) (1989) First international symposium on aviation turbine fuel specifications. Their history and future development, Guildhall, Bath, 19–21 September 1988. Cranfield/MOD
3. Datschefski G, Lewis C, Walters MB (1997) Jet fuel specification requirements. Strategic review. DERA SP-101, Copyright DERA
4. Edwards T (2007) Advancements in gas turbine fuels from 1943 to 2005. J Eng Gas Turbines Power 129(1):13–20 (published online Feb 2006)

5. Clark AQ (2007) Aviation gasoline: history and future. In: IASH 2007, the 10th international conference on stability, handling and use of liquid fuels Tucson, Arizona
6. Ogston A (1981) A short history of aviation gasoline development, 1903–1980 (SAE 810848). In: Trends in aviation fuels and lubricants, pp. 1–14, SP-492, SAE, Warrendale, PA
7. EN 590. Automotive fuels-diesel-requirements and test methods, British Standards Institution
8. Defence Standard 91-091. Turbine fuel, aviation kerosine type, Jet A-1 NATO code: F-35, Joint Service Designation: AVTUR, Issued by the United Kingdom (U.K.) Ministry of Defence. Glasgow, UK. https://www.dstan.mod.uk/
9. ASTM D1655. Standard specification for aviation turbine fuels. ASTM International, 100 Barr Harbor Drive, PO Box C700, West Conshohocken, PA 19428-2959, United States
10. CAN/CGSB 3.23 aviation turbine fuel (Grades JET A and Jet A-1), by the Canadian General Standards Board, Ottawa, Ontario K1A. https://www.publications.gc.ca
11. DCSEA 134. French specification for F-34 and F-35, available from the French Army Fuel Service. www.defense.gouv.fr/essences
12. ASTM D6615. Standard specification for Jet B wide-cut aviation turbine fuel. ASTM International, 100 Barr Harbor Drive, PO Box C700, West Conshohocken, PA 19428-2959, United States
13. CAN/CGSB 3.22. Wide-cut type aviation turbine fuel (Grade JET B), published by the Canadian General Standards Board, Ottawa, Ontario K1A
14. World jet fuel specifications with avgas supplement, ExxonMobil Aviation. 2008 edition can be downloaded from, e.g., https://docslib.org/doc/11927340/world-jet-fuel-specifications-with-avgas-supplement
15. Aviation fuels technical review, Copyright 2007 Chevron Products Company, 6001 Bollinger Canyon Road, San Ramon, CA 94583; https://www.chevron.com/-/media/chevron/operations/documents/aviation-tech-review.pdf
16. The AeroShell Book, Shell Global Solutions, 2021 edition. https://www.shell.com/business-customers/aviation/aeroshell/knowledge-centre/the-aeroshell-book.html
17. IATA. Aviation turbine fuels specifications guidance material. https://www.iata.org
18. ASTM D7566. Standard specification for aviation turbine fuel containing synthesized hydrocarbon. ASTM International, 100 Barr Harbor Drive, PO Box C700, West Conshohocken, PA 19428-2959, United States
19. GB 6537. The People's Republic of China government specification for No 3 Jet Fuel. Partial translation can be found online, with tables in [10]
20. GOST 10227-1986 (revised 1995; 2013 version cancelled). Jet fuels. Specification, maintained by the Euro-Asian Council for Standardization, Metrology and Certification (EASC). Covers TS-1 and RT grades
21. GOST R 52050-2003. Aviation turbine fuel Jet A-1 specification, maintained by the Euro-Asian Council for Standardization, Metrology and Certification (EASC). Covers a Jet A-1 grade using ASTM as well as GOST methods
22. MIL-DTL-83133. Detail specification: turbine fuel, aviation, kerosene type, JP-8 (NATO F-34), NATO F-35, and JP-8+100 (NATO F-37)
23. Standardisation Agreement STANAG 3747—guide specification (minimum quality standards) for aviation turbine fuels (F-34, F-35, F-40, AND F-44)
24. NATO AFLP-3747. Guide specifications (minimum quality standards) for aviation turbine fuels (F-24, F-27, F-34, F-35, F-37, F-40 AND F-44)
25. Defence Standard 91-087. Turbine fuel, aviation kerosine type: containing fuel system icing inhibitor NATO code: F-34 joint service designation: AVTUR/ FSII, Defence Equipment, UK Defence Standardization, Kentigern House, 65 Brown Street, Glasgow G2 8EX, UK
26. CAN/CGSB 3.24. Aviation turbine fuel (military grades F-34 and F-44), published by the Canadian General Standards Board, Ottawa, Ontario K1A
27. CAN/CGSB 3.23. Aviation turbine fuel (Grades JET A and JET A-1), published by the Canadian General Standards Board, Ottawa, Ontario K1A
28. Defence Standard 91-088. Turbine fuel, aviation: wide cut type, containing fuel system icing inhibitor; NATO code: F-40; Joint Service Designation: Avtag/FSII, Defence Equipment and

Support, UK Defence Standardization, Kentigern House, 65 Brown Street, Glasgow G2 8EX, UK
29. MIL-DTL-5624. Detail specification: turbine fuel, aviation, kerosene type, Grades JP-4 and JP-5
30. Defence Standard 91-086. Turbine fuel, aviation kerosine type: high flash type, containing fuel system icing inhibitor NATO code: F-44 JSD: AVCAT/FSII, Defence Equipment and Support, UK Defence Standardization, Kentigern House, 65 Brown Street, Glasgow G2 8EX, UK
31. EI/JIG 1530. Quality assurance requirements for the manufacture, storage and distribution of aviation fuels to Airports. http://www.jigonline.com; https://publishing.energyinst.org/featured/1530
32. ATA-103. Standard for jet fuel quality control at airports. Air Transport Association of America, Inc. (ATA) d/b/a Airlines for America, 1275 Pennsylvania Ave. NW, Suite 1300, Washington, D.C. 20004. http://www.airlines.org
33. IP 475. Petroleum liquids—manual sampling (ISO 3170:2004). Energy Institute, London
34. ASTM D4057. Practice for manual sampling of petroleum and petroleum products. ASTM International, 100 Barr Harbor Drive, PO Box C700, West Conshohocken, PA 19428-2959, United States
35. EI 1533. Quality assurance requirements for semi-synthetic jet fuel and synthetic blending components (SBC). Energy Institute, London.
36. ASTM D4175. Terminology relating to petroleum products, liquid fuels, and lubricants. ASTM International, 100 Barr Harbor Drive, PO Box C700, West Conshohocken, PA 19428-2959, United States
37. Manual on significance of tests for petroleum products, MNL 1. ASTM International, 100 Barr Harbor Drive, PO Box C700, West Conshohocken, PA 19428-2959, United States
38. Totten GE, Westbrook SR, Shah RJ (eds) Fuels and lubricants handbook: technology, properties, performance, and testing, MNL 37. ASTM International, 100 Barr Harbor Drive, PO Box C700, West Conshohocken, PA 19428-2959, United States
39. CRC handbook of aviation fuel properties, Coordinating Research Council Inc. 5755 North Point Parkway Suite 265, Alpharetta, GA 30022
40. ASTM D156. Standard test method for Saybolt color of petroleum products (Saybolt chromometer method). ASTM International, 100 Barr Harbor Drive, PO Box C700, West Conshohocken, PA 19428-2959, United States
41. ASTM D6045. Standard test method for color of petroleum products by the automatic tristimulus method. ASTM International, 100 Barr Harbor Drive, PO Box C700, West Conshohocken, PA 19428-2959, United States
42. ASTM D5452. Standard test method for particulate contamination in aviation fuels by laboratory filtration. ASTM International, 100 Barr Harbor Drive, PO Box C700, West Conshohocken, PA 19428-2959, United States
43. IP 565. Determination of the level of cleanliness of aviation turbine fuel—portable automatic particle counter method. Energy Institute, London
44. IP 577. Determination of the level of cleanliness of aviation turbine fuel—automatic particle counter method using light extinction. Energy Institute, London
45. IP 354. Determination of the acid number of aviation turbine fuels—colour-indicator titration method. Energy Institute, London
46. ASTM D3242. Standard test method for acidity in aviation turbine fuel. ASTM International, 100 Barr Harbor Drive, PO Box C700, West Conshohocken, PA 19428-2959, United States
47. IP 156. Determination of hydrocarbon types in petroleum products—fluorescent indicator adsorption method. Energy Institute, London
48. ASTM D1319. Standard test method for hydrocarbon types in liquid petroleum products by fluorescent indicator adsorption. ASTM International, 100 Barr Harbor Drive, PO Box C700, West Conshohocken, PA 19428-2959, United States
49. IP 436. Determination of aromatic hydrocarbon types in aviation fuels and petroleum distillates—high performance liquid chromatography method with refractive index detection. Energy Institute, London

50. ASTM D6379. Standard test method for determination of aromatic hydrocarbon types in aviation fuels and petroleum distillates—high performance liquid chromatography method with refractive index detection. ASTM International, 100 Barr Harbor Drive, PO Box C700, West Conshohocken, PA 19428-2959, United States
51. IP 336. Petroleum products—determination of sulfur content—energy-dispersive-X-ray fluorescence method. Energy Institute, London
52. ASTM D3227. Standard test method for (Thiol Mercaptan) sulfur in gasoline, kerosine, aviation turbine, and distillate fuels (potentiometric method). ASTM International, 100 Barr Harbor Drive, PO Box C700, West Conshohocken, PA 19428-2959, United States
53. IP 30. Detection of mercaptans, hydrogen sulfide, elemental sulfur and peroxides—Doctor test method. Energy Institute, London
54. IP 123. Petroleum products—determination of distillation characteristics at atmospheric pressure. Energy Institute, London
55. ASTM D86. Standard test method for distillation of petroleum products and liquid fuels at atmospheric pressure. ASTM International, 100 Barr Harbor Drive, PO Box C700, West Conshohocken, PA 19428-2959, United States
56. IP 170. Determination of flash point—Abel closed-cup method (ISO 13736:2021). Energy Institute, London
57. IP 365. Crude petroleum and petroleum products—determination of density—oscillating U-tube method. Energy Institute, London
58. ASTM D4052. Standard test method for density, relative density, and API gravity of liquids by digital density meter. ASTM International, 100 Barr Harbor Drive, PO Box C700, West Conshohocken, PA 19428–2959, United States
59. IP 16. Determination of the freezing point of aviation fuels—manual method. Energy Institute, London
60. ASTM D2386. Standard test method for freezing point of aviation fuels. ASTM International, 100 Barr Harbor Drive, PO Box C700, West Conshohocken, PA 19428-2959, United States
61. IP 71. Section 1: Petroleum products—transparent and opaque liquids—determination of kinematic viscosity and calculation of dynamic viscosity, Section 2 Glass capillary kinematic viscometers—specifications and operating instructions. Energy Institute, London
62. ASTM D445. Standard test method for kinematic viscosity of transparent and opaque liquids (and calculation of dynamic viscosity). ASTM International, 100 Barr Harbor Drive, PO Box C700, West Conshohocken, PA 19428-2959, United States
63. IP 598. Petroleum products—determination of the smoke point of kerosine, manual and automated method. Energy Institute, London
64. ASTM D1322. Standard test method for smoke point of kerosene and aviation turbine fuel. ASTM International, 100 Barr Harbor Drive, PO Box C700, West Conshohocken, PA 19428-2959, United States
65. ASTM D1840. Standard test method for naphthalene hydrocarbons in aviation turbine fuels by ultraviolet spectrophotometry ASTM International, 100 Barr Harbor Drive, PO Box C700, West Conshohocken, PA 19428-2959, United States
66. IP 154. Petroleum products—corrosiveness to copper—copper strip test. Energy Institute, London
67. ASTM D130. Standard test method for corrosiveness to copper from petroleum products by copper strip test ASTM International, 100 Barr Harbor Drive, PO Box C700, West Conshohocken, PA 19428-2959, United States
68. IP 323. Determination of thermal oxidation stability of gas turbine fuels. Energy Institute, London
69. ASTM D3241. Standard test method for thermal oxidation stability of aviation turbine fuels. ASTM International, 100 Barr Harbor Drive, PO Box C700, West Conshohocken, PA 19428-2959, United States
70. IP 540. Determination of the existent gum content of aviation turbine fuel—Jet evaporation method. Energy Institute, London

71. ASTM D3948. Standard test method for determining water separation characteristics of aviation turbine fuels by portable separometer. ASTM International, 100 Barr Harbor Drive, PO Box C700, West Conshohocken, PA 19428-2959, United States
72. IP 274. Determination of electrical conductivity of aviation and distillate fuels. Energy Institute, London
73. ASTM D2624. Standard test methods for electrical conductivity of aviation and distillate fuels. ASTM International, 100 Barr Harbor Drive, PO Box C700, West Conshohocken, PA 19428-2959, United States
74. ASTM D5001. Standard test method for measurement of lubricity of aviation turbine fuels by the ball-on-cylinder lubricity evaluator (BOCLE). ASTM International, 100 Barr Harbor Drive, PO Box C700, West Conshohocken, PA 19428-2959, United States
75. ASTM D1660. Standard test method for thermal stability of aviation turbine fuels (suspended). Method of test for thermal stability of aviation turbine fuels (withdrawn 1992). ASTM International, 100 Barr Harbor Drive, PO Box C700, West Conshohocken, PA 19428-2959, United States
76. ASTM D910. Standard specification for leaded aviation gasolines. ASTM International, 100 Barr Harbor Drive, PO Box C700, West Conshohocken, PA 19428-2959, United States
77. ASTM D7547. Standard specification for hydrocarbon unleaded aviation gasoline. ASTM International, 100 Barr Harbor Drive, PO Box C700, West Conshohocken, PA 19428-2959, United States
78. Defence Standard 91-090. Gasoline aviation: grades UL91, 100/130 and 100/130LL. Joint Service Designation: AVGAS 100 and AVGAS 100LL. Defence Equipment and Support, UK Defence Standardization, Kentigern House, 65 Brown Street, Glasgow G2 8EX, UK

Chapter 2
Manufacture of Jet Fuel in Refineries

Amiram Groweiss and Joanna M. Bauldreay

Abstract Aviation fuels are produced in oil refineries around the world mainly by atmospheric-pressure distillation of crude oil. Refinery activities that take account of crude properties and quality requirements for their complete range of products are essential to produce problem-free, fit-for-purpose aviation fuels. This chapter focusses on jet fuel production, from distillation and heavy stream upgrading processes that set most bulk properties, to finishing processes that control cleanliness and stability. It also addresses briefly more activities in the refineries, such as additivation, sampling, testing, and product release. The slightly different requirements for military grade fuels are discussed as are trends in refinery operations toward production of future fuels.

1 Introduction

1.1 General Introduction and Scope

The manufacture of jet fuels is more conservative than for other transportation fuels. Nevertheless, over the last 25 years the industry has made significant progress; it has approved new components to be mixed with conventional jet fuel, such as synthetic paraffinic kerosenes (SPKs) and, to help mitigate aviation's contribution to carbon dioxide (CO_2) emissions, increasingly favouring sustainable aviation fuel (SAF) components. Despite this, most jet fuels are still prepared by the distillation of petroleum and will be for the immediate future. The most recent generation of aircraft, including the "Dreamliner" (Boeing 787), Boeing 737 Max, Airbus A350, and Airbus A321neo, which reflects the quick technological developments of our

A. Groweiss (✉)
Ashdod Oil Refinery, Ashdod, Israel
e-mail: amiramg@zahav.net.il

J. M. Bauldreay
Bauldreay Jet Fuel Consulting Limited, Chester, United Kingdom
e-mail: Joanna@Bauldreay.com

M. Rabaev et al. (eds.), *Aviation Fuel Problems and Solutions: From Refinery to Wingtip*, https://doi.org/10.1007/978-3-032-05514-9_2

era, will be dependent on similar hydrocarbon-based aviation fuels for the duration of their lifespans, which could be at least 20–30 years. Liquid hydrocarbon fuels will need to be available for use for several decades to come.

This chapter deals with the production of aviation fuels from conventional crudes in oil refineries, focussing specifically on jet fuels for both civil and military users. Aviation gasoline production, which takes place at very few refineries nowadays, is covered in Chap. 4. Relevant and useful reading matter is compiled in the bibliography at the end of this chapter.

The crude or crudes at a refinery are separated by distillation in towers to provide several products, such as light gases, naphtha, kerosene,[1] gasoil (e.g., diesel), fuel–oil and bitumen. Those initial distillates and residual oils are then treated in secondary processes and are provided with additives to produce fuel products that are fit-for-purpose. The predominant aviation fuel, jet fuel, is prepared almost exclusively from kerosene, which is a middle-distillate fraction. This chapter will go into more detail about the various steps from distillation to final jet fuel manufacture and release; this falls into the two main areas of processing and oil movements. Making—including storing and blending—a good jet fuel is the first step in delivering a fuel suitable for the end users, airlines.

Production of synthetic or sustainable jet fuel components (the two categories overlap but are not synonymous) is outside the scope of this chapter. The reader is directed instead to Chap. 3, which covers Fischer–Tropsch (FT), hydrogenated esters and fatty acids (HEFAs), alcohol-to-jet (ATJ), and other approved synthetic components that can be blended with conventional crude-derived jet fuels and streams.

1.2 Specifications and Kerosene

Jet fuel specifications are very explicit about the materials that may—or may not—be used in their manufacture. Defence Standard 91-091 (Def Stan 91-091) [1] says in its materials section that jet fuel "shall consist predominantly of refined hydrocarbons derived from conventional sources including crude oil, natural gas liquid condensates, heavy oil, shale oil and oil sands, and qualified additives…"; a similar requirement is set out in ASTM D1655 [2]. Both specifications (see more in Chap. 1) refer to other sections to spell out how and when material from other sources may be used.

Kerosene consists of paraffinic, naphthenic, and aromatic hydrocarbons, predominantly having carbon numbers in the C8 through C16 range [3], and it mainly boils in the range of 150–290 °C (302–554 °F). It may also contain less than 0.1% w/w of

[1] The preferred English spelling is kerosine, but this chapter will use kerosene. This avoids confusion when mentioning synthetic jet fuel components, which are in the ASTM D7566 standard specification for synthetic kerosenes [4]. Likewise, S is spelled out as sulfur rather than sulphur, the former now being adopted for English scientific usage rather than the latter, traditional spelling.

performance additives.[2] This broad definition covers the most commonly accepted aviation turbine fuel known as Jet Fuel A-1 or, more usually, as Jet A-1.[3] Jet A-1 is defined by many properties, but the two key ones for the refiner are the freezing point, −47 °C maximum, and the flash point, +38 °C minimum. There are other aviation fuel grades differing in those two properties like Jet-A (freezing point below −40 °C and the primary fuel made for the USA market), Jet B (a mixture of kerosene and naphtha with a lower freezing point, −60 °C maximum and greater volatility) and JP-5 (a military aviation fuel with minimum flash point of +60 °C for shipboard safety purposes); see more in Chap. 1. Based on these properties, the refiner has the freedom to choose between grades of products to make; this includes selecting the proper crude oil, the distillation conditions and how to make the "cuts" between fractions, and then subsequent treatments that will make aviation fuel that meets the specification completely and is fit for purpose.

1.3 Refining and Crudes

Most jet fuels are produced in refineries by the atmospheric distillation of crude oil, often referred to as petroleum crude or petroleum. The source of crude oils may also be oil sands, bituminous rocks, or other resources of fossil organic materials, though not coal, but still petroleum from oil wells is the main supply for all fuel products including kerosene. Worldwide there are hundreds of crude oil types, usually recognized by their names, which often reflect their geographical origin. Everyone knows Brent crude oil from northern Europe, West Texas Intermediate (WTI) from the south of the USA, Ural from central Russia, Arabian Light, and many more.

Crude and refinery process units need to be compatible. A refinery's configuration reflects the crude or crudes being processed and the products that the refinery can sell. Historically refineries often used a single, local crude but increasingly they now must process a range of crudes. The more complex the refinery, the more flexibility it will have in its choice of crudes and thus the more scope to buy cost-effective crudes. Technically driven choices will depend on information, provided by crude assays, which impacts the quality and quantity of jet fuel that can be made.

Each crude oil has different chemical and physical properties: density, type of hydrocarbons and their relative abundance, content of sulfur, salt, water, residue of metals, and other elements, etc. Crude assays describe these properties over temperature ranges covering the entire distillation range and, therefore, the entire product

[2] In Europe, chemical substances must be assigned CAS and EINECS numbers. (Similar requirements, e.g., TSCA numbers, exist for US systems). There are at least 6 kerosene range products that can be found in jet fuel although few safety data sheets recognise this, with one CAS number 8008–20-6 being most referenced.

[3] This document concentrates on Jet A-1 made to Def Stan 91–091 requirements. Where relevant, the differences between this and Jet A-1 made to ASTM D1655 are noted, as are differences between the Jet A-1 s and other grades of fuels. Chapter 1 addresses specifications and key quality requirements in detail; this includes mention of Russian and Chinese jet fuels.

portfolio; they are not discussed in detail in this chapter, except as far as they affect jet fuel production.

Initially, the type of the hydrocarbons is a very important factor for the selection of a proper crude. Different categories of crudes include:

- paraffinic, where at least 75% of the petroleum is comprised of straight ("normal paraffins") or branched chains ("iso paraffins"),
- naphthenic (cycloparaffinic), which contains at least 70% of naphthenic hydrocarbons, with five-, six- or seven-membered non-aromatic rings, asphaltic crude, and mixtures thereof,
- aromatic, with at least 50% being aromatic rings.

Aromatic-type crudes are not ideal for jet fuel production because standards put a limit of 25% (v/v) total aromatics for jet fuel. Higher total aromatics in jet fuels may cause unacceptable levels of soot formation and incomplete combustion. Paraffinic light crudes tend to be preferred, though naphthenic ones can provide excellent low temperature properties. The final choice will be a compromise that optimises the refinery's slate of products and the overall economics.

Assays carried out for many crude oils have shown that the kerosene fraction in many of the paraffinic light crudes, namely the maximum portion of hydrocarbons that boils between 150 and 290 °C (302 and 554 °F), may reach up to 20%, however, in practice, the potential content of jet fuel in the total crudes is usually between 9 and 12% by weight. Table 1 provides a comparison of crudes' yields in an indicative kerosene relevant range of 150–230 °C. (Data from assay reports of major oil companies like BP, ExxonMobil, and Shell.)

Table 1 Kerosene contents in assorted crudes

Crude	Gravity API	Distilled between 150–230 °C (%)	Sulfur (% weight) (%)
Arabian light	33.0	15.6	1.77
Azeri (Ceyhan)	38.2	16.9	0.17
Bakken	43.3	17.2	0.07
Brent	38.6	15.1	0.39
Dalia	23.0	12.2	0.51
Kirkuk	34.2	15.2	2.24
North Alaska Slope	33.0	12.9	0.91
Osberg Blend	39.6	16.2	0.20
Ural	31.8	11.6	1.35
West Texas intermediate (WTI)	40.8	15	0.34

1.4 *Important Crude Aspects and Allowed Starting Materials for Jet*

As well as carbon and hydrogen, fossil sources of petroleum also include elements such as oxygen, nitrogen, sulfur, silicon, sodium, and other metals, e.g., nickel and vanadium. Many are separated out during distillation and do not appear in the jet fuel product. There are also organic compounds, like carboxylic acids, amines and other nitrogenous compounds, sulfides, disulfides, mercaptans and other sulfur-containing compounds, oxygenates, and so on. Not all can be separated out during distillation and may be found in the jet fuel product. The refiner needs to know the relative levels of these compounds in different crudes and how they can be controlled to acceptable levels to make the final product.

1.4.1 Acids

Some crude oils are more "acidic" than others and therefore may be responsible for corrosion of metallic parts. Acidity is measured by titration, with the total acid number, TAN, reported in terms of the amount of potassium hydroxide solution needed to neutralise the fuel. Comparing, say, Azeri Light crude (TAN: 0.28 mg KOH/g) with Brent blend (TAN: 0.064 mg KOH/g) clearly demonstrates the acidity of the former crude. The high acidity is not related to strong mineral acids, like hydrochloric or sulfuric, which may be separated easily from crude oil, but to the presence of organic acids containing a carboxyl functional group –COOH. A special group of organic acids, naphthenic acids, can remain even after mild hydrotreatment. Their structure involves a naphthenic ring, a chain of few carbon atoms and a terminal carboxylic group; a typical compound is shown in Fig. 1.

Naphthenic acids are one of the reasons for jet fuel acidity, which is limited to 0.015 mg KOH/g maximum in Def Stan 91-091. As mentioned in Chap. 1, ASTM D1655 has a significantly higher limit of 0.10 mg KOH/g but the North American distribution system often has downstream clay treating options, in place to remove surfactants and other impurities that also may reduce the TAN well below this level before it reaches aircraft. High acidity of kerosene streams intended for jet fuel production may be reduced in the refinery by caustic treatment with NaOH or by hydroprocessing; see later.

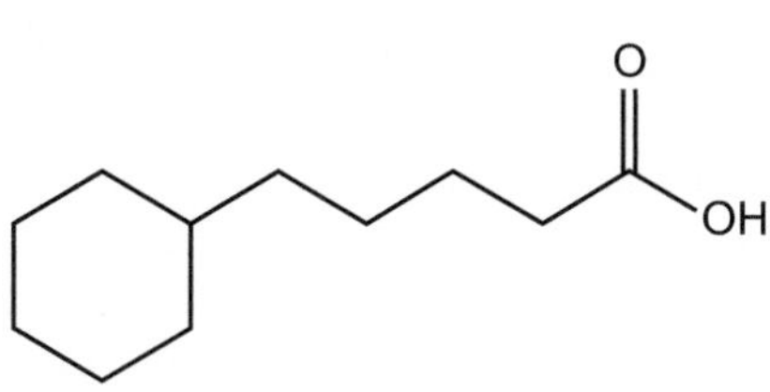

Fig. 1 A typical naphthenic acid, $C_{11}H_{20}O_2$

1.4.2 Sulfur

Sulfur is a common element in all crude oils. There is a distinction between low content (say, below 0.50% by weight) which is called "sweet crude", and high content of sulfur, "sour crude". Sulfur occurs in light gases like the toxic hydrogen sulfide (H_2S) or carbonyl sulfide (COS), in the acidic, corrosion-forming mercaptans or thiols (R-SH), sulfides (R-S-R), disulfides (R-S-S-R), and aromatic heterocyclic compounds like derivatives of thiophene, etc. The main issue with sulfur is its combustion to sulfur oxides such as SO_2 and SO_3, a major source of air pollution and of acid rain. Nevertheless, unlike the limitation of sulfur in gasoline and diesel to 10 ppm in those fuels in many parts of the world, the jet fuel limit is 3000 ppm (0.30 wt%), though the average sulfur level is often well below this.

Note: with gasoline and diesel fuels making up the biggest proportion of their product slate, refiners nowadays favour the use of low sulfur crudes and more severe processing conditions to make ultralow sulfur fuels. This will impact the kerosene qualities too because it is needed as a diesel blending component. The refiner may alter the processing for the kerosene when making jet rather than diesel blending components, or, in more modern refineries, the diesel and kerosene will be treated in a single unit (reducing the number of process units) and then be separated.

Therefore, the total sulfur content in jet fuel is not a real limitation, but mercaptans must be reduced to levels below 30 ppm because of their ability to induce corrosion in metallic aircraft components or to attack elastomers such as may be found in seals. There are several processes in the refinery to reduce the total content of sulfur, such as the hydrodesulfurization process (HDS), which converts sulfur into H_2S by catalytic reaction with hydrogen. Additionally, mercaptans can be reduced by HDS, or otherwise converted into less corrosive disulfides via processes such as the Merox™ process, though the total sulfur level is not reduced.

1.4.3 Other Heteroatomic and Undesirable Elements

Various oxygen and nitrogen species occur naturally in crude oils and are not considered extraneous. As such, there are no direct measurements for their content in jet fuels, although their amounts, especially nitrogen, are reduced during hydrodenitrogenation, which occurs simultaneously during HDS, or other chemical treatments.

Fuel streams can also contain other undesirable elements including silicon and metals, such as nickel, vanadium, sodium, iron, copper, and aluminium. These metals may remain in the fuel from the crude oil or, more probably, be carried over from caustic washing, salt drying, mercaptan oxidation or other kerosene upgrading processes. Their presence in jet fuel should be avoided because such elements, particularly Cu and Fe at ppb levels or Zn at more like 200 ppb, may harm the thermal stability of the fuel (as measured by ASTM D3241/IP 323) or be deleterious for turbine blades (vanadium in particular). Organometallic compounds are present in crude oils, but these are either decomposed by high temperatures in the distillation tower furnace or distill at higher temperatures than those required for kerosene. Most inorganic salts are removed by "de-salters" before the distillation tower, and in any case, those compounds are not distilled at temperatures below 300 °C, as required for kerosene. No metal Ions are expected in the fuel, but if found, are the result of a catalyst carryover event or process upset in the refinery, or from metal corrosion in some point of the refinery or downstream. Metals or nitrogen are not measured in specifications written for crude-derived jet fuels, but they are sometimes identified when problematic fuels are tested.

1.4.4 Additives from Upstream Supply and Refinery Activities

Refiners need to understand and manage the full range of additives that could be in the crude used to make jet fuel. Crudes inevitably contain some additives by the time they reach the refinery, such as H_2S scavengers and pipeline drag reducers, PDRs.[4] These may be added during production or while the crude is in transit, the former to mitigate H_2S when there is potential for human exposure and the latter to reduce pumping constraints in the crude supply system. Once at the refinery, chemical and refinery slops and "marketing returns" can sometimes be added to the crude, bringing other finished fuel additives and metals into the mix. The refiner will have no control over the pre-refinery additives, though should be aware that they will change if the crudes change. However, the refiner can and should assess the risks associated with adding in the slops and returns; many refineries have done this and have decided not to allow them to be used when jet fuel is being made.

Refinery process additives may also be introduced ahead of the distillation of the crude petroleum or during other refining processes used to manufacture the finished fuel. Those functional compounds may be added to the crude at tens to hundreds of ppm levels to separate emulsions of oil and water ("demulsifiers"), to neutralize acids (by use of ammonia and amine compounds), to prevent corrosion in the distillation tower (by using filming amines), and so on. These process additives are industrial chemicals, and the refiner should take all necessary means to prevent those chemicals—or any deleterious by-products—from reaching the jet fuel product. The refiner has responsibility for process additives, even if a third-party additive company controls the additives needed to achieve suitable process unit performance. The specification explicitly requires that the risks be assessed for any changes, to type or level of additive, as part of a Management of Change (MOC) process. This requirement became part of the Def Stan 91-091 specification in 2002 after major issues in Australia in December 1999, grounding a significant portion of the light aircraft fleet that were caused by an increased level of a refinery process additive used in aviation gasoline production.

Since many refinery process additives have surfactant properties, a special test for the finished product is used to prove that the jet fuel does not contain problematic surfactants (ASTM D3948: "Determining water separation characteristics of aviation

[4] Also known as drag reducing additives, DRAs.

turbine fuels by portable separometer", MSEP, is the reference water shedding test used at point of manufacture). Some surfactants are deleterious since they can disarm downstream filters which remove water from jet fuel.

2 Initial Refinery Processes

The production of fuels in a modern oil refinery involves a few major steps or considerations after the preliminary crude treatment in a desalter to remove water, dirt, and salts, see Fig. 2.

Separation of fractions by atmospheric and vacuum distillation. Differences between boiling points of the various hydrocarbons are used to separate them into products of varying boiling point ranges.

Cracking and reforming processes are used to increase the amount of the desired products and add to their value. Feedstocks for these processes are often heavier, high boiling point materials with lower inherent value. Chemical changes brought about by cracking and reforming these heavy molecules can yield higher value, lighter products, improving the quality and yield of these streams above those levels achieved simply by the initial distillation of crude oil.

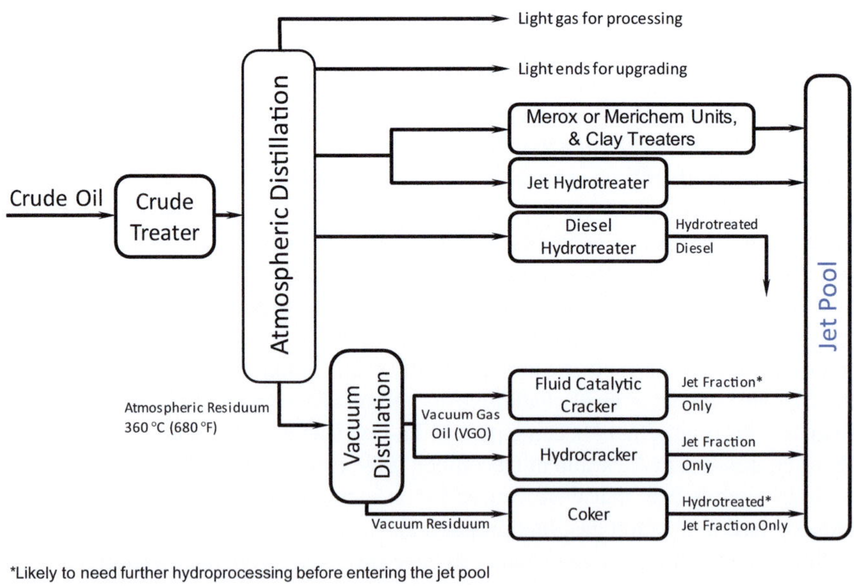

Fig. 2 Refinery jet fuel production schematic

Upgrading and Purification processes, which are intended to improve the quality of the materials that are obtained by distillation or cracking and bring them to meet specifications. An example is hydrodesulfurization to bring about the reduction of sulfur in fuels.

Finalization steps allow the products to meet finished fuel specifications. These steps may include improvement of the colour, treatment of acidity, complete elimination of water, and the use of permitted additives to prevent oxidation, to improve lubricity, or to increase electrical conductivity and offset the risks of electrostatic discharges during product movement.

Ecology: the production of fuels should be carried out in environmental-friendly processes, as much as possible, with minimal emissions to the atmosphere and minimal waste generation.

All the above are involved and considered during the production of jet fuel and are now discussed in more detail.

2.1 Separation Step—Primary Distillation

After treatment in the desalters units, crude oil is heated in a crude furnace and pumped into the atmospheric distillation unit[5] where it undergoes separation into fractions based on molecular boiling point differences. Large amounts of heat and steam cause the hydrocarbons to evaporate and then travel upwards in the tower. There is a gradient of temperatures in the tower, so when a fraction of molecules arrives at a point where the temperature is below their boiling points, they condense to liquids, and exit the tower as a fraction of different compounds that share similar boiling range, and in our case, a stream of kerosene. The decision about the range of temperatures over which kerosene should be collected, or "where to make the cut?", is specific to each refinery, though is driven by some common points. The specifications for Jet A-1 allow the laboratory distillation (according to ASTM D86) to reach a final boiling point of 300.0 °C, so the higher cut point can be made at, say, 290 °C. Jet fuels having a final boiling point that approaches 300 °C may be considered 'heavy', and usually have a density at the higher end of the permitted range approaching 840 kg/L. However, these properties are also greatly influenced by crude type as discussed previously. Aromatic and naphthenic crudes could be used to produce a jet fuel having similar final boiling points, but significantly different densities based on total aromatic content. A concern for the refiner though is that as jet fuels approach the permitted final boiling point limit, their low temperature performance (e.g., freeze point or viscosity) may be poor and higher aromatic levels may be present; the choice of crude may alleviate such concerns, with naphthenic crudes tending to produce kerosenes with very good low temperature properties. On

[5] Also called a crude distillation unit (CDU), crude unit, primary distillation or, particularly in the USA, a pipe still.

the lower side of the temperature range, the refiner may decide to make the cut deep into the naphtha fraction, say below 140 °C. Such a decision will increase the kerosene fraction at the expense of gasoline yield but may harm the flash point of the kerosene fraction. A low temperature will increase the amount of volatile, more flammable compounds in the kerosene fraction, which could lower the flash point below the 38.0 °C minimum. Therefore, the quality demands of the finished jet fuel product limit the full potential of the kerosene fraction, and the simultaneous play between flash point and freezing point will direct the refiner to select the range between, for example, 145 and 260 °C, as the two boiling points between which kerosene is taken. In the USA, which is more of a gasoline driven economy than elsewhere, the main jet fuel is Jet A, which has a higher freezing point (−40 °C maximum) than Jet A-1 and cuts more into the diesel range (maybe to 280 °C or higher) to meet demand requirements. Jet A therefore tends to have a higher final boiling point and density than Jet A-1. If a refinery makes kerosene for various jet purchasers, with different freeze or flash point requirements, they usually achieve this with a single kerosene product that meets all the base fuel specifications simultaneously. For instance, JP-5 is just a higher flash point version of Jet A-1, and Jet A-1 is a lower freeze point version of Jet A.

The initial atmospheric distillation often produces straight run kerosene that differs from the clear, bright, and almost colourless jet fuel in the later stages of the process. The stripping of light and volatile compounds by steam may leave enough water to create a turbid product. It may have a very intense bad smell due to the presence of light mercaptans above the allowable 30 ppm, may be too acidic and may have a yellowish colour, such as shown in Fig. 3. The most important qualities at that point are a laboratory distillation range that typically does not exceed 290 °C, a density between 0.775 and 0.840 kg/L at 15 °C, a flash point above 38.0 °C and a freezing point below −47.0 °C for Jet A-1 (−40 °C for Jet A). Most other specification needs can be met using further downstream processes.

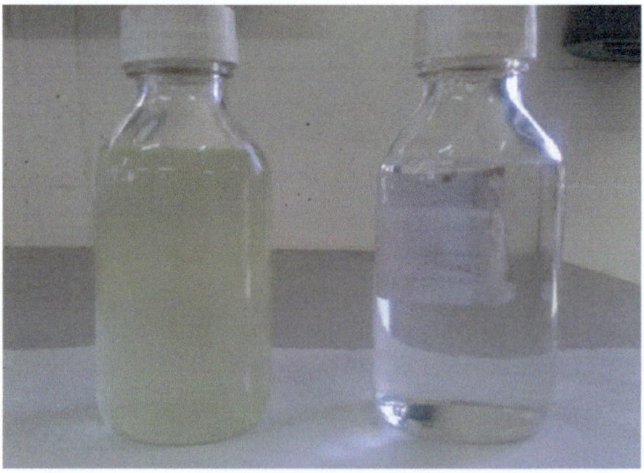

Fig. 3 Kerosene following atmospheric distillation (left) and Jet A-1 after Hydrodesulphurisation treatment (right) (Image courtesy of A. Groweiss)

2.2 Cracking Processes

Kerosene may also be produced in a refinery through various cracking processes. As the demand for air travel increased, there was a need to generate higher yields of jet fuel from a barrel of oil. The increased demand accompanied the growing price of petroleum, making use of "the bottom of the barrel" an obligation for the profitability or even survival of the oil refining industry. Cracking processes enable the heavier, lower value portion of the petroleum barrel to be upgraded to higher value products. There are several kinds of cracking processes, but this discussion focusses on the kerosene fractions from catalytic cracking, hydrocracking and delayed coking. The feeds for cracking are heavy streams of hydrocarbons that otherwise may find only limited use by customers. One example is VGO (vacuum gasoil) or HVGO (high-vacuum gasoil), a stream of aliphatic, naphthenic, and aromatic hydrocarbons that normally boils above 370 °C, so cannot be used in diesel or gasoil. The cracking processes convert or break these larger molecules into smaller ones having lower molecular weights and shorter carbon chain lengths and, therefore, lower boiling points.

Catalytic cracking takes place in a unit known as an FCC (fluidized catalytic cracker) where the molecules of VGO are broken at high temperatures (520–550 °C), low pressure and with the use of large amounts of aluminosilicate catalysts. Most products of the FCC are light gases and cracked naphtha, which are directed to products like LPG (domestic cooking gas), propylene for the plastics industry and, mainly, gasoline. Another fraction is heavy (cat) cracked naphtha, some of which can be directed, after proper treatment as discussed below, to be blended with other kerosene component streams.

Hydrocracking is another process where a VGO fraction is broken in a fixed-bed plant using higher pressures (35–200 Bar), an excess of hydrogen gas at somewhat milder temperatures than catalytic cracking (300–480 °C), and catalysts usually comprised of metal pairs (e.g., nickel-molybdenum, cobalt-molybdenum) on activated alumina. Fractionation of the hydrocracked products by atmospheric distillation gives high quality kerosene needing minimal further treatment.

Delayed coking takes even heavier feeds such as the residual materials from vacuum distillation of crude oil having a normal boiling point above 520 °C and breaks the molecules in a catalyst-free thermal process at temperatures that may reach 500 °C. The kerosene fraction from a delayed coker is not recommended for use in jet fuel, largely because of its olefin content, but potentially may be upgraded in downstream processes in a similar matter to the heavy naphtha from the FCC. Olefins may cause the jet fuel to fail the thermal stability test requirements if significantly above 1–2% volume.

2.3 Upgrading and Purification

Following the initial production of kerosene via processes such as atmospheric distillation or cracking, the next steps improve the quality of the kerosene fractions to the necessary fit for use quality expected of jet fuel. These upgrading steps may be achieved by various wet treating processes (Sect. 3) or by treatment with hydrogen (Sect. 4). The choice is often governed by the sulfur content of the kerosene, with the less expensive wet treating options not being able to reduce a high sulfur content kerosene down to jet fuel specification limits.

3 Wet Treating Processes

3.1 Caustic Washing

The simplest wet treating process caustic washing reduces the acidity of jet fuel to acceptable levels helping mitigate corrosion to the fuel distribution system and aircraft components. The main source of acidity in kerosene is usually traces of organic acids, like naphthenic acids from the crude oil and mercaptans (organic compounds containing S–H groups). The acidity content must not exceed 0.015 mg KOH/g of fuel (or 0.10 mg KOH/g in ASTM D1655).

In the refinery, caustic washing involves mixing of the fuel with an aqueous solution of caustic soda (sodium hydroxide at 5–15%), which converts all the acidic species into neutral or even slight-basic compounds. This treatment should be carried out immediately following the distillation step, because contact with air may form free sulfur, which is corrosive and difficult to remove. The reaction with sodium hydroxide converts organic acids into the sodium salts of the acids and some lighter mercaptans into organic sodium sulfides, which are more water-soluble. There are potential issues with caustic treaters (see more in Sect. 9.2.1) but in all cases the product needs to be: (1) water washed to remove caustic and water-soluble products, (2) passed through a salt dryer, to eliminate all traces of water from the fuel, and (3) treated with clay or diatomaceous earth to remove traces of the caustic solution, coloured materials, and precipitates. This should produce a clear and bright kerosene product that is an acceptable jet blending component.

Caustic washing primarily eliminates lighter mercaptans but does not significantly reduce the total sulfur content of the fuel. The next level of upgrading complexity involves the conversion of "sour" mercaptan components into "sweeter" sulfur compounds. As with caustic upgrading, the mercaptan oxidation approaches usually involve subsequent water washing, salt drying and clay treatment.

3.2 Older Sweetening Processes

Several technologies were used in the 1950s to 1970s to reduce the levels of mercaptans in kerosenes.[6] The best known were copper sweetening, the Doctor process and the Bender process. With the advent of more modern approaches to sweetening (e.g., Merox™ and Merichem™ oxidation technologies, addressed in Sects. 3.3 and 3.4, respectively) and the greater use of hydrotreating (Sect. 4), most if not all these older sweetening processes have been replaced. Details about them are limited,[7] as are details about any associated problems.

Copper sweetening[8] involves the use of copper (II) chloride to oxidise organic mercaptans to disulfides. Several stages are involved to produce a dry, H_2S-free feed, but the overall process can be described by a sweetening stage and a catalyst regeneration stage.

$$4CuCl_2 + 4RSH \rightarrow 2RSSR + 2Cu_2Cl_2 + 4HCl \tag{1}$$

$$2Cu_2Cl_2 + O_2 + 4HCl \rightarrow 4CuCl_2 + 2H_2O \tag{2}$$

This process was widely used at one stage, surviving at least into the late 1970s in China if not elsewhere. Final jet fuels always required use of metal deactivator additive to reduce the impact of trace copper on the thermal oxidative stability of the fuels.

Bender process achieves mercaptan reductions using a lead sulfide catalyst in a fixed bed configuration. Air, sodium hydroxide, and sulfur are added to the stream that is then passed over the catalyst. As with copper sweetening, the mercaptans in the feed are converted to more benign disulfides.

$$PbS + 0.5O_2 \rightarrow PbO + S \tag{3}$$

$$PbO + 2NaOH \rightarrow Na_2PbO_2 + H_2O \tag{4}$$

$$Na_2PbO_2 + 2RSH \rightarrow Pb(RS)_2 + 2NaOH \tag{5}$$

$$Pb(RS)_2 + S \rightarrow PbS + R_2S_2 \tag{6}$$

[6] Kerosene for domestic heating, lighting, and cooking prior to this time may also have benefitted from these and other approaches to mercaptan control but these are not in the scope of book.

[7] Handbook of Petroleum Refining, Speight (ed), for example, provides more information on these older sweetening processes [5].

[8] More details about copper sweetening are in Goodger and Vere's book, Aviation Fuel Technology [6].

Doctor process also uses a lead-based catalyst, sodium plumbite, formed in-situ, in the presence of free, powdered sulfur. The feed is usually caustic washed and dried, to remove any H_2S and lower molecular weight mercaptans, before contacting the sodium plumbite. Fuel mercaptans are eventually removed as complex alkyl polysulfides, so there is some reduction of the fuel sulfur level as well as removal of mercaptans.

$$PbO + 2NaOH \rightarrow Na_2PbO_2 + H_2O \quad (7)$$

$$2RSH + Na_2PbO_2 \rightarrow Pb(SR)_2 + 2NaOH \quad (8)$$

$$Pb(SR)_2 + S \rightarrow PbS_{(s)} + RSSR \quad (9)$$

With no added sulphur, the disulfides can be produced (and the catalyst regenerated) by a slower reaction with atmospheric oxygen:

$$2Pb(SR)_2 + 4\,NaOH + O_2 \rightarrow 2RSSR + 2Na_2PbO_2 + 2H_2O \quad (10)$$

3.3 Merox™ Process

Merox™ (MERcaptan OXidation) is a proprietary catalytic process that was developed by Universal Oil Products (UOP), as shown in Fig. 4. It is used in oil refineries and natural gas plants for the removal of mercaptans from LPG, natural gas, light naphtha, kerosene, and jet fuel by converting them to liquid disulfides that, in the cases of naphtha and kerosene, are dissolved in the fuels. In addition to the corrosive aspects of mercaptans previously mentioned, light mercaptans have a very offensive and unpleasant smell. The allowed content of total mercaptans in finished jet fuel shall be below 0.0030 wt% (30 mg/kg). The Merox™ process results in products which no longer have the sour, foul odours of mercaptans and hydrogen sulfide. It does not require complex and sophisticated plants and is therefore less expensive than the alternative process of catalytic hydrotreatment. In the Merox™ process liquid hydrocarbon disulfides remain in the sweetened kerosene and the total content of sulfur in the jet fuel is not significantly reduced. Though the allowable content of total sulfur in jet fuel is 0.30% by weight (3000 ppm), the current trend in the world to reduce the emission of sulfur oxides (SO_x) is acting against the use of Merox™ or similar sweetening processes and favours treatment with hydrogen.

The Merox™ process for fuels includes two chemical reaction steps. In the first, a small amount of basic aqueous solution, usually caustic soda, is injected into the fuel feed. This provides the required alkalinity to convert the mercaptans into mercaptide negative ions (RS^-). This reaction also takes care of any residual H_2S if present:

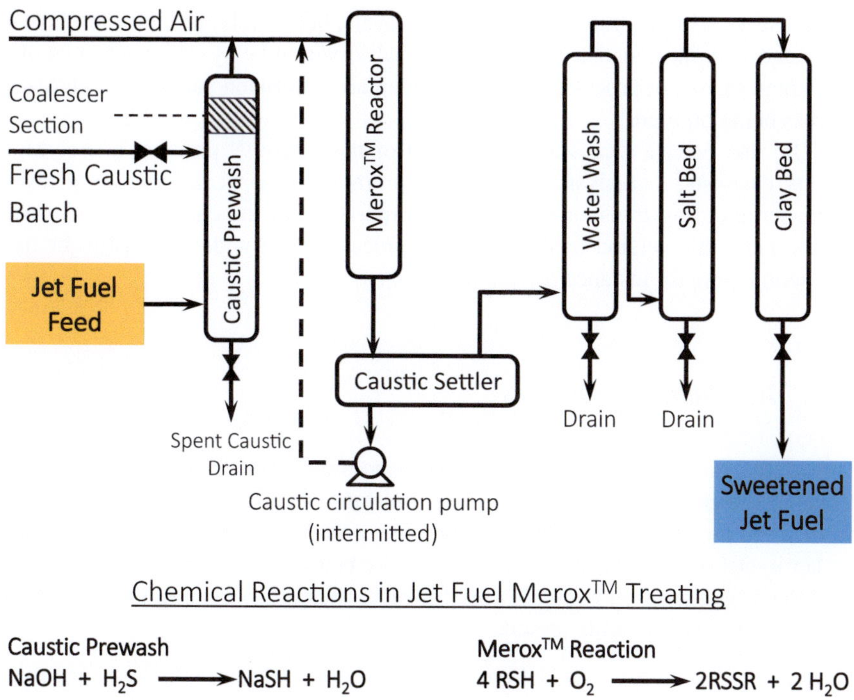

Fig. 4 Schematic of Merox™ process

$$RSH + NaOH \rightarrow RS^{-}Na^{+} + H_2O \tag{11}$$

In the second reaction, the mercaptides are oxidized to disulfides (RSSR) by air over the catalyst, leading to the net overall equation of the process being:

$$4RSH + O_2 \rightarrow 2RSSR + 2H_2O \tag{12}$$

Meanwhile, the Merox™ process for kerosene involves only one-stage, in which the feedstock, after pre-wash to extract out H_2S, is turned alkaline, mixed with compressed air which is dissolved in the liquid phase and then flows through a fixed bed of catalyst in a reactor vessel. The Merox™ catalyst is comprised of charcoal granules that have been impregnated with UOP's proprietary catalyst, which is listed in the material safety data sheet as cobalt phthalocyanine. The caustic fraction is separated in a settler, and the sweetened product is purified by flowing through a water wash vessel to remove any entrained caustic, followed by a salt bed vessel to remove water, and finally passes through a clay treater. The clay treater removes oil-soluble substances (many of which have surfactant properties), organometallic compounds (especially cobalt with its deeply blue colour) and particulate matter, which might prevent the jet fuel meeting product specifications.

Kerosenes with naturally low mercaptan levels, but with high acidity levels, can be treated with the caustic, water-washing, salt dryer and clay treatment units of a mercaptan oxidising process, with the mercaptan conversion step of the Merox™ process being omitted.

UOP also license a caustic-free version of the Merox™ process. It uses high activity catalysts, a catalyst activator, and ammonia as a weak base. As a result of concerns during transportation and at airports about the carry-over and smell of ammonia, this version of the Merox™ process is not widely used for jet fuel production. See, for instance, EI/JIG 1530 [7].

3.4 Merichem™ Processes

The Merichem Corporation has two kerosene sweetening processes targeted at reducing mercaptan levels: Mericat™ II (first used in 1986) and the more recent Mericat™ J process [8]. As with all sweetening processes, they are only appropriate for kerosenes streams whose sulfur contents are below 3000 ppm. Some customers have them retrofitted into vessels used for older mercaptan treating units, but they can also buy specially built vessels.

Mericat™ II mercaptan oxidation process, like the Merox™ process, involves kerosene mixing with air, caustic, and a catalyst on a fixed carbon bed. The sweetened stream passes to an Aquafining™ water wash, a salt dryer, and a clay treater. Next it can pass to a Napfining™ unit, with "FIBER-FILM® contact" (FFC) technology, upstream of the mercaptan treatment; this step of the processing targets naphthenic acids and leads to a reduced TAN in the feed to the mercaptan removal step. The lower TAN feeds, with fewer surfactant species than the neat straight run kerosene, have fewer problems related to emulsion formation in the mercaptan unit. A typical line up is shown in Fig. 5. FFC technology reduces issues with emulsion formation and carryover that are seen with conventional caustic and water washing steps.

Mericat™ J process shown in Fig. 6 uses a special solution, JeSOL®-9, mixed with air to oxidise heavy mercaptans. There is no fixed carbon bed, but similar reactions take place as in a fixed bed technology, at a higher rate. Merichem is responsible for the removal and replacement of the spent solution, not the refinery.

Note: When mercaptan levels in the straight run kerosene are low enough, the Mericat™ II or Mericat™ J step can be bypassed, with the treatment in the polishing steps (water wash (such as Aquafining™), salt dryer and clay treater) being sufficient to produce an acceptable product.

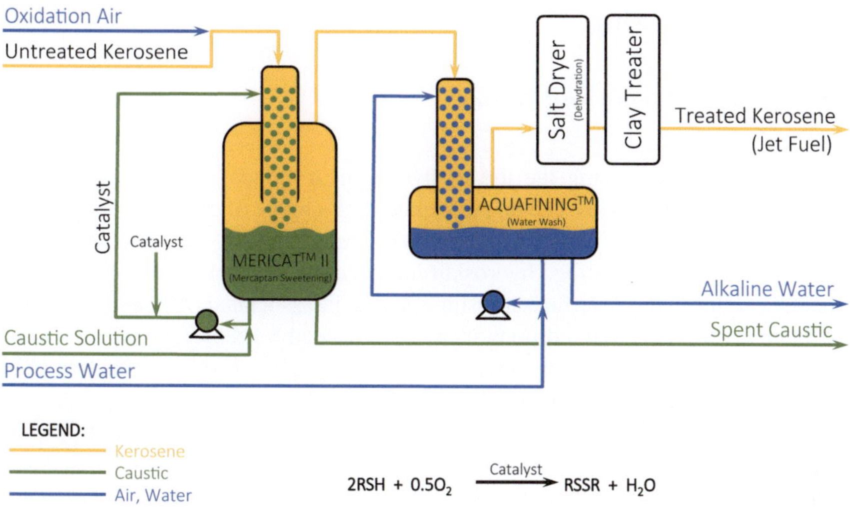

Fig. 5 Merichem kerosene treating unit with Mericat™ II

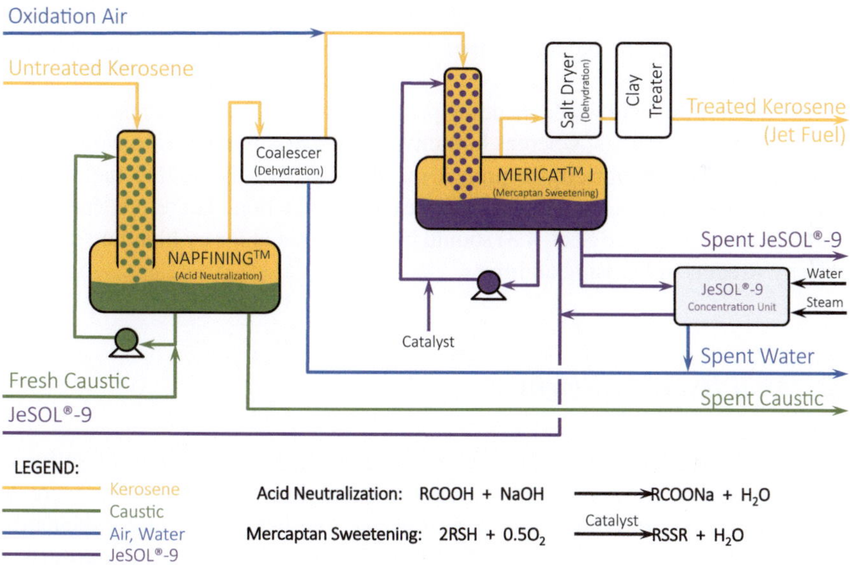

Fig. 6 Merichem kerosene treating unit with Mericat™ J

4 Hydroprocessing

4.1 General Terms and Issues

Hydroprocessing, involving the use of hydrogen to reduce sulfur, oxygen, some nitrogen, and olefin species, can be performed over a range of pressures and temperatures. The terminology is varied, dependent on the process provider and the streams being treated. Hydrotreating, hydroskimming, and hydrofining tend to be mild processes, while hydrodesulfurization units tend to be associated with higher pressure and more severe process conditions such as are needed for diesel/gasoil upgrading; the name used by a given company may reflect the degree of sulfur reduction to be achieved, with most diesel now made to 10–15 ppm limits.

Hydrogen gas is usually produced in a refinery by one or two of the following processes. A general method, which is not specific to refineries, involves methane steam reforming in which natural gas is transformed into a mixture of hydrogen and CO_2. In the second process, which is specific to refineries, a stream of heavy naphtha undergoes aromatization in units such as Continuous Catalytic Reformers (CCR). The main product of the chemical reaction is a stream of aromatic hydrocarbons that is needed for the making of high-octane gasoline, and the by-product is hydrogen. The latter gas is used for the treatment of sulfur in various refinery streams, including kerosene.

Thermal stability and water shedding properties tend to improve as the hydroprocessing severity increases, whereas lubricity is a property that tends to worsen. The fuel needs sufficient lubricity to protect metallic parts of the engine fuel system, such as the fuel pumps, against wear and excessive friction. Therefore, fuels with insufficient lubricity (see Sect. 4.3) should be compensated for by blending in better lubricity streams or by using additives.

4.2 Mild Hydrotreatment

Hydrotreatment is a catalytic process intended to reduce the content of sulfur in various streams by treatment with hydrogen gas at high temperature and pressure. In contrast to 10–15 ppm sulfur limits for gasoline and diesel in many current standards, the allowable limit in kerosene is 3000 ppm (0.30 wt%). Very little kerosene is produced anywhere near this limit and many refineries achieve, say, 200–600 ppm of sulfur, with a standalone (mild) kerosene hydrotreater (KHT) fed with straight run kerosene.

Mild hydrotreatment is carried out normally at temperatures between 300 and 400 °C under the catalysis of metal oxides like cobalt-molybdenum or nickel-molybdenum on active alumina in fixed bed reactors. The sulfur-containing hydrocarbons react on the catalytic surface with hydrogen, which bonds to the sulfur atom and converts it to hydrogen sulfide. The latter gas is separated in the process from the

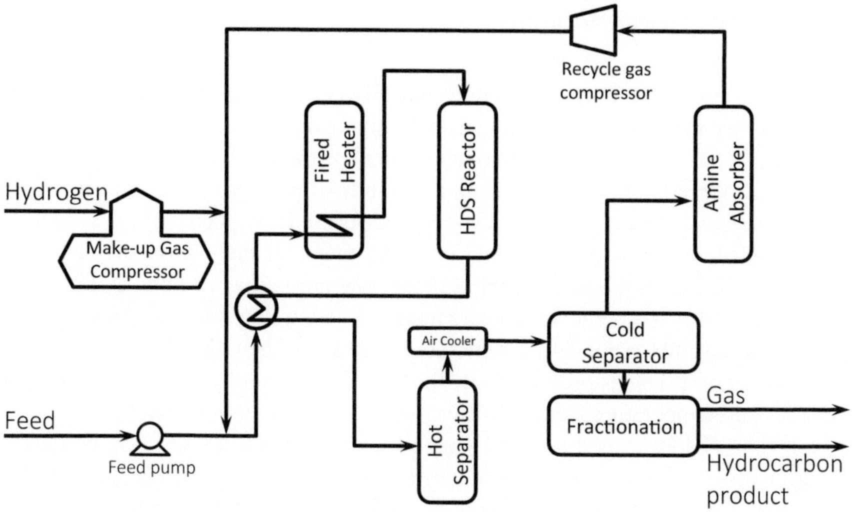

Fig. 7 An example of a mild hydrotreatment process line up

"sweet" hydrocarbon and converted downstream in the refinery to molecular sulfur. The partial pressure of the hydrogen gas in this process is an important issue. If the pressure is below 70 Bar (1015 psi) then the reaction will treat most of the sulfur content but not all of it. The treated hydrocarbon may still contain 50–600 ppm of sulfur, sometimes lower, depending on the pressure/temperature combination and feed kerosene's initial sulfur content. Catalyst performance and Liquid Hourly Space Velocity (LHSV) will also affect the sulfur levels that can be achieved.

Mild hydrotreatment brings more benefits than just sulfur level reductions. Together with the other sulfur species, mercaptans are reduced to well below the 30 ppm limit. Olefins will be saturated and oxygenate species will be removed. The colour of the jet fuel is improved, and if process chemicals are carried over into the kerosene fraction, the high temperature of the reaction normally helps to decompose those chemicals and eliminate them from the product, e.g., some nitrogen-containing species may be eliminated as ammonia. Also, the process needs to be carried out under very dry conditions so water in the feed is eliminated completely from the product. On the negative side, hydrotreatment eliminates some natural antioxidants that are part of the kerosene, and therefore addition of antioxidant (at up to 24 ppm) is often recommended.[9] See Fig. 7 for a mild hydrotreatment process line up example.

[9] Until recently, antioxidants were mandatory in hydrotreated streams used to make Jet A-1 to Def Standard 91-091, but this has been relaxed, to align with ASTM D1655 requirements, though additive levels must still be reported. The operator should check the requirements of military specifications.

4.3 Severe Hydrotreatment

More severe conditions may exist in units like HDS ("Hydrodesulfurization") plants. Their presence in refineries supported a revolution in the entire field of transportation fuels, mainly gasoline and diesel, making those fuels more environmentally friendly, by reducing the content of sulfur from thousands of parts-per-million to less than 10 ppm. Increasingly, refineries have used more severe HDS plants for all, or at least part, of their kerosene production and therefore many jet fuels have, say, less than 100 ppm of sulfur, and a significant number now are below 15 ppm. A few units have the flexibility to operate in a severe mode when ultralow kerosene is needed for diesel blending and to switch to a more moderate mode when the kerosene is needed for jet production. This is probably not "the norm" as the catalyst choices, hydrogen pressures and temperatures tend to differ for the two different outcomes.

Def Stan 91-091 uses the term "severe hydrotreatment" to describe processes which involve using hydrogen gas at partial pressure which is higher than 70 Bar (1015 psi), like the regular physical conditions that are applied in a hydrocracker (see Sect. 4.4) or in a hydrodesulfurization (HDS) unit used for making ultralow sulfur diesel. In some newer refineries, with fewer process units, these severe HDS units treat a combined kero/gasoil mix, with the kerosene and gasoil cuts distilled after the hydroprocessing. The proportion of the end product using severely hydroprocessed kerosene (as defined by Def Stan 91-091) must be reported in the refinery certificate of quality, if released to Def Stan 91-091; such reporting is not required for ASTM D1655 fuels.

4.4 Hydrocracking

The process of hydrocracking involves heavy molecules of VGO (vacuum gas oil) being broken in a fixed-bed plant using high pressures (between 35 and 200 Bar), an excess of hydrogen gas and elevated temperatures (between 300 and 480 °C). To protect the unit's catalysts, the sulfur in the feed must be completely removed by an initial hydrotreating step. The catalytic hydrocracking process uses catalysts comprised of a pair of metals (e.g., nickel-molybdenum, cobalt-molybdenum) on activated alumina. In the chemical reactions the long chains of hydrocarbons in VGO are broken into shorter chains of normal- and iso-paraffins. Olefinic compounds are not produced since double bonds are hydrogenated on-the-spot into saturated hydrocarbons. Moreover, the harsh reaction conditions may allow some aromatic compounds to be reduced into naphthenes and even to saturated paraffins. Aromatic saturation can also occur in some severe hydrotreating units (depending on catalyst selection), as covered in Sect. 4.3, without appreciable cracking taking place. A pre-condition to the operation of a hydrocracker is a source of pure hydrogen that is usually produced by steam reforming of natural gas in a HPU (hydrogen-producing unit), also known as a steam methane reformer (SMR); "greener" approaches are being encouraged.

Hydrocrackers, where used, tend to produce middle distillates with excellent properties. The absence of olefins improves the thermal oxidation stability and the shelf life of the fuel, and the reduced aromatics contents improve the burning quality as measured by the smoke point method. The slightly lower density of jet fuel from a hydrocracker, due to the reduction in levels of aromatic hydrocarbons, may also contribute to improved freezing points. On the negative side, hydrocracking is a severe process and produces very poor lubricity kerosene.

Kerosene does not provide much hydrodynamic lubrication and severe hydroprocessing—as defined by Def Stan 91-091 or more broadly by refiners—removes those naturally occurring and largely heteroatomic species which give a fuel its boundary lubrication properties. This is a particular challenge for kerosene fractions from a hydrocracker, which in other respects are high quality streams. These "dry" or "hard" fuels with poor boundary lubrication properties cannot give protection with the residual poor hydrodynamic lubrication properties of kerosenes. Lubricity of kerosene is measured by the size of the scar on a metallic ball (in the ASTM D5001 BOCLE test), and if inadequate lubricity is shown (large wear scar diameter), this should be compensated for by blending with streams of better lubricity or by using jet allowed additives.

Def Stan 91-091 does not require jet fuel lubricity to be tested for every batch, unless the blend contains more than 20% of kerosene fraction that was "severely hydroprocessed" and, simultaneously, has less than 5% of non-hydroprocessed components; ASTM D1655 has no mandatory lubricity testing requirement. However, all jet fuels with synthetic components must be tested, and if necessary, a qualified LIA (Lubricity Improver Additive) should be added to the fuel.

Note that non-additivated fuels that have started to degrade, producing oxidised species, can often appear to have better lubricity than should be the case; this is a particular challenge when collecting and testing small volumes of "new" fuel components.

5 Blending Components

5.1 Blending Rules and Strategies

"Finished" jet fuel can be made as a straight-run fuel without any additional blending but is often produced via the blending of component streams, depending on the complexity of a given refinery. While it is simpler if all the individual components meet the main physical properties in the final specification, this is not essential. Refineries establish "blending rules" that allow them to make jet when all or only some of the streams are available. The refinery establishes the range of component proportions that delivers a final kerosene mixture meeting all the requirements of the relevant specification(s). This is done as part of a whole refinery optimization process, so that the amount of all the products produced matches the demand; this involves deciding how much "product giveaway" they are prepared to have to ensure that they meet their most blend-limiting properties all the time. Failing to meet the limits has costly implications for product supply and storage, with only so many tanks being available to store problem fuels until they can be reblended to come into specification.

Most but not all properties of blends follow linear by volume rules so the refinery will establish proportions that apply for the crudes (or synthetics[10]) it uses, with particular emphasis on properties such as thermal stability and lubricity which are driven by trace components. When new crudes or processes are introduced, these blending rules will need to be checked.

Jet fuel is the main, though not the only, output for refinery blended kerosene streams, and kerosene is also used in gasoil blends, particularly over winter months. Countries like Japan and the UK use kerosene for domestic heating and lighting and in these cases the refinery will make a dual-purpose kerosene, DPK. If it meets jet fuel specifications it will usually meet the heating and lighting requirements; a single base fuel will be made and only at the final stages of batching and additivation will the jet and non-jet products be separated. Likewise, if a refinery makes jet fuels to meet several different civil and military requirements, e.g., for Jet A, Jet A-1 and JP-5, it will not make 2 or 3 separate jet fuels but will make one product that simultaneously meets all the base fuel requirements, making kerosene that falls within the most stringent of the flash point and freeze point requirements.

5.2 Components in Blends

A finished jet fuel may contain one or more of the following components:

- A kerosene fraction from atmospheric distillation of crude oil that undergoes one or more of the "wet treating processes" (Sect. 3) such as caustic treatment, or mercaptan oxidation in Merox™ or Merichem™ processes.
- A kerosene fraction from atmospheric distillation of crude oil that undergoes mild hydrodesulfurization (Sect. 4.2).
- A kerosene fraction that is obtained from a hydrocracker or other severe hydroprocessing unit (Sects. 4.3 and 4.4).

The above are the most common streams, but the more complex refineries are starting to include small amounts of streams like cracked heavy naphtha, or coker-derived products which may be used after suitable upgrading; see more in Sect. 10.1. A small but growing number of refineries are also starting to include non-conventional synthetic components in their jet fuel production or to co-process synthetic streams

[10] This can cover synthetic blending components for final blending, or synthetic crudes blended with refinery crudes or intermediates for coprocessing.

with crude-derived streams. More details on these options are covered in Sect. 10.2 and in Chap. 3.

5.3 *Blending Process*

The blending, which may also include the in-line introduction of additives, is mainly achieved by the mixing that occurs as separate components enter the jet fuel tankage. This may be achieved with sequential blending as product enters a tank, or in-line blending of streams as they enter a tank; it is good practice to add the components in the order of their densities, the densest being first, or to add all components simultaneously. Alternatively, there may be scope to recirculate the tank's contents to produce a homogeneous batch of fuel, though this approach is more complicated. (See EI/JIG 1530 for more details about blending.) Density measurements at multiple heights in the tank are used to check that adequate mixing has occurred to achieve homogeneity as required by the manufacturing specifications. Where the specifications require it, the amounts of each component and additive are recorded and, where required, reported.

Def Stan 91-091 requires information about the streams, e.g., in terms of relative hydroprocessing severity, which have been blended and the percentage of each component; this means that the downstream users, on request, should be able to have access to such information though not to the individual processing routes involved. The Refinery Certificate of Quality (or more generally, the Certificate of Quality) shall include % (v/v) amounts of non-hydroprocessed, severely hydroprocessed (defined as involving a hydrogen partial pressure above 70 bar), and synthetic components in the blend.[11] The production of jet fuel containing synthetic components, particularly sustainable blending components for making sustainable aviation fuels (SAFs), is covered in Chap. 3.

6 Additivation

Permitted "additives" are all those outside chemical ingredients that are added to the kerosene to improve some key qualities; they either assist in meeting the specification requirements or are added at the request of the purchaser. There are few allowable additives for civil jet fuels (shown in Defence Standard 91-091 Annex A, and in ASTM D1655 Table 2). As mentioned in Chap. 1, specifications describe additives by their function and by their defined concentrations, but in a few cases the approved manufacturers and name brands are also specified. This chapter deals with refinery use of jet fuel additives; Chapter 5 discusses in more detail the additives,

[11] Previously the amount of mildly hydroprocessed component was also reported, so that the total %v/v of all components was 100%. This requirement may be introduced again.

their chemistry and usage levels, including permitted additive top-ups downstream of the refinery.

Antioxidants are widely used at refineries, mainly in hydroprocessed products, which may have little natural antioxidancy, unlike straight-run and wet-treated products. Usage was mandatory in Def Stan 91-091 until recently, but this situation has changed to align with the ASTM practice, which permits antioxidants but rarely uses them. Def Stan 91-01 also says, "If antioxidant is added post manufacture, report the total active material concentration in the final batched fuel on the Certificate of Analysis and the Release Certificate". The total active matter must not exceed 24 ppm. The antioxidant content in the fuel is not measured by a test method, so the exact amount is calculated from the rate of addition at the metering pump and/or additive tank level changes.

If fuel is being made for military purchasers, who are more likely to store their fuel for long periods, the more stringent, older style requirements for antioxidant may still apply. The manufacturer should check the latest military specifications to ensure compliance.

From a technical perspective, the recommended point of addition of antioxidants was always on run-down from the hydroprocessing stage, i.e., before the kerosene cooled or was exposed to air, but now there is more flexibility. Operators who have used antioxidants in the past should undertake a careful risk assessment that includes all the usage, storage, and distribution that their product will undergo and thus the chances of gum formation and peroxidation leading to off-spec product if they stop or change their use of antioxidants. This assessment should reflect the system into which the fuel is delivered, such as whether or not it enters a fungible system that may provide anti-oxidancy benefits from mixing many batches, and also that any importing or exporting of products could involve and must comply with these different systems.

Static dissipator additive (SDA) is added to most jet fuels outside the USA, for safety reasons. The addition does not relate to the fitness for purpose aspects of the fuel but is to increase the static charge dissipation when used in fuel distribution systems that do not incorporate sufficient charge relaxation into their designs; it reduces risks in fast flowing locations like ground handling systems at airports or in pumping activities on and off ships. Def Stan 91-091 and other standards demand that the fuel supplied to aircraft will have electrical conductivity between 50 and 600 pS/m, when SDA is added to the fuel; this range applies in ASTM D1655 if SDA usage is requested. The two qualified additives are: Stadis® 450 (from Innospec Ltd.) and AvGuard SDA (from Afton Chemical Ltd.). The allowed level of initial doping is 3.0 ppm, though more typically is 1.0 ppm, as the stabiliser part of the SDA formulations can have a negative impact on water separation (as measured by water separation methods, if not on actual water separation in a filter/water separator). The practice is to add or top up SDA further down in the supply chain, and—if possible—close to the airport. Therefore, it is a matter of contractual agreement to decide where and when to add SDA, and jet fuel may be pumped, under proper notice, without the additive and therefore with conductivity that does not meet the specification at that point (this shall be mentioned on the CoQ).

Metal deactivator additive (MDA) may be added to ensure that the fuel passes the thermal stability requirements of ASTM D3241/IP 323 at the minimum test temperature of 260 °C. It should not be added pre-emptively, only when a fuel fails D3241/IP323 test at 260 °C but then passes at 245 °C, and specifications now say that MDA may not be used for over 5% of batches released in a 12-month period.

The initial doping must not exceed 2.0 mg/L. Failures in thermal stability testing may result from the presence of cracked compounds or olefins, traces of metals and other reasons. Traces of metals may enter the fuel through corrosion of tanks, pumps, or hoses, from residual catalysts like cobalt or copper from the Merox™ process or other catalytic processes. Usage is not common nowadays, and the expectation, from both Def Stan 91-091 and ASTM D1655, is that the manufacturer sorts out the underlying issues.

Biocides may be used periodically at the refinery. They are designed to kill viable microorganisms and most typically are used only during tank maintenance operations and only if necessary; overuse of biocides is not recommended as it could lead to more biocide-resistant fuel microorganisms. The tanks should be thoroughly cleaned afterwards to prevent any potential cross-contamination, and any biocide-containing kerosene should be downgraded to non-aviation use. Chapter 12 provides a thorough discussion of microbiological activity in jet fuel systems, and how best to manage it.

Other additives. There are a few other permitted additives for jet fuel, but they are mostly used downstream of refineries or other points of manufacture. These include but are not limited to Lubricity Improver Additive (LIA) and Fuel System Icing Inhibitor. Additives are discussed in Chap. 5.

6.1 Additive Storage and Shelf Life

Additives in the refineries are supplied and stored in iso-tanks of approximately 20 MT, in 200 L barrels and also in intermediate bulk containers (IBCs) of 1000L. They should be stored in covered or shaded areas under proper conditions. Their use should not exceed the manufacturers' recommended shelf life.

6.2 Additive Injection and Product Mixing

There are two separate issues: point of injection of the additives, and how to disperse the additive and keep the products homogenous, because in most cases, the additives are more viscous than the fuel. The points of injection depend on the type of blend that makes the batch of the fuel. Antioxidants, which are added only to the components that undergo treatment with hydrogen, normally are pumped immediately into the outgoing streams in the line from the plant to the storage tank. The same additions, namely pumping the right amounts of additives into the outgoing products, are carried out for additives like LIA and SDA. If the jet fuel is transferred by ships, addition of SDA can be made manually into the ship tanks or during discharge. A different case is MDA, which is typically added—only when necessary and with usage restrictions at a refinery—directly into storage tanks. Many jet fuel tanks are equipped with pumps that can circulate the entire content, so homogenous mixing and equal spread of the additive can be achieved inside the tanks.

Additive information (function, code, name, amount) shall be reported on the Refinery Certificate of Quality for Def Stan 91-091 but is not required for ASTM D1655.

7 Component and Finished Product Storage

7.1 Tankage and Settling

Following production, blending and additivation, the jet fuel is stored for a short period of time in the refinery tank farm, and then pumped downstream. Storage tanks and pipelines should be designed, constructed, and maintained according to recognized industry standards for jet fuel. It is highly recommended to use dedicated facilities for jet fuel, and not multi-purposed, though this is not always possible. Segregation from other fuels, especially diesel and biodiesel, should be enforced to prevent potential cross contamination. The use of metals such as copper, zinc, or cadmium in a fuel distribution system for the construction of tanks and pipelines and usage of zinc silicate linings for them including in points where fuel samples may be grabbed, may harm thermal stability of the fuel (as measured by thermal oxidation stability test, IP 323/ASTM D3241) and should therefore be avoided. For full details, the reader is encouraged to check standards and manuals referenced by ICAO Doc 9977 [9], in particular consulting with EI/JIG 1530 Standard.

After introduction of the fresh batch of kerosene in the storage tank, the fuel may contain traces of water and suspended solid particles. To eliminate these, the fuel shall settle in the tank for 24 h, or dependent on the height of the fuel in the tank, to at least three hours per each metre height of liquid. Then, water shall be drained from the bottom of the tank (the "sump"), and the jet fuel may be pumped downstream to the next station, preferably by passing through proper filter(s) to eliminate large particles and some water. Adherence to good housekeeping practices reduces the potential for future problems.

7.2 *Filtration*

Filtration may be necessary to eliminate solid particles, including residual catalysts or clay from the processes, unwanted products of metal corrosion or dusts ingested by the tank. Filtration may also remove sludge of all sorts, though it should never be considered as the ultimate option to remove microbial sludge. At the point of manufacture, the specification demands total particulates of no more than 1.0 mg/ L (by ASTM D5452). Over the last decade there has been a quality shift from just reducing total particles towards the determination of particle size distribution, capturing the populations in ranges from 4 to >30 μm (microns). To reduce the particulate content, it is a common practice to use filters of 5.0 μm, although the use of 1.0 and 0.5 μm is much preferred. The use of coalescers to simultaneously separate water and particles during filtration is also common, although in many cases it is carried out downstream in the supply chain and not in the refinery. Some crudes will lead to higher water levels in the fuel, and this may require tank and filter unit drainage frequencies to be adjusted. Unlike the intrinsic properties of the jet fuel (like density, freezing point, or amount of sulfur), the content of particles and water change constantly during the flow from the refinery to the wing of the airplane as a function of dirt in tanks and pipelines accumulating through atmospheric vents. Filtration is, therefore, a common demand along the entire supply chain. It is discussed in more detail in Chap. 10.

7.3 *Sampling*

Like other industrial and commercial products, the fitness-for-purpose of jet fuel shall be determined and approved as meeting all specifications by a series of careful tests in a qualified laboratory. A crucial step for obtaining reliable testing results is proper sampling, or in other words: making sure that a uniform, homogenous and representative sample is drawn from a storage tank and brought to the lab. The same rules, as briefly described herein, are applied to all fuels and the handling of kerosene is no exception.

Most jet fuel storage tanks are equipped with either floating or fixed roofs. The practice of manual sampling (ASTM D4057, ISO 3170) involves either drawing samples manually from a pre-arranged hatch on top of the roof or by tap sampling from side outlets on the side of the tanks (Figs. 8 and 9).

Samples shall be taken only when the content of the tank is at rest, namely after filling and letting the fuel settle for a minimum of 30 min prior to sampling. For determining batch homogeneity, upper, middle, and lower samples as defined in ASTM D4057 shall be taken, and their density determined. Homogeneity for aviation products is defined as having a density difference between samples of no more than 3.0 kg/m^3. Aviation fuels at the point of manufacture shall be homogenous prior to certification and release. If a batch tests as non-homogenous, additional

Fig. 8 Line sampling arrangement on side of tank—lines from different tank locations (Image courtesy of A. Groweiss)

Fig. 9 Array of taps allowing line sampling from various depths in fuel tank (Image courtesy of A. Groweiss)

mixing processes shall be commenced until homogeneity is achieved. Once a batch is identified as homogeneous, a composite or single spot sample may be used for assessing specification compliance.

When manual sampling is carried out from the roof of the tank, the collection devices are submerged to a certain height of the liquid in the tank, e.g., the middle section, determined by the length of the rope in use, and the fuel is collected in glass bottles or weighted metallic cans. Ropes should be made from natural materials like linen, not synthetic, due to the risk of formation of electrostatic charges. Another method of sampling is called "all levels" where fuels from all zones are collected during one pass of the collecting device from the lower to the upper part of the tank. A 'running' sample is like an all-levels sample but is taken during a pass of the container from the top of the tank to the bottom and back to the top. It can often be difficult to collect all levels and running samples without significant practice, as these sampling methods require an ullage to remain in the top of the sample container after sampling to ensure the container did not fill part-way during the procedure. Sampling from a roof hatch through a stilling well can present challenges. Stilling wells shall be slotted, to ensure the product in the well is representative of the tank. There shall also be sufficient slots to ensure samples from differing tank levels are representative of the levels at the same height within the tank. In general, upper samples are taken prior to the middle sample, followed by the lower sample. This is to prevent disturbance in the stilling well as the sample container moves through the liquid column. Taking tap samples from side outlets is less complicated, but it is still important to drain the content of fuel in the short sample lines before the actual sample is collected to ensure that only fuel representative of the tank is drawn.

7.4 In-Process Testing

Aviation fuels shall meet all the physical and chemical properties that are required by the specifications referenced in any contract of sale. The most referenced international jet fuel manufacturing specifications are ASTM D1655 and Defence Standard 91-091. While batch certification against a recognized specification is necessary for commercial sale of a fuel, in-process testing of the daily refinery production, whether the distilled fraction from the crude unit (Sect. 2) or the treated kerosene from other production plants in the refinery (Sects. 3–4), is important to track quality problems even before the fuel streams reach the large storage tank and potentially harm the entire batch. Crucial tests at this point are intended to confirm that the density at 15 °C is between 0.775 and 0.840 kg/L, the flash point is >38 °C, freezing point below −47 °C (for Jet A it must be below −40 °C) and the final boiling point in laboratory distillation according to ASTM D86 is below 290 °C, because those properties can only be fixed by re-distillation of the batch. Each refinery should decide on and use the most important in-process tests for their production set up; see the list of periodical checks that appears in EI/JIG Standard 1530.[12] The latter document suggests checking the physical appearance of the fuel (colour, free and total water), corrosiveness to copper due to mercaptans (ASTM D130), thermal oxidation stability (ASTM D3241/IP 323) and the ability of the fuel to separate water. Having a baseline set of results helps with any future troubleshooting work. Other recommended periodical tests include parameters not in the final product specifications, like total and basic nitrogen, trace metals and any other qualities that each refinery defines as important for ensuring process control in the manufacture of the aviation fuel components.

7.5 Final Product Testing

A segregated batch of jet fuel is only ready to be dispatched after all the quality demands have been fulfilled in the production process, in storage and in additivation. All those requirements together define a specification for the fuel, as described by Def Stan 91-091 or ASTM D1655 or others as covered in Chap. 1 of this book. The final product testing should confirm that the product indeed is "on spec" and is fit-for-purpose. The tests are conducted in a chemical laboratory which is defined as "QC (quality control) lab". There is no mandatory requirement about a lab's quality system and what kind of certification should be held, from a specification perspective. It is, however, highly recommended that laboratories that release products like jet fuel should be accredited to ISO 17025 [10] or to an equivalent standard, e.g., one from the National Measurements Accreditation Service, NAMAS; purchase contracts may set out specific requirements. Such standards, which involve periodic laboratory audits, require that the tests be conducted by qualified and trained personnel, using validated test methods, traceable standards, and instruments that are calibrated periodically. All those demands provide assurance that the lab can produce reliable results for reliable samples.

After testing a batch, the laboratory shall issue a Certificate of Quality (COQ) which states the results that were achieved for each specification parameter. Approximately 20 different tests are conducted and reported but several of those, like distillation or particle size distribution, provide multiple results, so the RCQ may contain over 40 different entries as well as bulk processing compositional information. All the parameters must be "on spec" so even if only one result is out-of-specification (OOS) then the entire batch shall be rejected. The only exception is conductivity, which is allowed though must be stated, if SDA is to be added downstream of the point of manufacture.

> Note: trending of specification data can highlight potential issues in unit operation, test equipment or product quality, which the producer can act upon before product failures occur.

[12] EI/JIG Standard 1530, Quality Assurance Requirements for the Manufacture, Storage and Distribution of Aviation Fuel to Airports.

The tests are divided into groups that may include several properties: appearance and colour of the fuel, composition, volatility, fluidity, combustion, corrosion to metallic parts, thermal stability, presence of contaminants, water separation and electrical conductivity. Lubricity (ASTM D5001) is evaluated only if the fuel contains a large portion of component that is prepared under severe hydrotreating conditions in a hydrocracker, as discussed earlier in Sect. 4.4. Several test properties are demanded for the fuel only at the "point of manufacture", namely in the refinery or in a plant that blends components into jet fuel. Those include MSEP, colour and particle size distribution, because those properties are prone to change during transfer downstream.

8 Product Release

A "batch" is defined by a certain quantity of homogenous fuel that is confined in one tank and identified by one set of representative test results. The documents and paperwork that accompany a batch relate to the entire quantity that is pumped downstream from the "point of manufacture". The "release" is the process in which the manufacturers check and confirm that the material was produced and stored properly, that it was handled as required by all the procedures and passed all the tests as required by a standard like EI/JIG 1530,[13] and in summary is fit-for-purpose.

8.1 Physical Aspects

Once a batch has been certified and settled according to agreed specifications and standards, the last task prior to release is to perform a tank dewatering to verify the batch is clear, bright, and visibly free of water and particulate. A retained sample of every batch should remain in the refinery (usually in the laboratory) for a pre-defined period until it is no longer needed; minimum retention times are given in standards such as EI/JIG 1530 or may be agreed in supply contracts.

8.2 Documentation and Accreditation

Several documents shall accompany a batch. A release certificate (RC) should mention the grade of fuel (e.g., Jet A-1), batch number, name and address of the refinery, date and time of release, quantity of fuel in the batch, and tank number from which the batch was drawn. It is common to add such details as fuel density

[13] While this is a voluntary standard, it is increasingly being used or considered for use in contracts of sale.

and the visual appearance as observed or tested. Reporting of fuel conductivity and temperature at time of release is also important. The RC should be signed and dated by an authorized person.

Another document is the certificate of quality (COQ). This paper shall contain all the test results, and a declaration that the tested fuel has met the requirements of the governing specification. In addition, the COQ should contain the name and address of the laboratory, the dates of sampling and performing of the tests and tank number. The RCQ should be signed by the lab chief or quality manager or another authorized person. If the lab is accredited, it is advised to put the logo of the accreditation body on the paperwork.

In addition to the COQ, material safety data sheets should also be sent with all cargos leaving the point of manufacture.

9 Problems With Fuels and Process Streams

A refinery may commonly have challenges with jet fuel production through the normal but infrequent process of unit shutdowns/turnarounds when essential maintenance and catalyst changeovers occur. In such circumstances a refiner may not be able to make any jet fuel, or they may have to alter the blend recipe, and deal with the predictable changes in the product quality. During such periods, any kerosene made may be better routed to an alternate disposition. The refinery will also need to decide how to manage the kerosene product and jet fuel blending when the systems start up again, in the period before they are "back to normal". These are predictable activities and should be covered in product quality processes adopted for jet fuel production.

But problems with jet fuel quality can arise during normal operations, related to final blends, individual streams, storage systems, etc. Routinely running product and stream checks may flag issues before they occur and should help identify the most probable causes of the issues.

9.1 Problems Seen with Final Product

A variety of issues can be experienced during sampling or testing of the final jet fuel product. Table 2 starts by summarising some problems seen most often and indicates probable causes of the issues, or the processes to be checked. It is indicative of issues, not aiming to be very detailed or exhaustive. The sections that follow it expand on the most common problems.

Colour problems. There is no colour specification for jet fuel, but at the point of manufacture there is a requirement in Def Stan 91-091 and some other specifications to report the colour, on the Saybolt scale, or to indicate if the fuel has some other colour. This provides the basis for declining a fuel further down the distribution

Table 2 General problems for fuels and probable causes

The observed problem	Potential reason	Potential solution ("what to do")
Dark (Saybolt) colour	Old material	Confirm all properties, then mix with a fresh batch or do nothing
	Degraded product	Check nitrogen content and levels of cracked products in the blend (and olefin contents)
High acidity	Acidic crude Insufficient caustic treatment	If possible, blend with another batch May need to stop using the crude if caustic process units cannot cope with the feed, but check caustic is not exhausted first
	Sulfate-reducing bacteria (SRB)	Test for SRB and ensure housekeeping is in line with standards
High mercaptan sulfur levels (and associated smell)	Insufficient processing	Check unit throughput and catalysts in the treater
	Sulfate-reducing bacteria	Housekeeping, possibly including use of biocides
Sludge, water, and dirt	Water drainage (in both main tanks and filter water separators (FWS), if used), poor maintenance	Leave to settle for extra time, then drain the sump. Filter assess if filter water separator drainage periods need to be changed to accommodate new streams/crudes with different water shedding characteristics
High content of solid particles	Filtration, maintenance	More settling time, then filter
Failure of water microseparometer test	Process chemicals, e.g., filming amines, demulsifiers	Try longer settling Change out (refresh) clay treater and check salt dryer Check any process chemical changes
	Too much SDA	Reduce SDA content and add more later in distribution system
Failure of oxidation stability test	Traces of metals, corrosion, cracked components (high olefin levels), high N levels?	Filter, use MDA or re-process
High freezing point	Heavy cuts, diesel	May be fixed by mixing with another batch, otherwise—re-process
Low flash point	Mix with light naphtha, other volatiles	Either add heavier batch or re-process

system if there has been an unacceptable level of colour deterioration, indicating product degradation or contamination (EI/JIG 1530 indicates the acceptable level of colour change downstream of production.) Colour is not explicitly included in the ASTM D1655 specification requirements. Anyone handling jet fuel that looks different from normal, particularly if coming from the same refinery as usual, should consult with their supplier to understand any reason for the colour change.

Contamination with other materials, e.g., fuel and lubricant streams, at refineries is very unlikely, if they adopt positive segregation between product lines. Most refinery colour issues relate to wet-treated fuels though hydroprocessed fuels can raise challenges from time to time. Some crudes are known to produce more coloured or off-Saybolt coloured fuels that are perfectly on-specification. UOP holds a list of crudes that lead to coloured Jet made using the Merox™ process; this list is proprietary information, and the refiner should discuss any concerns directly with UOP.

Oddly coloured fuels can also be made when the rate of flow of kerosene through the mercaptan removal upgrading process is different from that in the design. Longer or shorter residence times mean that the reactions on the catalysts may lead to different chemistries being produced than under "normal" conditions. Very small amounts of colour bodies can lend a blue or green tint to a fuel that meets all specification requirements; this is not a common occurrence, but purchasers have challenged the fuel quality and rejected it. The issues should disappear when the flow rates and residence times are back to normal.

Hydroprocessed fuels are sometimes known to be made with unusual colours, for instance, pink fuels have been reported. These colours are usually linked to the antioxidant added on rundown from the hydroprocessing unit; there is no indication of any specification parameters being impacted but customers are wary. Changing the antioxidant, maybe changing to a different formulation/dilution of the same active ingredient, can prevent the unusual colour developing.

Particulate Problems. The problem of excess particles in jet fuel may be created from residual solids of catalysts, dirt accumulated in tanks (from dust or sandstorms in parts of the world), metallic particles from corrosion processes and even slime derived from microbiological activity. There is a limit of 1.0 mg/L on the total amount of particles in the outgoing fuel, and the immediate solution to the problem is to drain the bottom of the tanks very often, let the fuel stand for hours to allow separation of water and large size particles, and to keep filtering the fuel again and again throughout the supply chain. Those steps will decrease the level of total particles but may not solve a problem of very small particles. If necessary, check the composition of the particles as failing salt dryers and clay treaters may also produce unexpectedly high levels of particulates. Clay treaters are deactivated by water (see Sect. 9.2.1), and one may see channelling of slugs of saline from a salt dryer that is not working correctly. Analysis of the chemistry and morphology of any particulates seen in the product may confirm that salt comes from the dryers (is it NaCl or $CaCl_2$, is it brick or pellet) and that clay matches clay treater materials rather than deriving from other particulates and sediments.

Microbiological Problems. Most refinery processing steps are at high enough temperatures to sterilise the fuel, though as the fuel cools substantial amounts of water can be shed so if housekeeping is poor, this could lead to significant microbiological contamination of run-down and batching tanks, as they are exposed to air which inevitably carries microbiological species. Potential problems for fuel in final storage are usually controlled by good housekeeping measures (including regular drainage) and tank design. See Chap. 12 for a comprehensive overview.

MSEP Problems. There are three methods now that assess the water separation characteristics of aviation fuels, specifically related to the fuel's ability to shed water in a filter water separator unit: ASTM D3948, ASTM D7224 and ASTM D8073. ASTM D3948, which uses a portable microseparometer (hence the name MSEP) has been the main method used at the point of manufacture, but the other two methods are now appearing in specifications. These set limits at the point of manufacture depending on the presence or absence of static dissipator additives, and the method chosen. All three methods are mentioned in the "MSEP protocol"[14] issued as a bulletin by JIG (Joint Inspection Group) [11], which applies strictly downstream of the point of manufacture; the document provides a good comparison of the methods.

In a manufacturing location, poor MSEP, i.e., poor water separation, results mostly occur because of the presence of surface-active agents that disturb the natural fuel/water interfacial tension and impact the ability of the fuel to shed water. The problem can be fixed by preventing surface-active materials like demulsifiers, which are regularly used in distillation towers or in hydrotreating processes, to reach the final kerosene fraction. Other potential causes for low result of water shedding test result include the presence of nitrogenous compounds from the crude oil, traces of metals, biodiesel, or compounds surviving from clay treatment, though not all of these apply at the refinery. A poor result at the refinery should, however, prompt actions to resolve the issues, such as changing the clay in the clay treater.

Thermal Stability Problems. Thermal oxidation testers assess the potential deterioration of jet fuels due to oxidation which occurs during operation of the aircraft fuel systems and engines. Oxidation may occur because of the presence of trace levels of specific metals, heteroatoms, the use of cracked component streams (particularly the olefins they contain), or other adverse and uncommon elements that may find a way to be mixed into the fuel. During the test, filtered fuel passes over an electrically heated metallic rod. Any particles formed during oxidation may increase the pressure on filter downstream of the rod to more than 25 mmHg. More frequently the heated oxidized fuel produces a lacquer on the heater tube, the colour of which is evaluated in a comparative manner and is graded as a number between "0" and "4". To pass the test the number shall be less than "3", and at the same time, the pressure on the filter shall be below 25 mmHg. Alternatively, quantitative metrological methods, e.g., ellipsometric or interferometric measurements, can determine the deposit thickness on the heater tube, with a passing result being less than 85 nm.

---

[14] Now being referred to as "Testing Water Separation Properties of Jet Fuel (Revised MSEP protocol)".

Another reason for an automatic test failure is a "peacock" coloured lacquer, as highlighted in the circle in Fig. 10, and a fuel like that should be treated as Out of Specification (OOS). The main reason for peacocks is the presence of very low levels of residual metals, such as copper and iron. Copper could be found due to corrosion of brass fittings (which are not recommended in any jet systems) and historically came from carry over from some wet treating processes; iron derives from other corrosive processes, like those in that happen in inadequately lined tanks, particularly after salt dryer problems. Prevention of those metals is by maintenance of tanks and pipelines or by good separations following the Merox™ reaction. Such depleted jet fuel may be successfully treated using Metal Deactivator Additive (MDA), as described above in Sect. 6. High nitrogen levels, often associated with filming amines used in distillation units or with cracked products (nitrogen is more common in those parts of the barrel that are heavier than kerosene), may also produce peacock-like failures.

A failing MSEP may be the first sign of actual or potential problems with thermal stability, though poor MSEP and poor thermal stability do not necessarily correlate. The result may point to issues with the clay treater (and units upstream); a poorly performing clay treater will not remove surface active (surfactant) materials and some of the metals that it would normally remove.

Another reason for a thermal oxidation test failure, namely the use of cracked material (kerosene fractions from FCC or Coker), may be prevented by either not using such fractions or limiting their amounts to a minimum, unless they receive some hydroprocessing to remove olefins, etc.

Note: the specification requirement to measure olefins was dropped in the late 1990s after work by BP proved to be redundant; fuels at the olefin limit (5% volume) always failed thermal stability tests.

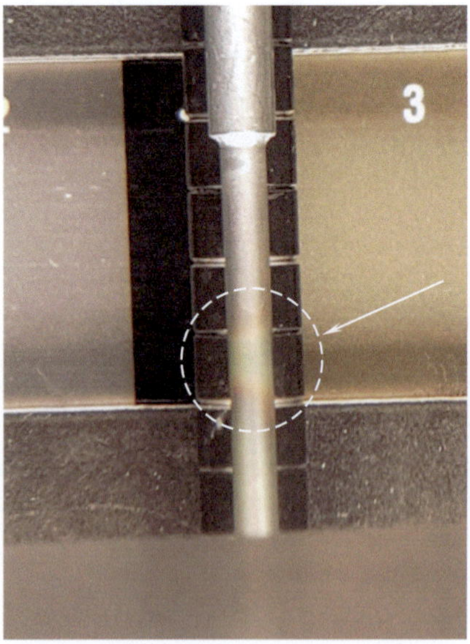

Fig. 10 Heater tube with peacock deposit (Image courtesy A. Groweiss)

Odour problems. Bad fuel smells are usually associated with the presence of H_2S or light mercaptans (compounds containing the sulfur group –SH). Those compounds are separated by distillation and the maximum allowable content of total mercaptans in the fuel shall not exceed 30 mg/L, and even those should not be light compounds and therefore not volatile enough to impart the bad smell. Sometimes poor separation, especially after hydrodesulfurization, may leave H_2S and light mercaptans in the fuel. There is no solution for the smell of a certain batch except aeration, which is highly unrecommended for the environment. The way to prevent those smell problems is by control over the distillation of the kerosene, and a proper operation of the Amine Units in hydrodesulfurization plants, which are expected to absorb all H_2S into ethanolamine and extract it out of the fuel for further treatment.

Ammonia smells have been associated with products from caustic-free Merox™ units, which is why such units are not recommended for jet fuel production.

9.2 Problems Seen with Wet Processing Streams and Units

Refineries monitor and inspect process units on a regular basis. Normally this prevents issues in one unit working through the following units but in extreme cases, the final product is impacted. For wet processing operations the following risk areas need to be mitigated and managed, with the process technology provider being consulted wherever necessary.

9.2.1 Subsidiary Units

Distillation unit additives: Corrosion control in the distillation steps can involve injection of corrosion additives such a filming amine, often by a third-party additive company. Whilst these and their primary degradation products should not end up in kerosene, this is not guaranteed. The refinery remains responsible for the product and should ensure the additive company is engaged in all Management of Change activities related to changes in levels or types of process additives. Changes in rate of usage of caustic downstream may be one indicator of issues and *in extremis*, neat or partially converted additives can end up as surfactant species in the kerosene, impacting MSEP, the levels of static dissipator needed to achieve a given conductivity and possibly thermal stability.

Caustic units: As well as monitoring the rate of caustic usage, which can change reflecting a change of crude or an increase in unwanted additive carry over, the refiner should be alert to issues relating to excessive emulsion formation in this two-phase reaction unit. This is encouraged by any vigorous mixing and the operator may be advised to remove "scum" at fuel/water interfaces. The newer FIBER-FILM™ contactors in the Merichem Napfining™ process, if operated correctly, do not require vigorous mixing of oil and water phases so minimise emulsion formation and separation difficulties compared to most caustic units. Manufacturer maintenance recommendations relating to emulsions should be followed.

Water washing units sit downstream of caustic units and, most of the time, a Merox™ or Merichem™ unit. They are intended to remove excess caustic and water-soluble contaminants. They will be impacted by any emulsions not stopped upstream and will be subject to corrosion if the pH is outside the appropriate range.

Salt dryers follow the water washing steps and are intended to dry the fuel before it goes into a clay treater. They must be checked to ensure that the salt is still present and active. If slugs of water pass through the dryers, not only will they deactivate the clay treater, but they may also take salt downstream into the storage tanks. As well as salt precipitating out as an unwanted solid in the fuel, over time, that salt will corrode even stainless steel. This can lead to the catalytic degradation of jet fuel held in those tanks, seen as thermal stability test failures.

Clay treaters are used at the end of most wet treating processes and are designed to remove most surfactants and some metals. Over time, the clay surfaces become saturated with surfactants and can no longer remove the problem species. Operators may notice this drop off in performance as increasing pressure drop (dP), degrading water separation test (e.g., MSEP) results, degrading thermal stability performance and increased particulates. However, most refineries focus on thermal stability as a pass/fail result and may not act on anything other than MSEP. Additionally, not all the issues are picked up by dP. If a water slug hits the clay treater, even one with fresh clay, the clay becomes useless. Water channels through the clay, there is no dP registered, and all the salt, dirt, etc. with the water is carried through too.

9.2.2 Main Mercaptan Treatment Units

Copper sweetening: The primary problem known with this older process is the inevitable presence of small amounts of copper in the final product. This catalyses fuel oxidation, so such fuels always needed metal deactivator additive to be able to pass the thermal stability requirements.

Merox™ and Merichem™ II: These carbon-bed based catalytic processes have some issues in common. These include a susceptibility to temporary fouling and permanent poisoning of the catalyst[15]; the latter is caused by certain impurities (assorted upstream additives and their degradation products, maybe). Following manufacturer advice on how to control and recognise these issues and minimise off spec fuels is essential. Control of upstream emulsions (from caustic unit) and emulsion formation in the main units is also necessary.

Merox™: Product made using the Merox™ process may have an unusual colour, i.e., not fall on the usual Saybolt Colour scale. There is no indication that the product

[15] Merichem J paper.

fails any specification requirements, but end customers may need assurance. The colour issues appear to relate to the choice of crude, and the technology provider, UOP, holds information on this.

Caustic-free Merox™: This version of the Merox™ process uses ammonia and there have been issues with the final product, at and beyond the refinery, still releasing ammonia. See EI/JIG 1530. For this reason, it is not a recommended route for jet fuel production.

Merichem™: No colour-related issues have been reported back to Merichem relating to either Mericat™ II or Mericat™ J technologies. The Mericat™ J technology has fewer issues with emulsion formation, fouling and poisoning than carbon-bed technologies. Merichem note that some issues can be linked to poor end point control; too much diesel in the kerosene can be associated with high TANs, the higher molecular weight acids also being more prone to emulsion formation. Most of the issues they are aware of are linked to not following maintenance instructions properly, not including unit maintenance during refinery shut down periods, or to running flows through units above design conditions, which both increases the chances of pushing material between units and increases the vigour of the mixing, exacerbating any emulsion formation and complicating phase separation steps. Merichem provide advice on emulsion control, clean up and removal.

9.3 Problems with Hydroprocessing

Conventional kerosene range products from hydroprocessing units have very good properties, and the units tend to have a smaller range of issues than wet processing options. As a greater range of inputs are added to the kerosene production, checks will need to be performed to ensure that olefins, aromatics, and nitrogen levels remain acceptable.

The limited issues currently encountered relate to odour (poor operation of stripper units that remove the H_2S from the product), non-ideal residence times (partial conversion of unwanted components) and depleted catalysts.

The more severe the hydroprocessing, the more likely it is that the kerosene stream will have poor lubricity performance. This is the main challenge with hydrocracker products, which are excellent materials in most other respects (e.g., stability, water separation, smoke point). Lubricity additive and/or blending options may be needed to release a satisfactorily performing jet fuel.

10 Trends

10.1 Changes in Current Crude Refinery Operations

Refineries produce a slate of products, to meet all quality and statutory requirements with the best possible payback. They are sensitive to feedstock prices, the costs of upgrading and to demand for products they can sell at a given time. This can include a concern about security of supply and their crude flexibility: their choice of crude or crudes will depend on the process units at the refinery, which usually reflect the original crudes processed and products made. Having more types of units, i.e., being a more complex refinery, broadens the range of crudes that can be processed.

As traditional crude sources dry up, as is happening with the North Sea, refineries may need to source alternative crudes. The more problematic the crude, the cheaper it is. Typical challenges would be with high acidity, high sulfur content, high viscosity, trace metal content, etc. If a refinery can upgrade such "opportunity" crudes, with changes to distillation protocols and upgrading processes, then it can still make acceptable product. In the USA another set of "new" crudes are those obtained by fracking, from such locations as the Permian fields in Texas or Bakken in North Dakota. Many of these have high nitrogen contents, and not all units can remove enough nitrogen to make a stable jet product; hydroprocessing catalysts usually target sulfur removal so more severe hydroprocessing units may be needed. This may require products to be treated with LIA to overcome lubricity issues with high severely hydroprocessed blends.

The changing demands for "road" fuels, to low sulfur fuels in the 90s and 00s, and now because of a push to electrification, affects how refineries process their crudes to meet the revised quality and quantity requirements of their slates of products. Peak gasoline demand has happened in the West and peak road diesel is approaching. Meanwhile, apart from the temporary, albeit dramatic reduction in demand due to Covid-19 in 2020–2021, the demand for kerosene remains strong and is expected to grow. Aviation fuel is the main outlet but, depending on country and climate, it is also used for domestic applications (heating, cooking and light) and as a component for winter grade diesel fuel blends. One option to meet the demand is to import kerosene, but refineries can also expand the type of refinery streams they use in their kerosene pool. This amounts to upgrading cracked streams that would historically been in the gasoline and diesel (gasoil) pools, from Fluidized Catalytic Cracker (FCC) or (Delayed) Coker units. This includes heavy catalytically cracked gasoline (HCCG) and light light cycle oil (LLCO); terminology for these streams vary as do the cut-points and is unlikely to be consistent across all refineries. These tend to have high levels of olefins and/or aromatics, the former leading to stability issues for kerosenes if used "as is" and the latter leading to unacceptable smoke points. However, with appropriate upgrading, including relatively severe hydroprocessing, it is possible to make final blends with these that meet jet fuel requirements.

The key to changing crudes and/or expanding the allowed streams is to understand and mitigate all possible problems in advance. Management of Change (MoC) exercises, ideally including outside expert views, should identify all the fuel properties and process unit operations that may need controls beyond those previously used for the jet fuel production.

Another trend is that newer refineries, rather than producing and processing kerosene and gasoil streams separately, will co-process a single, combined stream then distil off separate kerosene and gasoil cuts. This cuts down on process units, but the refinery will still need to address potential MoC issues for the feedstocks and processing they adopt. In particular, the refiner will need to assess how to deal with the requirement for trace amounts only of sulfur in ultra-low sulfur diesel (ULSD) and the impact such processing would have on the jet fuel's lubricity, particularly if the refinery does not have another kerosene stream with which to blend to make jet.

10.2 Future-Proofing Changes

The aviation industry is under pressure to reduce its environmental impact. So far, this has focussed on reducing its carbon footprint, but non-CO_2 effects may be greater. Flying with sustainable aviation fuels is an important tool for achieving CO_2 reductions. Refineries can also contribute to these; any efficiency gains they can make automatically reduce their CO_2 footprints (and that of the conventional jet fuel they make) but there may be scope to phase out their current sources of hydrogen and to bring in "blue" and "green" hydrogen. The kerosene would be the same as now, just with significantly lower CO_2 footprints.

Refineries can also help address the non-CO_2 impacts. They can take further steps to reduce the sulfur content of the fuel (helped by having access to lower sulfur content crudes or by increased hydroprocessing/reduced wet treating). Additionally, reducing the aromatic content of the kerosene may reduce the levels of particulates in the final exhaust products from the aircraft into the atmosphere, potentially reducing the degree of contrail formation. This is an area attracting more interest and research is on-going to understand the relationships between detailed fuel composition and exhaust products. Crudes all contain similar levels of aromatics so aromatics reductions would require new or repurposed processing units. Some improvements could be achieved by reducing the levels of diaromatic species (possibly by reducing the end point with associated decrease in yield) but significant reductions in aromatics contents would require at least one of the following:

- Dearomatisation (more specific hydroprocessing technology).
- Blending conventional jet with synthetic (not just sustainable) blending components with zero aromatics.
- Increased levels of coprocessing of synthetic fuel intermediates with zero aromatics. This is already happening commercially: new, renewable components are being brought into traditional refinery operations, primarily to increase the

amount of renewable diesel being produced. Western jet specifications currently permit the coprocessing of several streams, including Fischer–Tropsch waxes, bio-derived mono-, di-, and triglycerides, free fatty acids, and fatty acid esters with conventional hydrocarbon diesel streams. Further additional synthetic streams are being assessed, as is the scope to raise the permitted levels of synthetic component being coprocessed. It is probable that future jet fuels will include other sustainable feedstocks being blended with refinery streams for upgrading to kerosene. (Chapters. 1 and 3 provide more information about the specification changes and the new synthetic stream approvals.)

Also, as the need to produce road fuels decreases and less petroleum warrants being processed, refineries may also convert into biorefineries. The trend has started recently toward biorefineries, especially in Europe and America but less in other continents.

Notwithstanding the above changes, the conservative thinking of the aviation fuel industries, and the policy of not taking risks, will limit the changes expected in most of the production of jet fuel in refineries. As a bare minimum, refineries will continue to accommodate ordinary specification changes (new tests methods and new testing requirements, such as a switch from filterable deposits to particle counting, which may require additional measures to keep the fuels clean).

11 Military Specialist Fuels

The military require a range of fuels for their aircraft and other vehicles. Most of the fuel is the same as commercial ones, other than in the usage of additives. Variants on the Jet A/Jet A-1 like fuels require higher flash points (for aircraft carriers) or wider distillation ranges (for cold environments or to ensure sufficient fuel in a crisis). Very specialist applications, like ramjets, rockets, etc., require specialist fuels; none of these has a commercial equivalent, at least in Western operations, and they are made at very few locations, using proprietary processes and special additives. Consequently, this chapter does not include the production of these kerosene or non-kerosene specialist fuels.

Table 3 summarizes the most common of these three groups of fuels.

The main military fuel is based on Jet A-1 type fuels made to Def Stan 91-091 or derivative standards,[16] with extra additives to create the JP-8/F-34 fuel that NATO and linked air forces use for most operations. (In the USA, the military have transitioned to using a Jet A-based fuel but outside USA, they use JP-8/F-34 grade, based on Jet A-1). NATO's Single Fuel Policy, intended to support military operations when a single fuel supply system helps supply logistics, extends to having aviation fuels being used in ground vehicles. Once the JP-8/F-34 grade base fuel (Jet A-1) has been produced, which usually requires the manufacturer to add antioxidant on run-down

[16] As set out in STANAG 1135 (5) Amd 5 [12], used by the Armed Forces of NATO, and STANAG 3747 Amd 12 [13].

Table 3 Main classes and examples of military fuels and their commercial equivalents, where relevant

Category	Commercial equivalent	Examples[17] (US, UK, NATO)	Variant
Commercial	Jet A basis	F-24	–
	Jet A-1 basis	JP-8, Avtur + FSII, F-34 JP-8+100, F-37	F-63 is single battlefield fuel for road vehicles and land-based diesel equipment, made by adding S-1750 package to Jet A-1/JP-8 to improve lubricity and cetane number (replacing diesel-based F-54)
	Jet B basis (wide-cut gasoline/kero range)	JP-4, Avtag/FSII, F-40	–
Commercial, modified	Jet A-1 high flash	JP-5, Avcat/FSII, F-44	–
Specialist	None	JPTS	Thermally stable, low freeze point fuel in the lower density/distillation end of kerosene range
	None	JP-7	Low volatility (high flash) turbine fuel, with low aromatics, falling within the kerosene range
	None	JP-10	High density hydrocarbon fuel, composed solely of exo-tetrahydro dicyclopentadiene
	None	RP-1 RP-2	Rocket grade kerosene propellants. Low freeze point products in kero range. RP-1 has low sulfur and RP-2 has ultralow sulfur content

from any hydroprocessing step, the customer is usually responsible for the other JP-8 and diesel specific additives being put into the fuel. There are no extra refinery-related steps and challenges beyond those encountered in making Jet A-1 and JP-8.

As indicated in Chap. 1, little is known about the fuels used by the military in Russia/CIS and China. Russia appears to have specialist fuels, e.g., T-6 and T-8B, for some undefined applications, as indicated by different distillation, freeze point and thermal stability requirements, but they probably use additivated commercial grade fuels for most of their operations. The RT grade of civil fuel is described as being suitable for supersonic flight and has better thermal stability than the TS-1 grade. For both Russia and China, economics and security of supply probably preclude the adoption of specialist fuels for all operations. Any quality issues would therefore

[17] By default, listing the equivalent US, UK, and NATO specifications together. There are other national standards (issued by, e.g., CGSB for Canada or DCSEA for France) but these generally follow the US and/or UK specifications.

match those of the commercial fuels or would be related to additivation and handling downstream of the refinery.

Acknowledgements Thanks to William Echt, Merichem Company, for references and specific feedback on the issues and controls involving the Merichem II™ and Merichem J™ jet fuel processes (private communication). We also thank our reviewers Naphtali Brodsky and Daniel E Kadlecek for their significant contributions to improving the content of the chapter.

Glossary of Terms in this Chapter

Acronym/Term	Meaning	Comment
ADU or CDU	Atmospheric Distillation Unit or Crude Distillation Unit	First stage in crude separation at refinery
ASTM	American Society for Testing and Materials, International	Standards and methods organisation
Avcat	Aviation turbine fuel with high flash point	Kerosene, suitable for use onboard aircraft carriers. JP-5 grade
Avtag	Wide-cut kerosene + gasoline fuel for aviation turbine engines, with low flash and low freeze points	JP-4 or Jet B; made in very few locations
Avtur	Aviation turbine fuel	Jet fuel; terminology in Def Stan Jet A-1 and similar specifications
BHT	Butylated hydroxy toluene	A permitted antioxidant
CAS	Chemical Abstracts Service	Used as Substance (component) identification number
CI/LI	Corrosion inhibitor/ lubricity improver additive	Also known as LIA (see below)
CCR	Continuous Catalytic Reformer	Refinery unit making gasoline components and hydrogen
EINECS	European Inventory of Existing Commercial Substances	–
FCC (U)	Fluid Catalytic Cracker (Unit)	Upgrading unit for VGO and similar heavier cuts, mainly used for gasoline production
FFC	Fiber Film Contactor	Merichem process technology

(continued)

(continued)

Acronym/Term	Meaning	Comment
FSII	Fuel System Icing Inhibitor	–
FWS	Filter water separator	Dirt and water removal system common in aviation fuel supply systems
GOST (ГОСТ in Russian)	gosudarstvennyy standard, = state or governmental standard	Covers specifications and methods, maintained by the Euro-Asian Council for Standardization, Metrology and Certification (EASC), for the Commonwealth of Independent States (CIS)
HCCG or HCCN	Heavy catalytically cracked gasoline or naphtha	A heavy gasoline range cut normally from an FCC, the heavier part of which overlaps with the kerosene boiling range
HDS	Hydrodesulfurization	–
IBC	Intermediate Bulk Container	Storage option for fuels and additives
ISO	International Organization for Standardization	For aviation fuel, most relevant for quality assurance standards rather than test methods or specifications
JIG	Joint Inspection Group	–
KHT	Kero hydrotreater	Mild hydrotreatment unit used for kerosene streams
LHSV	Liquid Hourly Space Velocity	Measurement of residence time in a reactor
LIA	Lubricity improver additive	Also known as CI/LI
LLCO	Light light cycle oil	A light diesel feedstock cut from a Fluid Catalytic Cracker unit, usually with high aromatics content
MDA	Metal deactivator additive	–
MoC	Management of Change	Process to assess risks to product quality before making crude, process, or additive changes
MSDS	Material safety data sheet	Safety-related document to accompany all fuel movements
MSEP	Microseparometer	–
NAMAS	National Measurement Accreditation Service	Quality Assurance system used by some laboratories in addition to or instead of ISO standards
OOS	Out of Specification	–
REACH	Registration, Evaluation and Authorisation of Chemicals (REACH)	European system for chemical substance and harms data. All chemical materials made in, imported into or exported from Europe need their component parts to be REACH registered

(continued)

(continued)

Acronym/Term	Meaning	Comment
RT (PT in Cyrillic)	One of the two main grades of jet fuel made in Russian and some CIS countries to GOST 10227	Higher quality fuel (probably 100% hydroprocessed) for subsonic and supersonic aircraft
SDA	Static Dissipator Additive	–
SRB	Sulfate-reducing bacteria	Formed in anaerobic environments, they generate H_2S
TS-1 (TC-1 in Cyrillic)	One of the two main grades of jet fuel made in Russia and some CIS countries to GOST 10227	Lower flash point and lower freeze point than Jet A-1. Suitable for subsonic aircraft
TSCA	Toxic Substances Control Act	TSCA numbers are given to components in products listed in safety data sheets, for safety reasons
VGO/HVGO	(High) Vacuum Gas Oil	Heavy stream fed to upgrading units

References

1. Defence Standard 91-091. Turbine fuel, aviation kerosine type, Jet A-1 NATO code: F-35, Joint Service Designation: AVTUR, Issued by the United Kingdom (U.K.) Ministry of Defence. Glasglow, UK. https://www.dstan.mod.uk/
2. ASTM D1655. Standard specification for aviation turbine fuels. ASTM International, 100 Barr Harbor Drive, PO Box C700, West Conshohocken, PA 19428-2959, United States
3. Aviation fuels technical review, Copyright 2007 Chevron Products Company, 6001 Bollinger Canyon Road, San Ramon, CA 94583
4. ASTM D7566. Standard specification for aviation turbine fuel containing synthesised hydrocarbon. ASTM International, 100 Barr Harbor Drive, PO Box C700, West Conshohocken, PA 19428-2959, United States
5. Speight JG (ed) Handbook of petroleum refining. CRC Press (© 2017 by Taylor & Francis Group, LLC)
6. Goodger E, Vere R (1985) Aviation fuels technology, in particular Chapter 4.Macmillan Publishers
7. EI/JIG 1530 standard (2019) Quality assurance requirements for the manufacture, storage and distribution of aviation fuels to airports, Second Edition. Energy Institute (Great Britain) & Joint Inspection Group
8. Bussey K, Chachere S, Mericat J. A new kerosene treating technology to meet jet fuel specifications, Publication from Merichem. www.merichem.com
9. ICAO manual on civil aviation jet fuel supply, Doc 9977 AN/489, First edition 2012, International Civil Aviation Organization, 999 University Street, Montréal, Quebec, Canada H3C 5H7
10. ISO/IEC 17025 (2017) General requirements for the competence of testing and calibration laboratories. Available from ISO at https://www.iso.org/standards.html

11. JIG Bulletin 150. Testing water separation properties of jet fuel downstream of point of manufacture (Revised MSEP Protocol), Joint Inspection Group Limited, February 2024. www.jigonline.com
12. STANAG 1135 Amd 5. Interchangeability of fuels, lubricants and associated products, NATO unclassified
13. STANAG 3747 Amd 12. Guide specifications (minimum quality standards) for aviation turbine fuels, NATO unclassified

Chapter 3
Building Safety and Performance into Synthetic Aviation Turbine Fuels

Mark Rumizen

Abstract Concerns about the security of the supply of aviation fuels, and more recently the sustainability of aviation fuels, have challenged the aviation industry to expand the allowable components used in jet fuel manufacture. This has led to the ASTM D4054 guidance material that sets out how to confirm the suitability of candidate synthetic components. The ASTM D7566 specification then defines requirements for new components and blends made with conventional jet fuel components that can then drop into jet fuel distribution systems as Jet A or Jet A-1 meeting ASTM D1655, Def Stan 91-091, or similar existing specifications. The evolution of these changes is described, as well as more recent changes that are now allowing some synthetic materials to be coprocessed with conventional refinery streams to create jet fuels with sustainable content. Industry jet fuel product quality guides, such as EI 1533, provide guidance for manufacturers and suppliers of semisynthetic jet fuels that supplement the specification controls used to ensure fit-for-purpose jet fuels.

1 Introduction

The production of fuel from non-petroleum raw materials was initially driven by supply constraints on petroleum resources. In the 1940s, during World War II, Germany produced diesel fuel from coal using the Fischer–Tropsch (FT) process as they were losing access to petroleum sources. In the mid-1990s, South Africa embarked on a challenging effort to gain approval to operate aircraft with jet fuel produced from coal using the same Fischer–Tropsch process. This was done to compensate for inadequate domestic petroleum reserves in the face of increasing demand for jet fuel. This effort was led by the South African petroleum company SASOL, with technical support provided by Dr. Clifford Moses from Southwest Research Institute in San Antonio, Texas. By 1999, SASOL's efforts led to the inclusion of the SASOL coal-to-liquid jet fuel in the major jet fuel specifications Def Stan

---

M. Rumizen (✉)
ASTM International, Subcommittee D02.J for Aviation Fuels, West Conshohocken, PA, USA
e-mail: mark.rumizen@gmail.com

M. Rabaev et al. (eds.), *Aviation Fuel Problems and Solutions: From Refinery to Wingtip*, https://doi.org/10.1007/978-3-032-05514-9_3

91-091[1] and ASTM D1655 [2]. The interest in these synthetic aviation turbine fuels (SATF) gained momentum in the mid-2000s from geo-political instability in the oil-producing regions along with the perceived threat of the exhaustion of petroleum reserves. The United States Air Force (USAF) and Defense Advanced Research Projects Agency (DARPA) provided funding to support the further evaluation and approval of FT fuel, along with the evaluation of conversion of plant and other natural oils to jet fuels.

The commercial airline community was also seeking alternatives to petroleum-derived jet fuel to counter the price volatility of these fuels. In 2006, the Commercial Aviation Alternative Fuel Initiative (CAAFI) was formed by a coalition of airlines, engine and aircraft manufacturers, trade associations, fuel producers, and other stakeholders in the aviation fuel community. The CAAFI leadership and members recognized at the time that the use of SATFs could also contribute to the mitigation of the environmental impact of jet aircraft operations by offsetting carbon emissions with renewable feedstocks. More recently, the concern with the role of carbon emissions on climate change has taken precedence over supply security and price volatility as evidenced by government-sponsored policy initiatives such as the United States' Sustainable Aviation Fuel (SAF) Grand Challenge [3] and the research funding and tax incentives included in the Inflation Reduction Act of 2022 [4].

The ultimate approval to fly with these new SATFs and integration into the existing aviation fuel supply infrastructure initially was perceived as a major risk factor within both the USAF and CAAFI. This was due to both technical and regulatory concerns. It was therefore agreed that the new SATFs had to be drop-in fuels that could seamlessly enter the supply chain infrastructure and be safely used on the existing fleet of aircraft without any design changes or the need for airworthiness authority approval.

The technical approach relied on a comprehensive and rigorous evaluation by the aircraft and engine technical experts of new fuel composition, properties, and performance for comparison with those of conventional jet fuel. With financial assistance from the U.S. Federal Aviation Administration (FAA), testing was conducted by third parties such as the USAF laboratories, universities, and at engine and manufacturer facilities to ensure integrity and objectivity of the technical evaluation. In addition, stringent production quality controls were incorporated into the resulting specifications for these fuels to further ensure acceptable performance and quality throughout their journey from production facility to the aircraft.

The regulatory framework was based on the fact that these drop-in fuels would be essentially the same Jet A or Jet A-1 fuels to which the existing fleet of aircraft were certificated to operate with and would therefore fit within the existing airworthiness authority approvals for these aircraft. This was accomplished by integrating the specifications for the new SATFs into the existing specification framework for conventional jet fuel. This required a collaborative effort from the aviation fuel community working with the standards development organizations (SDOs) that wrote the current jet fuel specifications such as ASTM International and the UK Ministry of Defence (MOD) Aviation Fuel Committee (AFC). These organizations rely on industry experts across the supply chain to ensure the final, approved SATF specifications provide the controls necessary for aviation.

2 Rigorous and Comprehensive Technical Evaluation of Candidate SATFs

The groundbreaking work conducted by SASOL and Dr. Moses was used as a basis to develop the technical evaluation approach for the approval of proposed SATFs [5]. The first phase of the SASOL project was directed at adding provisions to the Def Stan 91-091 specification to permit the production of jet fuel from coal using the FT process. Because their primary concern was the safety of flight operations, the original equipment (engine and aircraft) manufacturers (OEMs) demanded sufficient data to prove the new fuel was suitable, or fit-for-purpose, along with verification that this coal-derived jet fuel was essentially identical to petroleum-derived jet fuel relative to composition, properties and performance. The initial approval in 1999 was limited to specific product streams at SASOL only, and no other producers were covered by this approval. Over several years SASOL worked to expand this approval, though it still only applied to synthetic components made at a single refinery and blended into conventional jet fuel at only that single refinery. The Sasol FT approval was used as a building block towards a more generic approval of synthetic blending components (SBC) produced with the FT process from any producer at any facility. This led to a Coordinating Research Council (CRC) project for a generic FT approval process based on Moses' earlier work. It required extensive testing (mainly by US Air Force) of FT fuels from natural gas feedstocks as well as Sasol coal-based products. While the CRC work was still underway, the emphasis began to move away from security of supply towards increased sustainability. CAAFI, the Commercial Aviation Alternative Fuels Initiative, came into existence. As well as supporting R&D, CAAFI pushed for certification and qualification of the new fuels. The USAF and CRC work was then instrumental in being able to expand in 2009 the scope of ASTM International standard practice D4054 from a solely additive evaluation protocol to an additive and new fuel evaluation protocol. ASTM D4054 is now entitled "Standard Practice for Qualification and Approval of New Aviation Turbine Fuels and Fuel Additives" [6]. This scope expansion relied on significant input from the OEMs to ensure the document addressed the full slate of jet fuel properties and other characteristics necessary for the safe operation of all types of gas turbine engines and aircraft powered by those engines.

The scope of D4054 is designed to support an "up front" evaluation of the composition, properties, and performance of synthesized aviation turbine fuels (SATFs) produced from non-petroleum raw materials to ensure these fuels meet the demanding safety and performance standards of aviation. The document is organized in four tiers, each with a progressively more demanding set of tests.

The Tier 1 testing is based on the property and compositional criteria in the most broadly used conventional jet fuel specifications, such as ASTM D1655 and Def Stan 91-091. Because historically these specifications have only been applicable to conventional jet fuel produced from petroleum using well-understood refining processes, these properties only reflect a subset of the properties that are critical for safe operations of engines and aircraft. Other critical properties of interest are captured in the CRC Aviation Fuel Properties Handbook, which is used extensively by OEM engineers [7]. Therefore, this is considered an initial screening step where this limited set of properties is examined. Properties evaluated in Tier 1 include bulk properties typically measured at one temperature or condition, or a limited number of temperatures, to provide a snapshot of the fuel's properties. Some examples of these bulk properties are density, viscosity, freezing point, flash point, and thermal stability, which are measured at a specific temperature, and distillation which is measured at four boiling points. If the Tier 1 properties fall within the specification limits, then it only indicates that the SATF is a viable candidate for further evaluation. Then additional testing is required to fully evaluate the SATF's properties to ultimately determine if the fuel would provide acceptable performance across the entire operating envelope of the diverse fleet of existing gas turbine powered aircraft.

The Tier 2 testing delves more deeply into the physical and chemical properties of the candidate SATF. These properties are called fit-for-purpose properties and are defined in D4054 as "characteristics of an aviation fuel or aviation fuel additive in the fuel that are not listed in the specification criteria but are inherent in petroleum-derived jet fuel".[1] It is necessary to measure fit-for-purpose properties because the slight compositional or property shifts or the introduction of new and novel trace materials that have not been present in petroleum-derived jet fuel must be evaluated. These shifts or materials may impact on engine operation or the aircraft fuel system functioning over the broad range of the operating envelope, or over the lengthy operating intervals typical of today's commercial aircraft. For example, Tier 1 testing only requires the measurement of density at 15 °C to provide a snapshot of the fuel's density, but this does not provide any information regarding how the density of the fuel will change as the temperature in the fuel tank changes. An unanticipated change in density could impact the accuracy of the capacitance fuel gauging sensors resulting in misleading fuel level indications to the pilot; there may not be enough fuel to complete the planned flight! Consequently, Tier 2 testing requires that the density be measured across the temperature range from −20 to 60 °C. Other bulk physical properties such as viscosity and distillation are also measured over broad ranges. Tier 2 testing also includes fuel properties not specified in the jet fuel specifications such as surface tension, vapor pressure, specific heat, thermal conductivity and water and air solubility; many of these properties are covered for conventional fuels in the CRC Handbook [7].

The compositional evaluation included in Tier 2 analyzes both the bulk hydrocarbon compound classes and the trace organic and inorganic elements in the fuel. The bulk hydrocarbon composition is analyzed, typically using advanced two-dimensional gas chromatography (2D GC) to provide a detailed report of the hydrocarbon compound classes present in the fuel, along with the distribution by type and percent volume and weight. This will highlight if the fuel introduces any shifts in the typical composition of current jet fuel population, relative to size (molecular

[1] ASTM International Standard D4054. "Standard Practice for Qualification and Approval of New Aviation Turbine Fuels and Fuel Additives", paragraph 3.2.3.1, ASTM International, West Conshohocken, PA, USA.

weight) or type (isoparaffin, cycloparaffin, normal paraffin, aromatic) of hydrocarbons. The gas turbine OEMs look very closely at this composition to determine if any shift or difference relative to petroleum-derived jet fuel will impact the operability, durability, or safe operation of their engines.

It is accepted that petroleum-derived jet fuel will contain small amounts of non-hydrocarbon trace materials. ASTM D1655 includes this statement in the Materials and Manufacture section of the specification; "Conventionally refined jet fuel contains trace levels of materials that are not hydrocarbons, including oxygenates, organosulfur, and nitrogenous compounds".[2] Candidate SATFs are analyzed to ensure that the amounts of these trace materials are no greater than seen in typical petroleum-derived jet fuel, and that no new harmful trace materials are introduced into the jet fuel supply. For example, the advanced, specialized coatings that engine OEMs apply to blades and nozzles in the turbine section to withstand the extreme temperatures experienced in the turbine are susceptible to attack by certain trace elements, such as sodium, potassium, lithium, and vanadium. And copper can impact the thermal stability of jet fuels thereby increasing the possibility of fuel nozzle deposits and associated blockage of fuel flow as well as leading to less effective cooling of lubricants in fuel cooled oil coolers.

At the conclusion of Tier 2 testing, a team of fuel specialists from the OEMs and airworthiness authorities will evaluate the data from the Tier 1 and 2 testing to determine if additional testing is required on actual aircraft or engine components, systems, or hardware. This is called Tier 3 testing, and it includes testing such as combustor rig, fuel nozzle rig, turbine hot section rig, or materials compatibility testing. For example, if there is a concern regarding turbine engine combustor operability due to a slight compositional shift of the hydrocarbon compounds, a combustor rig test would be conducted to evaluate performance characteristics such as cold starting, lean blowout, aerial restarting after a flame-out event, and turbine inlet-temperature distribution. A shift in the aromatic composition, and/or the cycloparaffinic composition might lead to materials compatibility testing of the fuel system elastomeric seals (see more in Chap. 6), or evidence of certain trace materials may lead to testing of aircraft/engine fuel-wetted metallic materials.

The OEM team will also determine if Tier 4 testing is required, which includes testing of complete turbine engines or auxiliary power units (APUs) on test stands, or even flight testing of a complete aircraft. This testing would be required in a case where there is OEM interest in validating the test results from the lower tiers under actual operating conditions. Because of the analytical rigor applied in Tiers 1 to 3, Tier 4 testing is not typically required. However, in some cases the OEMs or military organizations form partnerships with the candidate SAF producers and offer to conduct engine or flight testing to augment the final data report.

OEMs recognize that conducting Tier 3 or Tier 4 testing imposes a significant expense and resource commitment from the candidate SATF producer due to the large volume of fuel required and the need to arrange for use of OEM assets to conduct

[2] ASTM International Standard D1655. "Standard Specification for Aviation Turbine Fuels", Note 1 to paragraph 6.1.1, ASTM International, West Conshohocken, PA, USA.

the testing. This, along with increasing confidence in the analytical techniques used in Tier 1 and 2 testing gained from previously conducted D4054 SATF evaluations (particularly for FT and Hydroprocessed Esters and Fatty Acids (HEFA)[3] SBCs), has led to less frequent demand for Tier 3 and 4 testing.

The data generated from the D4054 testing contribute to the development of the specification annex criteria for the specific SBC. While the criteria are similar across all the annexes, the D4054 data may reveal the need for additional tests or other requirements necessary to ensure a fit for purpose jet fuel is produced.

3 Ensuring Integrity and Objectivity of Technical Evaluation

As discussed above, the OEMs and the airworthiness authorities decided that the technical evaluation of candidate SATFs would be built around the existing industry consensus-based standards development process used by ASTM International to develop and maintain aviation fuel standards. Because this was an industry led effort, and not a regulatory compliance program, the airworthiness authorities did not have direct oversight of the testing and test facilities. In addition, the cost of the D4054 testing presented a challenge to the companies developing SATF pathways that were new ventures with lower capital resources. Initially, the US military took the lead to financially support the FT (USAF) and HEFA (US Navy) testing and evaluation efforts, but the FAA took a decisive step in 2014 to mitigate the OEM and industry concerns regarding the oversight of the testing, along with the SATF producer challenges with funding the testing, by establishing the D4054 Clearinghouse.

The FAA established the D4054 Clearinghouse under its Aviation Sustainability Center (ASCENT), also known as the Center of Excellence for Alternative Jet Fuels and Environment. ASCENT is a cost-sharing program that the FAA established with academia and industry to conduct research on alternative jet fuels and the environmental impact of aviation. The D4054 Clearinghouse is being managed by the University of Dayton Research Institute (UDRI). The Clearinghouse provides coordination and management of the testing and OEM review of the D4054 data generated for candidate SATF pathways. The use of an objective third-party to manage the process has instilled confidence with the OEMs and airworthiness authorities regarding the testing results, and the engagement of the OEMs and airworthiness authorities in the data review process has lent credibility to the data with the ASTM International aviation fuel subcommittee members, and other stakeholders in the aviation fuel community.

The Clearinghouse asks candidate SATF producers to provide 100 gallons of test fuel at the start of the process. This fuel quantity requirement is considered a screening step for validation of commercial viability because companies must advance beyond laboratory scale to be able to produce this much fuel. The Clearinghouse will then

[3] Annex A2, ASTM D7566.

conduct the Tier 1 and 2 testing using both in-house and third-party test facilities. As this data is being generated, the Clearinghouse will arrange for periodic engagement of the candidate SATF producer with the OEM team of fuel specialists to provide guidance and advice regarding the testing. Upon completion of the Tier 1 and 2 testing, the Clearinghouse assists the candidate SATF producer in the drafting of a Phase 1 Research Report, and then coordinates the review of that report with the OEMs. After an iterative period of questions and answers, the candidate SATF producer will either move on to Tier 3 or 4 testing or advance to the ASTM International specification development process. In either case, the Clearinghouse remains engaged to support the candidate SATF producer until final specification issuance.

The success of the US D4054 Clearinghouse was recognized in the United Kingdom and European Union, where each followed with the establishment of their own Clearing Houses. Their detailed remits and funding differ from the FAA-funded Clearinghouse based in the USA, but they are being run as complementary facilities to expand early testing options and support to candidate SAF producers engaged in the D4054 process. Most importantly, all three Clearing Houses are collaborating to avoid any overlap or duplication of effort to result in an overall improvement in efficiency of the D4054 process while providing more financial support to candidate SAF producers.

4 Stringent Production Quality Control–ASTM D7566

Because of the lack of commercial-scale production experience with SATF, the OEMs, airworthiness authorities, and the aviation fuel community demanded more stringent production quality control for these fuels. This was accomplished by issuing a dedicated ASTM International specification for SATF, separate from the D1655 conventional jet fuel specification. This specification was issued as ASTM D7566 in 2009.

The use of an ASTM International specification was agreed upon for two reasons. First, the existing conventional jet fuel specification, D1655, is issued by ASTM International. Second, the ASTM International specification development process is designed to leverage the widest possible segment of the aviation fuel community in the review both of the proposed SATF pathway and the specification criteria, to ensure a technically robust standard is produced that ensures consistently safe fuel.

ASTM D7566 builds upon the conventional jet fuel specification D1655 by including all the fuel property criteria in that specification, but adds additional properties considered necessary for SATF. D7566 further enhances the level of control by establishing separate annexes for each of the conversion pathways; these contain even more property criteria that must be met by the neat synthetic blending component (SBC). For example, the annexes include a requirement to conduct thermal stability test of the fuel, as measured by ASTM D3241/IP 323 (JFTOT) at higher temperatures (325 °C versus 260 °C for conventional jet fuel), and distillation slope requirements that are not imposed on conventional jet fuel. Each annex also includes unique criteria for defining the key elements of that conversion process to ensure producers are limited to using a similar process to that used to produce samples during the D4054 review stage.

Once the neat SBC is certified as meeting the annex criteria, it must then be blended with conventional jet fuel at less than or equal to a specified volume percentage limitation (either 10% or 50%). The requirement to blend with conventional jet fuel is imposed to ensure aircraft do not operate with solely SATF at this early stage in the maturation process. After blending, the resulting SATF must be tested again to properties specified in the standard. Additional and unique property criteria are applied at this stage also, such as a lubricity requirement and a viscosity requirement at −40 °C (versus −20 °C for conventional jet fuel).

5 Regulatory Endorsement and Global Acceptance

The approach described above was designed to provide technical validation that a candidate SATF is essentially the same Jet A or Jet A-1 jet fuel as is produced from petroleum feed stocks. Both D7566 and D1655 have acknowledged this with specific provisions in each document that allow SATF produced to D7566 to be re-certified as D1655 Jet A or Jet A-1 fuel without further testing. This then enables the SATF to make a seamless transition from production into the distribution and operational infrastructure, including operation on the aircraft. This process essentially does not approve a new type of jet fuel, but rather relies on the OEM technical experts and the ASTM subcommittee to make a determination that the new pathway produces Jet A or Jet A-1 fuel. Consequently, additional airworthiness authority approval is not required, because the existing fleet of turbine engine powered aircraft are all approved to operate on Jet A and Jet A-1 fuel.

The FAA has confirmed this with issuance of Special Airworthiness Information Bulletin (SAIB) NE-11-56 (latest revision) [8] which communicates that SATF meeting ASTM D7566 is acceptable to use on aircraft and engines approved to operate on Jet A or Jet A-1 fuel. The European Aviation Safety Agency (EASA) has also issued Certification Memorandum EASA CM—PIFS—009 [9] which describes the policy for reliance on appropriate specification control bodies for approval of aviation fuels and the inclusion of these fuel specifications in the aircraft flight manual limitations. The United Kingdom's Ministry of Defence has referenced ASTM D7566 in their Def Stan 91-091 specification for acceptance of SATF as has the Canadian General Standards Board (CGSB) for the Canadian civil and military specifications. Other airworthiness authorities, such as Brazil's Agência Nacional de Aviação Civil (ANAC), Japan's Civil Aviation Bureau (JCAB), and the Civil Aviation Administration of China (CAAC) have all engaged at various levels in the use of the D7566 SATF specification. Military specifications, which are developed independently from ASTM specifications, have also moved to refer to D7566, with the understanding that the final fuels will still need to meet any additional criteria unique to military grade jet fuels.

The International Civil Aviation Organization (ICAO) has adopted specific ASTM D7566 annexes for use in its regulatory policy for the Carbon Offset and Reduction Scheme for International Aviation (CORSIA). The CORSIA policy specifies default carbon emission values for SATF that meet selected D7566 annexes. This policy will be used globally by airlines to meet ICAO and national carbon emission limits. Finally, the International Air Transport Association (IATA) has incorporated the above SATF approval concept in its documents and in panel sessions at its semi-annual Aviation Energy Forums.

6 Supply Chain Quality Control Framework is Emerging

Historically, quality control of the movement of jet fuel through the supply chain has been accomplished by a collaboration of industry standard bodies and other industry organizations. Because this approach has been so successful, as evidenced by the outstanding safety record of jet fuel, the aviation industry is leveraging these existing organizations and associated standards to oversee the production and handling of SATF and SBCs. It should be noted however, that because SATF is simply Jet A or Jet A-1 fuel, very little unique guidance is required for the final, blended SATF. However, it is anticipated that neat SBC may be transported and handled separately from SATF, and consequently the aviation industry is taking a proactive approach and has developed guidance for this. The Energy Institute has published EI 1533, "Quality assurance requirements for semi-synthetic jet fuel and synthetic blending components (SBC)" [10], to provide procedures and other guidance for ensuring SBCs and SATFs are of good quality and fit-for-purpose when they arrive at the aircraft.

EI/JIG Standard 1530, "Quality assurance requirements for the manufacture, storage and distribution of aviation fuel to airports", was developed by industry stakeholders in 2013 to provide recommended quality and handling practices and procedures for jet fuel as it travels from the producer to the airport. The handling procedures described in this document reflect the combined knowledge and experience of a team of industry experts with extensive jet fuel production and handling experience. The rapidly expanding SAF market prompted the Energy Institute to apply the same expertise to develop EI 1533 in 2022 as a supplementary document to EI/JIG 1530.

EI 1533 was prompted to some degree by the knowledge that many of the companies producing or planning to produce SATFs or SBCs are new technology companies lacking in experience manufacturing or handling of aviation fuels. Additionally, the unique provisions of ASTM D7566 related to production, handling and blending of SBCs necessitated some form of procedural guidance to ensure fuels delivered to airports are of aviation-grade quality. EI 1533 is a comprehensive document that provides important background information such as definitions, description of specification properties and limits, along with guidance and procedures specific to SBCs and SATF on transporting, testing, storage, and blending operations. EI 1533 exemplifies the pro-active nature of aviation to implement risk-reduction tools at the introduction of new technology such as SATF and SBCs. Quality control issues that might arise in this new aviation fuel supply environment that EI 1533 is intended to address include the following:

- A misinterpretation of the materials and manufacture requirements of a particular ASTM D7566 annex by an SBC producer could lead to the use of an unapproved feedstock or process design, which in turn can lead to compositional differences or introduction of unknown trace materials. Neglecting to add antioxidant additive after production can lead to gum formation, and specification testing errors or skipped tests could lead to a multitude of operational issues.
- Marine or rail transport of SBCs introduces many opportunities to impact the properties or quality of the blending component. Inadvertent mixing with other SBCs, fuels or chemicals can impact key performance properties that impact engine operability or can introduce trace materials that damage aircraft materials or systems. Improper management of the transmix region during pipeline transport can also result in the same issues, and neglecting to conduct recertification testing can result in an off-spec SBC being blended with a conventional jet fuel.
- The blending of SBCs with conventional jet fuel introduces a new operation, along with associated personnel and facility requirements, into the jet fuel supply chain that requires adherence to blend percentage and property limitations, along with certification testing after blending. Not adhering to these limitations, or not testing after blending, can result in off-spec jet fuel being delivered to the airport. Also, due to density differences that may exist between the SBC and the conventional jet fuel, mechanical mixing should be used during blending. If this is not done, a layered or stratified batch would result in differing physical properties across the fuel batch that could impact engine or aircraft operability.
- The SBC and SATF supply chain introduce additional documentation requirements beyond the conventional jet fuel supply chain that the industry is used to operating in. Both the SBC manufacturing step and the blending step introduce new certification steps, along with new test requirements that can either be overlooked or improperly conducted. Operational issues such as fuel system seal leakage resulting from insufficient concentration of aromatics, or engine operability issues due to inadequate distillation slope can result if aromatics are not recorded properly or if the distillation slopes are not calculated properly.

IATA, the Joint Inspection Group (JIG), and Airlines for America (A4A) are also incorporating guidance for handling of SATF at the end of the supply chain where the SATF is uplifted to the aircraft.

7 Seeking Greater Production Efficiencies with Co-processing

The impending imposition of government policy to mandate renewable content in aviation fuel spurred traditional petroleum producers to investigate the incorporation of biomass or other renewable materials into their jet fuel production processes. It was realized that co-processing of these renewable feedstocks in parallel with conventional petroleum feeds at the front-end of the refinery could be done at lower costs than establishing separate SBC production facilities and blending the two finished products at the back-end of the production processes. This has led to the incorporation of three co-processing pathways in the conventional jet fuel specification, D1655 beginning in 2018. These are co-processing of free fatty acid and fatty acid ester feedstocks or Fischer–Tropsch synthetic crude feedstocks at up to 5%, or co-processing of hydroprocessed biomass at up to 24%. Once manufactured and tested to ASTM 1655, including the additional requirements of Annex A1 of D1655, these fuels are handled in the same manner as conventional jet fuel.

8 Building on SATF Experience with Non-Drop-in Fuels

The above overview has focused on the progress of the industry for the incorporation of drop-in SATF into the supply chain and aircraft operations. Limiting the scope of these environmentally friendly jet fuels to only drop-in fuels has greatly simplified this task by relieving the requirement to create a new airport infrastructure and to recertify existing aircraft to operate on a new fuel. However, interest is growing in environmentally friendly jet fuels that differ from petroleum-derived jet fuel relative to composition, properties, or performance. These are called non-drop-in fuels. Specifically, fuels that are compositional subsets of petroleum-derived jet fuel, such as a paraffinic hydrocarbon fuel lacking the aromatics that are found in conventional jet fuel, are being investigated. These fuels do provide operational advantages, such as reduced particulate emissions and improved combustor durability, however they may require some aircraft and engine design changes and must be recertified for use on any particular aircraft and engine model before they can be used. Also, there is interest in hydrogen for use in turbine engines or fuel cells, which requires a much greater aircraft redesign and certification effort, along with the creation of an entirely new and separate distribution and supply infrastructure.

As with drop-in fuels, the aviation industry is leveraging the current approaches used to control conventional jet fuel for these new non-drop-in fuels. For example, an ASTM task group has been formed in the aviation fuel subcommittee to develop a new specification for a paraffinic SATF, and the ASTM gaseous fuels committee is developing a specification for hydrogen aviation fuel. Both efforts will build upon knowledge and experience base gained by the OEMs, airworthiness authorities, and other aviation fuel stakeholders during the D4054 evaluations conducted for the D7566 SATFs. This experience has significantly contributed to the understanding of the relationship between fuel properties and composition and aircraft design requirements that can be applied to these non-drop-in jet fuel development initiatives.

9 Concluding Comments and Future Prospects

It is important to note that we are only at the early stages of SATF development and commercial scale-up. For example, of the approximately 25 billion gallons of jet fuel consumed in the United States in 2022, only 5 million gallons, or 0.2%, were SATF. This reflects the economic challenges to designing and building large-scale production facilities, and the challenges of creating a renewable feed stock supply and distribution infrastructure. However, even with this limited experience base, we are seeing the SATF industry propose refinements to the specification test requirements to align better with our improved understanding of any potential risks introduced with these new fuels.

For example, two SBC test requirements originally included in the annexes were a test for the Fatty Acid Methyl Ester (FAME)[4] concentration in SBC from the HEFA process, and a requirement in all the annexes to conduct the thermal stability test at a more severe condition (a higher temperature) than required for conventional jet fuel. However, after gaining limited production experience and a better understanding of the projected handling and distribution of SBCs, in combination with a recognition of the original intent of these requirements to ensure production quality control, it was realized that these requirements should only be required at production, and not downstream of production in the supply chain, and the specification has been revised to incorporate these improvements. In addition, the compositional and purity test requirements in the HEFA annex were originally required on a per batch basis, but this requirement has been relaxed to require these tests only at process start-up or in the event of a major change to the process. This reflects data collected that confirmed satisfactory experience with the 75 million gallons of HEFA produced (globally) at the time the change was approved.

Another example of a re-alignment of testing as a risk mitigation is the issuance of the D4054 Fast Track annex in 2019. The collective aviation fuel community experience with the review of a significant amount of test data from completed SATF evaluation projects, in combination with an understanding of the properties and performance of fuel possessing nominal jet fuel hydrocarbon compositions, led to this Fast Track provision, which allows a reduced test regime and associated reductions in cost and timing of the review process for candidate SATF producers. This provision, however, also limits the approved SBC blend percentage to 10% to offset the reduced testing requirements.

---

[4] More commonly referred to as "biodiesel".

In response to the risk of initiating the costly and time-consuming D4054 process without an understanding of the probability of successfully completing it, CAAFI established the Pre-Screening process [11]. The Pre-Screening process evolved from the National Jet Fuel Combustion Program (NJFCP) research program established to investigate the relationship between jet fuel properties and composition, and combustor operability requirements [12]. The CAAFI Pre-Screening process, along with the EU's comparable JETSCREEN program, applies advanced analytical chemistry techniques to very small fuel samples to measure and predict key fuel properties. This allows the candidate SATF producer to refine their conversion process at the laboratory scale to improve the likelihood of successfully completing the much more expansive D4054 process, thereby mitigating the risk of commencing that process.

Other noteworthy developments that reflect progress towards commercialization include the current effort to incorporate provisions in ASTM D7566 to enable the production of fully synthetic aviation turbine fuel, i.e., not blended with conventional jet fuel. In addition, two new conversion pathways, isobutene in Annex A5 and Alcohol to Jet with Aromatics as Annex A8, were incorporated into D7566 in 2023. As with the above examples, these efforts were built upon the knowledge and experience base gained from the past D4054 evaluation projects.

The industry standards that guide and control the evaluation and production of SBCs and SATF will evolve and be refined as we gain more experience and knowledge with these fuels. However, it should be emphasized that the safety of flight will always take precedence over efforts to rationalize testing requirements, and that the aviation fuel community has established a process with the key industry stakeholders that will rely on data, experience, and objective assessment to ensure this.

Acknowledgements The author would like to thank Chris Lewis for his helpful comments and additional insights during the preparation of this chapter.

References

1. Defence Standard 91-091. Turbine fuel, aviation kerosine type, Jet A-1 NATO Code: F-35, Joint Service Designation: AVTUR, Issued by the United Kingdom (U.K.) Ministry of Defence. Glasglow, UK. https://www.dstan.mod.uk/
2. ASTM International Standard D1655. Standard specification for aviation turbine fuels. ASTM International, West Conshohocken, PA, USA
3. FACT SHEET: Biden administration advances the future of sustainable fuels in american aviation (2021). https://www.whitehouse.gov/briefing-room/statements-releases/2021/09/09/fact-sheet-biden-administration-advances-the-future-of-sustainable-fuels-in-american-aviation/
4. Inflation Reduction Act of 2022, United States Public Law 117–169, August 16, 2022
5. Moses CA (2007) Coordinating Research Council, Inc. Technical Report AV-2- 04, Development of the protocol for acceptance of synthetic fuels under ASTM D1655
6. ASTM International Standard D4054. Standard practice for qualification and approval of new aviation turbine fuels and fuel additives. ASTM International, West Conshohocken, PA, USA
7. Aviation fuel properties handbook, 4th edn. Coordinating Research Council, Inc., 5755 North Point Parkway Suite 265, Alpharetta, GA 30022, USA (2014)

8. Special airworthiness information bulletin (SAIB) NE-11-56 (latest revision), Engine fuel and control - semi-synthetic jet fuel. Federal Aviation Administration. https://drs.faa.gov/browse/SAIB/doctypeDetails
9. EASA certification memorandum, CM No.: EASA CM—PIFS—009 Issue: 01, Issue Date: 28th of February 2013. European Aviation Safety Agency. https://www.easa.europa.eu/en/document-library/product-certification-consultations/easa-cm-pifs-009
10. EI 1533. Quality assurance requirements for semi-synthetic jet fuel and synthetic blending components (SBC). Energy Institute, 61 New Cavendish Street, London, United Kingdom, W1G 7AR. https://publishing.energyinst.org/topics/aviation/aviation-fuel-handling/1533
11. CAAFI prescreening guidance for alternative jet fuels. https://caafi.org/tools/Prescreening_Guidance.html
12. Colket MB, Heyne JS, Rumizen M, Edwards JT, Gupta M, Roquemore WM, et al (2017) An overview of the national jet fuels combustion program, Reston, VA. AIAA J 55

Chapter 4
Aviation Gasoline and Other Fuels for Piston Engine Aircraft

Alisdair Clark, Roger Gaughan, and Melanie Kimble-Thom

Abstract Aviation gasoline represents the first fuel specifically developed for flight in an era when the spark ignition piston engine ruled the sky. While the industry largely transitioned to the aviation turbine engine many years ago, the piston engine fleet still performs a useful and important role in global transport. Light aircraft represent a unique solution where ground transport is challenged due to adverse terrain, for example by desert, jungle, water, mountains and snow. In addition, the fleet offers pilot training and recreation to promote aviation across the world, foundations on which the wider industry is built. Aviation gasoline is the primary fuel for these aircraft and, in this chapter, the background to the fuel is presented together with its unique attributes, challenges and solutions. Further, details for aviation compression ignition (diesel) piston engines are also included to promote understanding of alternative power sources.

1 Introduction

Aviation gasoline represents a high-quality hydrocarbon fuel specifically developed for aviation spark ignition piston engines. The product is distinctly different from Jet A/A-1, to meet the demands of the intermittent combustion cycle used within such engines and the related aircraft fuel systems. Aviation gasoline has a long history, being the first fuel specifically developed for flight. An insight into this history is important to understand the role of various properties, how these can be challenged in production and distribution, and potential solutions. Within this chapter these aspects

A. Clark (✉)
BP International Limited, London, United Kingdom
e-mail: alisdair.clark@bp.com

R. Gaughan
RGG Aviation Consulting Services, Biltmore Lake, North Carolina, USA

M. Kimble-Thom
Baere Aerospace Consulting, West Lafayette, Indiana, USA
e-mail: melanieathom@BaereAerospace.com

M. Rabaev et al. (eds.), *Aviation Fuel Problems and Solutions: From Refinery to Wingtip*, https://doi.org/10.1007/978-3-032-05514-9_4

are examined, moving through the supply chain while highlighting knowledge built up in the Industry over many years. A section covering compression ignition (diesel) aviation engines is also included to offer an insight into this technology.

1.1 History of Aviation Gasoline

The success of the Wright Flyer in 1903 brought the world an understanding of the principles of controlled flight and the need for engines which offer high power-to-weight ratio and good reliability. This was achieved through use of a spark ignition gasoline piston engine, a concept developed 27 years earlier in 1876 by Nicolaus Otto. The Flyer featured such an engine, built by Charles Taylor [1], with four cylinders and weighing 179 lbs. The 12-horsepower output would be considered very low by today's standards. However, in its time, this engine gave the proof of concept which was to dominate aviation power-plant design for decades to come. The fuel utilized was straight run gasoline, a low boiling hydrocarbon fraction obtained by distilling crude oil for use in automobiles.

The combustion quality of such a product is very variable, as illustrated in Table 1 where straight run gasolines are compared in terms of Research Octane Number ASTM D2699 [2]—a measure of the fuel's ability to "wait for the spark" rather than explode uncontrollably within the engine.

Early aviators used such fuels in ever more powerful engines as aircraft performance grew. Any link between engine operation and fuel quality remained unexplored, and it was not until the First World War, with the need for reliability and higher performance, that this critical factor became apparent.

Engines tended to suffer over-heating and loss of power. This resulted in a preference for "rotary" designs such as the Gnome, Clerget, Le Rhone and Bentley, where the whole engine rotated with the propeller, gaining favor due to superior cooling and hence reliability.

To help resolve operational problems, two studies were undertaken by the Industry. One in 1915 by A. H. Gibson, Royal Aircraft Factory UK, and a second in 1917 by E. W. Dean and C. Netzen, National Bureau of Mines US. Both the UK and US studies concluded that a separate specification for aviation gasoline was important to ensure operational performance. A US Navy Department specification was introduced on the

Table 1 Example variance in straight run gasoline octane quality versus crude oil source

Crude oil	Research octane number
Ardjuna (light cut)	75
East Indies	65–70
Hoops Blend (light cut)	67
Lower Zakum (light cut)	61
Pennsylvania	45–55

Table 2 1918 US government aviation gasoline specifications

Grade	Export	Fighting	Domestic
General	Free from acid, water and suspended matter		
Color	Water white		
Doctor test	Negative		
Corrosion	No gray or black corrosion in a copper dish and no weighable gum		
Distillation			
5% distillate °C	50–65	60–70	50–75
50% distillate °C	95 max	95 max	105 max
90% distillate °C	125 max	112 max	155 max
96% distillate °C	150 max	125 max	175 max
End point °C	165 max	140 max	–
Total distillate % v/v	96 min	96 min	96 min
Loss % v/v	2 max	2 max	2 max

1st September 1917 "Aeroplane Gasoline" followed by US Government specification in 1918, Table 2, and later British specifications [3].

These early specifications focused on the cleanliness, volatility, reactivity and corrosion properties of the fuel as important. However, with the publication of the book "The Internal Combustion Engine" by Ricardo [4] the need for an aviation fuel with a high-octane quality for thermal efficiency and performance was identified. This represented a key finding and resulted in the Industry adopting use of the additive tetraethyl lead, with ethylene dibromide scavenger (TEL-B) to develop a range of products, covering unleaded fuel of 80 motor octane number MON, ASTM D2700 [5] to leaded Grades of 99.6 MON and higher [6]. Today these have been largely rationalized to 4 fuels, mid-octane unleaded Grades UL91, UL94, and leaded high-octane Grades 100LL, 100/130 with specifications detailed below. A very low lead Grade, 100VLL, has also been developed as a sub-set of 100LL.

1.2 International Specifications

ASTM (ASTM D910, D7547 [7, 8]) and Defence Standard (Defence Standard 91-090 [9]) specifications cover the majority of aviation gasoline sold in the world today with Chinese (GB 1787—2018), Russian (GOST 1012) and other minor specifications being in broad alignment. Table 3 summarizes the laboratory analysis required for the four major Grades. In addition, it should be remembered that aviation fuels must meet the full scope of specifications and the importance of this will become apparent in the manufacturing section.

Table 3 Summary of major aviation gasoline grades and laboratory analysis. From ASTM D910, D7547, Defence Standard 91-090 (June 2023)

Test	Units	Grade 100LL	Grade 100	Grade UL91	Grade UL94
Appearance	Visual	Clear, bright and visually free from solid matter and undissolved water at ambient temperature			
MON	ON	99.6 min	99.6 min	91 min[a]	94 min
Supercharge	PN	130 min	130 min	[b]	
TEL-B	gPb/l	0.56 max[c]	1.12 max[a]	0.013 max	
Color		Blue	Green	Red[d]	Purple
Blue dye[a]	mg/l	2.7 max	2.7 max	0.2 max	1.2 max
Yellow dye[a]	mg/l	None	2.8 max	None	
Red dye[a]	mg/l	None	None	2.3 max	0.9 max
Density	kg/l	Report			
Volatility					
Initial boiling point	°C	Report			
T10% evaporated	°C	75 max			
T40% evaporated	°C	75 min			
T50% evaporated	°C	105 max			
T90% evaporated	°C	135 max			
End point	°C	170 max			
Loss	% v/v	1.5 max			
Recovered	% v/v	97 min			
Residue	% v/v	1.5 max			
T10% + T50% evaporated	°C	135 min			
Vapor pressure	kPa	38–49			
Copper corrosion 2 h 100 °C	–	1 max			
Sulfur	% w/w	0.050			
5 h oxidation stability[a]					
Potential gum[a]	mg/100 ml	6 max			
Lead precipitate		3 max			
Net heat of combustion	MJ/kg	43.5 min			
Water reaction					

(continued)

Table 3 (continued)

Test	Units	Grade 100LL	Grade 100	Grade UL91	Grade UL94
Volume change	ml	2 max			
Freeze point	°C	−58 max			
Conductivity (when requested)	pS/m	50–600			

[a]Defence Standard 91-090:

- UL91 additional RON (ASTM D2699) requirement of 95 minimum
- Grade 100 0.85 gPb/liter maximum
- Lovibond Color:
  - Blue, Grades 100LL and 100/130 1.7–3.5
  - Yellow Grade 100/130 1.5–2.7
  - Red Grade UL91 6.7–9.1
- 16-h oxidation stability with lead precipitate of 2 mg/100 ml maximum
- Existent gum requirement of 3.0 mg/100 ml maximum

[b]Optional supercharge of 98 ON minimum if agreed between purchaser and supplier
[c]Reduced to 0.45 gPb/l for Grade 100 Very Low Lead (100VLL)
[d]Included in Defence Standard 91-090 Issue 6

2 Manufacture

2.1 Typical Manufacturing Process

The high-octane quality, volatility, energy content and stability characteristics required for aviation gasoline limit refinery manufacture to a small sub-set of components, typically:

- Alkylate for MON and energy content distilled to meet the 170 °C final boiling point requirement.
- Aromatics for supercharge, either from a chemicals unit or distilled from catalytic reformate.
- Isomerate/isopentane/straight run gasoline for T10% and vapor pressure.

Alkylate represents the primary component, a high-quality gasoline stream rich in iso-octanes primarily synthesis from a mixture of isobutane and isobutene in the presence of a strong acid catalyst [10]. This is used with various concentrations of aromatics, isomerate and related streams as necessary to enhance specific properties. Additives may include tetraethyl lead (TEL-B), dye and antioxidant dependent on the Grade in production. Compared to automotive fuel, aviation gasoline represents a more challenging product and various problems and solutions have been observed by the industry over the years which are detailed below.

2.2 Refinery Anti-corrosion Additives

The necessity to distill alkylate to meet aviation gasoline volatility requirements can lead to problems of corrosion in production units. To mitigate this issue good engineering, monitoring and maintenance of the distillation tower is required. In some cases, anti-corrosion additives may be injected but an incident in year 2000 highlighted that great care is required to understand additive chemistry and fate down-stream [11]. In this case an amine, ethylene diamine, was being used to control acidity. Due to various events, an excess of the additive entered product. This was not detected by copper strip analysis and product moved down-stream to airports. Unfortunately, when the fuel flowed across fine tolerance brass carburetors used in piston engine aircraft, a black sticky deposit formed causing malfunction and necessitating overhaul. This incident resulted in a detailed review of both aviation gasoline and turbine fuel specifications. As mentioned in Chap. 1, a section on refinery additives and Management of Change is now featured to avoid reoccurrence.

Given the above, any facility manufacturing aviation gasoline or turbine fuel should remain vigilant regarding the use of anti-corrosion and other additives. Good control and documented Management of Change are important to ensure product integrity.

2.3 TEL-B Content and Determination

The octane enhancement additive tetraethyl lead (TEL) remains necessary for the production of aviation gasoline Grades 100LL, 100VLL and 100/130 due to the exceptionally high octane quality required, greater than or equal to 99.6 Motor Octane Number (ASTM D2700) and 130.0 Supercharge (ASTM D909 [12]) respectively. In addition to TEL, it is important the additive package contains the correct concentration of ethylene dibromide scavenger to help remove lead deposits from the engine after combustion. This is calculated on a molecular basis of two bromine atoms per atom of lead, the aim being to form lead bromide ($PbBr_2$). The manufacture and handling of the additive is a specialist area and normally provided as a package "TEL-B" containing:

- Tetraethyl lead
- Ethylene dibromide
- Dye
- Antioxidant
- Solvent

Mixing the additive into product and determination of the lead concentration for the Certificate of Quality can be problematic. For mixing, specialist guidance is required due to the reactivity and highly toxic nature of the pure additive. This is best achieved by contacting the supplier where extensive knowledge is available.

Table 4 Precision of lead additive analysis methods for aviation gasoline

Method	Reproducibility mg Pb/l
IP 270/ASTM D3341	0.027
IP228 [13]/ASTM D5059[a] [14]	0.044 (method A) 0.032 (method B)
IP 428 [15]	0.062

[a]For 0.56 gPb/l

Differences in the concentration of the additive in the finished fuel are more easily managed and often relate to the precision of the test method for the high-octane grades, Table 4.

Where a fuel has been found to contain an unexpected quantity of additive, re-testing by the Defence Standard Referee titration method (IP 270 [16], D3341 [17]) can be beneficial. In addition, testing of Upper, Middle and Lower samples can also be beneficial to ensure a tank has successfully completed a mixing cycle.

2.4 *Motor Octane Quality*

There have been instances of confusion regarding aviation gasoline Motor Octane Number (MON D2700) values. By way of example, the same fuel, same test result, might have octane reported as:

- 100.0
- 101.1

This is due to 2 possible reporting scales being available:

(i) Motor Octane Number

Direct reporting of the result as a motor octane number by ASTM D2700.

(ii) Aviation Number

Conversion of the result to the historical Aviation Method D614 (ASTM D614 [18]) by use of Table 8 in ASTM D2700. In the example the former is MON and the latter Aviation Number. This can lead to misunderstanding when re-certifying fuel and raise concern about possible contamination. The industry is seeking to standardize on MON but verification on which method was used with the testing laboratory is important.

2.5 *Color*

The use of color to differentiate between leaded aviation gasoline grades can have some unexpected implications for manufacture. This usually relates to facilities distilling catalytic reformate to obtain an aromatic fraction for blending. If the distillation point is set too high within the reformate, conjugated aromatics rings can give the component a yellow tinge. For motor gasoline this would not be an issue but for Grade 100LL or 100VLL, where blue dye is used, the resultant product may appear green and be mistaken for Grade 100/130. Understanding and oversight of the reformate fraction boiling range is therefore important.

2.6 *Volatility*

When compared to automotive fuel, aviation gasoline has a narrow, flat distillation range with relatively low vapor pressure. This can lead to problems during laboratory testing where an automated ASTM D86 (ASTM D86 [19], IP 123 [20]) unit is used and the:

- fuel is inadvertently tested using a program for the incorrect product Group.
- heating rate is inappropriate for the flat distillation curve.

Errors of this type are usually reflected in T10%, and/or Loss and Residue results for the sample. The former is often too high, the heating rate having "rushed on" without giving due time for the correct volume of fuel to distill into the receiving vessel. Loss and Residue may also be excessive, the lighter components being driven out of the product without time to condense leaving traces of the heavier material with no fluid to help transfer. Such anomalies are corrected by ensuring Group 2 is selected for aviation gasoline samples, vapor pressure ≤65.5 kPa End Point ≤250 °C, and making sure the heating profile meets D86 requirements. For automatic instruments the latter may require a specific program to be developed for the product.

2.7 *Aromatics*

The aromatic concentration in aviation gasoline can influence distribution system and aircraft elastomeric seals, hoses and diaphragms. This is discussed further in Sect. 4. While specifications do not feature a limit for total aromatic or benzene content this is controlled by the 43.5 MJ/kg minimum energy requirement where significant differences in this property are present between saturates and aromatics as illustrated within Table 5.

Prior to the discovery of chemical carcinogenic properties benzene was widely used in aviation gasoline, offering good octane and appropriate volatility. The total amount was controlled by energy content, similar to other aromatics, and interestingly the freezing point which is very poor versus the specification maximum of −58 °C compared to other hydrocarbons, Table 6.

Today benzene is a highly controlled fuel constituent and aviation gasoline must comply with national health and safety requirements. Many industry members have

Table 5 Energy content of typical aviation gasoline components [21, 22]

Component	Net energy content MJ/kg
Iso-Pentane (2-methylbutane)	45.249
Iso-Octane (2,2,4-trimethylpentane)	44.651
Toluene	40.60

Table 6 Freezing point of typical aviation gasoline components

Component	Freezing point °C
Iso-Pentane (2-methylbutane)	−159.8
Iso-Octane (2,2,4-trimethylpentane)	−107.4
Benzene	+5.5
Toluene	−95.0

taken a proactive approach and reduced benzene content to below 0.1% w/w rendering the fuel non-carcinogenic for ease of distribution and use.

ASTM D910 section X1.8.1 offers a good summary of aromatic content in aviation gasoline with the observation that the minimum energy specification of 43.5 MJ/kg typically limits these components to about 25% v/v maximum. Further discussion is offered in the section below.

2.8 *Energy Content*

The minimum net energy content for aviation gasoline is set at 43.5 MJ/kg to ensure appropriate aircraft take-off weight and range for a given fuel load. A number of laboratory analysis methods are offered to measure this property, for example: IP 12 [23] Determination of Specific Energy, ASTM D3338 [24] Estimation of Net Heat of Combustion of Aviation Fuels, ASTM D4809 [25] Test Method for Heat of Combustion of Liquid Hydrocarbon Fuels by Bomb Calorimeter (Precision Method) and ASTM D4529 [26] Estimation of Net Heat of Combustion of Aviation Fuels.

For speed and simplicity, estimation methods are often used for fuel Certification. These generally rely on product density, distillation, aromatic and sulfur content. Problems can occur with this approach as a fuel approaches the specification limit, the estimation being exactly that, an estimation of energy content. The best policy in this situation is to revert to the referee method of bomb calorimetry to determine the true value. This requires good experimental technique to avoid loss of the lighter components. As seen earlier, aviation gasoline is a volatile fuel, and these components offer a high contribution to the total MJ/kg.

2.9 Oxidation Stability—5 Versus 16 Hours and Antioxidants

On a global scale, aviation gasoline may be stored and used in remote regions where there is little ground infrastructure and light aircraft represent the best, possibly only, form of convenient transport. In these circumstances the storge stability of the fuel is important to ensure product remains fit-for-purpose even under arduous conditions of temperature and time. Provision for this situation is made at point of manufacture with a 5- or 16-hour severe oxidation stability test, IP 138 [27] or ASTM D873 [28]:

- IP 138—Determination of Oxidation Stability of Aviation Fuel Potential Residue Method.
- D873 Test Method for Oxidation Stability of Aviation Fuels (Potential Residue Method).

The 5-hour test is appropriate for populated areas where fuel turnover is high. Where product may enter regions for storage of 6 months or more, the 16-hour test is advantageous to minimize the risk of oxidation and gum formation. Moving from one test scenario to another can give problems at point of manufacture. Aviation gasoline components may offer sufficient inherent stability to pass the 5-hour test without resorting to additives. When the 16-hour test is selected, the use of antioxidant becomes important to control reactions within the fuel and maintain quality. Depending on the test time and specification selected, up to 12 or 24 mg/l antioxidant may be used.

This completes the insight into the manufacture of aviation gasoline, the product now being ready to enter the distribution system as discussed in the next section.

3 Distribution

This section provides general information on how aviation gasoline is handled and distributed in the United States and other Western European countries. It should not be considered definitive, nor replace other legal and contractual requirements, nor does it attempt to address all regions of the globe. At the time of this writing, most aviation gasoline contains lead, although at least two grades of unleaded aviation gasoline are available.

Depending on where the refinery is located, and the airfield to which the fuel is destined, aviation gasoline is moved by marine tanker, railcar, road tankers or in drums. It may be stored in large distribution tanks, smaller on-field tanks, or in individual drums. The unique nature of leaded aviation gasoline requires a dedicated distribution system which is completely segregated from all other liquid fuels.

3.1 Segregation

As the last remaining liquid transportation fuel containing lead, and as an aviation product, all current grades of aviation gasoline must be segregated from any fungible distribution system. This means in countries like the United States and the United Kingdom, aviation gasoline is not transported using the multi-product pipeline system in which automotive fuels such as gasoline or automotive gasoil (diesel) and aviation turbine fuels are moved.

To maintain the segregation of the leaded aviation gasoline, dedicated systems are installed at refineries, aviation gasoline distributors and airports. Even at refineries which use the same distribution system within the refinery for all liquid transportation fuels, the requirement to prevent leaded fuel from entering unleaded products requires extensive controls when converting production from one product to another. Aviation gasoline distributors with sufficient volumes of fuel delivery may have dedicated tank trucks and rail cars. In other cases, these vehicles serve multiple purposes. In the U.S., these tanks are required to be constructed from stainless steel or epoxy-lined carbon steel. When tanks are used to transport multiple products, API RP 1595 [29] requires the tanks to be steam cleaned prior to loading with aviation gasoline if the tanks previously contained anything other than gasoline. Outside of the U.S., the Energy Institute guidance document Quality Assurance Requirements for the Manufacture, Storage and Distribution of Aviation Fuels to Airports provides information on the distribution of aviation fuels. While the document is primarily focused on jet fuel, it does include limited guidance specific to aviation gasoline [30].

Jetty and terminal operations, loading arms, manifolds and short distances of pipe may either be dedicated to aviation gasoline or shared with other fuels through necessity. Where facilities are shared, great care must be exercised to avoid contamination. For example, jet fuel or diesel rapidly increases aviation gasoline gum content, final boiling point and degrades octane quality rendering product unfit for use. Specific procedures are required to manage the receipt of fuel in these circumstances. These should be regularly reviewed to maintain best practice, particularly where a non-standard cargo, loading or discharge procedure is requested.

3.1.1 Marine Tankers

Depending on the location of a refinery producing aviation alkylate or the finished aviation gasoline, it may be necessary to move the fuel transoceanic. This requires the use of cargo ships, generically referred to as tankers. In most cases, ocean-going tankers are not dedicated to aviation gasoline transport alone, but to handle multiple flammable products in a range of tanks distributed along the vessel.

To safely transport flammable liquids, the tanker will typically be designed and equipped to minimize conditions which can start or sustain combustion. Readers may be familiar with the fire triangle which describes the conditions required for combustion as including an ignition source, fuel and oxygen. The fuel in this case is the flammable liquid and cannot be removed. The other two legs of the triangle are addressed by the design. While the removal of the sources of ignition is not trivial, it has little or no impact on the fuel itself and will not be addressed here. The use of inerting, that is making the environment so oxygen poor as to be unable to support combustion, can have an impact on the fuel.

In its simplest terms, aviation gasoline is flammable because of the ease with which the volatile vapors can form a combustible phase over the surface of the liquid fuel. One way to prevent the formation a combustible environment is to minimize the ullage space, the open space over the fuel. The less vapor space there is, the less flammable vapors that can accumulate. In theory, if the tanks were filled to the top, there would be no ullage space, and thus no flammable vapors. In reality it is neither practical nor safe to fill a tank completely. The vessel will experience temperature changes which cause the fuel to expand and contract. Either the tank must breathe with the expansion and contraction, or it must be completely sealed. If the tank is allowed to breathe, it will result in unacceptable spills during expansion, and the formation of an ullage space during contraction. If the tank is sealed, the result is a potential for explosion from the increasing pressure during expansion, or implosion during the contraction.

Alternatively, the ullage can be made inert by preventing the presence of sufficient oxygen content to support combustion. This level is generally accepted to be less than 8% oxygen in the tank. There are two methods available to inert a cargo tank; dilution or displacement [31]. Additionally, the safest manner to fill tanks, including cargo tanks is to fill from below the liquid level as opposed to from the top through the air space [29, 32]. To facilitate this, most cargo ships will have a drop line which exits near, but not at, the bottom of the tank. To allow the air to escape while the tank is filled, a separate vapor lines leaves from the top of the tank and goes to a vapor manifold. The cargo tank is also fitted with an inert gas (IG) line, and a purge pipe and valve. Depending on the time and the tank size, inert gas is used to dilute the air in the ullage space until it is inert, or it is used to displace the air in the tank with inert gas before loading [31].

In either case, what is important to the quality of the aviation gasoline is the choice of the inerting gas. In most cases aboard more modern tankers, the inerting gas of choice is nitrogen [31, 33], as it is easily produced both on-board ships and at shore-based loading facilities. However, there are cases where the inert gas is the cleaned and stored emissions from the ship's boilers, primarily carbon dioxide (CO_2) [34]. The presence of moisture in the emissions gas would result in an increase in the entrained and free water content of the fuel with the concomitant challenges of a wet fuel, e.g., fuel degradation, storage stability or corrosion, which would need to be addressed.

3.1.2 Rail and Road

In general, transportation of aviation gasoline in rail cars or over the road network using tanker trucks is handled under the same considerations as the ocean-going tankers. The tanks are to be prepared from stainless steel or epoxy-lined carbon steel. Steam cleaning of the tanks is to occur between fuel types when loading aviation gasoline unless the previous fuel was motor gasoline. When the previous load is gasoline, the only requirement is to fully drain the tank. API RP 1595 notes that with respect to the potential for fires during loading, the most dangerous case is the loading of a jet fuel into a tank previously loaded with aviation gasoline, referred to as switch loading. The jet fuel passing through the pre-loading filter may become electrically charged and will enter a flammable environment with associated risk of spark and ignition (see Sect. 3.9).

3.2 Storage

Once the aviation gasoline reaches the storage facilities, the fuel remains segregated not only by type (jet fuel versus aviation gasoline) but also by grade (for example 100LL, UL91), as commingling of high- and low-octane fuels would result in off-specification product. A batch system is used to ensure traceability [28]. Dependent on the location of the facility in the distribution chain, it may be possible to use common loading arms or pipelines for different products, but there must be a procedure for appropriately down-grading the interface material between batches. Typically, any shared hardware is over a short distance with the fuel returning to a segregated system for re-analysis to confirm quality before onward distribution to airport.

At larger airports with higher fuel volumes, the fuel is delivered into a settling tank where free water and dirt is allowed to settle before moving to the final refueling vehicle or bowser from which aircraft will be fueled. API RP 1595 specifies aviation gasoline be permitted to rest five minutes for every foot of fuel depth in the tank. In cases where fueling rates are high enough for operations to be impacted, API RP 1595 permits a tank which has outlet filtration and which uses a floating suction outlet to be released after only 45 minutes. Note that the use of API RP 1595 is not a requirement, and smaller airfields may not include a separate settling tank. At the smallest or most remote locations, fuel may be stored in large rubberized blivets or in drums.

While smaller and older airfields may still have underground storage tanks (UST) for aviation gasoline, most have transitioned to the use of above ground storage tanks (AST). This is because of Environmental Protection Agency (EPA) regulations regarding UST's enacted as part of the Energy Policy Act of 2005 which "significantly affected federal and state underground storage tank programs, required major changes to the programs, and is aimed at reducing underground storage tank releases to our environment." [35]. Aviation gasoline storage tanks are generally constructed of carbon steel and depending on the size, coated with epoxy. API RP 1595 requires the bottom and a minimum of the first three feet of the tank to be coated with an approved epoxy. All horizontal tanks and vertical tanks less than 30,000 gallons (114,000 L) are required to be entirely coated with a protecting coating system per API/EI 1541.

The piping on the storage tank is typically 304 or 316 stainless steel [36]. The use of galvanized steel is not recommended as it may react with the tetraethyl lead (TEL) and ethylene dibromide (EDB). Figure 1 provides a typical installation.

Vertical storage tanks are recommended to be constructed with a cone shaped roof and floor, a floating suction arm which draws product from the top, and either floating roofs or fixed roofs fitted with pressure-vacuum breather valves. The outflow fitting is recommended to be a floating suction arm which means the fuel will be suctioned from the top of the liquid and the arm will float up and down with the fluid level, Fig. 2. If the outlet is not floating, then the inlet of the outflow suction pipe is placed above the tank bottom and typically turned to draw from above, Fig. 3. This orientation and location help prevent drawing traces of settled dirt and free water from the bottom of the tank into the fuel flow. Note drawings are informational and not to scale nor complete.

Fig. 1 Typical horizontal aviation gasoline tank installation for refueling vehicle supply

Fig. 2 Schematic of floating suction arm

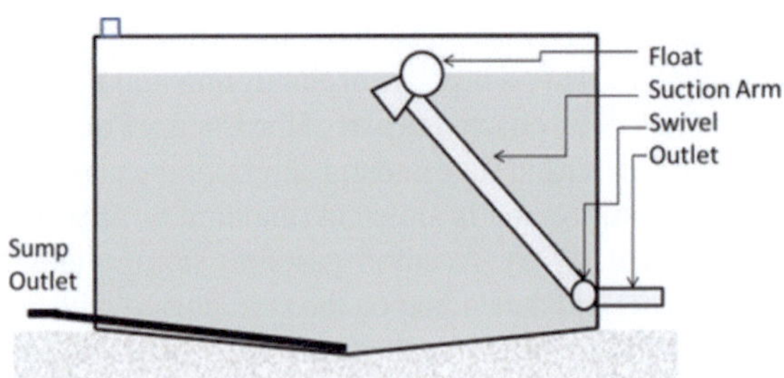

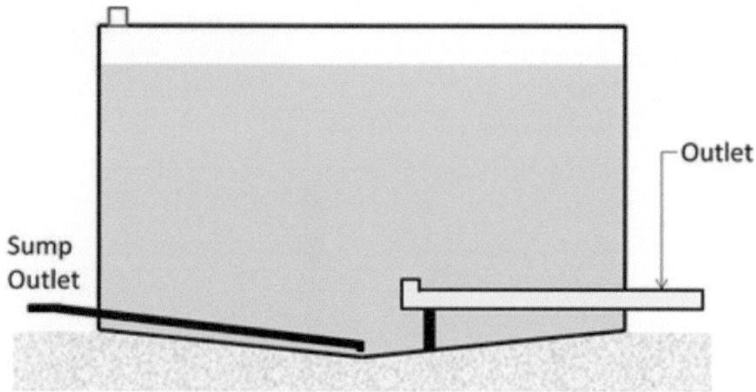

Fig. 3 Schematic of bottom mounted exit pipe

Within the distribution system, there are a number of considerations which directly affect the quality of the aviation gasoline. The following sections include an insight into various challenges and considerations but should not be considered an exhaustive discussion.

3.3 Tank Metallurgy

When preparing to install storage for aviation gasoline, especially at smaller airfields or private hangars, the use of galvanized steel tanks should be avoided. This is due to zinc acting as a catalytic metal causing the dyes used in aviation gasoline to react in the presence of trace water and change color. In addition, zinc oxide can precipitate as a white material and can cause both the tetraethyl lead additive and associated ethylene dibromide scavenger to be lost from the fuel into the precipitate [37].

3.4 Inerting Gases

Examples of inerting gases used on cargo ships have already been provided above. But the use of inerting is not limited to cargo ships and may be found in use on rail cars, road tankers and fixed storage tanks. In most cases, the inerting gas of choice is nitrogen. The use of dry nitrogen is not reported to have any negative impacts on the quality of the fuel. Similarly, the use of dry, clean carbon dioxide, typically the emissions gas from engine and boiler operations, does not have reported issues. However, the use of wet inert gases such as carbon dioxide from combustion does have the potential to result in elevated entrained and free water in the fuel. The presence of this water can result in issues, such as metal corrosion. Although traditionally the presence of the lead additive in aviation gasoline has acted as a biocide, moves to very low lead and unleaded fuels present an opportunity for microbial growth in the water phase provided by wet inerting gases. The best protection is to assure only dry inerting gases are used. With respect to use of engine exhaust gas, it is important that the vessel scrubbing system to remove particulates is operating correctly. Failure in

Fig. 4 Typical 100LL aviation gasoline and microfilter drain sample contaminated with inert gas soot

this respect can severely contaminate product with soot particles which do not settle from fuel due to low density and small size range. Where concentrated, for example in microfilter drainings, these can discolor product to a gray-green [38], Fig. 4.

3.5 *Corrosion*

Distribution systems handling aviation gasoline seek to avoid extended exposure to bare steel at low points due to the potential for corrosion from trace free water entrained in the fuel. For example, coating storage tanks to a height of 1 m with an aviation industry approved inert epoxy lacquer. Water entrainment occurs from the point of refinery manufacture due tank breathing and ambient conditions and may build or deplete as product moves through the distribution system. The literature reports that aviation gasoline leaves the refinery saturated with water and enters the distribution system which is at ambient temperature. When the fuel cools, the entrained water drops out of the fuel, forming free water. This water combined with oxygen, also provided by the fuel, reacts with the iron in the steel, potentially causing corrosion (rust) [37].

One of the ways to deal with pipeline corrosion is the use of fuel soluble corrosion inhibitors (CI). ASTM D910, *Standard Specification for Aviation Gasoline* includes a list of acceptable CI additives and their respective dosage rates. In the context of distribution of aviation gasoline by pipeline, this is now very rare on a global basis due to the small volume, leaded nature and high quality of the product making interface management practically impossible.

3.6 Pressure–Vacuum Valves

Aviation gasoline is generally be stored in tanks, whether at the refinery, the distributor, or the airfield. Larger storage tanks, smaller vertical, and horizontal fuel tanks require the use of breather valves to compensate for the expansion and contraction of the liquid fuels, and the formation of a vapor phase with temperature changes, a particularly important aspect for a volatile product. This results in the tanks "breathing", potentially drawing in moisture laden air. In order to minimize this influx of moisture, and to control exhalation of vapors, tanks which must breathe are recommended to be fitted with a pressure-vacuum valve. The tank may also be constructed to withstand greater pressure fluctuations to limit these effects. Where a tank is protected by a pressure-vacuum valve; the valve only actuates when the maximum/minimum pressure values within the tank are exceeded [39]. Owner/operators of the storage tanks are also encouraged to have tanks designed in manners to reduce the amount of temperature change experienced by the fuels. This can be done by the use of tank insulation, placing the tanks in a manner to minimize direct sunlight on the walls, particularly the roof [39], and the use of white/reflective paint [36, 37, 39]. The impact of losses of volatile components from aviation gasoline is discussed under weathering, Sect. 3.8.

3.7 ISO Tank Container and Drums

ISO tank containers are also popular in some regions for ease of transport and intermediate storage. Care is required to ensure these are fit for use with an appropriate cleaning procedure as previous cargos can be very variable, spanning chemicals, vegetable oils and other products.

At some locations, logistics require fuel to be stored in drums. These should be lacquer lined and of the correct design to receive aviation fuel, for example as detailed in ISO/ANSI MH2a or ISO 15750-2. The storage location should be selected to minimize temperature changes and associated expansion and contraction, recognizing drums will respond to temperature changes more quickly than large tanks. The drums should be stored on their sides in a barrel rack to protect the tops from standing water and contamination that could enter during expansion and contraction of the rim seal and bung. The drum should also be stored with the large bung below the fluid level to ensure the elastomer seal remains wet. If this is allowed to dry, there is risk of seal contraction/leakage. The use of drums for long-term storage is less desirable and good housekeeping is critical both to avoid hair-line corrosion with risk of drawing in pooled water and dirt during contraction, and to prevent fuel seepage during expansion.

One of the major risks of drum storage in remote locations is reuse for different products. There have been instances of aviation gasoline drums being refilled with jet fuel and resulting in inadvertent misfuelling of aircraft with serious consequences. The use of tamper-proof tab seals is vital with strict control of stock to avoid problems.

Further details for tank, iso-tank and barrel storage of aviation fuels are provided in EI/JIG Standard 1530, Quality Assurance Requirements for the Manufacture, Storage and Distribution of Aviation Fuel to Airports.

3.8 Fuel Weathering

The volatility of aviation gasoline, as measured by its distillation range and vapor pressure, directly influences the ability to start the engine (higher volatility desirable) while avoiding vapor lock and minimizing weathering [40] (lower volatility desirable). "Weathering" is a term used to describe the loss of the volatile fractions of the aviation gasoline. As the temperature of the fuel rises, the lighter, more volatile fractions essentially "distill" or "evaporate" away. This potentially results in two negative impacts to the quality of the fuel. The first is a loss of the more volatile fractions which can result in an out-of-specification product should the distillation curve be sufficiently changed. The second is a lowered ability to start the aircraft at lower temperatures. In addition to negatively impacting the fuel's absolute volatility, the loss of a portion of the fuel volume effectively increases the lead concentration in the remaining liquid. This potentially results in the lead maximum for the Grade being exceeded, changes (increases) the octane rating of the fuel and promotes lead fouling in aircraft.

To minimize weathering, storage tanks feature pressure-vacuum valves and seek to control bulk temperature changes (see above) to minimize heat rises which might cause loss of the more volatile fractions. The use of floating tank roofs [37] and maintaining fill levels [39] to minimize the vapor space above the fuel all help control fuel weathering. At locations where fuel is stored in drums, weathering may be reduced by storing the barrels on their sides out of direct sunlight. The sealed nature of drums effectively eliminates the risk of lost volatility. However, the risk of gum formation remains with product slowly oxidizing over time. As detailed in the manufacturing section, the use of antioxidant additives and the 16-hour oxidation stability test helps maintain quality for such severe environments.

3.9 Static Dissipator Additive (SDA)

One of the challenges from moving or flowing flammable liquids is the ability of the fuel to experience an increase in static electrical charge. Like rubbing one's stockinged feet across a carpet on a cold, dry day, the fuel will pick up a charge as it flows through pipes, filters and nozzles [41]. In use, if the charge does not have time to relax or "bleed away", it can release the stored charge all at once, creating a spark. This is much like touching a door handle after walking across a synthetic carpet in a dry environment. In the case of fuel, if there is a flammable vapor mixture present, this spark can have enough energy to ignite the vapors.

A fuel can generate this charge because it is non-conductive. Its relative motion across a conductive material, including air, water or metal, generates a surplus charge within the fuel ([37], p. 188). This charge must be given a path to equilibrate by bonding or grounding to prevent a spark, or by giving the charge in the fuel time to bleed away. Bonding and Grounding are different and warrant a brief explanation. Bonding is connecting two pieces of equipment together using an electrically conductive element, like a cable connecting a fuel truck to an aircraft. Grounding is connecting the fuel truck or aircraft to the physical Earth. Note that neither Bonding, nor Grounding directly relaxes the charge in the fuel; it only provides a path for the charge generated to move along when it reaches a metallic surface, such as a the conductive refueling hose or aircraft wing-tank wall.

The discharge of static electricity does not only occur outside of the hardware. Incidents have been documented over the decades in which the discharge occurs inside the hardware. Examples of locations of the discharges, at times sufficient to ignite the fuel, include within the aircraft on-board fuel tanks, within filter housings in the distribution system, and within the storage and distribution tanks. The loss of epoxy coating on fuel tanks is known to provide a location for charge accumulation within fuel tank. When derating filter housings, just removing one of the canisters may provide a condition for charging. The removed canister should be replaced by a blank and consideration be given to the increase in fuel flow rate through the remaining canisters. To eliminate these internal discharges, bonding and grounding have no effect if the charge generated within in the fuel cannot reach a surface connected to the bonding path. The only way to eliminate this risk is to assure the fuel can transfer the charge generated more effectively ([37], p. 170).

Charged fuel will only relax with time. The rate at which this relaxation occurs is related to the conductivity of the fuel. By its nature, fuel is not particularly conductive. In order to improve the conductivity of the fuel, that is to increase the rate at which the charge can dissipate from the fuel, the use of an electrical conductivity additive is permitted. In 2025 at the time of this writing, the latest revision of ASTM D910 permits the use of one product, Stadis® 450. Because the additive has a surface-active nature, the additive can be depleted during the distribution of the fuel because it cleaves to the surface of the equipment through which the fuel moves. Stadis® 450 may be added in single additions of concentrations up to 3 mg/L. The additive is a long-chain hydrocarbon with an anionic polar head group [42]. Where necessary, additional doses of the additive may be added as long as the total addition does not exceed 5 mg/L.

In practice, similar to Jet A, SDA is rarely used in aviation gasoline. This may seem surprising but is primarily due to the successful development of industry Bonding requirements to manage electrical discharge before the invention of such additives. Flow rates and volumes for aviation gasoline are also much lower than Jet reducing the accumulation of charge when passing through filters. Where use of additive may be considered is usually to meet National requirements or in very cold climates. The latter is due to the risk of low ambient temperature potentially making the headspace move from the over-rich to flammable region combined with a dry atmosphere promoting static accumulation. In addition, if there is the possibility of non-standard refueling operations, for example during military use, SDA may be mandatory for example as in Defence Standard 91-091 for Jet A-1 [43].

When used, SDA can have unexpected consequences. For example, the additive surfactant properties may begin to impact the fuel's ability to shed water. As the amount of additive increases, the ability for the fuel to demulsify and settle trace water decreases. Additionally, too much additive can begin to affect the accuracy of capacitance type fuel gauges. Two instances of SDA overdose in aviation gasoline and excess conductivity were reported in the market in recent years. In both cases the solution was to dilute the product with fresh fuel of lower conductivity. Clay treating might have been considered as it is successfully used in Jet fuel, but the use of antioxidant, dye and lead additive in aviation gasoline makes such an option more challenging.

Depending on the ambient temperature of the fuel when adding the SDA, differences in viscosity between the fuel and the additive can result in poor mixing and lack of homogeneity. Therefore, care must be taken to assure adequate mixing using appropriate injection equipment.

4 Aviation Gasoline Use

Having reviewed the history, manufacture and distribution of aviation gasoline in the previous sections, various observations, problems and solutions relating to use are now considered.

4.1 Seal Swell

Polymers or elastomeric non-metallic materials are used to make various parts of aviation gasoline distribution and aircraft fuel systems, like hoses, O-rings and tank bladders. While referred to descriptively as "rubber", rubber is a term specific to a material produced from naturally occurring "natural rubber" called latex. Natural rubber parts are not typically used in aviation as better synthetic polymers have been invented. Synthetic polymers or elastomers are long chains of repeating chemical units which are cross-linked together. The type of repeating units is specific to the polymer type, three of which are shown in Fig. 5.

The interaction of the fuel with these components is important to ensure correct engineering performance. As described earlier, aviation gasoline is a mixture of hydrocarbon components and these can be broadly divided into alkanes and aromatics Fig. 6.

The fuel may interact with the polymers in one of three primary ways; (1) Soak into the polymer structure like water into a sponge, (2) Solvate the polymer, causing the

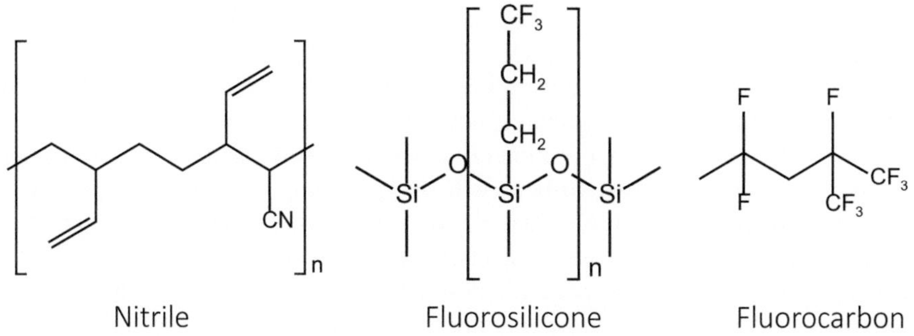

Fig. 5 Representation of repeating polymer units—nitrile, fluorosilicone and fluorocarbon (Viton™)

Fig. 6 Examples of alkane and aromatic hydrocarbons found in aviation gasoline

connections to break, (3) Extract component such as additives and other compounds. These are not the only ways materials react but are sufficient for this discussion. In the first case, the hose or aircraft part may just swell as the fuel gets in the spaces between the polymer units. In the second case the hose or aircraft part may become soft, sticky, gummy or slimy possibly abrading away, or even start to dissolve. In the third, the material may change properties, becoming harder or more brittle which could impact function or service life.

One of the considerations in the design and support of piston-powered aircraft is the interaction of the fuel with the non-metallic components within the fuel system, specifically the hoses and seals. As part of the development of the ASTM D7826 *Standard Guide for Evaluation of New Aviation Gasolines and New Aviation Gasoline Additives,* a survey of the current supported systems was made to identify the non-metallic materials used. Even after combining like materials, i.e., multiple types of nitrile, 30 different polymeric material types were identified. This does not include non-metallic sealants, paints or the plastic parts. The list is very extensive because of the wide range of materials used in fuel systems over the last 70 years. To rationalize the situation, five major classes of polymeric materials used for seals and hoses were selected to make up the core materials. The five material classes are: nitrile butadiene (Buna-N or generically nitrile), ethylene propylene diene terpolymer (EPDM), silicone, fluorosilicone and fluorocarbon (Viton™). While a discussion of polymer chemistry is beyond the scope of this section, it is useful to consider that each type of polymer was selected for a specific reason. This selection considers the requirements of the component and the trade-offs in the material properties. For example, Buna-N has good resistance to hydrocarbons like those used in the aviation gasoline alkylate, but poor resistance to alcohols and aromatics. Viton™ has exceptional resistance to aromatics and high temperatures but is not as flexible at colder temperatures.

The requirements of an application drive polymer selection, designers balancing the largest number of positive attributes with the lowest number of negative attributes, to identify an appropriate class of material for use. In other cases, the selection may be based on the most critical requirement of the application, even if it means dealing with a larger number of negative attributes. Within a class of polymer, the polymer designer can affect minor changes in attributes by modifying the specific formula. The result is a selection that is the best choice for the particular application. During this selection, the typical and extreme properties of the aviation gasoline used for the airframe and engine certification are considered as part of the system, for example, Grade UL91, 100LL or 100/130.

Herein lies the challenge. When the fuel composition is changed, these design assumptions and property balances may be altered. An aircraft originally certified for use with historical Grade 80 has a slightly different set of requirements than one designed for 100/130. This relates primarily to the difference in the typical aromatic content. The lower octane Grade 80 was more likely to have a lower aromatic content compared to 100/130. This means the designers of the Grade 80 aircraft had more latitude in their polymer selection than the designer of the 100/130 aircraft with respect to aromatics. Remember, nitrile polymers are not as resistant to aromatics as fluorosilicone for example. When the Grade 80 aircraft receives a steady diet of a higher-octane fuel, for example modern Grade 100LL, it may be receiving exposure to more aromatics on a regular basis than the designers anticipated.

Use of automotive gasoline (AutoGas), as permitted by a Supplementary Type Certificate, may also be challenging to elastomers as aromatic concentration can be significantly higher than aviation gasoline and alkenes (olefins) and alcohols are also present. This is part of the reason why many approvals to use AutoGas stipulate only alcohol-free gas may be used to help minimize issues. For example, the presence of the ethanol can cause excessive swell of some polymers up to and including destruction of nitrile parts.

Seal swell can also work in the other direction. Seal swell of a known amount is generally expected as part of the fuel system hardware design. In aircraft which have been in long-term storage or which have been allowed to sit dry for long periods, especially older or classic aircraft, polymers can actually shrink. Once brought into service the fuel system polymers may not readily swell again, or possibly attempt to swell and crack. The result can be failed joints and fuel system leaks.

Regular inspection and replacement of aircraft hoses and seals is important to avoid the problems noted above. This is also important for aircraft which may have been operating on a particular aviation gasoline composition. Introduction of a new supply with either low or higher aromatic concentration can lead to seal shrinkage or over swell/cracking with resultant maintenance requirement, an effect which has been observed in the field.

4.2 *Lead Deposits*

During the operation of the aviation piston engine, the fuel is combusted in the piston cylinders. Within the fuel are the tetraethyl lead (TEL) and ethylene dibromide (EDB) additives, which are also part of the combustion reaction. During this reaction, the lead in the TEL reacts with the EDB scavenger and is converted to lead bromide, a volatile compound which ideally exits the engine with the exhaust gases. However, this reaction is not always complete. Some of the lead is converted to lead oxybromide, which does not exit the engine, but will instead deposit within the engine [44], ([37], p. 72). During this combustion process, some of the lead may not react with the EDB and will instead form lead oxides and lead sulfates. These compounds will deposit on the engine surfaces, becoming very hard and are difficult to remove, with further potential problems detailed below.

4.2.1 Spark Plug Fouling

The lead compounds that are deposited in the engine are conductive. When they deposit on spark plugs, the deposits may cause misfires and low energy sparks ([37], p. 235). In a 1952 SAE study, Greenshields [45] reported observing three types of spark plug fouling:

1. Deposits on the plug bridging the gap and creating a short circuit.
2. Low shunt resistance because the deposits on the ceramic portion of the plug caused a resistance too high for the spark to overcome. Smith described this as the electricity drained away more quickly than the charge could be generated resulting in "no spark" ([37], p. 235).
3. Tracking caused by the highly conductive deposits discontinuously depositing on the surface and causing tracks and short circuits.

Lead fouling can be particularly problematic for aircraft using engines which were originally designed to operate on a lower octane number/lower lead content aviation gasoline such as Grade 80. Grade 80 was specified with a maximum of 0.14 g lead/L (0.02 oz/gallon). When the engine is fueled by necessity with currently prevalent Grade 100LL with up to 0.56 g lead/L (0.07 oz/gallon), there is nearly four times the amount of lead present than the design expected. The excess lead in the fuel combined with the lower combustion severity in such engines, can result in excess lead deposits and increased spark plug fouling. A similar situation can also occur for performance engines operated for excess time at reduced power output.

For environmental benefit and to assist the historical Grade 80 and similar low-octane requirement fleet, unleaded mid-octane Grades have been developed. These are Grade UL91 and UL94, the former being featured in Defence Standard 91-090 and both in ASTM D7547. With appropriate oil selection, engine manufacturers have recognized the benefits of such products on reduced lead fouling and possible increase in engine overhaul interval. Development of these Grades is discussed elsewhere.

4.2.2 Lead-Induced Corrosion

In addition to lead fouling of spark plugs, lead bromide deposits can exacerbate corrosion. When the engine cools off, moisture condenses on the surfaces of the engine, particularly in the exhaust. This moisture can then cause rust. When lead bromide and excess bromine from the ethylene dibromide scavenger deposit on the surfaces, the corrosion is further exacerbated. This corrosion is generically referred to as "cold corrosion" ([37], p. 233).

There have also been cases of "hot corrosion" reported over the years. In this situation, the lead exits the combustion cycle as lead oxide and lead bromide. Because the exhaust surfaces are so hot, the lead becomes molten metal. This molten lead then scours the exhaust surfaces, physically abrading away material and appearing as corrosion ([37], p. 234), [45]. This can be a significant problem for historically important museum aircraft using engines designed to operate on unleaded aviation gasoline. The supply of modern equivalent unleaded Grades of higher octane quality, UL91 or UL94, can be beneficial in such instances if approved for the engine model.

4.2.3 Lead Precipitates

There is a potential for the formation of lead precipitates in aviation gasoline due to the oxidation of the tetraethyl lead additive. This is not a common occurrence for bulk storage but is often seen due to poor sampling practice. In such instances, the aviation gasoline sample is contained in a clear glass bottle which has been left exposed to sunlight. Ultra-violet radiation perpetuates degradation of the additive which forms a fine white precipitate cloud at the bottom of the container. To rectify, a fresh sample is required in an amber glass bottle or metal can. Care is required when handling such precipitates, either from storage vessels, samples, aircraft fuel systems or engines as the lead content renders them poisonous and appropriate precautions are necessary.

4.3 Tricresyl Phosphate Additive

For aircraft which experience troublesome lead fouling, such as those originally designed to operate on aviation gasoline Grade 80, one means of reducing deposits is to seek an after-market additive to further scavenge lead during combustion. Studies performed in the 1950s determined that a reduction in lead fouling could be achieved with the addition of tricresyl phosphate (TCP), Table 7.

Studies performed with TCP demonstrated the additive as successful in reducing fouling, and potentially helping to clear existing lead deposits. It was determined that the lead oxybromide formed during combustion was converted to lead phosphate. The study found that; (1) metallic lead deposits, a result of high temperature exposure of the lead oxides and lead oxybromides, were never formed, (2) the lead

Table 7 Properties of tricresyl phosphate

Synonym	Tricresyl phosphate (TCP[a])
Molecular formula	$(CH_3C_6H_4O)_3PO$
Molecular weight	368.37 g/mol
Density	1.143 kg/l
Boiling point	265 °C °C @ 10 mmHg
Flash point	>110 °C

[a]Not to be confused with Trichlorophenol/pharmaceutical products

phosphates created had a low conductivity and (3) the lead phosphate deposits were soft and powdery instead of hard and glazed. The soft, powdery deposits could be easily dislodged during the exhaust stroke and exit the engine combustion chamber [37, 45, 46].

While TCP does successfully help treat lead fouling, its use is only recommended under aircraft manufacturer guidance and for non-turbocharged engines. This is due to potential side effects, for example turbocharger fouling. Alcor, the manufacturer at the time of this writing, recommends using one ounce per gallon of aviation gasoline and injecting the TCP directly into the aircraft fuel tank [47]. It should be noted that tricresyl phosphate is a hazardous chemical which can cause chemical burns to the skin and strip the paint from the aircraft when in its undiluted form. When using TCP, aircraft, engine manufacturer and additive supplier recommendations and procedures should always be followed as stated.

4.4 Fuel System Icing Inhibitors

Aviation gasoline is permitted to use fuel system icing inhibitor (FSII) in the form of either isopropyl alcohol or di-ethylene glycol monomethyl ether (DiEGME). Addition of FSII will normally be at the specific request of the owner/operator at the time of fueling and is not generally added at the time of manufacture. At the time of this writing, there are multiple DiEGME containing products sold as FSII. Within the Western aviation community, the owner/operators will generally refer to the FSII as "Prist". To be clear, Prist is a manufacturing company, one of whose products is a DiEGME-based fuel system icing inhibitor called Prist HiFlash™. Much like the terms Band-Aid™ and Kleenex™, the product brand name has become equated with the product itself regardless of the actual manufacturer. Here in lies one of the challenges in the use of the term "Prist" when referring to FSII. Prist the company produces FSII, but they also produce an aviation window cleaner, Prist Glass Cleaner™. Therefore, it is entirely possible to be at an airfield and ask for Prist and not be sure which product is being requested or which was received.

Another challenge in using a fuel system icing inhibitor in aviation gasoline is the successful addition of the DiEGME into the fuel as it is being loaded. DiEGME is only slightly soluble in fuel and very soluble in water. Its function is actually to partition into any trace free water in the fuel and act as an anti-freeze, preventing the water from freezing and blocking fuel lines and filters. It is this low solubility in fuel which causes the challenge. The person fueling the aircraft must introduce the FSII uniformly into the flowing stream over the entire fueling, to improve the likelihood of blending the additive throughout the fuel load. Just adding the permitted amount of additive (0.10–0.15% by volume) either before or after loading the fuel will not provide adequate mixing. The additive will predominantly pool at the low point of the fuel tank. While it may provide protection against freezing any water bottoms, it will not provide protection against freezing of the entrained free water in the fuel. Furthermore, this highly concentrated DiEGME may cause removal of the surface coating of integral fuel tanks, may attack the structure of polymer fuel bladders and may cause removal of external paint especially at the fuel filler and sump valves. To achieve acceptable integration of FSII additives, carefully follow the additive manufacturers' instructions.

4.5 Misfuelling

One of the most catastrophic and preventable failures for the owner/operator of a piston-powered aircraft is to misfuel the aircraft with jet fuel. The NTSB Safety Recommendation bulletin, A-86-135 reported "during the period 1978 through 1985, the NTSB investigated 18 accidents stemming from misfuelling" [48]. In the same article, the FAA reported five additional misfuelling incidents from 1980 to 1986. And these incidents represent only those cases where misfuelling resulted in a reportable incident. In cases where the misfuelling was not caught prior to take-off, the most likely outcome was catastrophic, meaning loss of aircraft, property damage, or injury was high.

The reason that a piston engine will not operate on jet fuel is that the spark plug will not ignite the product due to the low volatility and the engine will stop, or more typically where the mixture can be ignited, the octane quality will be insufficient and extensive detonation damage will occur. In cases where a full tank of fuel has been loaded, the pilot may be unable to start the engine or to keep the engine running, and as such will never reach the point of take-off. If there is some amount of aviation gasoline remaining in the tank or in the fuel lines and carburetor, it may be possible to successfully run pre-flight checks and taxi to the runway. Depending on the volume of this fuel, when the aircraft is put to take-off power, the jet fuel will reach the engine at the most critical point of the take-off. In cases where there is actually a blend of aviation gasoline and jet fuel reach the cylinder, this may still be sufficient to support combustion. However, the flame will not be smooth due to the degraded octane quality resulting in detonation and high temperature shockwaves reaching the piston crown and valves. Rapid failure of the upper piston usually occurs followed by loss of oil, engine seizure and catastrophic failure. If this occurs at altitude, the pilot may be able to safely land. If the destruction occurs during rotation, it is unlikely the pilot will be able to recover the aircraft.

Piston-powered aircraft are unique in that the pilot may self-serve, defined in NFPA 407 as fuel being dispensed by someone other than the facility owner/operator. This may be from a traditional fuel pump, or from drums using a hand pump. In other cases, fueling may be provided by a Servicing Tank Vehicle called a fueler, by which a tank of fuel is brought to the aircraft and the truck provides the transfer equipment. Because of the presence of multiple types of fuel at an airfield, e.g., aviation gasoline, jet fuel, diesel fuel, automotive gasoline and the existence of a number of human factors that have the potential to facilitate misfuelling, owner/operators of piston-powered aircraft should exercise a heightened level of due diligence to avoid or mitigate potential misfuelling. Various campaigns have been run over the years for example by the National Air Transport Association, Aircraft Owners and Pilots Association and Energy Institute. An example from EI 1597 Aircraft Misfuelling Prevention is given in Fig. 7.

Larger airfields may have implemented procedures to help prevent misfuelling. FAA AC20-43C [49] suggests aggressively labeling the equipment with large lettering on the refueling truck, identifying the product in the truck. The truck should be labeled on the sides and rear, and on the nozzles. The nozzles themselves should be marked and color-coded specific to the fuel. Fuel specific nozzles are available, with the nozzle for jet fuel being broader, generically referred to as a duck-bill nozzle as compared to a standard fuel nozzle, Fig. 8. The broader jet fuel nozzle is too wide to fit into the one-inch fuel openings on aircraft using aviation gasoline [50–52]. It is important to recognize that not all fuelers use the two different types of nozzles. Self-service locations may have only standard nozzles fitted to their fuel pumps. The hand pumps used to fuel from barrels may not be fitted with the larger nozzle for jet fuel. In addition, not all jet powered aircraft have the larger fueling ports, and as such the duck-bill nozzle may have been removed to service the aircraft and not replaced. Without the duck-bill nozzle, the jet fuel nozzle would now fit the smaller fuel port of a piston-powered aircraft.

4.5.1 Diesel Refueling Trucks

While not directly related to the topic of misfuelling, there are two safety considerations related to fueling an aircraft from a fueler. Newer fuelers are most likely to be diesel powered vehicles which include contemporary environmental controls specific to the vehicle itself. Two recent challenges which have presented themselves relate to the emission controls. The first is the potential for contamination of the fuel with the diesel exhaust fluid (DEF) due to the visual and package similarities between DEF and Fuel System Icing Inhibitor (FSII). The second challenge is with respect to the operation of the regenerator. Diesel regeneration is a process which literally burns the soot and ash accumulated on the diesel particulate filter (DPF). On standard vehicles, this process occurs when the engine control module senses a

Fig. 7 EI 1597 Three-Way Cross-Check for Misfuelling Prevention

rise in exhaust pressure indicating the filter is plugging. The system automatically initiates the burn-off cycle, completely combusting the deposits and exhausting them. Here in lies the challenge, this temperature at the DPF is raised to roughly 1000 °F (540 °C) and the exhausted gases remaining hot enough to be an ignition source for the aviation gasoline vapors. Depending on the design of the truck, this exhaust may be located close to the vehicles dispensing point. It is critical that operators of diesel powered fuelers actively confirm the regeneration cycle cannot occur during fueling or when the truck is located on the ramp (apron) near aircraft as it is a measurable fire safety hazard.

Fig. 8 Example fuel nozzles—jet (left) and aviation gasoline (right)

4.6 Water

Due to the low density and viscosity of aviation gasoline, free water will generally settle from the fuel with perceived ease, especially as compared to the settling time of jet fuel. This is illustrated by data from the Coordinating Research Council (CRC) Aviation Fuel Properties Handbook [53], Table 8.

Inspection of a fuel sample from the aircraft sump is a standard entry in pre-flight check lists. This is to confirm the fuel tank is not filled to some level with water and has the correct Grade on-board, high-octane aviation gasolines being dyed blue or green. General knowledge within the piston engine user community would suggest an expectation that the water settling from the gasoline during fueling will accumulate in the fuel sump within minutes of the aircraft being fueled. In 1976, the U.S. FAA released an Advisory Circular 20-43C in which a review of reported studies on water settling was provided. In summary AC 20-43C reported that when three gallons of water had been added to a half-full fuel tank on what was hypothesized to be either a Cessna 152 or 172, it required removing 10 oz (295 mL) of fuel before water was observed in the fuel sump. In another report reviewed by the FAA, an aircraft only described as a "tricycle gear craft", one gallon (3.79 L) of water was added to the half-full fuel tank, and in this case, more than a quart (0.98 L) of fuel was drained before water began to show up in the sump sample [49]. The point being made in the advisory circular and pertinent here is that it takes time for water to settle to the

Table 8 Typical free water droplet settling rates at 15.5 °C [53]

Droplet size (μm)	Free water droplet settling rates at 15.5 °C (m/h)	
	Avgas	Jet A
15	2.88	0.72
20	5.4	1.08
25	7.56	1.8
30	10.44	2.16

sump. Furthermore, the amount of time required is affected by local conditions such as the orientation of the aircraft and the ambient temperature. Therefore, it is due diligence for the owner/operator to assure that both sufficient time from fueling is allowed to pass and sufficient volume of fuel is drained to inspect the fuel load. The pilots operating manual and check lists provided with the aircraft should be consulted and followed.

4.7 Use of Automotive Gasoline (MoGas) in Aviation

4.7.1 Overview

In 1979, fuel researchers undertook a comparison of the specification requirements for aviation gasoline Grade 100LL and automotive gasoline (Autogas). The findings were, as reported by Zeisloft in Future Fuels [54], that with the exception of volatility and octane, the two specifications were both essentially spark ignition piston engine fuels and there might be some scope for interchangeability.

One of the observations was that Autogas had a wider volatility range based on seasonal requirements and market demand. In order to support starting in the winter months, the fuel was made more volatile [44, 55] and the vapor pressure was in general higher than aviation gasoline. For example, Autogas may have a Reid vapor pressure of 12 lbs/in^2 (83 kPa) maximum as compared to 7 lbs/in^2 (48 Kpa) for aviation gasoline. This raised concern that use of the product might cause vapor lock in aircraft. In an extensive research program, the Experimental Aircraft Association (EAA) tested a maximum volatility Autogas in multiple flight tests with operating temperatures up to 100 °F (38 °C) [54]. Another observation was that Autogas generally had a higher energy content by volume (BTU value); 124.33 BTU/gallon (34,700 J/l) compared to 120.190 BTU/gallon (33,500 J/l) for aviation gasoline [56]. This would benefit aircraft range limited by fuel volume, accepting some loss of load carrying capacity. It should be noted that the minimum Autogas octane value was greater than Grade 80, alleviating detonation worries for this portion of the fleet and providing an option to reduce lead fouling problems. However, the fuel could not be considered for any aircraft requiring high-octane aviation gasoline Grades 100LL or 100/130.

Based on the above research, in 1982 the EAA received approval from the Federal Aviation Administration (FAA) to issue supplemental type certificates (STCs) for the conversion of certain aircraft/engine combinations to operate on non-ethanol containing Autogas. This was limited to aircraft where Autogas had sufficient octane to allow safe operation and fuel systems capable of managing the higher volatility of the product without vapor lock.

When using Autogas, there are a number of considerations for its safe use. According to recommendations provided by the FAA in Advisory Circular 91-33a [57], the following actions should be taken.

(1) Do not store Autogas in aircraft fuel tanks for long periods of time due to the potential to form gum deposits.
(2) Carburetor icing can occur at warmer temperatures and form more quickly with Autogas than with aviation gasoline.
(3) After prolonged heat soaking, for example during pre-flight, confirm there is full power available prior to take-off.

While considered at the time a successful program for Grade 80 designed aircraft experiencing lead fouling, various challenges have been encountered due to the lack of aviation industry control over motor gasoline specifications, manufacture and quality control during distribution. In the United States and Europe motor gasoline is covered by specifications ASTM D4814 [55] and EN 228 [58], respectively. These have evolved to meet market demand and environmental legislation. Oxygenates such as ethanol are now in prevalent use in both regions, a situation not envisaged by the EAA at the time of the STC campaign.

While E10, automotive fuel containing 10% ethanol, began to be available in the 1970s, widespread use was not common until the 2000s. This occurred when the use of methyl tertiary butyl ether (MTBE), the octane improver of choice, was banned or partially banned in the U.S. By 2007, 25 U.S. states had some level of ban on the use of MTBE [59]. Instead of using MTBE, motor gasoline began to incorporate ethanol as the oxygenate of choice. By the end of 2010, over 90% of all gasoline sold in the U.S. contained ethanol [60]. The required inclusion of ethanol in automotive gasoline as an octane source was done specifically to reduce the per volume use of petroleum in automotive applications. This meant the standard motor gasoline distillate could be produced with a lower inherent octane value because the mandated ethanol addition would improve the octane rating of the finished fuel. Therefore, it was not enough to separate motor gasoline distillate for aviation use before the ethanol was added because the distillate would not have the necessary octane rating on its own.

The obvious question is "why not allow the use of Autogas containing ethanol?" There are a number of issues with ethanol that are unacceptable for aviation use. With respect to the physical properties of the fuel [61], the addition of ethanol results in:

(1) A loss of energy density per volume
(2) An increased risk of vapor lock
(3) The potential for phase separation.

As reported above, isopropyl alcohol is permitted at limited amount, in controlled application, as a fuel system icing inhibitor in aviation gasoline because it moves into the trace free water and acts as an anti-freeze. Ethanol will also move preferentially to the trace water but, because a significant amount is present, this may result in complete phase separation of the fuel into:

- A low-octane hydrocarbon layer
- An incombustible, corrosive, water–ethanol layer.

While the water–ethanol mixture remains in the aircraft sump, the remaining distillate will be at a lower octane rating and no longer able to provide sufficient protection against detonation for the engine. This attribute of fuel phase separation is promoted by low temperatures, and therefore a risk for any aircraft flying at altitude or during winter months.

There are also operational challenges caused by the presence of the ethanol. Ethanol can attack the nitrile polymers used for seals, hoses and fuel bladders. There are also suggestions the first-time use of ethanol may clean-up existing deposits due to the higher polarity of the liquid fuel, resulting in filter blockages and fuel starvation [62]. As noted, ethanol's propensity to attract water means a high propensity for facilitating corrosion, the alcohol also being gradually oxidized to form acetic acid.

4.7.2 Ethanol Test

The consideration for the owner/operator is how to confirm the motor gasoline supplied at an airfield does not contain ethanol. According to Petersen Aviation [63], Inc. it is relatively simple to check fuel for the presence of ethanol. Use a glass tube or bottle six to seven inches long (15–18 cm) on which is a permanent marking of level, two inches (5 cm) from the bottom. Test kits are available which include this tube. Fill the tube with still water to this mark and then fill with the Autogas of interest. Cap the tube and agitate the mixture. Allow the tube to rest upright until the water settles to the bottom. In order to more clearly see the difference between the fuel and the water, a water-soluble dye is provided in the test kit. If the water level increases, is higher than the permanent mark, the fuel contains ethanol and is not approved for use in the aircraft.

In response to the challenge of obtaining Autogas with an appropriate anti-knock index for Grade 80 STC aircraft which does not contain ethanol the aviation industry developed specification ASTM D6227 [64] for Grade 82UL. This was essentially equivalent to motor gasoline at the time but based on a hydrocarbon product with limited use of ethers. In March 2000, the FAA released a SAIB approving the use of ASTM D6227 as an alternative to automotive gasoline per ASTM D4814 [55] without further modifications to the STC [65] provided the minimum 82 MON was sufficient. The use of 82UL was not permitted as a direct alternative for all Grade 80 aircraft due to differences in volatility and (optional) ether content. Unfortunately, uptake of Grade 82UL, and a second higher-octane Grade UL87 added later, was never achieved. The product has now been overtaken by UL91 and UL94 detailed in the next section.

4.8 Unleaded Aviation Gasoline

To resolve the issue of lead fouling in engines, reduce lead emissions to the environment and offer a solution to Autogas/STC complexity, a range of unleaded aviation gasoline Grades have been successfully developed by the Industry. This started in 1981 with Hjelmco Oil™ in Sweden specifically promoting an unleaded Grade 80 later followed in 1991 by a mid-octane Grade, Hjelmco 91/96UL™. ASTM development followed with the introduction of D7547 Standard Specification for Hydrocarbon Only Unleaded Aviation Gasoline covering Grades UL91 and UL94, the former Grade then being also adopted in Defence Standard 91-090. In the European Union, EASA has released a Safety Information Bulletin (SIB) 2011-01R2 which permits the use of UL91 aviation gasoline in aircraft with lower octane requirements with no additional approval, meaning no STC is required. In the U.S., Grade UL94 is gathering market interest for environmental and operational benefits. Per EASA, a U.S.-based aircraft operating in the European Union and requiring a lower octane fuel would use ASTM D7547 unleaded aviation gasoline or the Defence Standard 91-090 equivalent.

Development of a high-octane unleaded aviation gasoline as a replacement for 100LL and equivalent Grades is also being evaluated under the Piston Aviation Fuels Initiative [66]. The program is a joint Government—Industry undertaking to assess potential fuel formulations and engineering solutions to allow performance engine aircraft to operate without the need for the lead octane enhancement additive TEL. This is a challenging program requiring detailed evaluation of full-size engine octane requirement. To assist, two agreed detonation testing protocols have been developed:

- ASTM D6424 [67] Practice for Octane Rating Naturally Aspirated Spark Ignition Aircraft Engines.
- ASTM D6812 [68] Practice for Ground-Based Octane Rating Procedures for Turbocharged/Supercharged Spark Ignition Aircraft Engines.

Studies have highlighted the influence of TEL on the combustion process in full-size engines [69] and led to the development of technical guidelines for aviation gasoline assessment through ASTM D7826 [70]. At the time of writing, work remains ongoing.

5 Aviation Diesel (Aerodiesel) Engines

While the Wright Flyer featured an Otto gasoline spark ignition engine for flight, setting a path which would dominate aviation for over 40 years, this was not the only type of piston engine considered for aircraft development. In 1893 Rudolf Diesel successfully proved the concept of a compression ignition (CI) engine in Augsburg, Germany. Such engines were to become very popular for ground vehicles and marine vessels, often referred to simply by the inventor's name, diesel engines. Adapting this technology for flight is challenging due to power-to-weight ratio and the need for sufficient charge for compression ignition at altitude, but possible.

5.1 History of Aerodiesel Engines

The first airplane diesel engine to power an aircraft was produced by Packard Motor Car Company (225hp Model DR-980 radial engine) in 1928 [71] and advertised as "the first oil burning engine to power an airplane". The specific aircraft, a Stinson SM-1DX "Detroiter" first flew in September of 1928. Interest in developing aviation diesel engines was prevalent in the 1920s, as demonstrated by the Bristol Aircraft Company Phoenix engine based on the Pegasus design [72]. Such engines were also being developed in France (Panhard) Germany (Junkers) and Italy (Fiat). "Furnace oil" was the typical fuel used at this time as it provided increased range compared to aviation gasoline. This, in addition to increased reliability (purportedly due to simplicity of design) and increased safety (low flash point reducing potential fire hazard), constituted the three main drivers for development of the aviation diesel engine, or aerodiesel as referred to in this chapter. These early engines were predominantly radial in design.

The most successful early engine came from Junkers in the 1930s with the Jumo series of aircraft diesels [73]. The Junkers Jumo diesel engine featured an unusual design of six cylinders, with two opposed pistons operating in each cylinder, under a two-stroke operating cycle. This allowed the 16.6 L unit to deliver a maximum of 600 hp and offer excellent fuel economy for Lufthansa's Do18 and Do26 transatlantic flying boats.

A Junkers/Lufthansa aviation diesel specification [74] was developed to ensure fuel quality, Table 9.

Table 9 Junkers/Deutsche Lufthansa diesel fuel specification

Property	Units	Limits
Saybolt viscosity @ 20 °F		39–57
Saybolt viscosity @ 50 °F		33 min
Cetane number	CN	45–60
Pour point	°F	−4 max
Cloud point	°F	10 max
Ash	%	0.02 max
Sulfur	%	0.1 max
Distillation		
Boiling point	°F	392 min
70% point	°F	572 min
90% point	°F	662 min
Recovery	% v/v	98 min
Specific gravity	kg/l	0.840–0.880
Calorific value	Btu/lb	19,380 min
Flash point	°F	150 min

[a] a number of impurity limits also included

While technically successful, the necessary ruggedness of these early aerodiesel engines detrimentally impacted power-to-weight ratio, thus hindering development compared to lighter spark ignition piston units. Moreover, the discovery of tetraethyl lead and availability of high-octane aviation gasoline allowed the latter to gain in thermal efficiency, reducing size and increasing maximum power output, leaving diesel development behind. In addition, from the late 1930s onward, the advent of the aviation turbine engine for military and commercial markets offered high-speed aircraft combined with turboprop engines for lower speed applications. With a superior power-to-weight ratio compared to both aviation spark ignition and diesel compression ignition piston engines, such technology now dominates flight as exemplified by other chapters in this book. Diesel powered aircraft essentially vanished leaving the spark ignition units using aviation gasoline to support light aircraft in General Aviation applications. The story might end here but a change in circumstances has led to the re-emergence of the aerodiesel.

5.2 Changing Circumstances and The Modern Aerodiesel Engine

Over time, conditions have changed with several factors now influencing the viability of the aerodiesel. In the US, introduction of the 1990 amendments to the Clean Air Act [75] precipitated an effort to replace leaded aviation gasoline. The CAA banned future production of "non-road" lead burning engines and it was unclear at that time whether aircraft were included. In the end, the apparent intent was to regulate aircraft separately. Currently, the Environmental Protection Agency (EPA) is studying lead emissions from aircraft and will likely result in an endangerment finding for aviation sourced emissions.

In Europe and other regions, a significant cost differential developed between aviation gasoline and jet fuel when the former fell into decline. This related to the simplicity of manufacturing turbine fuel compared to a high-octane gasoline combined with the loss of critical mass in the market. Remote locations would often have jet fuel readily available but aviation gasoline would be hard to find. The situation was further compounded by countries then applying duty tax to aviation gasoline as the piston engine fleet was no longer the mainstay of international travel.

The scene was set for the return of the aerodiesel burning readily available, lower cost, jet fuel. However, if diesel technology were to re-enter the GA market and compete with spark ignition flight, it would have to be competitive in terms of purchase cost and performance. This was realized when government regulators allowed upgraded automotive production line engines, with additional built-in safeguards, to be certified for aviation use. In essence, development and tooling costs were absorbed by the automotive industry, making the engine competitive with aviation spark ignition units. Major advances in automotive diesel engine technologies, for example, direct injection, FADEC, and turbocharging, also helped to resolve the poor power-to-weight ratio of earlier designs. The combination of these factors opened the door leading to the reintroduction of aerodiesels.

5.3 *Certificated Engines and Aircraft*

A brief insight into the current situation regarding certified aerodiesel engines and aircraft is provided below to build understanding before examining the problems and solutions facing this technology.

5.3.1 Certificated Engines

Modern aerodiesel development started in the late 1980s/early 1990s. However, none of the early projects progressed to the certification process. One example is the VM Motori company's development of four-, six- and eight-cylinder, liquid-cooled, turbocharged aerodiesels rated at 206, 315 and 424 horsepower respectively. Work was not pursued, and it was only at the start of the next decade that Thielert and SMA successfully progressed through the certification process to give commercial offers. The first engine, a Thielert TAE110-01, was certified in March of 2001 but never went into production being overtaken by a later model, the TAE 125-01. In April of that same year, SMA obtained certification for their SMA SR305-230 [76].

Further activity was not realized until 2004. From 2004 to 2017 seven additional engines were certified. Of the seven certified, engines Continental obtained TCDSs for their Centurion 4.0, TAE 125, TD-300-B and Centurion 3.0 series engines [77–81]. Austro obtained certification for their E4 series engines in 2009 [82], Diesel Jet in 2010 for their TDA CR engine [83] and RED for their A03 series engines in 2012 [84].

Substituting SI engines with aerodiesel engines constituted the first applications for these certified engines. The supplemental type certificate (STC) process provided the path for achieving the goal of introducing certified aerodiesel powered aircraft into the modern fleet. For example, the Continental CD-135 aerodiesel has STCs for the Cessna 172 and Piper PA 28 [85, 86]. At the time of this writing (2025), 6000 TAE 125 engines have been produced and delivered since certification in 2006. The number of active engines in today's fleet will likely be less due to older engines being retired. Considering the size of the worldwide fleet, these numbers are small.

Production and certification of new aircraft fitted with an aerodiesel as original equipment was the next logical step.

5.3.2 Certified Aircraft

Diamond Aircraft was the first modern manufacturer to obtain certification for aircraft fitted with aerodiesel engines, these being offered as an alternative to a conventional aviation gasoline powered spark ignition unit. For example, a Diamonds DA 42 twin engine aircraft fitted with two aerodiesel engines was certified in May of 2004 [87].

Subsequent to the DA 42 certification, Diamond set out to obtain approval for a basic entry level single engine aircraft designated as the DA 40 [88]. This fixed gear aircraft was certified with the Austro AE-300 aerodiesel engine in January of 2005. Again, a conventional gasoline powered version of the aircraft is also available.

Fifteen years passed before Diamond obtained a third certification only this time it was for an intermediate level aircraft designated as the DA 50 [89]. The DA 50 was certified with a 300 horsepower Continental CD 300 engine fitted with a constant speed propeller and a retractable gear in August of 2020.

The Technam P2010 represents the latest aircraft that has obtained certification including an aerodiesel engine as original equipment. It is a single engine plane fitted with a 168 HP Continental TAE 125 fitted with a fixed pitch propeller and is manufactured Costruzioni Aeronautich TECNAM S.P.A. [90]. The company obtained certification in October of 2020 and designated the aircraft the Tecnam P2010.

A key message from the above is that aerodiesels are typically offered as an alternative to spark ignition engines in the GA fleet. This has implications as will be seen later in the chapter.

5.4 Addressing Fit for Purpose Issues

Fueling aerodiesel engines with Jet A or Jet A-1 does provide some significant advantages. As previously mentioned, Jet fuel remains readily available worldwide. Moreover, the product is handled to aviation standards, following rigorous quality handling procedures that maintain fuel quality from refinery to the aircraft as detailed in other chapters. It also lends itself to the "one fuel" concept for global aviation. Unfortunately, Jet fuel is specifically produced to satisfy the requirements of turbine engines, discounting any properties that may or may not be necessary for the safe operation of an aerodiesel engine. This raises some fit-for-purpose challenges. These can be identified by comparing ASTM jet fuel and diesel fuel specifications, D1655 [91] and D975 [92] respectively, Table 10. As a first observation, the reliability and time before overhaul for an aviation turbine engine makes this more sensitive to a range of fuel properties compared to the piston diesel. The gray areas highlight a range of aspects where the properties are not listed in D975 rendering them unnecessary for the aerodiesel. There are, however, three properties where the variance in specifications is critical for the safe operation of aerodiesel engines—Cetane, Freeze Point and Lubricity. These are discussed in more detail below, highlighting the issue and possible solution.

These are discussed in more detail below, highlighting the issue and possible solution.

Table 10 A comparison of ASTM D1655 aviation turbine fuel (Jet A/A-1) and D975 automotive diesel specifications

Property	Max./Min.	D1655: Jet A	D975: Diesel (No.2)
COMPOSITION			
Acidity, Total (mg KOH/g)	Max.	0.1	
Aromatics (vol%)	Max.	25	35
Sulfur, Total (wt %)	Max.	0.3	0.0015
Sulfur, mercaptan (wt %)	Max.	0.003	
or Doctor Test		Negative	
VOLATILITY Distillation Temperature:			
10% Recovery (°C)	Max.	205	
50% Recovery (°C)	Max.	report	
90% Recovery (°C)	Max.	report	338
Final BP (°C)	Max.	300	
Distillation Residue (vol %)	Max.	1.5	
Distillation Loss (vol %)	Max.	1.5	
Flash Point (°C)	Min.	38	52
Density @ 15 °C (kg/m³)		775-840	
FLUIDITY			

(continued)

Table 10 (continued)

Freezing Point (°C)	Max.	- 40*	~ - 8**
Viscosity (cSt)	Max.	8 @ - 20	4.1 @40
LUBRICITY			
HFRR @ 60 °C (microns)	Max.		520
BOCLE (mm)	Max.	0.85	
COMBUSTION			
Net Heat of Comb. (MI/kg)	Min.	42.8	
Smoke Point (mm)	Min.	25	
or Smoke point (mm)	Min.	18	
and Naphthalenes (vol %)	Max.	3	
Cetane number	Min.		40
CORROSION			
Copper Strip (2h @ 100 °C)	Max.	1	3
STABIALITY			
JFTOT Dp (mm Hg)	Max.	25	
Tube Deposit Rating (Visual)	Max.	<3	
CONTAMINANTS			
Existent Gum (mg/100 mL)	Max.	7	
Microseparometer	Min.	85	

5.4.1 Cetane

Cetane quality (ASTM D613, [93]) is a measure of a fuel's ability to ignite in a compression ignition engine and burn in an appropriate manner to deliver power. Typically, a minimum cetane number of 40 is specified to allow the engine to start and operate under the compression ignition regime as indicated by automotive diesel specifications (ASTM D975). In contrast, the turbine engine has no such requirement as it operates under steady-state combustion conditions with no need for cetane or octane. After Jet fuel is introduced into the combustor of a turbine engine, the vaporized fuel is set alight with an igniter, generating a self-sustaining flame. Once lit, the igniter is turned off and the combustion section essentially functions as a burner. In a diesel engine, combustion is initiated during the compression stroke where adequate heat is generated to cause the fuel to auto ignite whereupon it burns and expands providing the power stroke. The fuel must possess enough cetane quality to reliably ignite each compression stroke, allowing for uninterrupted smooth running of the engine. Too low a cetane can result in an engine failing to start. For an aerodiesel engine this is a critical air safety issue, as it is unacceptable to be unable to restart an engine in flight should an engine-out situation occur, for example through failure to switch between fuel tanks.

The combustion portion of Table 10 shows there is no cetane requirement (in red) listed in D1655 for jet fuel while a minimum of 40 cetane is listed for D975 diesel fuel. Therefore, fueling an aerodiesel aircraft with jet fuel will not necessarily satisfy all the needs of the engine as, by definition, turbine fuel is not produced for that purpose. However, jet fuel broadly has appropriate properties and can be used in some diesel vehicles, for example as part of the Military single fuel concept. The aerodiesel seeks to take advantage of this aspect but lacks a supporting minimum cetane specification.

In 2009, an ASTM JetDiesel taskforce was formed to help resolve the complexity of fuel cetane in aerodiesel engines. US market surveys performed in 2010 and 2011 revealed that jet fuel cetane values ranged from 38 to 51 and estimated that roughly 10 percent of product possessed a value of less than 40. The taskforce also noted that new, highly isomeric synthetic jet fuels could have very low cetane numbers. The implications for aerodiesel operation were further emphasized following a report of an aircraft incident in the Far East where an engine failed to restart in flight. Investigations determined Jet cetane quality in the region could be as low as 28.

With the aviation industry unable to accommodate aerodiesel engine requirements in D1655 turbine fuel specifications due to possible ramifications, an engine and airframe manufacturer approach has been necessary. This requires the aerodiesel industry to diligently monitor changes to turbine fuel compositions and specifications and ensure operational performance. For example, following the incident in the Far East, one manufacturer adopted use of 28 cetane fuel at −32 °C in a climatic chamber to demonstrate cold-start performance.

5.4.2 Freezing Point

Clearly a fuel in an aircraft operating at high altitude must not to freeze but remain fluid and pumpable during flight. Fluidity properties required for aviation turbine engines compared to those for piston engines are different, the former must possess a freezing point no higher than −40 °C (Jet A) Table 10, and the latter −58 °C (aviation gasoline Grades UL91, UL94, 100LL, 100VLL, 100), Table 3. A brief explanation for this difference is warranted. Jet aircraft can fly close to the speed of sound (for example, Mach 0.8) and feature fuel tank temperature sensors. If the fuel temperature approaches −40 °C (−47 °C for Jet A-1), the pilot is alerted and may accelerate or move to a lower altitude to increase aerodynamic heating to keep the bulk fuel above its freezing point. Indeed, for very high-speed aircraft aerodynamic heating is a significant problem and roles are reversed, the fuel now having to adsorb as much heat as possible and act as a coolant before combustion. The low freezing point of aviation gasoline is necessary because flight speeds are much lower and the fuel volume relatively small, allowing the liquid to approach the external ambient temperature at altitude more rapidly.

Aerodiesel powered aircraft are not capable of utilizing the aerodynamic heating used by jets making the freezing point of the fuel an issue. To resolve this problem engine and aircraft manufacturers have sought to apply heat to the fuel where necessary to ensure uninterrupted flow to the engine. In addition, flight limitations may be applied to ensure ambient conditions are appropriate for operation. At this time there are no known incidences here freezing point has been linked to an off-field landing of an aerodiesel powered aircraft. It remains prudent for the Industry to monitor any changes to the turbine fuel freezing point specification and be realistic about aerodiesel applications, for example use in high altitude drones.

With respect to ground fuels, ASTM D975 does not list a freezing point and simply contains a footnote stating *"It is unrealistic to specify low temperature properties that will ensure satisfactory operation at all ambient conditions. In general, cloud point (or wax appearance point), Low Temperature Flow Test, and Cold Filter Plugging Point Test may be used as an estimate of operating temperature limits for"*... numerous grades ...*"diesel fuel oils"*. The value of ~−8 °C listed in Table 10 is only a rough estimate and would vary dependent on many factors such as region and season.

5.4.3 Lubricity

The ability of a fuel to also act as a lubricant for fuel system parts is termed lubricity. Straight run jet fuels possess sufficient natural lubricity to satisfy the turbine engine. However, when severely hydrotreated blend stocks are introduced into the pool, lubricity properties can be impacted as discussed in ASTM D1655 section X1.10. Looking to the lubricity portion of Table 10, the method used to measure lubricity for turbine fuels (Ball On Cylinder Lubricity Evaluator "BOCLE", ASTM D5001 [94]) is significantly different to that used for diesel fuels (High Frequency Reciprocating Rig "HFRR" ASTM D6079 [95]) due to the different mechanical action of engine parts, and therefore different specification limits are set. Additives are available to improve lubricity when required, but similar to the methods of measurement, these are not aligned. Diesel engine lubricity additives are not approved for use in turbine fuel and therefore will not be present for the benefit of aerodiesel powered aircraft. With aviation additive approval processes being necessarily rigorous, this situation is unlikely to change for the relatively small aerodiesel fleet. Similar to cetane and freeze point, engine manufacturers have had to address this issue by incorporation of materials that can stand up to worst-case turbine fuel properties. For some designs, this has featured ruggedness testing using 1000 μm HFRR product.

Similar to cetane, an off-airport landing occurred due to use of poor lubricity fuel [96]. On investigation, engine failure was attributed to the lack of adequate fuel pump lubrication causing seizure. After identifying the problem, the manufacturer upgraded the fuel pump's metallurgy to handle low lubricity fuels, protecting against any future incidents.

5.5 *Misfuelling Risk*

As discussed in Sect. 4.5 the risk of aircraft misfuelling is an on-going concern and occurs when the wrong grade of fuel is delivered into an aircraft. This may happen due to lack of training, complacency, or a momentary lapse in concentration. Even conscientious ground personnel can easily refuel an aircraft with the wrong fuel due to misplaced familiarity. For example, the Cessna 404 Titan and 441 Conquest appear very similar but the former requires aviation gasoline and the latter jet fuel. The odds of a misfuelling increase significantly when aircraft possess the same basic airframe but different engine types. If both aircraft were parked on the ramp and a request were to be that both aircraft be refueled, unless there is clear communication between the pilot ground crew there is a greater potential for error. Carefully examining the aircraft and experience with cowling/layout details can help, in this case, to determine which aircraft is fitted with an aviation gasoline fueled piston engine and which with a jet-fueled turboprop engine.

With the introduction of aerodiesel piston engine aircraft into the market, particularly as an option to aviation gasoline piston engines, such differences in appearance are lost. Aircraft cowling and exhaust pipe configurations essentially become the same, Fig. 9.

Fig. 9 A Diamond DA40 NG fitted with an aerodiesel engine (left) and a diamond DA40 XLT fitted with a gasoline powered SI engine (right)

While similar in appearance, the risk of misfuelling an aerodiesel with aviation gasoline is similarly as deadly as the spark ignition case. Aviation gasoline is formulated to resist compression ignition and would not necessarily ignite in the combustion chamber, causing engine failure with little hope of restart.

Physical measures like color coded filler caps, fueling nozzles, wing decals and fuel specific self-serve credit cards, can go a long way to alert line personnel and pilots to avoid a potential misfuelling. FAA Advisory Circular 20−43 C reiterates that FAR Part 23 requires that aircraft fuel filler openings must be marked to show the "fuel" and the minimum fuel grade or designation for the engines, Figs. 10 and 11.

While the wider Jet "duck billed" nozzle referenced in Sect. 4.5 helps prevent entry into small diameter spark ignition piston aircraft filler ports, the same is not true for the aviation gasoline nozzle and aerodiesel aircraft. Indeed, when first introduced aerodiesels were at significant risk of misfuelling with aviation gasoline. To help resolve this problem, the Industry developed a mechanical gate which could only be opened by the jet nozzle, Fig. 12, and details can be found in SAE AS1852D [97].

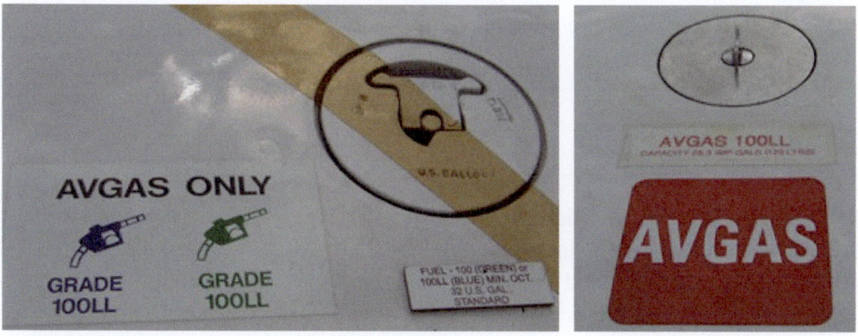

Fig. 10 Examples of aviation gasoline wing decals

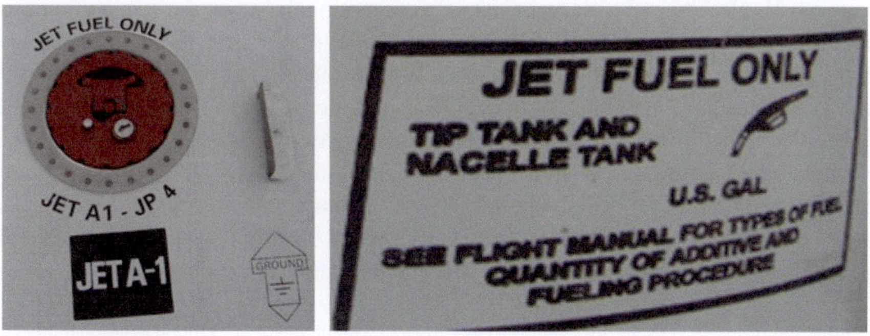

Fig. 11 Examples of aviation turbine fuel (jet) wing decals

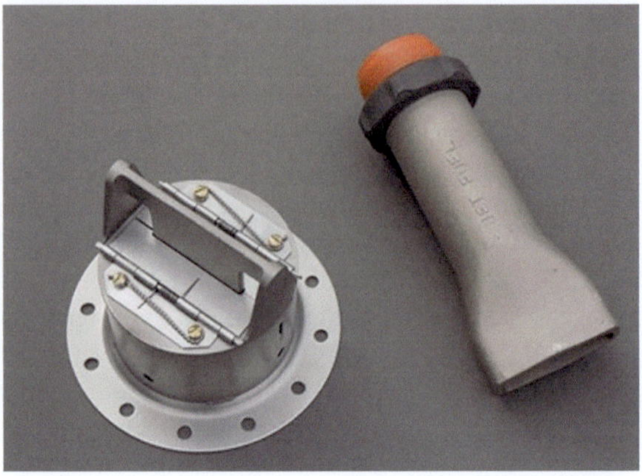

Fig. 12 Jet only filler port for aerodiesel aircraft

Human error is the one thing that all misfuelling incidents have in common and more vigilance is required with aerodiesels entering the fleet. In-depth training in combination with clear communication between line personnel and the pilot are necessary if misfuellings are to be prevented.

6 Summary

Aviation gasoline was the first fuel specifically developed for flight. The product is used in spark ignition aviation piston engines and is of exceptional quality and manufactured from a limited number of refinery components. Tetraethyl lead additive is necessary for production of Grades 100, 100LL and 100VLL due to the minimum octane requirement far exceeding automotive fuels to maximize thermodynamic combustion cycle efficiency. Looking to the future, the industry is seeking to develop unleaded aviation gasoline grades for environmental benefit, with two mid-octane fuels already in the market. Given this exceptional quality, the segregation and handling of aviation gasoline requires care and attention to detail similar to turbine fuels to ensure operational performance. With respect to turbine fuel, aerodiesel piston engines have also been developed for light aircraft use. This has highlighted further technical aspects to ensure flight performance in an alternative engine environment which has required industry research and understanding. With both aviation gasoline and aerodiesel piston engines now in the market and aircraft looking practically identical, a key message from this chapter is to ensure the correct fuel is uses. Both technologies are reliant on highly developed, aviation grade, products.

Acknowledgements The authors gratefully acknowledge members of the general aviation industry, engine and aircraft manufacturers, fuel producers, distributors, pilot's associations, the Federal Aviation Administration, the European Union Aviation Safety Agency and related groups for their insight into piston engine fuels. This knowledge has been shared at ASTM International, The Coordinating Research Council Aviation Committee, the Aviation Fuels Committee, Energy Institute and The

International Conference on Stability, Handling and Use of Liquid Fuels over many years to the benefit of the Industry. The wealth of historical data, field experience and laboratory test work has been invaluable in preparation of this chapter. The authors would like to thank Ron Wilkinson, Dave Atwood and Joanna M Bauldreay for their helpful comments and additional insights during the preparation of this chapter.

References

1. Ogston A (1981) A short history of aviation gasoline development, 1903–1980. Aeronaut J
2. ASTM D2699. Standard test method for research octane number of spark-ignition engine fuel. ASTM International, 100 Barr Harbor Drive, PO Box C700, West Conshohocken, PA 19428-2959. United States
3. Nash AW, Howes DA (1935) The principles of motor fuel preparation and application, Volume II
4. Ricardo HR (1923) The internal combustion engine. Blackie & Son Ltd, London and Glasgow
5. ASTM D2700 Standard test method for motor octane number of spark-ignition engine fuel. ASTM International, 100 Barr Harbor Drive, PO Box C700, West Conshohocken, PA 19428-2959. United States
6. Banks FR (1978) I kept no diary, air commodore. Airlife Publications, 7 S. John's Hill, Shrewsbury, England. ISBN 0 9504543 97
7. ASTM D910. Standard specification for leaded aviation gasolines. ASTM International, 100 Barr Harbor Drive, PO Box C700, West Conshohocken, PA 19428-2959. United States
8. ASTM D7547. Standard specification for hydrocarbon unleaded aviation gasolines. ASTM International, 100 Barr Harbor Drive, PO Box C700, West Conshohocken, PA 19428-2959. United States
9. Defence Standard 91-090. Gasoline, aviation, grades UL91, 100/130 and 100/130 low lead. JSD: AVGAS UL91, AVGAS 100 and AVGAS 100LL. Defence Equipment and Support, UK Defence Standardization, Kentigern House, 65 Brown Street, GLASGOW G2 8EX, UK
10. Goodger E, Vere R (1982) Aviation fuels technology. Macmillan Publishers Limited. ISBN 0-333-35787-6
11. Systemic investigation into fuel contamination. Australian Transport Safety Bureau (ATSB), Air Safety Investigation (2001). ISBN 0 642 27470 3
12. ASTM D909. Standard test method for supercharge rating of spark-ignition aviation gasoline. ASTM International, 100 Barr Harbor Drive, PO Box C700, West Conshohocken, PA 19428-2959. United States
13. IP 228. Determination of lead content of gasoline—X-ray spectrometric method. Energy Institute, 61 New Cavendish St, London W1G 7AR, UK
14. ASTM D5059. Standard test methods for lead and manganese in gasoline by X-Ray fluorescence spectroscopy. ASTM International, 100 Barr Harbor Drive, PO Box C700, West Conshohocken, PA 19428-2959. United State
15. IP 428. Liquid petroleum products—petrol—determination of low lead concentrations by atomic absorption spectrometry. Energy Institute, 61 New Cavendish St, London W1G 7AR, UK
16. IP 270. Petroleum products—determination of lead content of gasoline—iodine monochloride method. Energy Institute, 61 New Cavendish St, London W1G 7AR, UK
17. ASTM D3341. Standard test method for lead in gasoline—iodine monochloride method (withdrawn 2022). ASTM International, 100 Barr Harbor Drive, PO Box C700, West Conshohocken, PA 19428-2959. United States
18. ASTM D614. Test for knock characteristics of aviation fuels by the aviation method. ASTM International, 100 Barr Harbor Drive, PO Box C700, West Conshohocken, PA 19428-2959. United States

19. ASTM D86. Standard test method for distillation of petroleum products and liquid fuels at atmospheric pressure. ASTM International, 100 Barr Harbor Drive, PO Box C700, West Conshohocken, PA 19428-2959. United States
20. IP 123. Petroleum products—determination of distillation characteristics at atmospheric pressure. Energy Institute, 61 New Cavendish St, London W1G 7AR, UK
21. McAllister S et al. (2011) Fundamentals of combustion processes. Mechanical engineering series. Springer Science and Business Media, LLC. https://doi.org/10.1007/978-1-4419-7943-8
22. Heywood JB (1988) Internal combustion engine fundamentals. McGraw-Hill Book Company
23. IP 12. Determination of specific energy. Energy Institute, 61 New Cavendish St, London W1G 7AR, UK
24. ASTM D3338. Standard test method for estimation of net heat of combustion of aviation fuels. ASTM International, 100 Barr Harbor Drive, PO Box C700, West Conshohocken, PA 19428-2959. United States
25. ASTM D4809. Standard test method for heat of combustion of liquid hydrocarbon fuels by bomb calorimeter (Precision Method). ASTM International, 100 Barr Harbor Drive, PO Box C700, West Conshohocken, PA 19428-2959. United States
26. ASTM D4529. Standard test method for estimation of net heat of combustion of aviation fuels. ASTM International, 100 Barr Harbor Drive, PO Box C700, West Conshohocken, PA 19428-2959. United States
27. IP 138. Petroleum products—determination of oxidation stability of aviation fuel potential residue method. Energy Institute, 61 New Cavendish St, London W1G 7AR, UK
28. ASTM D873. Standard test method for oxidation stability of aviation fuels (potential residue method). ASTM International, 100 Barr Harbor Drive, PO Box C700, West Conshohocken, PA 19428-2959. United States
29. API RP 1595 (2019) Design, construction, operation, maintenance, and inspection of aviation pre-airfield storage terminals. American Petroleum Institute, Washington, DC. https://www.engineerdocuments.com/standards/API-RP-1595-R2019/
30. EI/JIG 1530. Quality assurance requirements for the manufacture, storage and distribution of aviation fuels to Airports. Energy Institute (Great Britain), & Joint Inspection Group. http://www.jigonline.com; https://publishing.energyinst.org/featured/1530
31. Jassal R (2018) Inerting of cargo tanks: how exactly to do it? https://www.myseatime.com/blog/detail/inerting-on-tankers
32. American Petroleum Institute (2018) API 2610 design terminal and tanks
33. Liquefied Gas Carrier.com (n.d.) Preparation for loading LNG cargo—inerting of cargo tanks. Retrieved May 17, 2021, from http://www.liquefiedgascarrier.com/inerting-tanks.html
34. ScienceDirect Topics—Elsevier (2018) Inert gas system—an overview. https://www.sciencedirect.com/topics/engineering/inert-gas-system
35. US-EPA (2017) Energy policy act of 2005 and underground storage tanks (USTs). https://www.epa.gov/ust/energy-policy-act-2005-and-underground-storage-tanks-usts
36. Personal communication true north steel (2021) In True north steel
37. Smith M, Foulis GT (1970) Aviation fuels
38. Anderson S, Clark AQ, Air BP, IASH (2015) Fuel cleanliness—the source and fate of contaminants from refinery to aircraft. In: The 14th international conference on stability, handling and use of liquid fuels
39. Royal Dutch Shell (1959) A petroleum handbook, 4th edn. The Asiatic Petroleum Company Limited
40. Barnett HC, Hibbard RR, August W (1956) NACA technical note 3276 properties of aircraft fuels
41. SAE (2018) API 2610 design, construction, operation, maintenance, and inspection of terminal and tank facilities. https://www.agilent.com/cs/library/applications/5990-9939EN.pdf
42. Wilhelm D, Huber U (2017) Determination of static dissipater additives in aviation turbine fuel and middle distillate fuels by HPLC. Agilent Technologies Application Note
43. Defence Standard 91-091. Turbine fuel, aviation kerosine type, Jet A-1 NATO code: F-35, joint service designation: AVTUR, Issued by the United Kingdom (U.K.) Ministry of Defence. Glasgow, UK. https://www.dstan.mod.uk/

44. Aviation fuels and their effects on engine performance—NAVAER-06-5-501. Ethyl Corporation (1951)
45. Greenshields RJ (1953) Spark-plug fouling studies. https://doi.org/10.4271/530206
46. Johnson CR, Harris DN (1953) Service experience with TCP fuel additive. https://doi.org/10.4271/530011
47. Get the lead out. Sales brochure. Alcor precision aviation products (2012). http://www.alcorinc.com/wp-content/uploads/2012/06/TCP_FAQ_06_12_no_bleeds.pdf
48. Johnson M, Kimble-Thom M (2019) Safety enhancements for aircraft fuel identification markings
49. FAA (1976) AC 20-43C—Aircraft fuel control—document information. https://www.faa.gov/regulations_policies/advisory_circulars/index.cfm/go/document.information/documentID/22581
50. ExxonMobil Aviation (2003) Aviation fuelling guide. ExxonMobil Aviation International Limited
51. Phillips 66 Aviation (2000) Tech tips—misfueling of aircraft. http://aviation.phillips66.com/misfueling.shmtl
52. Phillips 66 Aviation (2000) Tech tips—over wing nozzles. http://aviation.phillips66.com/overwingnoz.shmtl
53. CRC (2014) The coordinating research council handbook of aviation fuel properties, 4th edn. CRC Report No. 663. Coordinating Research Council, 5755 North Point Parkway, Suite 265, Alpharetta, GA 30022, USA
54. Zeisloft HC (1989) Future fuels for general aviation (Strauss K, Gonzalez C, eds.). ASTM International. https://doi.org/10.1520/STP1048-EB
55. ASTM D4814. Standard specification for automotive spark-ignition engine fuel 1. ASTM International, 100 Barr Harbor Drive, PO Box C700, West Conshohocken, PA 19428-2959. United States. https://doi.org/10.1520/D4814-21A
56. US EIA (2021) U.S. energy information administration energy explained. https://www.eia.gov/energyexplained/units-and-calculators/
57. FAA (1984) Advisory circular 91-33A use of alternate grades of aviation gasoline for grade 80/87, and use of automotive gasoline. https://www.faa.gov/documentLibrary/media/Advisory_Circular/AC91-33A.pdf
58. EN 228. Automotive fuels—unleaded petrol—requirements and test methods. European Committee for Standardization
59. US Environmental Protection Agency (2007) EPA420-B-07-013 state actions banning MTBE (Statewide)
60. RFA (2011) Building bridges to a more sustainable future
61. FAA (2006) Special airworthiness information bulletin CE-07-06. https://rgl.faa.gov/Regulatory_and_Guidance_Library/rgSAIB.nsf/dc7bd4f27e5f107486257221005f069d/6f3250f958b6a22286257259006d6dab/$FILE/CE-07-06.pdf
62. Visser B (2007) Own a certified airplane? Just say no to ethanol. General Aviation News, 24-undefined
63. Ethanol Fuel Testing | Ethanol in Gas (n.d.) Retrieved June 8, 2021, from https://www.autofuelstc.com/fuel_testing.phtml
64. ASTM D6227. Standard specification for unleaded aviation gasoline containing a non-hydrocarbon component. ASTM International, 100 Barr Harbor Drive, PO Box C700, West Conshohocken, PA 19428-2959. United States
65. FAA (2000) Special airworthiness information bulletin CE-00-19R1
66. PAFI (2021) Piston aviation fuels initiative (PAFI). Federal Aviation Administration. www.faa.gov/about/initiatives/avgas/
67. ASTM D6424. Practice for octane rating naturally aspirated spark ignition aircraft engines. ASTM International, 100 Barr Harbor Drive, PO Box C700, West Conshohocken, PA 19428-2959. United States
68. ASTM D6812. Practice for ground-based octane rating procedures for turbocharged/supercharged spark ignition aircraft engines. ASTM International, 100 Barr Harbor Drive, PO Box C700, West Conshohocken, PA 19428-2959. United States

69. Atwood D (2007) High-octane and mid-octane detonation performance of leaded and unleaded fuels in naturally aspirated, piston, spark ignition aircraft engines. DOT/FAA/AR-TN07/5. National Technical Information Services (NTIS), Springfield, Virginia 22161
70. ASTM D7826. Standard guide for evaluation of new aviation gasolines and new aviation gasoline additives. ASTM International, 100 Barr Harbor Drive, PO Box C700, West Conshohocken, PA 19428-2959. United States
71. Meyer RB (1964) The first airplane diesel engine: packard model DR-980 of 1928. Smithsonian Institution—National air Museum, Washington, D.C.
72. Rolls Royce Heritage Trust. "Bristol". B Gunston (1998). ISBN 1 872922 13 9
73. Deutsches Museum, Museumsinsel 1, 80538 Munich, Germany. Inv.-No.: 68302
74. Wilkinson PH (1940) Aircraft diesels, pp 207–216
75. CAA (1996) US environmental protection agency (EPA) clean air act prohibition on gasoline containing lead or lead additives for highway use
76. Type-Certificate Data Sheet, April 20, 2001, No. EASA E.076, SR305 Series engine, SMA Aero Engines GmbH, Flugplatz, 63329 Egelsbach, Germany
77. Type-Certificate Data Sheet, March 8, 2001, No. E.019, TAE 110 series engines, Continental Aerospace Technologies GmbH, Platanenstrasse 14, D-09356 Sankt Egidien, Germany
78. Type-Certificate Data Sheet, March 24, 2006, No. E.005, TAE 125 series engines, Continental Aerospace Technologies GmbH, Platanenstrasse 14, D-09356 Sankt Egidien, Germany
79. Type-Certificate Data Sheet, October 28, 2004, No. E.014, Centurion 4.0 series engines, Continental Aerospace Technologies GmbH, Platanenstrasse 14, D-09356 Sankt Egidien, Germany
80. Type-Certificate Data Sheet, December 19, 2012, No. E00017AT, Continental TD-300-B engine, Continental Motors Inc., 2039 Broad Street Mobile, Alabama
81. Type-Certificate Data Sheet, June 20, 2017, No. E.104, Centurion 3.0 series engines, Continental Aerospace Technologies GmbH, Platanenstrasse 14, D-09356 Sankt Egidien, Germany
82. Type-Certificate Data Sheet, January 28, 2009, E.200, E4 series engines, Austro Engine GmbH, Rudolf-Diesel-Strasse 11, A-2700 Wiener Neustadt, Austria
83. Type-Certificate Data Sheet, June 10, 2010, E.079, DieselJet TDA CR, DieselJet s.r.l., Via Marabini, 11, 40013 Castel Maggiore (BO), Italy
84. Type-Certificate Data Sheet, December 19, 2014, E.150, RED A03 series engines, Raikhlin Aircraft Engine Developments GmbH, Am alten Wehr 23, D-53518 Adenau, Germany
85. Supplemental Type Certificate SA01303WI, January 31, 2005, Cessna 172 models, Thielert Aircraft Engines GmbH, Platanenstrasse 14, Lichtenstein, Germany
86. Supplemental Type Certificate SA03303AT, July 24, 2006, Piper PA-28-161, Thielert Aircraft Engines GmbH, Platanenstrasse 14, Lichtenstein, Germany
87. Type-Certificate Data Sheet, May 13, 2004, EASA.A.005, DA 42, Diamond Aircraft Industries GmbH, N-A-Otto Strassa 5, A-2700 Wiener Neustadt, Austria
88. Type-Certificate Data Sheet, January 21, 2005, EASA.IM.A.022, DA 40, Diamond Aircraft Industries GmbH, N-A-Otto Strassa 5, A-2700 Wiener Neustadt, Austria
89. Type-Certificate Data Sheet, August 27, 2020, No. EASA.A.639, DA 50, Diamond Aircraft Industries GmbH, Nikolaus-August-Otto Strassa 5, A-2700 Wiener Neustadt, Austria
90. Type-Certificate Data Sheet, October, 2020, EASA.A.576, P2010, Costruzioni Aeronautiche TECNAM S.P.A., Via S. D'acquisto, 62, 80042 Boscotrecase, Napoli, Italy
91. ASTM D1655. Standard specification for aviation turbine fuels. ASTM International, 100 Barr Harbor Drive, PO Box C700, West Conshohocken, PA 19428-2959. United States
92. ASTM D975. Standard specification for diesel fuel oils. ASTM International, 100 Barr Harbor Drive, PO Box C700, West Conshohocken, PA 19428-2959. United States
93. ASTM D613. Standard test method for cetane number of diesel fuel oil. ASTM International, 100 Barr Harbor Drive, PO Box C700, West Conshohocken, PA 19428-2959. United States
94. ASTM D5001. Standard test method for measurement of lubricity of aviation turbine fuels by the ball-on-cylinder lubricity evaluator (BOCLE). ASTM International, 100 Barr Harbor Drive, PO Box C700, West Conshohocken, PA 19428-2959. United States

95. ASTM D6079. Standard test method for evaluating lubricity of diesel fuels by the high-frequency reciprocating rig (HFRR). ASTM International, 100 Barr Harbor Drive, PO Box C700, West Conshohocken, PA 19428-2959. United States
96. Melaranta H (2007) Emergency landing on Porvoo motorway on September 28, 2007, Finnish Accident Investigation Board, Sornas strandvag 33C, 00500 Helsingfors, Finland
97. SAE AS1852 (2012) SAE aerospace standard. Nozzle and ports—gravity fueling interface standards for civil aircraft

Chapter 5
Additives in Aviation Fuels

Dietmar Posselt, Joanna M. Bauldreay, Amiram Groweiss, and Patrick Bosmans

Abstract Aviation fuel specifications set out minimum requirements for a given liquid to be accepted as *fit for purpose* for use in aircraft. These requirements are a combination of compositional restraints, physical properties and performance criteria that, historically, have proven to produce a fuel that is acceptable for Aviation. Additives have been incorporated into the fuel specifications to help address inherent functional and performance limitations with the fuel. Aviation fuel additives, whether deliberately added or present because of crossover from other fuels during storage and transportation, are tightly controlled in jet fuel specifications. The chapter describes additive usage as well as the approval processes involved for aviation additives to be accepted into aviation fuels, focussing on jet fuel applications. Challenges with additive usage are also addressed and ways to understand, prevent or mitigate the issues are discussed.

1 Introduction and Definition of an Additive

This chapter concentrates on jet fuel additives, while Chap. 4—aviation gasoline—covers additives that are specific for aviation gasoline. Some additives are common to both types of fuel, though may be used at slightly different concentrations, and

D. Posselt (✉)
Aviation Fuel Projects Consulting GmbH & Co. KG, Prien, Germany
e-mail: dietmar.posselt@afpconsulting.org

J. M. Bauldreay
Bauldreay Jet Fuel Consulting Limited, Chester, UK
e-mail: Joanna@Bauldreay.com

A. Groweiss
Ashdod Oil Refinery, Ashdod, Israel
e-mail: amiramg@zahav.net.il

P. Bosmans
Central Europe Pipeline System (CEPS) Programme Office (PO), Versailles, France
e-mail: patrick.bosmans@nspa.nato.int

M. Rabaev et al. (eds.), *Aviation Fuel Problems and Solutions: From Refinery to Wingtip*, https://doi.org/10.1007/978-3-032-05514-9_5

where possible they are noted in this chapter. There are common features, some of which are addressed first.

What is a fuel additive? Fuel additives are chemical substances that are specifically added to fuels in a range of a few up to several hundred parts per million (ppm—either by weight or volume) to improve desired qualities or maintain specific fuel properties, meet specification requirements and making the fuel fit-for-purpose. Products which are added to aviation fuels at higher treatment rates, e.g., at percentage levels, are normally called 'components' rather than additives. For instance, Synthetic Blending Components (SBCs) are fuel components, which are blended with petroleum derived fuels, often to create Sustainable Aviation Fuels (SAFs) (see Chap. 3).

Aviation fuel additives are organic chemicals, typically being hydrocarbon-like with small amounts of oxygen, nitrogen and/or sulphur. Metal-containing additives are not approved for use in turbine aviation fuels, i.e., jet fuels, but are used in Avgas. The additive may be a single chemistry, but it can also be a mixture of chemicals, with carrier fluids and diluents, as necessary. See more in Sect. 7.2.

They are purposely added to improve functional and performance properties of a fuel. Additives improve functional properties such as product oxidation stability during storage with antioxidants, or product safety during handling with static dissipator additives. They can address performance properties that impact the attributes of the fuel during use in the airplane (main frame or engine). These include lubricity additives to protect fuel pumps, metal deactivator to enhance fuel thermal stability, fuel system icing inhibitors to diminish the propensity of ice formation in the fuel during flight.

The world of aviation is very conservative to ensure safety of flight. The fuel composition and additives used in aviation are highly controlled. Compared to automotive gasoline and diesel which contain detergents, dispersants, friction modifiers, middle distillate flow improvers, wax anti-settling additives or cetane improvers, aviation turbine fuel has a relatively concise list of allowable additives. Defence Standard 91-091 Annex A [1], List of Qualified Additives and in ASTM D1655 Table 2 [2] have a similar list of allowable additives. These specifications not only mention the allowed additives by their function and by their minimum and maximum concentrations, but in few cases also list the approved manufacturers and brand names. Additives approved in Defence Standard specifications are usually marked as RDE/A/xyz, where x, y, z are numeric figures that are unique to the chemistry or brand name.

Specifically with respect to jet fuels, additives are used in the range of about 1 ppm up to 1500 ppm, depending on the type of additive and the quality requirements of the respective base fuels. Individual additive characteristics and properties of each additive type are described in Sect. 5.

Jet fuels used in other applications, such as diesel engine-powered aircraft, may require additional additives but these are outside the scope of aviation specifications and this book.

2 General Overview of Additive Usage in Relation to Aviation Fuels

In the ideal world, only Original Equipment Manufacturer (OEM) approved aviation additives would be present in jet fuels. However, additives may be introduced at many points in the lifecycle of producing an aviation fuel. There are two categories of additives that while unintended to be present can be considered as trace contaminants in jet fuel. First there are the *unintended materials*. These are non-aviation additives used in drilling, stimulation and production and transport of crude oils, and additives used in the refining processes for converting of crude oil to jet fuel. These materials find their way into jet fuel mainly because they are chemicals that distil in the jet fuel range. The presence and composition of these materials are generally unknown, and they are expected to be in amounts that generally do not impact jet fuel quality.

The second category are termed in the specifications as '*identified incidental materials*'.[1] These tend to be picked up after manufacture, in the distribution systems used for both jet and non-jet fuels. The two listed in the fuel specifications are Fatty Acid Methyl Esters (FAME) used in diesel fuels also known as Bio Diesel, and Drag Reducing Agent (DRA) used in the transport of petroleum products (aviation gasoline is transported in dedicated systems so is less prone to such additives).

Table 1 provides a non-exhaustive list of most classes of additives from production to end use of jet fuel. Some but not all additives are controlled explicitly in the aviation fuel specifications. The extent of that control varies between specifications and can change as specifications are updated. The default assumption is that if the specification does not mention an additive, it is not allowed. However, listing of an additive in the fuel specification (Def Stan 91-091 or ASTM D1655) does not ensure the additive is approved for use in aviation. For an additive to be approved for use it must not only be listed in a national fuel specification, but more importantly it must be listed in OEM documents. It is important that the fuel producer or supplier ensure that the additive they place in the fuel is approved and listed by the major OEMs. The specifications listing should be regularly checked for any new restrictions imposed by the and OEM or changes in the relevant specifications to which they are producing.

2.1 *Upstream of Refineries and Refinery Process Units*

As mentioned in Chap. 2, additives may be present in crude oil production and distribution; they may include surfactants, corrosion inhibitors, drag reducing additives and H_2S scavengers. Additionally, additives are used in various refinery process units upstream of the main kerosene production and upgrading units. ASTM D1655 Section X2.1[2] recognises that traces of these additives, frequently unknown/

[1] Defined in ASTM D1655 as 'chemicals and compositions that have defined upper content limits in an aviation fuel specification but are not approved additives'.

[2] An information section of the specification.

Table 1 Broad classifications of additive usage in aviation turbine fuels

Location	Sublocation	Classes of additives	Specification controls
Upstream of refinery	Drilling	Drilling muds	None. Refinery will treat these as part of the 'crude' and apply processing experience to ensure they do not enter the produced jet fuel
	Fracking	Various polymers and surfactants	
	Production	Demulsifiers, H_2S scavengers, scale inhibitors, tertiary oil recovery additives (usually surfactants)	
	Transportation of crude to refinery	Drag reducing additives (DRAs) H_2S scavengers corrosion additives pour point depressants (usually heavy crudes/winter time)	None. Refinery will treat these as part of the 'crude' and apply processing experience to ensure they do not enter the produced jet fuel. Worth checking potential impact in MOC process
Refinery	Crude tanks	Marketing return additives chemical slops	None, though usage should be challenged in MOC process to avoid complications from performance additives (e.g., dispersants and detergents) found in non-jet fuels
	Distillation unit	Corrosion additives (filming amines)	None, though any changes (amounts and/or additives) should be subject to MOC process
	Jet process units	Antioxidants	Yes
	Jet blending, storage and release	Static dissipator additive (SDA) metal deactivator additive (MDA) Lubricity improver additive (LIA)	Yes. Should not see regular use of MDA at refinery but get to root cause of thermal stability issues and fix
Jet fuel storage and transportation	Pipelines, trucks, trains, ships	DRAs, FAME	DRAs and FAME are not allowed in jet above specified very low levels (coming from other products)
	Terminals	SDA MDA	Both may be added to maximum total levels in specifications

(continued)

Table 1 (continued)

Location	Sublocation	Classes of additives	Specification controls
Airport and airfield	Maintenance	Leak detector, biocide	–
	Performance	Military packages	–
Aircraft	Maintenance	Biocide	Biocides in aircraft tanks, only used during maintenance. Must be flushed out before flights
	Continual use	SDA, LIA, fuel system icing inhibitor (FSII), water scavenger KEROJET® AQUARIUS, performance packages for military	–

Table 2 Identified incidental materials and limits (ASTM D1655 and Def Stan 91-091)

Incidental material	Maximum permitted level	Test method(s)
FAME (as per EN14214 [10] or ASTM D6751 [11])	50 mg/kg (50 ppm); 100 mg/kg is permitted on an emergency basis only	IP 585 [12], ASTM D7797[13]/IP 583 [14], IP 590 [15], IP 599 [16]
DRA/PDR	72 μg/l of active polymer	ASTM D7872 [17]

undisclosed, and/or their degradation products, can cause problems even though they cannot be detected by specification Table 1 tests. It is not possible to identify appropriate analytical detection methods or set 'no harm' levels for these varied and unknown additives. It is, however, incumbent on each manufacturing location at least to control known additive usage and to manage changes of crudes (and associated unknown additives) through the implementation of robust management of change (MOC) processes.

The refiner may not control the use of process additives directly, having a third-party additive company (AddCo) oversee their usage to achieve agreed unit performance criteria. It is, however, the refiner's responsibility to ensure that the additives used are approved and listed by the OEMs and in the specification to which they are produced, and the used additives do not cause problems for the jet fuel produced. They must therefore ensure that the AddCo is an active participant in any MOC process that should be undertaken if there is a change in levels or types of additives being used.

Another potential source of unidentified additives is in the slops that get processed with crude oil at a refinery. These can be from chemical units (so should NOT be used) or off-specification products. Refineries may also process 'marketing returns', which include fuel cuts from pipeline operations and off-spec fuel from terminals and road fuel forecourts. Most road fuels contain additives that are not permitted in jet fuel and are used at significantly higher levels than in jet fuel. The refinery must decide through a risk-based MOC process if any of the additives, now significantly diluted, have the potential, as is or after degradation, to cause issues in jet fuel.

2.2 Jet Fuel Production

Specifications concentrate on the deliberate use of additives during the final production and the blending of one or more kerosene streams to create jet fuel. They vary in whether they mandate the usage of specific additives but generally align on the dosage levels and the additives (named or chemistries) that may be used; see more in Sect. 3.2, which also addresses reporting requirements on the appropriate quality certificate, such as a Certificate of Quality (COQ) or Release Certificate (RC).

These intended additives are used to prevent problems in transit to the airport and in the aircraft fuel system. In civil fuels the additives are usually limited to antioxidants, static dissipator additives and metal deactivator additives (though not on a continual basis). More detail on why and where these additives are introduced are given in Chap. 2.

2.3 Jet Fuel Storage and Transportation

Aviation fuel is expected to meet specification both at the point of jet fuel production and at the wing tip, when delivered to the customer. However, jet fuel specifications do not and cannot define all the quality assurance testing and handling procedures needed to maintain the quality of the fuel throughout the distribution system. Industry standards have developed to provide guidance for jet fuel handling, which cover more than keeping the fuel 'clear and bright and clear from particulates and free water'. Many are listed in the specifications and are increasingly being written into fuel purchasing contracts. The most notable of these are ICAO 9977 [3], EI/JIG 1530 [4], EI 1533 [5], JIG 1[6], JIG 2 [7], API 1543 [8] and API RP 1595 [9]. They cover 'production to pre-airfield', 'airfield to wingtip' or specific subsections of those storage/transportation stages in the life of a jet fuel. Changes in these fuel handling systems are subject to risk and management of change assessments.

After manufacture, usually at a refinery, jet fuel is stored and moved in various systems. Some of the permitted additives may need to be topped up after the refinery release, usually at a terminal or an airport tank, to maintain specific fuel properties. Specifications concentrate on SDAs and MDAs in this respect, with a limit on the total loading of the given additive type.

Pipelines are an important part of the distribution system. Their operators control jet fuel product quality by a variety of means, depending on the type of pipelines and related storage systems involved. Close to airports, dedicated pipelines only carry jet fuel but earlier in the transportation process the jet fuel may move in multi-product pipelines (MPPL). These transport jet fuel but also transport fuel products like gasoline, road diesel and heating fuels.

In North America, particularly in the US, jet parcels in MPPLs are frequently a mixture of separate batches all conforming to ASTM D1655. The system is termed *fungible* because of lack of traceability of each batch of fuel and allowance of mixtures of batches of jet fuel during transport and storage. Outside this region jet fuel batches tend to be discrete, with properties and additisation history more readily traceable back to the point of manufacture.

Non-jet additives may be found in jet fuels passing through these systems, largely resulting from fuel movements in MPPLs. Different grades of fuel are carefully segregated, and product sequencing is an important part of the controls, but additional measures are required. Operations aim to prevent cross contamination of fuel grades, including the 'trail back' of additives from products that lead the jet fuel cargo. There is always some risk of crossover of low levels of additives that have the potential to contaminate the jet fuel. They can be surface active or non-surface active and include those classified in aviation specifications as 'identified incidental materials' (DRA and Bio Diesel, containing FAME) as well as more general performance additives needed in other products, in particular diesel fuels. See Table 2.

Pipeline and terminal operators must have control systems and management of change procedures to ensure that the levels of these identified incidental additives are kept below the levels set out in the jet fuel specifications. Testing is only required at points in the system identified as having a potential risk of incidental materials being present, and any mitigating actions must be taken immediately because airports are not equipped to deal with such contaminants.

Non-surface-active additives normally can be satisfactorily controlled by interface cutting and additive injection control. Additives like DRA and dyes (such as in heating oil) are detrimental to jet fuel because of their chemistry (see more in Sect. 8) and where possible should be excluded for use in MPPLs. If exclusion cannot be avoided, products containing non-approved additives shall be separated from jet fuel by a buffer of acceptable fuel or the injection of those additives into the product preceding jet fuel shall be stopped sufficiently in advance of the jet fuel batch.

If a batch of jet fuel is to follow a non-jet batch containing DRA, the DRA injection shall be stopped before the end of the non-jet fuel batch (e.g., at least 500 m^3 for an 8″ and 12″ line). If a jet fuel batch precedes a batch that will be doped with DRA, the injection shall start a few m^3 after the beginning of the non-jet fuel batch (e.g., at least 50 m^3 for an 8″ and 12″ line). Before doing so, a risk assessment shall be done, typically comparing full specification analysis on both neat and suitably doped jet fuel samples (e.g., 5% of maximum additive dose rate recommended by the DRA manufacturer), possibly backed up with pipeline tests using method ASTM D7872 to test for DRA.

Note: There is a strong and increasing commercial incentive to use DRAs in MPPLs. Adding 15 ppm to a fuel allowing its usage could increase the flow rate up to 70% in a pipeline, potentially reducing electricity bills (and CO_2 footprint) for pumping but more importantly allowing more product to pass through the system. Numerous major pipelines are running close to capacity which may impact the delivery of jet fuel and other fuels in required volumes.

A 15-ppm limit is often imposed for pipelines carrying motor gasoline as the large DRA molecules can lead to the formation of deposits inside internal combustion engines (ICE), negatively impacting engine performance. One option to overcome this problem is the inclusion of deposit control additives (DCA).

Surface active additives must be listed in EI 1535 [18] before acceptance in any MPPL. Only a limited number of diesel performance additives are admitted in pipelines that transport jet fuel. These surface-active additives may by adsorption/desorption on the pipeline walls result in trail back of the unapproved additive into jet cargos. A list of allowed non-jet additives in MPPL is contained in EI 1535 protocol, the majority of which have gone through the EI 1535 protocol. However, some of the listed additives are 'grandfathered' in as they have field history of use without reported incident predating the EI 1535 protocol.

2.4 Military Versus Civil Systems

Pipelines mainly move civil fuels, but some systems, including those operated by NATO in Europe, also carry fuel for military customers. For instance, the Central European Pipeline System (CEPS) transports (not just during a crisis or in times of war) fully formulated military jet fuel meeting F-34 requirements in some of its spur lines; several CEPS depots routinely store F-34 for airbases. In this way, its Jet A-1 banking system can be most easily managed.

Most of the time Jet A-1 is turned into F-34 at the connected military airbases by remote injection systems that introduces LIA and FSII at a fixed ratio. However, LIA and FSII are forbidden to be in the Jet A-1 at the ingress to the CEPS system to avoid possible overdosing when making F-34. An overdose of LIA can negatively impact MSEP results as well as the filtration downstream. This ban on LIA usage may become an issue for SAF blenders who need to use LIAs to get their blends on specification for lubricity. Also, not all filter elements, such as water separation/coalescing systems including the filter vessels, are compatible with military jet fuel containing FSII. Where the FWS is exposed to fuel containing FSII, only category M elements shall be used.

Another feature of the military systems is that most lined tanks shall not be used for F-34 storage because FSII is incompatible with tank liners. FSII resistant epoxy coatings such as Tankard Special Ultra may be suitable (some military aircraft have seen FSII-related material degradation of wing tank coatings and linings).

2.5 Airport Operations

Additive usage at or near to airports and military airfields is generally controlled by the purpose of the additive, dosing capability of the location and the possible impact of the additive on fuel handling and transport system.

Additives, which are injected at the point of fuelling onto the aircraft, include: FSII (due to requirement of shear to disperse, and water pick up during storage), military additive packages, and lubricity additives (impact on fuel storage and transport systems as well as filtration, and water scavenging additives (not approved for all aviation equipment).

Leak detector additives are used to ensure that hydrant systems and tanks are not leaking fuel. Fuel leakage would compromise airfield safety and result in undesirable environment impact. Fuel treated with leak detection additive should not be uplifted onto aircraft.

Biocides may be used in fuelling facilities and in aircraft fuel systems, in a manner discussed in more detail in Chap. 12, to limit potential microbiological contamination issues including but not limited to blocked filters, impacting aircraft gauges and resulting in microbial induced corrosion.

3 Additive Usage and Specification Requirements

3.1 Additives in Use and History of Their Introduction into Specifications

The first and oldest fuel additive was discovered in 1921: Tetraethyl Lead. It was further improved and introduced to the market in the 1920s for the use *inter alia* in aviation fuels. At that time, only aviation gasoline was used in aviation to power airplanes with spark-ignited piston engines. Gasoline qualities in aviation and road transportation with quite low octane ratings often suffered from poor combustion characteristics due to severe engine knocking. Tetraethyl Lead in combination with ethylene dibromide scavengers helped effectively to overcome that problem. A first specification for leaded aviation gasoline was defined by the US Army Air Corps in 1930 with a maximum limit of 3 ml/US gal Tetraethyl Lead. This additive is still in use within many aviation gasoline grades, though there is considerable effort to find a suitable octane improving additive with better HSSE credentials.

Turbine aviation fuels also have a long tradition regarding the application and use of additives. The introduction and more frequent use of cracking refinery processes produced more reactive species in the fuel such as olefins, leading to fuel oxidation and gum formation. This was the birth of antioxidant developments in the 1930s to combat the tendency of fuel degradation. Another example for an additive introduction in the mid of last century was the need to protect fuel wetted steel tanks and pipes from corrosion. With the knowledge from the lubricant industry, fuel specific corrosion inhibitors (known as CI/LI) were also developed and launched for the aviation industry. Thus, one after another various types of additives have found their way into civil and military aviation fuel specifications.

These specifications have incorporated fuel additives to maintain or even to improve fuel performance characteristics and to increase safety aspects regarding fuel storage and transport.

Proper additisation, using the right additive at the right dosage, has always been seen as the most cost-effective solution for the industry as compared to a complete re-formulation of a base fuel.

3.2 Additives Used in the Main Specifications

The use of additives in aviation fuels is carefully controlled and limited because of the potential for undesirable side effects. This is strictly reflected in all relevant aviation fuel specifications. Under certain circumstances additives may affect the ability to maintain fuel cleanliness during shipment and handling or may adversely impact fuel system and turbine engine operation and maintenance. Therefore, only qualified additives of defined composition and amount which are approved and listed by the airframe and engine manufacturers and cited by the relevant fuel specification authority may be used. The strict control over the use of additives is clearly expressed by the classification in the following categories:

- **Mandatory** additives—Shall be present between defined minimum and maximum concentration or property limits.
- **Optional** additives—May be added up to a maximum concentration or property limits.
- **Additives by agreement**—May be added only with agreement of the user/purchaser within specified limits.

All other additives must not be used in aviation fuels. This also means that refineries and distributors must ensure that the fuel cannot pick up any intentionally added additives used for refinery processes or from other products carried in the pipeline.

Whether an additive is classed as, e.g., mandatory or optional, depends on the specification. New additives can appear in specifications (and/or qualified product lists) after they have been officially approved (please also refer to Sect. 4). Generally, more additives are used in military jet fuels than in civil. Table 3 gives an overview of the known jet fuel additive types, and their classifications related to some major global aviation fuel specifications. **The latest version of any specification shall always be used to ensure correct additive usage and reporting**. Those additive classes that may also be used in aviation gasoline are marked with an asterix *. The Avgas specifications should be checked for other permitted chemistries (CI/LIs) and for the mandated use of Tetraethyl Lead/ethylene bromide and dyes needed in most Avgas grades. Chapter 4 provides more information on aviation gasoline additives.

Table 3 Overview of aviation fuel additives in major civil and military jet fuel specifications

Additive type	Jet A/Jet A-1 ASTM D1655	Jet A-1 DEF STAN 91-091	JP-5/F-44 MIL-DTL 5624 [19]	JP-8/F-34 MIL-DTL 83133 [20]
Explicit reporting on CoQ	No[a]	Mandatory	By agreement[b]	By agreement[c]
Antioxidant*	Optional	Optional[d]	Mandatory	Optional[e]
Metal deactivator	Optional	Optional	Agreement	Agreement
Static dissipator/ electrical conductivity improver*	Optional	Mandatory	Agreement	Mandatory
Corrosion inhibitor/lubricity improver*	Optional	Optional	Mandatory	Mandatory
Fuel system icing inhibitor*	By agreement	By agreement	Mandatory	Mandatory
Biocide	By agreement	Not listed	Not listed	Not allowed
Thermal stability improver	Not allowed	Not allowed	Not allowed	Not allowed[3]
Water scavenger	Agreement	Not allowed	Not allowed	Not allowed

[a]The quantity of FSII or biocide (not at point of production) shall be declared by the fuel supplier and usage must be agreed by the purchaser
[b]If specified in the acquisition documents, information concerning the type and amount of each additive used shall be reported
[c]The type and amount of each additive used shall be made available when requested by procuring activity or user, so including details on CoQ helps to fulfil this requirement
[d]Previously mandatory for hydroprocessed components. Now only mandatory in fuels with synthetic components
[e]Previously mandatory for hydroprocessed components. May still be required in synthetic components to be blended to make F-34

3.3 Additive Reporting Requirements

In principle, in Def Stan 91-091 aviation fuel additives shall be identified by their appropriate RDE/A/XXX numbers. In ASTM D1655, they are defined in Table 2. A proper documentation of using the right additive products in combination with the right amounts is essential and required. This means that refineries and other storage facilities at which additives are injected shall have a system in place that ensures that only approved additives[4] are used and that the correct dosage rates are applied. A

[3] Removed from list of additives in 2024 edition of MIL-DL-83133.

[4] While OEMs play a key role in the approval of additives, the OEMs do not immediately approve new additives listed in the specifications for their equipment; it may take over a year for the OEM customer bulletins/specification to list the newer additives. This has been the case with AvGuard SDA.

Table 4 Example of additive reporting

Antioxidant	RDE/A/608	20 mg/l
SDA	RDE/A/621	1.1 mg/l

method of additive addition shall be covered by detailed on-site procedures such as control of quantity and type of material used as well as timely monitoring of volume used to confirm the final addition rate.

Additionally, a system of additive batch recording shall be in place to allow traceability of additive batches in the finished aviation fuel. These records must be managed as follows: refineries or other manufacturing locations making jet fuel to meet Def Stan 91-091 shall include the relevant information on their Certificate of Quality (CoQ) or Release Certificates (RCs). Specifically:

> Additives shall be identified by the appropriate RDE/A/XXX or RDE/A/DSFA/XXX number as shown in Annex A (of the specification). The amount of all additive additions shall be reported to the purchaser on batch quality certificates or as otherwise directed by the purchaser and/or contract'

For instance, this could be as Table 4.

Where re-batching fuels at other supply installations, the amount of additive product added shall be recorded on the Certificate of Analysis (CoA) or Recertification or Release Certificate as required. When additive is injected as fuel is dispatched from an installation, the amount added shall be recorded on the Release Certificate.

4 Approval Processes for Aviation Additives

4.1 Historical Approach

Historically, approving a new aviation fuel additive was a very lengthy process, often overseen by military laboratories, and commonly taking up to a decade to reach completion. Timelines and costs deterred additive companies from proposing new candidate additives. Consequently, most aviation fuel additive types have been in use for decades and no compositional changes to petroleum-derived fuels were made for many years. This situation changed in the 1990s when the US Air Force in close cooperation with a team of jet engine and aircraft OEMs developed and introduced a thermal stability improving additive for use in selected military airplanes.

All the experience from that first approval work was used to conduct the OEM approval evaluation for some additional thermal stability improvers in the 2004–2010 period. These years were the birth of the guidelines for the evaluation of new jet fuels and additives.

The first fuels-containing version of the protocol designated as ASTM D4054 was issued in 2009 [21]. This early ASTM D4054 was mainly focused on requirements for evaluation new fuels. The extensive document (over 40 pages) had little coverage (two paragraphs) focused on evaluation of additives. The subsequently the additive evaluation protocol was included in ASTM D4054 and was first published in 2014. Commercial and military jet engines rely on certified aviation fuels and whenever a new additive product is intended to be introduced the additive proponents must follow ASTM D4054—Standard Practice for Evaluation of New Aviation Turbine Fuels and Fuel Additives. This practice is under the jurisdiction of ASTM Committee D02 on Petroleum Products, Liquid Fuels, and Lubricants and is the direct responsibility of Subcommittee D02.J0.04 on Additives and Electrical Properties. Since then and until today, a static dissipator/electrical conductivity improving type of additive as well as a water scavenging type of additive have been evaluated using D4054 guidelines. The following Sect. 4.2 presents an overview about the principles of ASTM D4054.

4.2 ASTM D4054—Overview and Principles

ASTM D4054 Standard Practice for Evaluation of New Aviation Turbine Fuels and Fuel Additives covers and provides a framework for the OEM evaluation and approval of new aviation fuel additives for use in commercial and military aviation gas turbine engines. This practice was developed as a guide by the aviation gas-turbine engine OEMs in collaboration with ASTM International member support. The OEMs are solely responsible for approval of an additive in their respective engines and airframes.

Standards organisations such as ASTM International, United Kingdom Ministry of Defence and the US Military try to list only those additives that are mutually acceptable to all OEMs.

The OEMs will consider a new aviation fuel additive based on an established need or benefit attributed to its use. Airline advocacy for the candidate fuel additive can be important to warrant consideration for qualification. The OEMs need strong support from the airlines to justify committing internal resources to evaluating a new fuel additive for use in an aircraft. Interested airlines or other users (e.g., US Military and air cargo) must submit formal written requests to the OEM customer service groups expressing a need and requesting that the additive be evaluated for OEM qualification and approval. Military participation in the evaluation process is important because many commercial engines have military derivatives.

The additive will be considered approved for aviation use after completion of ASTM D4054 protocol, balloting of the inclusion of the additive into ASTM D1655 (and subsequently into Def Stan 91-091, US Air Force MIL-DTL-83133), and most importantly approval and listing of the additive in OEM service bulletins and manuals. While some additives are listed in ASTM D1655 and in Def Stan 91-091, they are not universally listed by all the major OEMs and thus are not considered to be approved for use in aviation fuels.

The whole evaluation and approval process has been coordinated with airworthiness and certification groups within each OEM, the Federal Aviation Administration (FAA) and the European Aviation Safety Agency (EASA).

One of the goals of ASTM D4054 guidelines is to guide a producer and offer an evaluation and approval route that is cost effective and completable in a timely manner. This practice also provides a basis for calculating the volume of additive needed for evaluation and gives insight into the cost associated with taking a new fuel additive through the evaluation process. The protocol offers a clear path forward for introducing a new technology for the benefit of the aviation community.

A schematic overview of the approval process is given in Fig. 1. The purpose of the test programme is to ensure that the candidate additive will have no negative impact on engine safety, durability or performance. This is accomplished by investigating the impact of the candidate additive on fuel specification properties, fit-for-purpose properties, component rig tests or engine tests. Figure 2 lists elements of the test programme and should be considered as a guideline. It is unlikely that all the tests shown in Fig. 2 will need to be performed. OEMs provide guidance on the required tests and applicability will be based on chemical composition of the new fuel additive, similarity to other approved additives and engine/airframe manufacturer experience. Departure from engine/airframe manufacturer experience on additive composition and use requires more rigorous testing.

The additive evaluation section of D4054 has protocols that are in addition to testing required for evaluations of new fuels. Specifically, ASTM D4054 requires the additive to be tested for *performance for its intended function* (related to additive purpose in the fuel) and no harm/impact on the performance of existing approved additives.

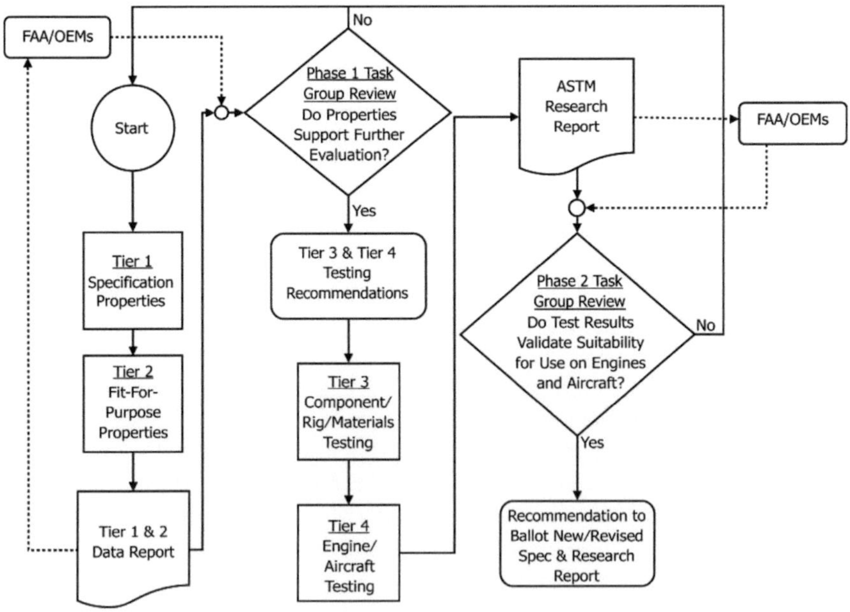

Fig. 1 Overview of fuel and additive approval process (Copyright ASTM; used with permission)

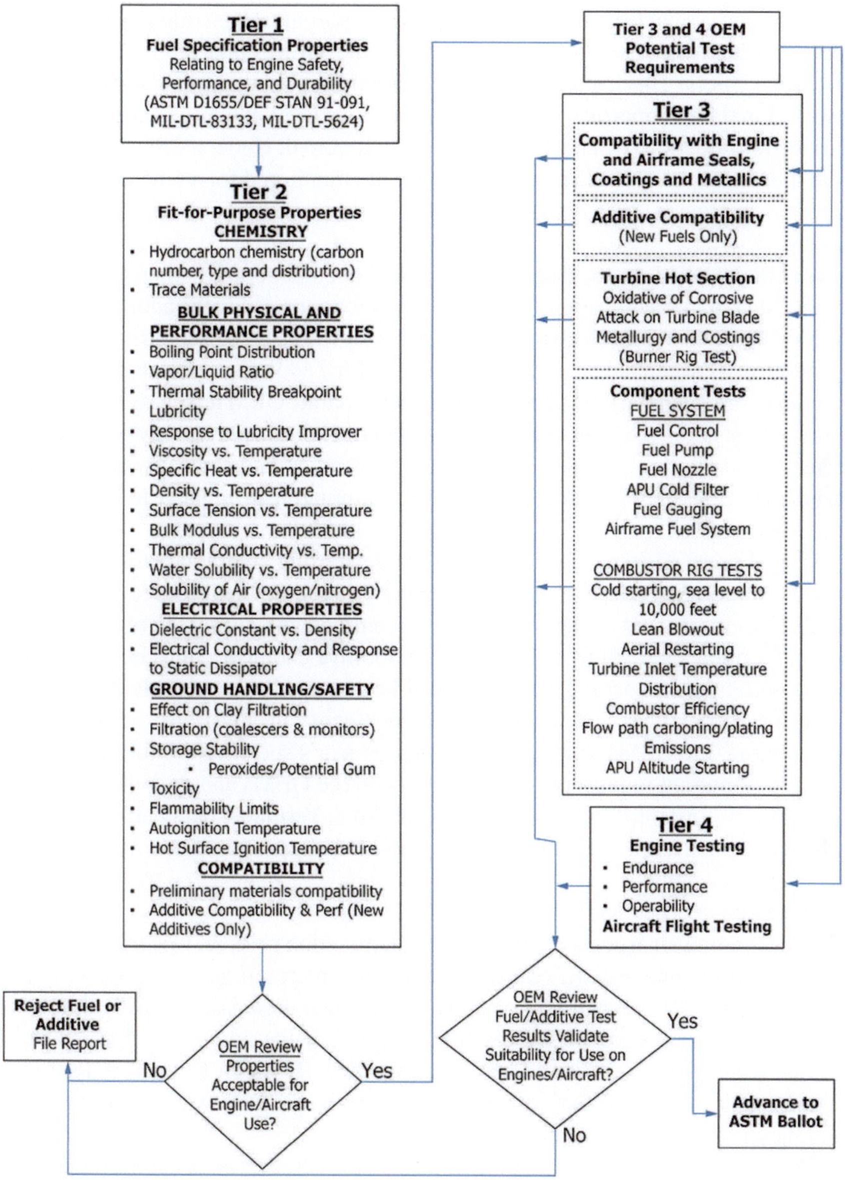

Fig. 2 Additive approval process outlined in ASTM D4054 (Copyright ASTM; used with permission).

At the end of that test programme a research report must be written by the additive sponsor which then is carefully reviewed by the OEM chief engineers and their discipline chiefs. This exercise can result in requests for additional information and testing. Final OEM approval is made at the executive level. It may be that a Controlled Service Introduction (CSI) will be additionally required. Under a CSI, engines in the field that are exposed to the new fuel additive are monitored for an increased level of fair wear and tear. The CSI is directed at identifying possible long-term maintenance effects.

An official OEM approval and listing in OEM documents and listing of the additive in ASTM D1655 specification are both required for approval of the additive for use in aviation but follow separate pathways. Changes to ASTM D1655 could include listing the additive as acceptable for use, changes to published limits, special restrictions or additional precautions. OEM approval requires separate OEM review and subsequent listing of the additive in OEM documents. Figure 2 gives an overview of the ASTM review and balloting process, which is quite rigorous and typically goes through several iterations before a ballot is successful, culminating in a change to specification ASTM D1655.

Several prerequisites must be fulfilled before starting an official test programme. A new aviation fuel additive formulation must be well defined prior to approaching ASTM; compositional changes cannot be accommodated during the review process without written approval by the OEMs. The additive chemistry must be provided. If it is proprietary, a generic description shall be provided to the OEMs. Information about the carrier fluid and recommended treatment rate must also be mentioned. Furthermore, the OEMs shall agree upon a producer's specification for the additive. It is also important to present a technical case to the OEMs and the relevant ASTM Subcommittee establishing a need for the additive. Verifiable data performed by an industry-recognised laboratory shall be presented supporting performance for the specified application.

The additive proponent seeking permission to use the new fuel additive is responsible for funding all aspects of the additive qualification process. Costs include laboratory, rig or engine tests (if required) as well as interpreting, communicating and reporting data. The US Military or other government organisations will sometimes participate in a Cooperative Research Programme if the additive is deemed to be of significant benefit to the military. Coordinating Research Council (CRC) may organise task forces and may solicit its members to perform work using available funding within their organisations.

The test programme for evaluating impact on mainframes and engine components is comprised of four tiers.

- Fuel specification properties
- Fit-for-purpose properties
- Component, rig and compatibility tests
- Engine test.

Each tier consists of a distinct set of tests focused on critical consideration that impacts engine and airplane design, safety, durability, performance and reliability.

In most cases, testing of a fuel additive candidate shall be performed at four times (4×) the concentration being requested for qualification. This allows for evaluation of possible overtreats during field operations and gives more flexibility in selecting the baseline fuel. Fuels can vary in their sensitivity to a particular additive. The testing at 4× negates the need to spend resources searching for a sensitive fuel for use as the baseline test fuel.

Another required section of the approval related work are evaluations to investigate:

- Performance of an additive's intended function and
- its compatibility with additives which are currently permitted in ASTM D1655.

The 'performance' section is geared to ensure the additive enhances or corrects the fuel property for which it is being added to the fuel. The 'compatibility' evaluation encompasses testing to evaluate physical properties of the additive to including solubility of the additive in fuels, and the propensity for adverse interaction between the candidate additive and the currently approved additives.

The specific additive task force, the OEMs or the Sub J committee as a whole may with technical justification request additional other test(s) to be performed or other requirements incorporated into the qualification process. There may also be instances where testing not detailed in ASTM D4054 is required. Examples include an additive with a completely new function or chemistry, or where specific concerns regarding the additive impact on unique engine or airplane designs features.

The process seems, and is, lengthy, but the ASTM D4054 updates in the last decade mean that it is becoming much quicker and clearer to all parties involved than historically used processes for additive evaluation.

Note: Additives for aviation gasoline go through a similar evaluation process, ASTM D7826 [22].

4.3 *Other Considerations*

Additives can be composed of a single chemical or a blend of multiple chemicals, with or without solvents and carrier fluids to allow them to be stored and readily used in the intended applications.

Additives procurement shall be accompanied by the supplier's quality certificate and MSDS. The markings on the containers shall be compared and correspond with the information on the supplier's quality certificate and delivery papers. Every container shall be examined for damage or possible contamination during transit. If a container is leaking, it shall not be used.

As well as ensuring that the correct level of additive is used, anyone handling the additive during blending or addition should take precautions that deal with the risks highlighted in the safety data sheet (SDS) that accompany any delivery. The SDS will provide a generic description of the additive, referring both to CAS numbers and to REACH (or similar) registration details for the main components.

Note that the obligations with REACH or similar registrations are extensive and expensive. On top of the additive approval processes, it is one of the factors that prevents the development of many new additives for aviation fuels market.

4.4 Additive Package Reformulation

From time to time, additive packages may need to be reformulated. This means that a slightly different package of active ingredient, solvent and carrier could be created (see more about additive formulations in Sect. 7.2). Even if the change is just the solvent or carrier, the additive supplier cannot make the change without first contacting the necessary specification bodies, who may require that the additive be submitted for reapproval.

5 Currently Approved Additives

Different classes of additives have distinct functions and depending on their composition may require different handling, storage and dosing. The next section builds on the general classifications discussed in Sect. 2, to explain the functions of the different classes and identify specific additive chemistries for the various functions. It focusses on the additives that appear in 'Western' civil and military specifications, highlighting where rules of usage differ, and briefly discusses the additive situation for less well-known specifications issued by Russia and China.

6 Western Specifications

'Western' is a shorthand way to describe major specifications produced by ASTM and the UK's Ministry of Defence, e.g., ASTM D1655 and Def Stan 91-091, and all those closely related specifications used in Europe, the Americas, Africa and Middle East, Asia and much of Oceania.

6.1 Civil Fuels

6.1.1 Antioxidants

During storage and handling jet fuel is in contact with air where oxidation of fuel components is a possibility. As oxidation proceeds, radical species and peroxides are formed. Peroxides remain dissolved in the product and may attack and shorten the life of some elastomers in the aircraft and fuel systems. Further the peroxides may carry out subsequent reactions in the fuel leading to darkening and increased viscosity of the jet fuel due to the formation of soluble gums and the build-up of polymeric, fuel-insoluble particles. The resulting oxidation products may clog fuel fine filters and deposit on the surfaces of aircraft fuel systems, restricting flow in small-diameter passageways. The oxidation process may also form organic acids that are corrosive to fuel handling and storage as well as aircraft components.

These instability reactions occur faster, and to a greater extent, at higher ambient temperatures. However, fuel oxidation is not considered to be a major problem in aviation as jet fuel that has been properly manufactured, stored and handled should remain stable for many years.

Antioxidants, whether naturally present or added during production, delay peroxide formation and the subsequent chemistries of oxidation like formation of gums or organic acids during long-term storage. Petroleum-based distilled and wet treated kerosene resists oxidation by air due to the presence of sulphur-, oxygen- and nitrogen-containing compounds (hetero atom containing compounds) from petroleum that act like natural antioxidants, so generally the use of external antioxidants in those fuels is not necessary. However, the use of hydrogen in high pressure and high temperature refinery processes like hydrodesulphurisation and especially in hydrocracking processes eliminate those hetero atom containing compounds and may leave the kerosene unprotected against oxidation. The use of synthetic antioxidant additive products is, therefore, allowed though not mandatory, and is usually limited to the fraction of kerosene that undergoes hydrotreatment. The specification standards Def Stan 91-091 and ASTM D1655 allow the use of commercial synthetic antioxidants that are known as 'butylated phenols' and butylated hydroxy toluenes (BHT) see Fig. 3). Six chemistries or mixtures of chemistries are listed. All are aromatic compounds, which include phenolic groups (aromatic O–H), tert-butyl substituent(s) and methyl ($-CH_3$). This kind of chemical functionality allows the interference with the radical oxidation mechanism and delays the oxidation process.

Historically, amine-type antioxidants were listed in jet specifications (and they may remain in some Avgas specifications, which is where they were first used) but were removed from most jet fuel specification by the early 1980s, possibly because of issues with colour instability, and/or precipitate formation following antioxidant polymerisation (see Sect. 2.3, Chap. 7).

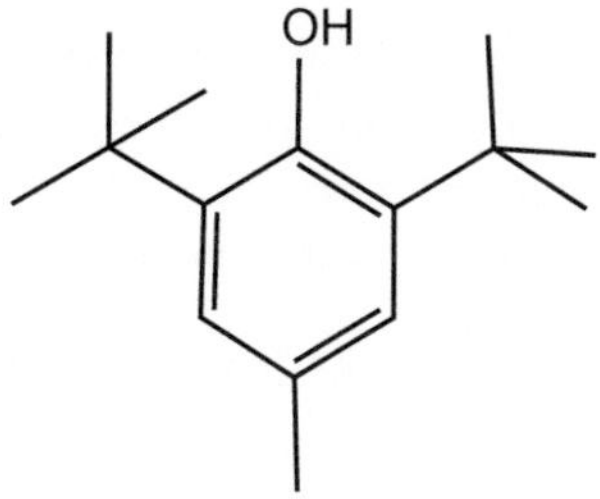

Fig. 3 Chemical formula of BHT

Approved antioxidants are used in the range of 17–24 ppm (g/m^3). The antioxidant additives can either be introduced as a pure compound or as mixture of various butylated isomeric compounds. Antioxidant additives may be pumped into the fuel at various points during production or storage, but the most recommended point is immediately after leaving the hydrogenation plant before entering the finished jet fuel tank. The proper use of antioxidants in jet fuels thus prevents gum formation, sedimentation and discolouration and improves the storage stability of aviation fuels.

There is no obligatory test method for the content of the additive in the fuel, so the exact amount should be calculated from the rate of addition at the metering pump.

6.1.2 Static Dissipator Additives

Static dissipator additives (SDAs) are intimately linked to fuel conductivity. Kerosene is a flammable and combustible liquid that consists of very poorly conducting hydrocarbon compounds. As produced in the refinery the electrical conductivities of hydrocarbon fuels are extremely low, close to 0 pS/m.

The risk associated with SDA-free fuels with low electrical conductivity/high resistivity is their low potential to dissipate any electrical charges that can build up because of friction during movements. Pipe joints, elbows, diameter reducers and especially filters contribute to increased friction. Fine filters (up to 25 μm) can give very high electrostatic charging owing to their very large surface area compared with pipes and fittings.

If the fuel conductivity is inadequate to transfer charge there is the potential for a fire or explosion to occur if the in an accumulation of generated charge in flowing aviation fuels which results in a spark or electrostatic discharges that can ignite flammable fuel/air mixtures, particularly if the fuel's flash point is near or lower than ambient temperature.

The mechanism of Static Discharge Ignition (SDI) can be briefly described as follows:

> Dissociated ions are formed in fast flowing fuels. One charge (e.g., +) is selectively adsorbed at the inner surface of the hose while the remaining charges (e.g., −) is carried by the flowing fuel as a 'net charge'. This can lead to a potentially hazardous situation when the liquid with the 'net charge' only relaxes slowly and is discharged into a receiving tank that is electrically isolated, and flammable vapours exist at the same time. The resulting relaxation time for discharge might be then too short resulting in an increased risk of Static Discharge Ignition.

To diminish electrostatic problems if no SDA is present in the fuel, the filters should be located several minutes flow time upstream of the loading point. The engineering requirement is generally difficult to achieve due to site design constraints and is easily alleviated using SDAs to reduce the relaxation time from minutes to seconds.

The risk of SDI is minimised by using SDAs which increase the conductivity of the fuel to impart greater ion mobility in the fluid and thus allowing the fuel to relax faster. That is the reason SDAs are often referred to as 'conductivity improvers' in the fuel industry.

Def Stan 91-091 demands that the fuel supplied to aircraft will have an electrical conductivity between 50 and 600 pS/m (pico-Siemens per meter), as measured by ASTM D2624 [23] or IP 274 [24]. There are two listed additives in ASTM D1655 and Def Stan 91-091: Stadis® 450 from Innospec LLC and AvGuard SDA from Afton Chemical Ltd. The maximum allowed initial concentration of SDA is 3.0 ppm (mg/L) though a target of closer to 1.0 mg/l is generally adequate to meet specification conductivity requirements. The additive level may be increased further downstream if losses in conductivity are observed in the aviation fuel infrastructure, but the cumulative level shall not exceed 5.0 ppm (mg/L).

In a clean jet fuel, the conductivity response will generally be linearly related to concentration of SDA: if 1.0 mg/l gives a conductivity of 150 pS/m, then 2.0 mg/l will commonly give 300 pS/m. Since the treat rate is very low, dilution may be necessary to bring the injection rate within the capability of the pump used to inject the additive. The amount of additive added must be recorded on a certificate at entry into a logistics system. Any additional doping downstream shall also be properly recorded on the certificate.

Static dissipator additives are composed of a mixture of chemistries that are proprietary to the additive producers. SDAs are very stable as neat additives and generally have a storage life of 10+ years.

Fluids treated with SDA are evaluated for conductivity response (the pS/m derived by ppm of additive) and conductivity retention (how long the initial conductivity goes unchanged). Both conductivity response and retention are highly sensitive to fuel impurities. The response is also dependent on fluid composition, and more importantly the temperature of the fluid. A clean (e.g., hydrotreated Jet) will give 100–200 pS/m for 1 mg/l of SDA measured at Room Temperature and will only lose 10% of initial conductivity in 3–4 weeks of storage. There are, however, other factors that impact conductivity retention such as the type of container used to store the treated fuel (preference is Pyrex/non soda glass bottles), the limitation of exposure to sunlight (preference is fuel is stored in the dark or use of amber bottles or in bottles covered with aluminium foil).

SDA chemistry can behave as a weak surfactant and can impact water separation properties of a fuel (as measured by micro-separometer instrument according to ASTM D3948 [25] or D7224 [26]). Global jet fuel specifications take into consideration this impact and set lower MSEP limits for SDA-treated fuels.

Most fuel supplied to Def Stan 91-091 has SDA added at the point of manufacture, but where and when to add SDA can be changed based on producer/purchaser contractual agreements. Risk assessments may define when SDA is necessary.

Jet fuel may be pumped, under proper notice, without the additive and therefore at a conductivity that does not meet the specification, with SDA being added later in the supply chain, closer to end use at the airport. Regardless of when additised, Def Stan 91-091 requires 50 pS/m conductivity in fuel delivered to the aircraft.

Fuels with sufficient electrical conductivity should lead to a relatively fast relaxation of existing charges before the fuel ends up in an isolated vessel. The proper use of static dissipator additives (SDA) does not prevent static charge generation

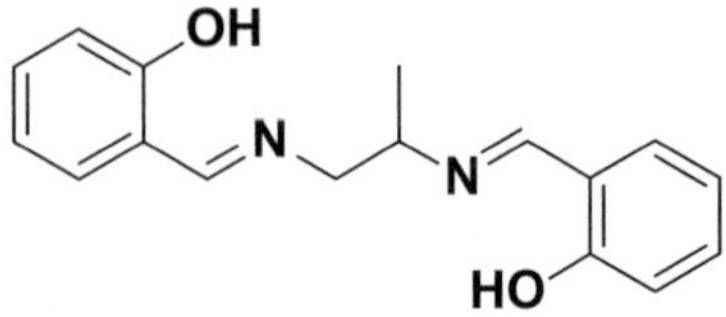

Fig. 4 Chemical formula of MDA

but rather enables rapid dissipation of electrostatic charge generated during fuel movement, diminishing the risks associated with fuel handling and transport.

6.1.3 Metal Deactivator Additives

One of the most important demands on jet fuel is for the fuel to be thermally stable. Historically accepted chemical reasons for failing the thermal stability test (ASTM D3241 [27] or IP 323 [28]) include the presence of cracked or olefinic compounds, or traces of metals that catalyse fuel thermal degradation. There have also been studies showing that trace non-metallic unidentified impurities or chemicals from crude production processing can also impact jet fuel thermal stability.

The presence of olefinic compounds can be minimised by hydrotreatment of the fuel. Kerosene products like Jet A, Jet A-1, F-44[5] or F-34 may, however, be exposed to and pick-up traces of metals during storage and transport. These metals may enter the fuel through corrosion of unlined tanks (especially if exposed to brine), pumps or hoses, from residual catalysts from the Merox™ process or introduced during a copper sweetening process to remove mercaptans, or other catalysts (e.g., from hydrodesulphurisation).

Copper, being the most problematic metal, may be solubilised in a nascent form and may form a fuel-soluble salt. The higher the total acid number of the fuel, the greater probability that metals will be solubilised. These solubilised metals can negatively affect thermal stability properties of the fuel and promote oxidation of the fuel with potential subsequent gum formation, deposit build-up in the fuel system. Even traces of metals in the ppb level for can be deleterious for fuel stability, e.g., copper as low as 20–30 ppb. However, many methods cannot detect the presence of copper <100 ppb.

Only one compound, N,N′-di-salicylidene-1,2-propanediamine, is approved as a jet fuel MDA (RDE/A/650). Figure 4 shows the chemical structure. This is an additive that can be supplied at various dilution levels, i.e., the active matter content of the additive may differ from 100%. Operators at refineries and downstream should check to ensure that correct active matter is used and reported for the final fuel.

MDA works by forming chelates with solubilised metals, thus preventing the metal ions from catalysing fuel oxidation that would lead to polymerisation reactions and gum formation. It is most effective with copper but can also bind iron, cobalt,

[5] This chapter uses NATO codes for jet fuel grades. In the US Military, F-34 is known as JP-8 and F-44 as JP-5.

cadmium and zinc to varying degrees. However, if the metal levels are high (say, above 1200 ppb), MDA will be ineffective. See more about the chelation process and MDA functionality in Chap. 7.

The proper use of a metal deactivator (MDA) is a good option to prevent situations where jet fuels can be destabilised by trace metal components. Most specifications permit the use of MDA when deemed necessary. In the past, the use of MDA was allowed only after showing presence of metal contamination. It is no longer a necessity to analyse the fuel for trace levels of copper cadmium, iron, cobalt and zinc before using MDA. However, MDA must not be used on a regular basis at the point of production to obtain thermal stability pass results. ASTM D1655 has instituted criteria (listed in Table 2 notes in ASTM D1655) upon which MDA can be used at production or when thermal stability issues occur downstream of production.

The main criteria for use at production are: no more than 5% of the jet fuel batches produced in a 12 month period may be treated with MDA to meet the thermal oxidative stability requirements 260 °C, the batch of fuel requiring MDA shall pass thermal oxidative stability requirements at a test temperature of 245 °C prior to any MDA addition, and after MDA addition (2.0 mg/L maximum MDA) the fuel batch shall pass thermal oxidative stability requirements at a test temperature of 275 °C.

Initial addition of more than 2.0 mg/L MDA to jet fuel already meeting 260 °C thermal stability requirements before addition of MDA is also permitted when fuel will be transported in supply chains where copper contamination can occur.

MDA may also be added to jet fuel in the distribution system to recover thermal oxidative stability performance lost during distribution (after refinery release). The maximum cumulative addition for MDA in fuels is 5.7 mg/L.

The Certificate of Quality shall show the initial thermal oxidative stability test result, the result after the addition of the MDA and the concentration of MDA added. Locations considering addition of MDA to deal with thermal stability problems should prepare a lab blend of the fuel with the proposed level of MDA and perform a JFTOT (thermal stability) test to confirm the effectiveness of the treatment. A dosage rate of 1.0 mg/L or less is usually sufficient to recover thermal stability—successive higher treat rates can be used as necessary subject to the limitations imposed in ASTM D1655 and Def Stan 91-091.

6.1.4 Lubricity Improver Additives

This type of additive can fulfil two separate intended functions in the fuel. It can improve fuel lubricity to meet component parts (fuel pumps) operating requirements, and it can perform its original role to protect against corrosion in fuel tanks and in transport lines. Therefore, lubricity improver additives (LIAs) can also be referred to as corrosion inhibitor/lubricity improvers (CI/LIs) or simply corrosion inhibitors (CIs).

Some hardware components in aircraft are designed requiring the fuel providing boundary lubrication to maintain proper performance. (A similar requirement is also found with fuel injection pumps of diesel passenger cars and light duty vehicles which need lubrication by the diesel fuel itself.) However, kerosene-range hydrocarbons with relatively low viscosity do not provide much hydrodynamic lubrication.

Intrinsic fuel lubricity is impacted by the amount of polyaromatic hydrocarbons as well as by N- and O-containing heteroatomic compounds in the fuel. If these chemical species are removed from the fuel by severe hydrotreatment or during processing by cutting off the distillation profile at a reduced final boiling point, or if the fuel contains a high blend percentage of synthetic fuel blend component (which has inherently poor lubricating capability), the resulting fuel may require a lubricity improver additive, LIA, to meet the aviation fuel lubricity requirement of 0.85 mm maximum wear scar evaluated using ASTM D5001 lubricity test [29].

The fuel producer also may not always have direct access to a good lubricity blending component to add to a synthetic or severely hydrotreated component thus would require the use of an LIA to meet specification requirements.

Historically, CIs were used to minimise corrosive rusting of steel pipelines and tanks in contact with fuels in which traces of water were sometimes present. However, it has been decades since major pipeline systems like CEPS have allowed CIs to be used in Jet A-1 as pipeline corrosion inhibitors before entry into the system.

Chemically these additives are derived from dimerisation of organic mono acids. While the exact composition offered by the various additive producers vary, most of the additives are based on the same dimer acid chemistry.

These semi-synthetic compounds, from terrestrial plants or animal-fat sources, are known to impart high lubricity to a fuel. Similar compounds are also used to improve the lubricity of ultralow sulphur diesel, ULSD.

CI/LI additives that are used in aviation have historically been qualified by US Military and separately by the UK MoD. They are listed in the qualified product list (QPL) which is accepted by all NATO members. Def Stan 91-091, being both a civil and a military specification, adopts all the CI/LI additives listed in the QPL. Currently, there are six approved lubricity improver additive products listed in Def Stan 91-091 made by several manufacturers and listed in QPL 68-251 [30]. However, ASTM D1655 contains a smaller list of approved additives (Innospec DCI-4A, Nalco 5403) as not all the additives listed in the QPL and in Def Stan 91-091 are approved by all the major OEMs, thus not all 6 listed in Def Stan should be used in civil fuels.

CI/LI additives should be added to jet fuel at somewhere between 9 and 23 ppm; see specifications for the actual minimum and maximum allowed limits for the given additive as dosage levels differ. The dosage is based on military studies, the minimum being Minimum Effective Concentration and the maximum denoted as Max Allowable Concentration. The Max Allowable Concentration is based on the highest concentration where there is no impact on MSEP rating of a fuel.

Usage tends to be optional in Jet A and Jet A-1 specifications, but mandatory in military grades of jet fuels such as F-44, F-34 and F-24.

There is currently no accepted industry test method for determining the amount of CI/LI in Jet fuel. However, the US Airforce is working to develop a method to evaluate CI/LI concentration in jet fuel. Currently, LIA levels in jet fuel are determined by a combination of metered injection and stock level checks.

6.1.5 Fuel System Icing Inhibitor

Water is omnipresent, and water contamination of hydrocarbon fuels is a constant challenge in fuel handling and use. Water tends to freeze in aviation fuels at altitude, a fact which might lead to negative impacts like ice crystals blocking important fuel handling airplane parts, e.g., unheated on-board fuel filters.

This technical and safety challenge plays a more important role in Military and General Aviation (Avgas) than in large commercial aircraft operations. In Civil aviation FSII is used in Jet A/Jet A-1 by agreement between purchaser and supplier; this is most likely to occur when the purchaser is operating a business jet. The requirement for and mode of fuel system icing inhibitor (FSII) addition is usually set by the business jet maker, e.g., Cessna or Honda, and the driver for FSII use is often due to biostatic properties associated with FSII.

Military operations are more challenging than civil thus FSII is mandatory for F-44 and F-34. FSII should never be used when filter monitors are involved. However, monitors are now obsolete in military fuelling operations and are in the process of being removed from commercial fuelling operations.

FSII is a glycol ether with the structure depicted in Fig. 5. The chemical composition imparts solubility in both fuel and water. When a hot, wet fuel containing FSII is cooled at elevation during flight, the water in the fuel coalesces and converts from dissolved water to free water and drops out of the fuel. The FSII in the fuel will then preferentially partition to the water phase and effectively suppress the freeze point of free water. This reduces the occurrence of suspended ice crystals and ensures that fuel will flow freely through aircraft fuel lines and filters. (Russian and Chinese specifications may allow other chemistries.)

There have been several approved chemistries as a FSII with varying operational and HSSE profiles, but currently the only approved material listed in Def Stan 91-091 and MIL-DTL-83133 is diethylene glycol mono methyl ether (DiEGME) (see Fig. 5). It is also known in some quarters by the trade name of PRIST. The approved treatment rates are 0.10–0.15 Vol. % in civil and 0.07–0.10 Vol. %, in military fuel specifications.

In special cases it is opportune to use LIA and FSII together, e.g., in military aviation fuels. Then it may be possible to add the LIA in a mixture with FSII. Def Stan 68-150 [31] then controls such a combined additive concentrate.

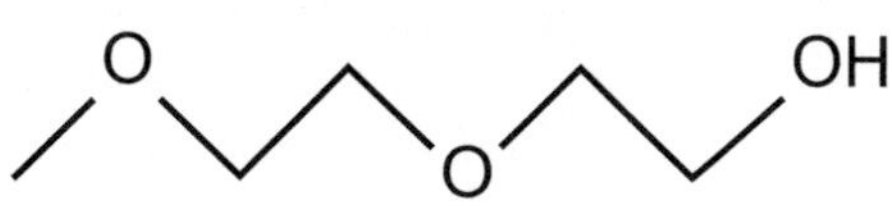

Fig. 5 Chemical formula of DiEGME

6.1.6 Leak Detection Additives

Airport hydrant systems consist of tankage, filtration units and pipework, much of which is underground. Any leakage of product is wasteful, harmful to the environment and potentially dangerous. Periodically systems may need to be checked for leakage; physical methods exist but may not work for all systems. In such cases, one can inject a leak detection additive and monitor for its presence. Tracer A (LDTA-A)[6] is the only additive approved for this application. It can only be used in the supplier's recommended manner for the leak detection process at airports and other storage and delivery locations but should not be used in any fuel being sent to aircraft.

6.1.7 Biocides

Microorganisms, such as yeasts, moulds and fungi, are omnipresent and, if adequately controlled, pose no threat to aviation operations. Left unchecked they can proliferate in fuel, water bottoms and fuel/water interfaces, causing assorted problems is storage and use. Appearance of such microbiota may harm the quality of the fuel by formation of slime that may block filters downstream. Another problem may be sulphate reducing bacteria, SRBs, that are favoured in anaerobic conditions and may form H_2S and mercaptans and increase the acidity of the fuel, impact on corrosivity and thermal stability of the fuel. These and other issues are discussed in more detail in Chap. 12 of this book.

Microorganisms rarely cause problems in refinery production, where the high temperatures of production processes, often above 200 °C, serve to completely sterilise fuel of any microbial contamination. Downstream of these hot units, good housekeeping and proper maintenance is required to prevent most microbial problems in wet treatment units, fuel storage tanks, filtration units, pipe lines, aircraft tanks, etc. For tanks, this starts with daily water drainage of tank bottoms. Steam injection into fuel raises temperatures and can sterilise systems and reduce issues. As a last resort, biocides may be used to control the growth of microorganisms though they do not deal with the problems caused by dead microorganisms which may require filtration to remove dead microbial matter.

Routine and non-routine inspections of storage systems and aircraft tanks may identify unacceptable levels of microbial activity, in which case biocides may be used as part of the remedial actions. See more in Chap. 12. Biocides may be used during maintenance of fuel storage and aircraft tanks, but the tanks should then be thoroughly cleaned to prevent any potential cross contamination. Fuels containing biocides should not be used for normal aircraft operations. If a batch of kerosene contains biocides it should be downgraded to non-aviation use.

There are two biocides that have been referenced by ASTM D1655 as being available for controlled aviation usage. These are Kathon™ FP 1.5 and ®Biobor JF, though only ®Biobor JF is listed in the most recent edition. Kathon™ had been

[6] Tracer A (LDTA-A) is a registered trademark of Praxair Services, Inc., Tucson, AZ 85705.

used universally in aviation for decades but has been delisted due to a recent field incident. New biocide chemistries are not expected to enter the market in part due to high costs of REACH and National registration, and the time and cost required for aviation approval for new materials. Defence Standard 91-091 no longer lists any biocides; it expects their usage downstream of the point of production to be controlled by aircraft maintenance manuals (AMMs) and fuel handling guidance documents. The end users shall be informed and agree to the biocide additive usages.

In addition, the fuel system icing inhibitor additive known as Di-EGME, imparts biostatic properties and at times is used to retard microbial growth. It is not routinely used in most civil aircraft, although is recommended by some business jet manufacturers (see previous section).

6.1.8 Water Scavenger Additive

With reference to the section above discussing fuel system icing inhibitor (FSII), it is evident that water contamination is a challenge within the aviation fuel distribution system and in aircraft. Regular draining operations of water and fuel-water mixtures are necessary in commercial aviation due to the specifications in aircraft maintenance manuals. This is to avoid some of the following negative implications with water as a contaminant:

- An excessive amount of water may interfere with the capacitance gauging probes in aircraft fuel tanks resulting in wrong readings.
- Water will form ice on cooling and this ice may detach and may restrict or block the fuel supply to the engine.
- Water may lead to additional corrosion in aircraft fuel tanks.
- The presence of water may further enhance the growth of microbiological proliferation with all its negative effects.
- Water needs to be drained and disposed of, leading to increased maintenance and downtime costs.

There are precautions taken during up lifting of fuel to an aircraft to remove free water from the fuel. Filter water separators and new technologies to replace filter monitors are used to reduce the free water in fuel to very low levels, typically below 15 ppm.

However, fuel itself always contains dissolved water, and the total amount depends on ambient temperature and humidity as well as fuel composition. For example, in hot climates the concentration of dissolved water could be up to 100 ppm. With reduced ambient temperatures during flight, the dissolved water will separate from the fuel-water solution and become free water. This free water can later freeze during flight. Another source of water during flight, up to litres per tank, can come from humid air sucked into the airplane fuel tanks during descent. Frozen water in the tanks from prior flights will add to the free water level on thawing and if not drained can accumulate in the wing.

A water scavenging additive solution was developed to deal with the formation of free water in aircraft tanks. It binds water molecules into the fuel using a micelle structure (or molecular cage) rendering the water molecule physically inert. Thus, this type of additive will maintain the solubility of the dissolved water in the aircraft down to temperatures of the fuel's freeze point. Additionally, it can capture free water of the bottom of fuel tanks and then expel the captured water during normal fuel burning by the engines. Salts and bacteria or other contaminants that might usually be found in the water with fuel are not sequestered, only the water molecule itself. This type of additive must be metered into the fuel at each refuelling site. In 2020, after going through a D4054 approval process, a water scavenger additive called Kerojet® Aquarius® was included into ASTM D1655 with an allowed concentration of maximum 250 ppmv (mL/m^3). However, the additive is not approved for use by the major airframes. Restriction for its usage under controlled (both for intended usage and in the event of possible defueling) conditions is detailed in in the specification.

6.2 Military Fuels

Main fuels. Most military jet fuels are those based on civil Jet A and Jet A-1 with the inclusion of mandatory additives. The use of un-additised civil fuels keeps costs down and aids in availability of fuels.

Conversion of two civil fuels to F-24 and F-34 for use in air force and army aircraft is achieved by the addition of required military additives. Some go in at the point of manufacture (e.g., antioxidants) while additional additives are normally added close to the point of use and no longer added at point of manufacture. Naval aircraft use a higher flash point base fuel to make F-44. The higher flash point is usually achieved by the refiner using a specific distillation range. F44, while having a higher flash point, will still meet other Jet A and Jet A-1 requirements. This is more cost effective than intermittently making special base fuel for F-44.

Most of the additives that are mandatory in these kerosene military fuels are also permitted in civil fuels. Table 2 of ASTM D1655 and Defence Standard 91-091 both give an overview of generic additive classes. The exact additives and dosage requirements for military fuels are listed in the current version of the military specifications and on the Military Qualified Products Lists (QPLs).

Special application fuels. The US Air Force also has special grades of fuels for aircraft operating under more extreme conditions. The main example of this is JPTS, a high thermal stability fuel (MIL-DTL-25524) used in some US Military aircraft. This fuel must pass the JFTOT test at 335 °C. It is a light kerosene cut and contains an additive package, JFA-5®, made by Innospec LLC, which is used at 8.6–11.4 mg/L. The package contains metal deactivator, antioxidant, and dispersant components to enhance fuel thermal stability. As JPTS is a specially produced fuel, it is expensive. This was one of the reasons for the initiation of the '+100 programme' described below.

+100 additives. The thermal stability of aviation fuels plays a critical role in proper and safe operation of aircraft and is one of the most important required and controlled properties for jet fuels. Such fuels perform other functions apart from being burned for propulsion. The fuel is, for example, used as a coolant for engine oils, electronics and avionics; this impacts the maintenance and performance of on-board fuel pumps and fuel controls, and, particularly in military applications, the fuel may act as a hydraulic fluid for engine control actuators. Generally, thermal stability requirements in military aviation are even more demanding because military jet fuels like F-34 are more stressed than Jet A/Jet A-1 used in commercial aviation; some of the extra fuel used with military afterburners/augmentors is used to cool these systems before returning to the tanks as a stressed fuel.

In the 1990s the US Air Force started an intensive programme to explore and develop a jet fuel additive which would allow F-34 to match the high thermal stability of JPTS at significantly reduced costs. Specifically, the programme set out to improve the JP-8/F-34 thermal stability performance by 100 °F.

One additive package was identified which improved thermal stability by 100 °F and was therefore called the '+100' additive. It was primarily developed for the then future generation of F-22 fighter aircraft, which were expected to need a fuel with increased heat sink properties. After the introduction it was also used successfully in other fighter and trainer jets, light transportation aircraft and helicopters. Main benefits were:

- A reduced coke build-up in fuel nozzles,
- less unscheduled engine removals,
- improved engine flameout rates,
- improved flight safety by reducing afterburner malfunctions,
- sustained engine power levels,
- reduced particulate emissions.

The '+100' additive is composed of the following effective components: a proprietary dispersant, an antioxidant and a metal deactivator (the two latter selected from the already approved aviation fuel additives). The approved treatment rate is 256 g/ m^3. F-34 fuel with a thermal stability improver additive at the required concentration shall be designated as 'JP-8+100' (NATO code F-37). Until recently, there was one additive listed as a qualified thermal stability improver additive in MIL-DTL-83133 with two different viscosity versions: SPEC AID 8Q462 and SPEC AID 8Q462W, both from Suez Water Technologies and Solutions. It has been removed from the US Air Force specification but may still appear in the specifications used by other NATO countries.

Several other formulations were also offered to US Air Force by BASF and Infinium chemical companies, but none were ever fully approved for usage. This additive, while approved and used after initial introduction, is now rarely used by the military, and is considered obsolete. The additive while performing technically never met the cost reduction drivers, mainly due to the very high treat rate of 256 g/ m^3 compared to the JFA-5 treat rate of 8.6–11.4 mg/L in JPTS.

6.3 Russian and Chinese Specifications

There is an increasing willingness to align all the major international jet fuel specifications, championed by the International Jet Fuel Specification Liaison Group, (IJFSLG), but the usage of additives remains an area where differences remain. This is particularly the case for specifications overseen by Russia and China. The additive roles are similar, but the lists of specific permitted additives differ, either containing additives not approved in Western specifications or not containing all the additives in the Western lists. Table 5 provides a summary of the current situation; Chinese authorities have discussed having their additives go through the ASTM D4054 process, but no research reports have yet been produced. The additive codes are provided but not all the chemistries are disclosed.

The authors do not know if additives used upstream of refineries and later, in the distribution systems, are like or different from those found in 'western' operations.

Table 5 Summary of readily available information about additive usage in Russian and Chinese specifications

Additive class	Russian fuels (GOST) TS-1 and RT (GOST 10,227, AM 1) GSTU 320.001149943.007–97 (RT type) GSTU 320.001149943.011–99 (TS-1 type)	Chinese fuels GB 6537-2018 for No 3 Jet fuel [32]
Antioxidant	Substituted phenols, including BHT. May be listed as Agidol + number, with Agidol being a trade name	2,6-di-tert-butyl-p-methylphenol
Static dissipator additive	Historically used an ASA-3 lookalike, may now use Stadis® 450	Stadis® 450 or T1502; previously T1501 was another SDA listed in the Chinese specifications
Lubricity improver additive (anti-wear additive)	Naphthenic acids	Naphthenic acid type (T1602), the amount of addition is not more than 20.0 mg/L There is also T1601 (was in 2006 version of the Chinese spec)
Metal deactivator additive	Probably same MDA as 'western'	As allowed by the user, it may add N, N'-di-salicylidene-1,2-propanediamine. The amount of the first addition shall not exceed 2.0 mg/L, the cumulative amount of addition shall not exceed 5.7 mg/L
FSII (anti-freezing additive)	TriEGEE or PVK	As allowed by the user, it may add the ethylene glycol methyl ether or diethylene glycol methyl ether, the volume fraction of the added amount is 0.10~0.15%

Likewise, there are no details available about the use of additives in the military operations not falling under 'western' specifications.

7 Methods of Introducing Additives

7.1 Hardware for Injection and Mixing

Additisation: There are two separate issues when additising fuel: (1) how to add the additive to the fuel and (2) how to adequately disperse the additive and keep the products homogenous in the fuel. Most of the approved additives, while fully soluble in jet fuel, are delivered as a concentrate thus are more viscous than the fuel and require agitation for uniform dissolution.

There are three ways to additise a fuel—splash blending, slugging and injection. Splash blending is pouring an additive concentrate into a fuel tank. This is possible as many of the additives have very low treat rates and can be poured into large fuel holding tanks. As an example, SDA additives require about 1 gallon of additive to treat a million gallons of jet fuel. The disadvantage of splash blending is the requirement of mixing which is very difficult in large fuel tanks.

Slug dosing is sometimes used where the additive is placed into a transfer pipe or hose and followed by the fuel. The expectation is that moving fuel when entering the holding tank will mix the additive in to the fuel. This is also not a preferred way to additise a fuel as the level of mixing of the additive in the fuel is better than splash blending but does not achieve uniform distribution of the additive.

The preferred mode of additive dosing is by pulse injection where the additive is added to the fuel in controlled small increments to flowing product during transfer (loading or unloading a fuel cargo).

The point of injection of the additive will depend on the type of blend (one or more kerosene streams) that makes the batch of the fuel. Antioxidants tend only to be added to components prone to peroxide formation like streams that have undergone treatment with hydrogen, or to oxidatively sensitive synthetic blending components listed in D7566 [33]. They are ideally pumped immediately into the outgoing streams from the refinery to the storage tank and are added continuously. The same additions, namely pumping the right amounts of additives into the outgoing products, are carried out for additives like LIA and SDA. If the jet fuel is transferred by ships, addition of SDA can be made manually into the ship tanks (splash blending). A different case is MDA, which is added—only when necessary—directly into product or storage tanks. Many jet fuel tanks are equipped with pumps that can circulate entire contents of a tank, so homogenous mixing and equal spread of the additive can be achieved inside the tanks if metered injection into or out of tankage is not possible or adequate.

In pipeline systems, the injection equipment should be verified at regular intervals (typically every six months) for the dosage being delivered. Secondly, if feasible ASTM D2624 and ASTM D5006 [34] should be conducted downstream of injection equipment on regular intervals to ensure the fuel meets SDA and FSII injection range requirements. Automated in-line SDA injection systems should be designed to either automatically shut down or give an alarm if the conductivity downstream is near the maximum limit (600 pS/m). The injection should be stopped if there is no flow (which is not always the case at the ingress points). An injection pump that adapts the dose according to the flow is the best solution (e.g., Hyrolec).

Additive tanks and containers. FSII, LIA and SDA tanks and piping shall be of stainless steel, preferably type 316 that has a better corrosion and chemical resistance than type 304. Additional recommendations for LIA and SDA tanks are: cone-down bottom tanks with a low drain point, the tank volume should be optimised to maintain a high product refresh rate (particularly for FSII tanks) and the tank have regular inspections. The tank should have a stirrer to maintain homogeneity of the mixture, possess a readable volume indicator, flushing connection for additive piping, accessible sampling point and should have water/humidity exclusion devices.

In all cases, copper or tin couplings shall be avoided. Teflon seals should be used for FSII valves, since FSII is not compatible with Viton, neoprene, chloroprene or natural rubber. If Teflon is not an option, isobutylene isoprene rubber, ethylene propylene diene monomer or polytetrafluorethylene can be used. Make sure that there is no backflow of fuel into an additive tank. Valve failures cause jet backflow.

Every six years, additive tanks should be opened, visually inspected, and cleaned if required (if inspection is impracticable, cleaning every six years is advisable). In addition, they shall be cleaned immediately if there is evidence of accumulation of sediment as disclosed by bottom samples (or upstream of the injection point) or by the need to frequently clean strainers. Details of inspection, recalibration and cleaning shall be recorded and retained for at least seven years.

7.2 Additive Packages and Additive Dilution

Most additives are provided as single application products, though sometimes (mainly in military applications) a mixture of additives is used. For single application products, the additives can be specified chemical species used 'neat', or provided in a mixture, with a carrier fluid and possibly a diluent fluid. Fully prepared materials of multiple components and diluents are called additive formulations or additive packages.

Some additive packages require more than one type of chemistry to be effective. Due to the manufacturing processes and the application requirements in the field, the effective additive component(s) are always diluted with appropriate, compatible refinery hydrocarbon streams or solvents, either by the manufacturer or later at the site where they are to be added to the fuel.[7] If there is dilution required, the user must ensure that they achieve the correct level of active matter in the final fuel. The choice

[7] See guidance in ASTM D1655 Sect. 6.

of diluent should be guided by the advice from the vendor in the additive specifications (for example, using a refined hydrocarbon stream from a refinery that meets jet fuel materials and manufacture requirements). If the Additive Company (AddCo) performs the additive injections at the refinery, they will already have established suitable diluents.

7.3 Determining Additive Levels Added

Where additives are deliberately added to aviation fuels, the amounts added should be reported on the appropriate quality certificates, such as Certificates of Quality (CoQs) used at point of production or on downstream Release Certificates. Knowing how much additive has been introduced is determined by one of both of the following:

- Change in level of stock in the additive dispensing tank during the injection process.
- Tracking volume via metered pump injection.

Note: When additives may be added at two or more points in the production to airport supply system, such is the case with SDAs and MDAs, the documents must report the cumulative amount of additivation, not just the last level dosed into the fuel.

Fuels are not often analysed to confirm that the correct type and level of a given additive is present. This is in part because stock levels in additive tanks and tracking metering pump injection are adequate to measure the volume of additive dosed to the fuel, and in part because appropriate analytical methods do not exist or are too unwieldy requiring specialised equipment or interpretive knowledge or require excessive time to perform the evaluation. More methods are becoming available, with the potential to support product quality investigations or management of change exercises; see Sect. 8.1 for more information.

7.4 Reporting Requirements for Additives—Active Matter or Full Package

Specifications set out the amounts of additive that may be used. The additive may be used 'as received' or diluted with kerosene to aid injection; the user then corrects the amounts used to reflect any dilutions. In some cases, the limits require that the active matter in the additive is reported. If, for example, 10 mg/l of a 50% active matter additive were used, the reported amount of additive would be 5 mg/l. In other cases, the reporting reflects the amount of the named additive (e.g., SDA), with no corrections for any diluents or carrier fluids in the additive formulation.

8 Challenges and Issues (Problems and Solutions)

The purpose of specification-approved additives is to prevent problems and/or to improve aircraft performance. The approval process requires that the additives deliver their intended benefits without compromising other fuel properties or performance of other additives; to that end, the ASTM D4054 process concentrates on ensuring that there are no unintended side effects.

Specifications are not prescriptive, and challenges exist to ensure that appropriate additives are used correctly. Two types of approval exist for additives in aviation, one based on chemistry (AO, MDA, FSII, etc.) and the second based on product trade name (SDA, CI/LI, leak detection). As an example, anyone can produce a MDA or an AO provided the chemical meets the industry registered chemical composition; however, the trade name additives with undisclosed composition may only be produced by the company registered with that product name.

There are challenges to ensure companies do not change additive composition for trade name listed products and still retain the same RDE numbers used in Def Stan 91-091, and also in ensuring that unapproved additives are controlled to as low a level as possible.

8.1 *Challenges with Permitted Additives*

Antioxidants are usually added at the point of manufacture. There are instances where they cause colour changes through conjugation in the additive sometimes giving the jet fuel a red tinge. The colour change is due to quinone formation via dimerisation of the hindered phenol antioxidant. This should not be a concern for end users, but it is important to note the fuel colour at the point of manufacture, e.g., on the CoQ, to avoid downstream concerns about product contamination with a different, dyed product.

AOs are not surfactant materials, and they have no known side effects that adversely affect fuel performance properties.

Fuels, which may or may not have had AOs added at the point of manufacture, may need AO to be added later in the distribution system; this may be because the end customer needs long term storage (say, for strategic purposes) or if there are indications of fuel degradation during storage. Rather than checking on the level of AO in the fuel, which is not straightforward, ASTM D3703 [35] is an accepted method for measuring hydroperoxide levels. 8 mg/kg of hydroperoxide is generally considered the maximum limit for fuel based upon prior USA Naval research. If hydroperoxides are trending higher in a storage tank, an approved AO can be added to delay additional hydroperoxide formation.

MDA is not reported as causing problems when used correctly in civil applications. While MDA has minimal surfactant properties and can adsorb on solid metal surfaces as well as chelating solvated metals, there are no reports of MDA interfering with other additives or otherwise having a deleterious impact on jet fuel properties. (In the past, the old formulation of MDA did cause filter blockages due to the formation of poorly soluble metal chelates. The current MDA avoids this issue.)

Naval fuels use MDA more regularly and may have identified issues, in part related to their greater exposure to copper and brass fittings (not recommended for systems moving jet fuel).

Timely addition of MDA is needed. Adding MDA too late will not stop or reverse oxidation processes already underway in a metal contaminated fuel. Also, MDA will not be effective at concentrations of metals greater than 1200 ppb.

SDAs are added to provide sufficient electrical conductivity to reduce the risks related to Static Discharge Ignition. In practice, conductivity is one of the most sensitive parameters to control. It can remain constant, decrease or even increase over a matter of days or even hours. Adding more SDA to increase conductivity may create its own problems, such as conductivity being above the specification limit of 600 pS/m. Further SDA can act as a weak surfactant and negatively impact water separation characteristics of the jet fuel (seen as poor MSEP results). However, the specifications allow for lower MSEP rating of a fuel containing SDA.

Fuels free from co-additives and trace contaminants are most responsive to SDAs and have fewest problems. The response is better for hydrotreated fuels than wet treated ones because the Merox™ or Merichem™ type processes do not remove trace contaminants as effectively as hydrotreating. This can lead to a significant drop of conductivity level over time, to below optimal levels without proportionally increasing the MSEP. Clay filtration that removes contaminants prior to additive treatment improves conductivity response. After additisation, clay treating at low flow rates can also be used to remove SDA to reduce conductivity that is too high, particularly if the alternative of diluting the fuel is not an option.

Time-dependent conductivity loss is partly due to fuel composition (presence of trace materials such as amines) or absorption onto dirt or tank surfaces, and by water removal which can extract the partially water-soluble SDA from the fuel. Sulphur oxides and combustion particulates delivered to the fuel due to poorly operating inert gas systems (IGS)/scrubber systems will often lead to conductivity variability (see also Sect. 8.8) during ship transport. Excessive nitrogen oxides from ship's combustion processes may also cause similar problems but have not been studied.

Fuel temperature also affects the measured conductivity of a fuel. Conductivity decreases with temperature owing to the increased viscosity of the fuel resulting in lower mobility of charge in the fuel. The conductivity change at a given temperature can be approximated with the formula in Appendix X2 of ASTM D2624. This is not an absolute calculation and is highly dependent on fuel composition. Hence, the temperature of the fuel shall always be mentioned on a test report when the conductivity is measured.

These conductivity-related challenges can become important when the period between receipts into a tank exceeds one month. It is good practice to check the conductivity of jet fuel containing SDA at monthly intervals and record it along with the temperature of the fuel.

Apart from conductivity and MSEP concerns, because SDA has surface-active components, it may also clean up distribution systems by dispersing dirt or rust previously attached to the pipelines. In this way high levels of finely dispersed rust can be mobilised which could cause filtration problems downstream. Pipeline operators may choose to pig the lines periodically to remove dirt and rust and reduce the filtration burden.

LIAs for the purpose of lubricity improvement in aviation tend to be added late in the fuel delivery process. However, as these additives also perform as corrosion inhibitors, they are added by some pipeline operators to inhibit corrosion of their assets. CI/LIs are weak surfactants, i.e., they are surface active. They can plate out onto metal surfaces in the distribution system, most particularly on higher energy sites like grain boundaries, thus forming a partial or complete coating to protect against corrosion of the metal surface. But as such the desired lubricity benefits for the fuel may be diminished. This is overcome by continuous LIA usage to ensure a state of equilibrium has been achieved between pipe coating and fuel concentration. CI/LIs can affect water shedding properties of a fuel. MSEP results can be impacted if excess CI/LI is dosed into the fuel. Military qualification information list maximum allowable amounts of the CI/LI allowed to be in the fuel to ensure the additive has minimal impact on fuel MSEP rating. The max allowable range is also incorporated into civil specifications.

High treat rates of LIAs can at times have an impact on the conductivity of fuels containing SDAs.

FSIIs are only slightly soluble in fuel, and preferentially partition into any water present. One well-reported set of issues relate to the subsequent formation of a slightly viscous material referred to as 'apple jelly'. This is formed by FSII reacting with water and other contaminants in the fuel and is a corrosive material that has caused fuel filter disarming (leading to water and particulates being loaded onto aircraft), corrosion of fuel tanks and fuel tank linings, and erroneous capacitance gauge readings. This is mostly an issue for military fuel users as few commercial aircraft apart from some business jets and helicopters use FSII. FSII is added during fuelling, under the commercial name of PRIST. Issues are mostly prevented by rigorous and daily sumping activities.

8.2 Analytical Methods for Additive Measurement

There are very few reliable chemical analyses to establish additive concentrations from samples of additised fuel. Even if methods were available, it would be particularly difficult to measure additive concentration when the additisation occurs downstream of the refinery and the fuel is evaluated at a laboratory not equipped to run the analysis. Furthermore, most of the additives are present at very low levels, and many are surface active, and have the potential to absorb on and desorb from transport infrastructure surfaces like pipeline and tank walls, as well as sample container surfaces, resulting in misleading or variable measurement results. They may also be removed in units like clay treaters. This is one reason SDA is not often added to fuels in the USA; the fuels are sometimes clay treated at terminals close to airports, which would strip out any SDAs added upstream. The assumption must be that the correct amount of additive was added to the fuel at the point of manufacture and that enough has survived transport and storage to achieve the required performance. If this is the case, having precise additive level data may not be warranted for day-to-day operations though could help with incident investigations.

Nevertheless, methods are being developed that may be suitable for use at well-equipped refineries or that may eventually appear in portable (or at least transportable) versions. Military laboratories have made most progress on these latter approaches. In civil applications, usually a method will be developed, undergo validation and interlaboratory studies to obtain precision data, and then be offered to the specification writers to include in the fuel specifications. Table 6 provides a non-exhaustive list of additive quantification methods.

8.3 Storage of Additives and Shelf Lives

Additives in the refineries are supplied and stored in iso-tanks of approximately 20 MT, 200 L barrels or, at least for SDAs at refineries, small 16 kg (25L) barrels. The materials used in the construction of the tank and injection equipment shall be compatible and suitable for use with the additives. Stainless steel tanks are recommended for the storage of additives.

Additives should be stored in covered or shaded areas to minimise water ingress and exposure to elevated temperatures. Their shelf-lives should not exceed the manufacturers' recommendations, and best site inventory management practices should be followed—for example using 'first in, first out' concept to ensure the additive is used with in the recommended shelf life. Where appropriate, they may need to be tested on receipt and later to ensure that they meet specification and qualified product list (QPL) requirements.

If feasible, at least once a month a sample should be taken into a glass bottle from the additive tank low point or upstream of the injection point, to confirm that the sample is visually clear and bright. If the sample is not free from contamination, the system shall be flushed until a satisfactory sample is obtained. The monthly sampling may be extended to a longer period if supported by a risk assessment demonstrating satisfactory additive quality.

Long term storage some additives, for example, FSII is questionable. This is mainly because of FSII's hygroscopic nature. It is recommended that the stocks are rotated as frequently as possible as possible and if not feasible, retested for quality conformance (see EI 1530).

There are best practices worth adopting when the same additive tank is used for different aviation fuel additive products. This can happen either with:

Table 6 Indicative methods for additive quantification

Additive class	Method(s)	Status/In spec	Comment
Antioxidant	ASTM D8564 [35]	Recently approved but not in specification	Unlikely to be added to specifications until an interlaboratory study (ILS) has taken place
CI/LI or LIA	FTIR (Fourier Transform Infra-Red spectrophotometer)	Not in spec	Test method[a] is not validated and not mandatory. Presence of naphthenic acid in the Kerosene gives false positive results
SDA	ASTM D7524 [38], IP 568 [39]	–	Measures surfactant level in multi-component SDA package, so possibly misleading beyond point of manufacture as there are situations when the 2 components are exhausted or removed differently
MDA	ASTM D8454 [40]	Not yet listed in specifications	Method with interim precision data, awaiting ILS. Detects uncomplexed, not total, MDA
FSII	ASTM D5006 ASTM D8564	–	Measures DiEGME; other methods needed for FSIIs in non-jet or Russian and Chinese specifications See above antioxidant entry; new method, in theory covering broader range of FSIIs
FAME	ASTM D7797/IP 583, IP 585, IP 590 and IP 599	Listed in specification	Run after fuels have moved in MPPLs or been in systems with scope for FAME carryover
DRA	ASTM D7872	Listed in specification	Used periodically to prove DRA not detectable in jet above specification level

[a]See method feasibility study by Larkin [37]

(a) aviation fuel additive products with the same chemistry and for the same use but from different suppliers, or
(b) additive types with slightly different chemistry but for the same use are sourced from different suppliers.

Such additive products can also contain different solvents and/or dilution grades. Any kind of miscibility and demixing or even compatibility problems must be avoided. Hence, it is recommended to empty the additive tank and rinse it with, e.g., kerosene, first before a new product batch is filled into the additive tank. If complete emptying and rinsing is not possible, the residual amount of the previously used additive product should be minimal to avoid problems with mixed 'neat additives'. Once two different sales products of aviation fuel additives are dissolved in aviation fuel, incompatibility or miscibility problems are generally not a concern.

8.4 Single Additives and Additive Mixtures

Fuel additives are tailor-made products. Their chemical compositions match the nature and needs of individual fuel types, e.g., diesel, aviation fuel or motor gasoline. Final fuel additive products can be composed of either single individual effective additive component or be a multicomponent mixture of different effective additive substances. The type and chemistry of the additive is dictated by the required performance profiles for the specific fuel. The latter types are called additive formulations. Examples of such aviation fuel additive classes are the antioxidant BHT (single effective component) and the static dissipator additive (multi effective component mixture). Normally, effective additive components are manufactured in a solvent which remains in the final additive product. To guarantee optimum stability and effectiveness, all individual additive components are selected carefully and manufactured under stringent quality controls. Furthermore, additive formulations are well balanced which means that all components are added at their optimum quantities. Thus, an adequately formulated final product on the one hand must ensure satisfactory storage stability to maintain performance features, and on the other hand significantly reduced potentials for 'no-harm effects' on the fuel and other additives present on the fuel. Sometimes solubilising co-solvents are added to ensure low temperature stability and meet viscosity requirements of the additive for field use.

The above-mentioned balance of components in an additive package is a prerequisite for good storage stability and acceptable shelf life. In addition, storage temperature and ambient conditions like direct exposure to light and moisture may have an influence on those properties. Normally, it is recommended to store aviation fuel additives under dry conditions between 0 °C and maximum 40 °C. Storage tanks should be designed to minimise condensation. Typical shelf lives under these storage conditions are between 12 and 24 months.

8.5 Additive/Fuel Interactions in Use

Most additives, stored and used correctly, do not react, or deteriorate to create problems. Exceptions to this include problems seen by several users of F-34, including the Royal Australian Air Force, with filter blocking precipitates forming in poorly stable AL-48 premix additive (FSII/LIA). Another example is BHT antioxidant that has precipitated out of formulation while being stored in 1 m^3 plastic containers. Figure 6 shows the white precipitate in the container.

Fig. 6 BHT precipitated from additive formulation in storage container (Image courtesy of A. Groweiss)

8.6 Additive Availability

Manufacturers of additives may stop making specific additives, sometimes because they are not prepared to pay for expensive processes such as those required for REACH or TSCA registration, and sometimes for other financial and/or HSSE reasons. When this happens, the fuel specifications will remove the additive from the relevant qualified products list (QPL) or specification, only allowing older stocks to be used up if they do not exceed their shelf life.

Examples include:

- ASA-3, Shell's SDA, removed from the market in the 1990s.
- ®Biobor JF, a biocide, because it did not have REACH registration. It is still listed in ASTM D1655.
- Recent removal of Hitec 580 from list of approved CI/LI additives in Def Stan 91-091 and MIL-DTL-83133 (also balloted for removal by ASTM) because it is no longer made.

Additives may also be removed from the QPLs and specifications if they are found to have issues not identified at the point of approval. An example is the removal of Kathon™ FP 1.5 biocide due to salt in the formulation having impact on aviation equipment.

8.7 Issues Related to Upstream and Refinery Additives

Upstream additives, the nature and levels of which are rarely known or disclosed, can cause problems for aviation fuels if they are not destroyed or otherwise removed during refinery processing. Well documented examples of issues caused to jet fuel production are not often reported, either because the issues are identified to be upstream additive related by the refinery or because there is no clear link to the suspected additive. Problem materials are sometimes blended away or are released into fungible systems where there is no link of the source to aircraft or fuel storage and handling problem.

One instance reported related to the same crude being used at two refineries. The hydroprocessing refinery had no problems but the wet treating refinery had thermal stability failures. It was postulated that the problem was linked to the drag reducing additive present in the crude, which was transported by pipeline; it is postulated that trace components, possibly monomers from the DRA production, were removed by hydroprocessing but not by wet treating.

While the incidence of proven upstream additive problems is extremely low, it does highlight the need to consider the possibility of issues when changing crudes. Any information on additives put into the crudes after leaving the well (DRAs, corrosion inhibitors, H_2S scavengers, etc.) will help in related MoC processes.

Anti-corrosion additives known as 'filming amines' which are routinely used in distillation towers are the most common class of additive known to impact jet fuel. Under normal operations the expectation is that these are either removed through hydroprocessing steps (as ammonia or ammonium salts) or will be removed by water washing steps and clay treaters. If any of these processes is not running ideally, there can be issues which, for amine type additives, are likely to include poor MSEP, faster clay treater changeover rate, SDA disarmament and degraded thermal stability.

8.8 Issues with Additives in Distribution Systems

8.8.1 SDA

Fuel tanks on ships are 'inerted' with exhaust gas with the intention of maintaining the oxygen levels below 5%, reducing explosion risks. Most inert gas systems (IGSs) use the exhaust gas from fuel-burning boilers, which are used to generate steam to drive turbines for ship propulsion. In addition to a reduced oxygen content good for inerting, the exhaust gas also contains sulphur oxides, very fine particulates (soot) and water. Therefore, these gases need to be 'cleaned' (scrubbed) with seawater before to eliminate any trace of particulates and SO_2 or SO_3 in the inerting gas. The full suite of checks that should be done on IGSs before loading and discharging fuel products is out of the scope of this chapter. However, if exhaust gases of the IGS contain too much SO_x, it can dissolve into jet fuel and negatively influence the conductivity, thermal stability and total sediment. This phenomenon can occur in coastal storage terminals. A first sign of IGS problems is the much higher particle count will be seen at discharge than at loading. Soot (fine particulates) produced can block filters downstream, mainly ultra-fine filters at airports.

SO_2 and SO_3 in the fuel gas themselves do not affect the acidity, sulphur and copper strip values, but they may affect fuel thermal stability by producing sulphonated aromatic compounds by reaction of SO_3 with available aromatics compounds in the fuel. Multiple studies have shown that sulphur oxides are prone to cause temporary diminishment in jet fuel conductivity response. The presence of these sulphur oxides in the fuel (especially SO_3; amounts of SO_3 as low as 0.5 mg/l) can drastically diminish fuel conductivity response by 90%. Similar phenomena may occur with NO_x. Response rates below 100 pS/m per 1 ppm SDA should prompt caution.

The limitation of SDA response can result in repeat injection of the additive to achieve the minimum conductivity listed in the jet fuel specification. As the product moves through a pipeline network, an increase in conductivity can be noticed due to elimination of sulphur oxides through degassing or exposure to metal surfaces.

The increased SDA concentration due to repeat re-additisation will be reactivated and will now yield conductivity that is higher than the limit allowed in the fuel specification. The ways to correct the problem are either by dilution with a low conductivity fuel or by clay filtration and re-addition of SDA to meet specification conductivity range.

Operators shall pay close attention to imported jet fuels. Overall, it is recommended not to exceed a dose of 1.0 mg/l, even if it means that additive will have to be injected again before delivery.

8.8.2 Non-Aviation Additives

Dyes: A 10-year US FAA-funded test programme in the 00s was conducted to determine the effect of trace levels of red dye on jet fuel properties and on the performance of critical turbine engine components [41]. Results showed that trace amounts of red dye, present in untaxed or high sulphur diesel fuels, can cause a decrease in thermal stability and an increase in fouling rate in engine fuel system components. As a result of this testing, the restriction on red dye remains in effect; any visible pink or red colour resulting from dye contamination is grounds for rejection of the fuel.

DRAs: These are extremely high-molecular weight materials, usually derived from polyalphaolefins. Even when diluted in jet fuels and sheared, they impact the operation of some basic specification fuel tests like the thermal stability test—filters rapidly block. More importantly, work by CRC in the 00s (CRC report 642 [42]) concluded that use of DRAs was unacceptable, mainly because of safety-related ignition failures linked to poor atomisation of the additive-containing fuel at low temperatures.

FAME: These components of (first generation) biodiesel fuels received aviation industry attention in the 00s and shortly thereafter. Several programmes were undertaken by the Energy Institute and others to establish what impact trace levels of FAME would have on jet fuel properties (particularly thermal stability and handling). There was also parallel work developing methods that would be able to detect low levels of FAME. Work funded by CRC and US Military groups at Southwest Research Institute (CRC report AV-03-04 [43]) concluded that up to 100 ppm was unlikely to cause issues, but so far that level is only accepted in emergency situations. Currently, specifications allow 50 ppm of FAME in jet fuel, and OEMs expect that fuel handlers to control the FAME exposure of jet fuels to lower than 50 ppm.

8.9 Management of Change

A series of incidents with Avgas in the late 1990s that grounded about half the Australian fleet identified that the problems arose because the refinery had changed the levels of a processing additive previously which had been used without any field problems. The Australian Transport Safety Bureau Investigation Board (ATSB) requested that the production specification be modified to include a Management of Change (MOC) requirement, which cascaded into Def Stan jet fuel specifications and later into ASTM specifications. Put simply, if a manufacturer seeks to change their production methods, either by changing additives, additive levels or refinery streams (which may introduce further additives and byproducts), they must perform a risk management assessment to recognise and mitigate any potential problems. This can be performed 'in house' but should involve their additive suppliers (who often manage the additisation) and ideally a third-party challenger.

MOC requirements have been extended to include the production of SAFs, where the location of blending and variety of blending components may differ from those at a refinery.

Still further, MoCs apply to changes in operational procedures, including hardware changes, some of which can be directly related to additive handling and usage.

Jet fuel specifications and handling standards and guidance material, including EI/JIG 1530, provide examples of activities needing MoCs. It is up to the relevant companies in the fuel production, distribution and delivery systems to identify where their operations may need MoC oversight.

8.10 Incorrect Additives

While uncommon, there are occasional issues related to incorrect or inappropriate additives being used with jet fuel. This tends to occur at airfields, and since late 2017 FAA and others have reported several incidents of Diesel Exhaust Fluid (DEF) being added to jet fuel instead of the intended FSII additive [44]. DEF, which is a urea-based chemical, reacted with jet fuel to produce solids that caused Inflight fuel filter blockages and subsequent safety-related incidents. The first reports were from the Royal Canadian Air Force and incidents still occurred after the problem had been reported by FAA and JIG.

As a prevention method, industry bulletins—NOTAMs in the case of FAA—highlighting issues may be insufficient. Appropriate training and signages must be implemented at locations where both additives are used.

Acknowledgements The authors thank Paul Rawson and Jack Burgazli for their reviews and the editorial team for their time, comments and contributions to the production of this chapter.

References

1. Defence Standard 91-091. Turbine fuel, aviation kerosine type, Jet A-1 NATO Code: F-35, Joint Service Designation: AVTUR, Issued by the United Kingdom (U.K.) Ministry of Defence. Glasglow, UK. https://www.dstan.mod.uk/
2. ASTM D1655. Standard specification for aviation turbine fuels. ASTM International, 100 Barr Harbor Drive, PO Box C700, West Conshohocken, PA 19428-2959. United States
3. ICAO DOC 9977. Manual on civil aviation jet fuel supply. Available from ICAO Store https://store.icao.int/en/manual-on-civil-aviation-jet-fuel-supply-doc-9977
4. EI/JIG Standard 1530. Quality assurance requirements for the manufacture, storage and distribution of aviation fuel to airports. Published by Energy Institute, London and the Joint Inspection Group. Chapter 7 focuses on additives
5. EI Standard 1533. Published by Energy Institute, London
6. JIG 1—Aviation fuel quality control & operating standards for into-plane fuelling services (Issue 13, 2021). Published by the Joint Inspection Group, London
7. JIG 2—Aviation fuel quality control & operating standards for airport depots & hydrants (Issue 13, 2021). Published by the Joint Inspection Group, London
8. API 1543—Documentation, monitoring and laboratory testing of aviation fuel during shipment from refinery to airport. American Petroleum Institute
9. API RP 1595—Design, construction, operation, maintenance, and inspection of aviation pre-airfield storage terminals, 2nd edn. American Petroleum Institute
10. BS EN 14214:2012+A2:2019. Liquid petroleum products. Fatty acid methyl esters (FAME) for use in diesel engines and heating applications. Requirements and test methods. The British Standards Institution
11. ASTM D6751. Standard specification for biodiesel fuel blend stock (B100) for middle distillate fuels. ASTM International, 100 Barr Harbor Drive, PO Box C700, West Conshohocken, PA 19428-2959. United States
12. IP 585. Determination of fatty acid methyl esters (FAME), derived from bio-diesel fuel, in aviation turbine fuel—GC-MS with selective ion monitoring/scan detection method. Energy Institute
13. ASTM D7797. Test method for determination of the fatty acid methyl esters content of aviation turbine fuel using flow analysis by Fourier transform infrared spectroscopy—rapid screening method. ASTM International, 100 Barr Harbor Drive, PO Box C700, West Conshohocken, PA 19428-2959. United States
14. IP 583. Determination of the fatty acid methyl esters content of aviation turbine fuel using flow analysis by Fourier transform infrared spectroscopy—rapid screening method. Energy Institute
15. IP 590. Determination of fatty acid methyl esters (FAME) in aviation turbine fuel—HPLC evaporative light scattering detector method. Energy Institute
16. IP 599. Determination of Fatty Acid Methyl Esters (FAME) in aviation turbine fuel—gas chromatography using heart-cut and refocusing. Energy Institute
17. ASTM D7872. Test method for determining the concentration of pipeline drag reducer additive in aviation fuels. ASTM International, 100 Barr Harbor Drive, PO Box C700, West Conshohocken, PA 19428-2959. United States
18. EI 1535. Minimum criteria to determine acceptability of additives for use in multi-product pipelines co-transporting jet fuel, published by Energy Institute, London
19. MIL-DTL-5624. Turbine fuel, aviation, grades JP-5 (NATO F-44), DLA Energy, USA
20. MIL-DTL-83133. Turbine fuel, aviation, kerosene type, JP-8 (NATO F-34) and NATO F-35, DLA Energy, USA
21. ASTM D4054. Standard practice for evaluation of new aviation turbine fuels and fuel additives. ASTM International, 100 Barr Harbor Drive, PO Box C700, West Conshohocken, PA 19428-2959. United States
22. ASTM D7826. Standard guide for evaluation of new aviation gasolines and new aviation gasoline additives. ASTM International, 100 Barr Harbor Drive, PO Box C700, West Conshohocken, PA 19428-2959. United States

23. ASTM D2624. Standard test methods for electrical conductivity of aviation and distillate fuels. ASTM International, 100 Barr Harbor Drive, PO Box C700, West Conshohocken, PA 19428-2959. United States
24. IP 274. Determination of electrical conductivity of aviation and distillate fuels, Energy Institute
25. ASTM D3948. Standard test method for determining water separation characteristics of aviation turbine fuels by portable separometer. ASTM International, 100 Barr Harbor Drive, PO Box C700, West Conshohocken, PA 19428-2959. United States
26. ASTM D7224. Standard test method for determining water separation characteristics of kerosine-type aviation turbine fuels containing additives by portable separometer. ASTM International, 100 Barr Harbor Drive, PO Box C700, West Conshohocken, PA 19428-2959. United States
27. ASTM D3241. Standard test method for thermal oxidation stability of aviation turbine fuels. ASTM International, 100 Barr Harbor Drive, PO Box C700, West Conshohocken, PA 19428-2959. United States
28. IP 323. Determination of thermal oxidation stability of gas turbine fuels, Energy Institute
29. ASTM D5001. Standard test method for measurement of lubricity of aviation turbine fuels by the ball-on-cylinder lubricity evaluator (BOCLE). ASTM International, 100 Barr Harbor Drive, PO Box C700, West Conshohocken, PA 19428-2959. United States
30. Defence Standard 68-251. Fuel soluble lubricity improving additives for aviation turbine fuels, NATO Code: S-1747 Joint Service Designation: AL-61, Ministry of Defence
31. Defence Standard 68-150. Mixture of fuel system icing inhibitor and lubricity improving additive joint service designation: AL-48, Ministry of Defence
32. GB 6537-2018. PDF in English national standard of the People's Republic of China ICS 75.160.20 E 31. Replacing GB 6537-2006 No.3 jet fuel, Issued on July 13, 2018, Implemented on February 01, 2019
33. ASTM D7566. Specification for aviation turbine fuel containing synthesized hydrocarbons. ASTM International, 100 Barr Harbor Drive, PO Box C700, West Conshohocken, PA 19428-2959. United States
34. ASTM D5006. Test method for measurement of fuel system icing inhibitors (ether type) in aviation fuels. ASTM International, 100 Barr Harbor Drive, PO Box C700, West Conshohocken, PA 19428-2959. United States
35. ASTM D3703. Standard test method for hydroperoxide number of aviation turbine fuels, gasoline and diesel fuels. ASTM International, 100 Barr Harbor Drive, PO Box C700, West Conshohocken, PA 19428-2959. United States
36. ASTM D8564. Standard test method for determination of phenolic antioxidants and fuel system icing inhibitors in aviation turbine fuel. ASTM International, 100 Barr Harbor Drive, PO Box C700, West Conshohocken, PA 19428-2959. United States
37. Larkin C (2013) Quantification of corrosion inhibitor/lubricity improver in military fuels using infrared spectroscopy. US TARDEC Report #23814, U.S. Army TARDEC, 6501 East Eleven Mile Rd, Warren, MI 48397–5000
38. ASTM D7524. Test method for determination of static dissipater additives (SDAs) in aviation turbine fuel and middle distillate fuels—high performance liquid chromatograph (hplc) method. ASTM International, 100 Barr Harbor Drive, PO Box C700, West Conshohocken, PA 19428-2959. United States. See also IP 568. For use at point of manufacture
39. IP 568. Determination of the static dissipater additives (SDA) in aviation turbine fuel and middle distillate fuels—HPLC method. Energy Institute
40. ASTM D8454-23e. Standard test method for determination of the metal deactivator additive (MDA) (N,N′ disalicylidene 1,2 propane diamine) content in aviation turbine fuel by high performance liquid chromatography. ASTM International, 100 Barr Harbor Drive, PO Box C700, West Conshohocken, PA 19428-2959. United States
41. Moses CA et al (2010) The evaluation of red-dye contamination in jet fuel and the identification of a screening method for thermal stability issues. FAA Report DOT/FAA/AR-10/15
42. CRC Report No. 642 investigation of pipeline drag reducers in aviation turbine fuels (CRC Project No. CA−68−97, S Seto, October 2005)

43. Wilson G (2011) CRC report AV-03-04 diesel lubricity additive effect on jet fuel thermal oxidative stability with supplementary information on fatty acid methyl ester and jet engine nozzle performance
44. JIG operations bulletin no 124, DEF contamination in jet fuel, 29th October 2019. https://www.jig.org/

Literature for Additional Reading

45. Airbus technical magazine FAST; edition no. 60; October 2017
46. ASTM D4171. Specification for fuel system icing inhibitors. ASTM International, 100 Barr Harbor Drive, PO Box C700, West Conshohocken, PA 19428-2959. United States
47. ATA 103. Standard for jet fuel quality control at airports. Available from Air Transport Association of America, Inc. (ATA) d/b/a Airlines for America, 1275 Pennsylvania Ave. NW, Suite 1300, Washington, D.C. 20004. http://www.airlines.org
48. Aviation fuels technical review, Chevron Corporation (2007)
49. Defence Standard 68-252. Fuel system icing inhibitor, NATO Code: S-1745, Joint Service Designation AL-41, Ministry of Defence
50. JIG aviation fuel quality requirements for jointly operated systems (AFQRJOS)[8] Checklist for Jet A-1, available from www.jigonline as a Products Specification Bulletin.
51. JIG 4—Aviation fuel quality control & operating standards for smaller airports (Issue 4, 2021), published by the Joint Inspection Group, London

[8] Jet fuel that meets AFQRJOS is usually referred to as 'Jet A-1 to Check List' or 'Check List Jet A-1'.

Chapter 6
The Effects of the Fuel Composition on Non-metal Materials in Aircraft Systems

Melanie Kimble-Thom and Alan Fletcher

Abstract The effect of the composition of an aviation fuel, be it kerosene or gasoline, on the non-metallic components of the aircraft fuel system cannot be underestimated. Examples of unintended or undesirable effects of fuel compositional changes include leaking of fuel-wetted seals, swelling of polymeric fuel hoses, and removal of organic coatings from metal substrates. These events illuminate the importance of considering the impacts of changes in fuel formulation or composition. The complexity of the potential list of non-metallic materials, and determining the means of testing and evaluating data, is not immediately obvious. Considering the increasing contemporary efforts to develop alternative fuel sources, formulations, and compositions, the activities to support those challenges become more expansive, touching materials which may have a legacy of decades without being reviewed. While not an exhaustive discussion, the following chapter endeavors to provide a foundation for identifying challenges and developing the means for evaluating data.

1 Introduction

While the parts of an aircraft built from metal tend to get the greatest attention during the design, build, and maintenance activities, there are a great deal of components made of non-metals. Some materials are obvious, like paint or elastomeric O-rings. Others are less noticeable, like the coating on a fuel tank, or a polysulfide sealant, or an inked part number. When wetted with fuel, either intentionally or incidentally, the compatibility, functionality, and continued operability of the non-metallic components are just as integral to the system design as the metals, and often are more difficult to test.

M. Kimble-Thom (✉)
Baere Aerospace Consulting, West Lafayette, IN, USA
e-mail: Melanieathom@BaereAerospace.com

A. Fletcher
US Air Force Research Laboratory (Retired), Dayton, Ohio, USA

M. Rabaev et al. (eds.), *Aviation Fuel Problems and Solutions: From Refinery to Wingtip*, https://doi.org/10.1007/978-3-032-05514-9_6

One of the challenges to considering non-metallic materials lies in their pervasiveness and variety. Not only is there a large variety of non-metal materials, from O-rings and gaskets to inks and paints, but within a category there are seemingly endless formulations. Like the recipe for a favorite chocolate cake, every manufacturer may have their own proprietary formula. While it is beyond the scope of this text to cover non-metals exhaustively, the following represents a high-level introduction to major types, and considerations for their use and testing in fuel-wetted aviation systems. The reader is further advised to consider what the result of the contact might be; is the non-metal adversely impacted by contact with the fuel (more common) or will the fuel be negatively affected by contact with the non-metal (less common).

2 Non-metallic Materials

The following does not attempt to identify every type and class of non-metallic material found in an aviation system but presents several of the most common materials and those that have had a history of being more sensitive to changes in fuel chemistry.

2.1 Organic Coatings

Before exploring the category of organic coatings, it is useful to develop a working understanding of the term "organic." The McGraw-Hill Dictionary of Chemistry [1], defines organic as "of chemical compounds, based on carbon chains or rings, and also containing hydrogen with or without oxygen, nitrogen, or other elements." All of the coatings discussed in this section are based on some type of material that has an organic backbone. They may, and likely do, have other elements such as titanium, or manganese present.

Organic coatings used in aviation are typically cured coatings as opposed to dried coatings. The difference being the first undergoes a chemical change whereby the molecular structure in the coating becomes linked together. The latter is a film of material from which a solvating medium is evaporated. The cured types of coatings may result from a reaction of a multi-component blend, such as a two-part epoxy, which may be catalyzed by the presence of a catalyst, for example, the hardener added to paint, or it may be the result of a thermal cure in which the coating is actually baked, such as powder coatings. In each of these examples, how the molecules are bonded together changes after curing.

On the other hand, films from which a volatile solvent has been evaporated may undergo no further change in molecular structure other than consolidating on the surface. An example of this would be latex paint from which water evaporates as the paint dries or the ink markings used to identify parts. This type of coating is not common on fuel-wetted surfaces due to the ability to be dissolved and removed when

Table 1 Examples of approved organic coatings [2]

Purpose	Coating	Examples
Surface paint	Primer	Epoxy
Surface paint	Base coat	Polyurethane, Acrylic
Surface paint	Top coat	Polyurethane
Surface paint	Clear coat	Polyurethane
Fuel tank sealant/corrosion prevention coating	Primer	Epoxy
Fuel tank sealant/corrosion prevention coating	Top coat	Polyurethane
Surface color/surface protection	Powder coat	Proprietary polyester/epoxy resin Proprietary triglycidyl isocyanurate
Above ground storage tanks/ transportation tanks	Surface coating	Bis F epoxy phenol novolac epoxy

exposed to fuel. However, part markings are often ink, and as such may be removed by incidental fuel contact.

The benefits of a cured coating over dried coatings in aviation fuel systems include increased durability, resistance to abrasion, resistance to chemical exposure, and when properly applied, improved performance. The disadvantages include more complicated application processes, potentially greater challenges during repair and maintenance, and higher procurement and application costs. Dried coatings are generally less costly to procure and require less complex application systems, but they are generally less resistant to fluid exposure and abrasion.

Typical approved organic coatings encountered in both the aircraft and in the aviation fuel distribution system are shown in Table 1. The list of coatings is not exhaustive, and examples are not inclusive.

2.2 Potting Material, Conformal Coat, and Wire Insulation

It may not be surprising that an aviation or aerospace system has a number of electronics included in the design. What might be less obvious is that there are electronics which are actually fuel-wetted. Sensors, control units, switches, and printed circuit boards are all used within the fuel system and as such must be designed using materials which are compatible with the fluid in which they are submerged. This compatibility of the materials must occur while protecting the system from sources of ignition such as sparks or electrical discharge. The electronics in the aviation fuel system make use of both potting materials, which encapsulate a finished electronic component, and conformal coatings, which isolate and protect complex surfaces such as printed circuit boards. Potting materials are used to fill the space around an electronic connection, preventing fuel and moisture from entering the connection,

Fig. 1 Example of potted connector (Photo Dan McClelland (Image courtesy of Hapco International))

and providing strength and protection from vibration, see Fig. 1. The material used may include polyurethane, epoxy, silicone, polyolefins, and polyacrylates.

Conformal coating is a thin layer of polymer which when applied, conforms to the surface. Thus, any complex surface, such as a printed surface board, can benefit from the same benefits as the potted connection. Example polymers used for conformal coating in aviation and aerospace include polyurethane and polyimides. Specialty proprietary chemistries may also be used, necessitating collaboration with the vendor to assure proper compatibility with fuel. Because of the typical design hierarchy in aviation and aerospace, access to these sub-tier vendors may be challenging and require effort to achieve.

In addition to the potting and conformal coating, the coating on the wires should also be considered. The coatings on wires in aerospace system are typically made of polytetrafluoroethylene (PTFE, Teflon™), ethylene tetrafluoroethylene (ETFE), or polyimide (Kapton™). General aviation aircraft wiring may also include Kapton™ or polyvinylidene fluoride.

2.3 Sealants

Room Temperature Vulcanizations (RTV) Sealants—In general, RTV sealants are not used within the fuel system itself. This does not mean they never are, nor does it mean they are not used in locations where incidental fuel exposure (unintentional contact, or intentional but short-term contact) can occur. RTV sealants are silicone-based polymers which are made of a mixture of silicone, fillers, and catalysts. The reaction may be initiated using only the moisture in the air, or by using a two-part blend which initiates the reaction. RTV sealants are used because they do not require any other specialized equipment to use, are generally easy to use, are resistant to acidic and basic (alkaline) exposure, and age well. Because they are silicone, they benefit from the characteristics of any silicone polymer, such as heat resistance and lack of deformation. On the other hand, silicones tend to have low strength and are easily abraded. While they are easy to apply, if the bead becomes too thick, the high viscosity of the sealant may entrap air bubbles which can weaken the finished polymer. Some RTV's release acetic acid and this release can attack nearby components such as solder joints. Within an aircraft, RTVs are most likely to be used to seal air gaps or provide limited vibration protection.

Integral Tank Sealants—Both turbine and piston aircraft use integral fuel tank sealants. Integral fuel tanks, also referred to as "wet wings" are formed within the aircraft by sealing sections, typically within the wing, during manufacture, see Fig. 2. The joints between the wing skin sheets, and between the structural members are sealed with a flexible, 2-part sealant so the joints will not leak. This permits the entire space within the structure to be used as a fuel tank and reduces the aircraft weight through the elimination of separate fuel tanks. Joints between inspection covers and other routinely removed pieces, may be sealed with a low adhesion sealant to facilitate removal and replacement.

Sealants are controlled by industry standards, such as AMS-S-8802. A single standard may control multiple types of sealants. The first designation, when specified, is the type. Type 1 indicates the sealant uses a dichromate cure system. Type 2 indicates the sealant uses a manganese dioxide cure system. Each type of sealant may be blended with varying amounts of the two parts to affect changes in the viscosity and pot life. Depending on how the sealant is to be applied, a manufacturer may want a low viscosity blend so the sealant may be brushed into place, up through a thick paste which is applied with a caulk gun. The classes include Class A—suitable for brush (thin), Class B—suitable for application by extrusion, spatula, brush, or roller (medium), and Class C—suitable for application by gun or spatula (thick). Similarly, the pot life, or time to initial cure, may be as short as 30 min or as long as 4 h. The pot life is presented in hours; 1/2, 1, 2, and 4. Thus a B-1/2 is a medium thickness sealant with a pot life of 30 min.

Fig. 2 Inner view of integral fuel tank (Photographer: Faran Levi)

Table 2 Example tank sealants

Description	Cure type
Polysulfide with dichromate cure	Type 1—Dichromate
Polysulfide with manganese cure	Type 2—Manganese
Polysulfide	Type 2—Manganese
Polythioether	Type 2—Manganese
Polysulfide (light weight)	Type 1—Dichromate
Polysulfide with manganese cure (low adhesion)	Type 2—Manganese
Polysulfide, non-chromate	Non-chromate, not specified
Polyurethane	Not specified
Polythioether (sprayable)	Not specified
Polysulfide (low density)	Not specified
Polysulfide (non-curing)	None
Fluorosilicone (non-curing)	None

From a material compatibility perspective, a specified sealant will be the same regardless of its viscosity or pot life, albeit absolute test values will be class dependent. To execute industry standard testing, a Class B blend is typically specified. Compatibility testing should use the cure type specified when preparing specimens for testing.

The typical types of sealants used in aviation fuel tank systems are shown in Table 2 but the list is neither exhaustive nor inclusive.

2.4 Fuel Cells and Bladders

In some aircraft, instead of a "wet wing," see Sect. 2.3, a separate bladder is used. Within the industry, this polymer bladder may be seen referred to as a "fuel cell." This should not be confused with the similarly designated electrical term "cell," nor the class of devices that convert the chemical energy of a fuel directly into electricity by electrochemical reactions. These aircraft fuel cells are generally polymeric bladders constructed to fit into one or multiple locations in an aircraft, see Fig. 3. The fuel cells are used in both piston-powered aircraft that use aviation gasoline, and in turbine powered aircraft that use jet fuel (kerosene).

Unlike wet wings, fuel cells are completely self-contained tanks, which are essentially hung from the aircraft structure at attach points molded into the surface of the bladders. In many cases the bladder is fitted in between aircraft structure which may provide external support to the fuel cell. This would be like fitting a shaped balloon into a close-fitting box. While this positioning may provide support and minimize the ability of the fuel cell to move within the aircraft, it is the attach points which provide the actual mounting and are ultimately responsible for accepting 100% of

Fig. 3 Example fuel cells (Photo courtesy of Aircraft Owners and Pilots Association; *Source* https://aopa.org/-/media/Images/AOPA-Main/News-and-Media/2020/August/0826_fuel_bladder_01.jpg)

the load forces. These load forces include both the fuel cell mass as well as the contained fuel, at all points of a flight and at all aircraft orientations. The design of these attach points must be sufficient to withstand the forces from a fully loaded fuel tank at greater than one G-force due to flight maneuvers. These loads depend on the aircraft and its permitted flight profiles.

One benefit to the use of bladders is that they do not risk leakage due to flexing of the aircraft structure as can occur with the sealant in wet wings. As few as a single, large bladder, or many smaller bladders distributed throughout the airframe may be used. The disadvantages to the use of fuel bladders over wet wings include measurable increases to the total aircraft weight, challenges in inspections and leak identification, and effort levels required for removal and installation of fuel cells after original aircraft manufacture.

There are three basic types of fuel cell bladders. The first is simply for fuel containment and is generally composed of either nitrile or urethane with fabric reinforcement. There is also a nylon barrier in these cells to prevent fuel permeation through the bladder. These are used in wings, fuselage or external pods to contain the fuel going to the engines. In the case of military refueling aircraft, they are also used to store fuel to be offloaded to other aircraft. The second type is crash worthy fuel cells. These are generally used in helicopters and are strongly reinforced to withstand bursting during the high impact of a crash. These cells are required to withstand a drop test from 70 feet full of fuel or simulated fuel. The third type is used in the military to self-seal when hit by small arms fire. These contain a thick layer of material (usually natural rubber) that swells vigorously when exposed to fuel.

At the time of this text, two manufacturers were responsible for the majority of the original equipment fuel cells shipped in new aircraft built in the US. These fuel cells were constructed of multiple plies (layers) of urethane or nitrile polymer laid over a fabric layer with a proprietary innerliner material sprayed onto the assembled plies and then co-cured. These cells were most notable for their lighter construction (lower weight) while still being fuel resistant. The cells were compactable (able to be folded up tightly) due to their light mass, making them less challenging to install in the aircraft.

In the aftermarket, there were two main manufacturers responsible for most replacement bladders. In the case of the aftermarket fuel cells, construction of the fuel cells was from proprietary polymers with nitrile components, and nylon reinforcement. These fuel cells were most notable for their heavier construction and higher durability. However, these benefits came at the expense of greater weight and more challenges in replacement installation due to greater bulk.

All FAA (US Federal Aviation Administration) authorized fuel cells are required to undergo certification testing in compliance with FAA TSO-C80 which provides requirements on a variety of physical parameters. These parameters include fuel cell permeability (outgassing of the fuel through the bladder walls), material compatibility with test fuels both internally and externally, and physical durability under static and moving conditions while filled with fuel. With respect to the fuel's integrity, testing must also demonstrate a resistance to any chemical extraction from the bladder which can cause the fuel to have unacceptable gum residue when heated. Once a manufacturer receives certification on a construction (material and manufacture), no further testing to the Technical Standard Orders (TSO) is required unless changes to formulation or construction of the bladder are made. The TSO does provide a foundation for developing testing of fuel effects on the bladders as well as impacts on the fuel from materials extracted from the polymers.

2.5 *Gaskets, Packings, O-rings, and Molded Seals*

Any place two or more surfaces come together in a fuel system, a seal must be made to prevent fuel leakage. The design, development, and testing of "seals" are a large topic in itself and as such beyond the scope of this text. The following provides a general presentation of seals, and the reader is directed to references such as the Parker O-ring Handbook [3] for a more in-depth study of sealing.

When considering sealing, one must first consider the mating surfaces at which sealing is desired. In general, these surfaces will either have no relative motion between them and a single pressure, such as the mating surfaces between two engine block halves, or they will have one or more of the surfaces moving relative to another, such as a spinning shaft. The relative motion may be due to cyclic pressure changes, or to physical movement. The physical movement may be either reciprocating, oscillating, or rotational (see Fig. 4). In the first case with no relative motion, a static seal is required. In the second case, where there is some type of relative motion, caused by either physical motion of the surfaces, or by changing fuel pressure, a dynamic seal is required.

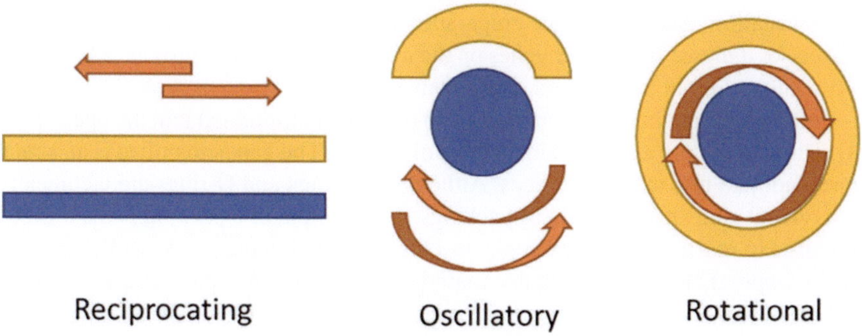

Fig. 4 Examples of relative motion (Figure courtesy of M. Thom)

Static seals include gaskets, packings, lip seals, and O-rings. Note that O-rings are also used as dynamic seals as discussed below. Gaskets are materials placed between two flat, mating surfaces. The shape may be quite complex, and the material used to produce the gasket may be from a variety of materials (Fig. 5). Piston engines still use impregnated paper, leather, and cork for gasket material. Silicone, fluorosilicone, and fluorocarbon are all commonly used in aviation applications.

Gaskets function by compressing the gasket between the mating surfaces under a single pressure. The pressure on the gasket is fixed and does not change in operation, i.e., two halves of an engine bolted together to a specified torque value. Gaskets are best suited for lower pressure applications and will fail when the pressure experienced from the medium that is being sealed is greater than the applied sealing pressure.

Fig. 5 Example aircraft gasket (Photos courtesy of Y. Cohen)

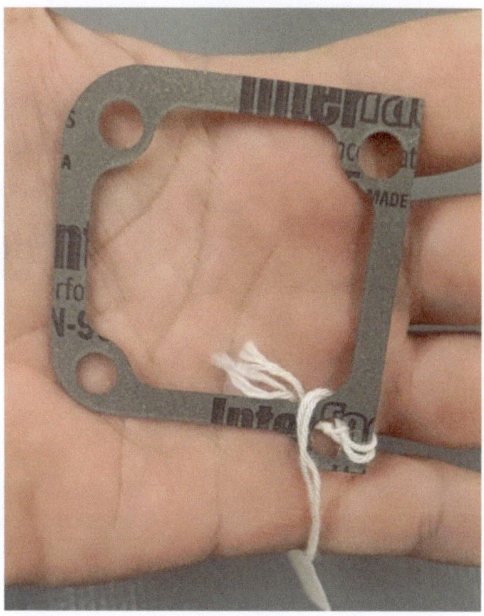

When installing gaskets, surface preparation is critical as any residue from previous gaskets, or damage such as scratches on the mating surfaces will provide a leak path, and sealing will not be obtained or maintained.

Packings are materials which plug a gap by being compressed into the open space [4]. This is typically a cord, braid, or bead of sealant. The gap being filled is an actual space, not just two mating surfaces. Although packings and O-rings are technically different, O-rings may be used as packings. An example of a packing is the braided cord used to seal a valve, for example, in a fuel distribution pump. Typically, the cord will be impregnated with an organic material or polymer. When installing a packing (except rings), it is necessary to cut the material to length with the cut ends being at a 45° angle to each other, called a skive cut. This prevents a leak path being generated by a straight cut (Figs. 6 and 7).

In other designs, the packing provides its own compression by expanding into the space. Examples of this type of packing include shaped packings made from polymers such as V-Lip and U-Lip seals (Fig. 8). Lip seals or molded seals are seals which only seal in one direction as the lip of the seal presses against the sealing surface. To increase the sealing force, lip seals may be energized by the inclusion of a spring. The installation of a lip seal generally involves an interference fit, so installation of the

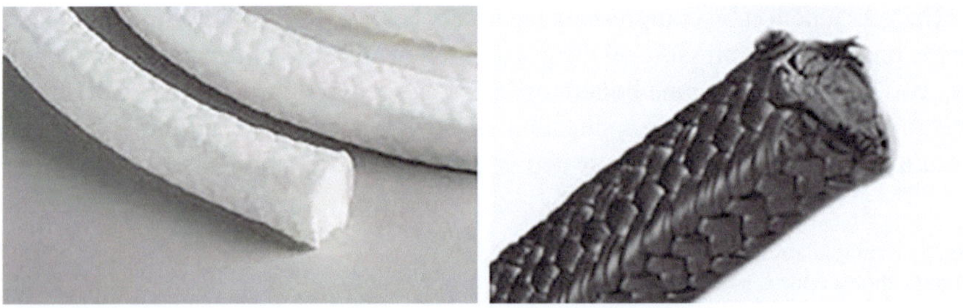

Fig. 6 Examples of "rope" packing (Photos courtesy of M. Thom)

Fig. 7 Demonstration of how a skive cut minimizes leak paths (Figure courtesy of M. Thom)

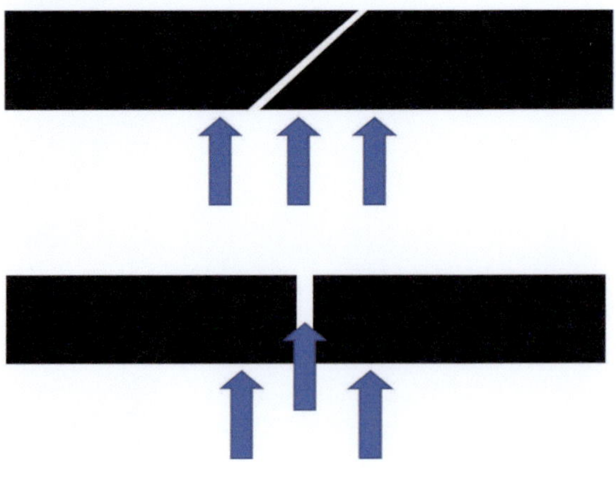

seals requires extra attention and care to avoid damaging or inadvertently twisting the sealing lip.

The last type of static seal to be presented here is the O-ring or occasionally D-ring (Fig. 9). Where all the other seals discussed to this point may be essentially any shape, O-rings are always rings. When used as a packing, the O-ring is pressed into the gap, filling the space.

O-rings are measured relative to the diameter of the ring and to the cross-sectional area of the physical tube (Fig. 10). The external diameter determines the O-ring size. The cross section determines the depth and width of the gland it will fill. O-rings are installed into a gland and the amount of the empty volume the O-ring fills at rest is referred to as *gland fill*. Typically, a cross section is selected that will result in 60–85% fill [3]. However, designers may choose a lesser or even greater percent fill, depending on the application.

In addition to gland fill, the other primary design parameter is the initial O-ring squeeze. This is how much force is applied to the O-ring when the seal is

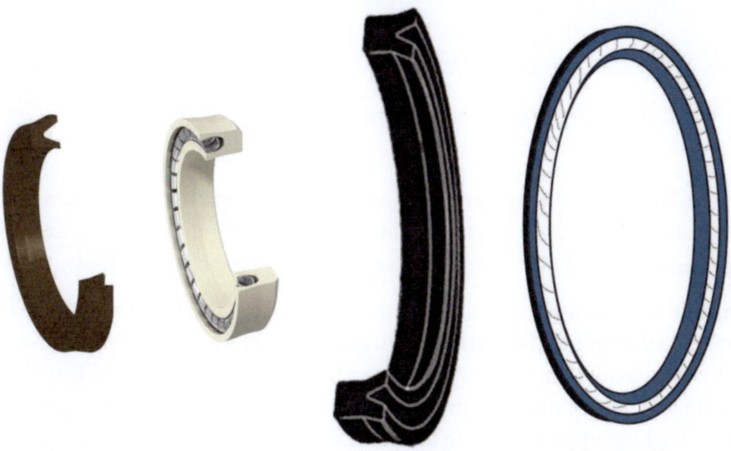

Fig. 8 Examples of lip seals; V-seal (L) and energized lip seal (R) (Figure courtesy of Parker)

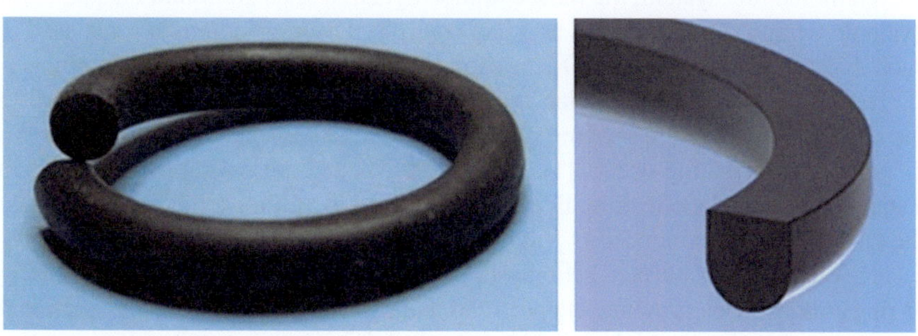

Fig. 9 Examples of shaped rings; O-ring (L) and D-ring (R) (Photo courtesy of M. Thom)

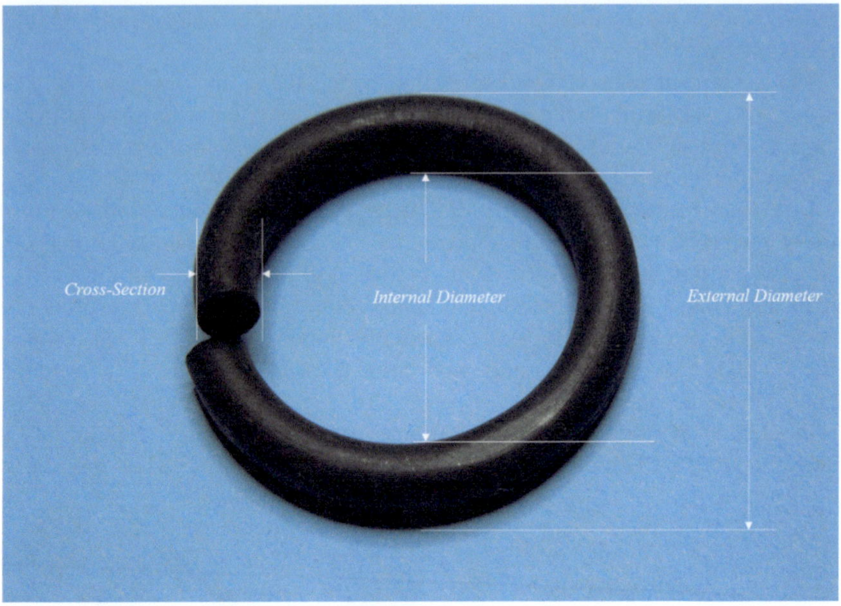

Fig. 10 O-ring dimensions (Figure courtesy of M. Thom)

assembled. Typically, the squeeze applied in static applications is 25–30% (Parker). Seal manufacturers warn that for O-rings less than 0.2 mm in cross section, the O-rings will take an almost immediate set when the squeeze is applied.

In dynamic applications, the most commonly used seals are O-rings, T-seals, face seals, mechanical face seals, and labyrinth seals (Fig. 11). Note that labyrinth seals are generally metallic and function by creating a tortuous path from one side to the other. This path is so convoluted that the media being sealed enters the path and "plugs" the space, backing up the media in the path, and sealing the gap. Labyrinth seals are found, for example, on the main shaft of turbine engines and seal the lubricating path. Given their predominant metal construction, they are not discussed further here.

Within the fuel system, the most common dynamic seals are found on shafts, for example, within the fuel pumps and fuel controls. The most common seals used in such dynamic applications are O-rings and shape seals (i.e., D-rings, V-rings).

The ring seal is a toroid which is installed into a gland and which seals due to the applied fluid pressure pushing the seal up into the leak path. Figure 12 shows a diagram of how sealing occurs [3]. This is true whether the O-ring is being used in a static or dynamic application. Because of the need to minimize the friction against the O-ring, the recommended O-ring squeeze in dynamic applications is only 16%.

It is only the seal surface actually in the gap providing sealing. The remaining O-ring area provides structure, support, and sealing capacity. A larger cross-sectional area means there is material available as the seal degrades or extrudes. When asked during a sealing conference, aerospace designers of fuel-wetted systems have been

Fig. 11 Example labyrinth seal (Photo courtesy of Dover Precision; www.Waukbearing.com)

Fig. 12 How sealing occurs—**a** O-ring as installed, **b** O-ring under pressure, **c** O-ring extruding, and **d** O-ring failure

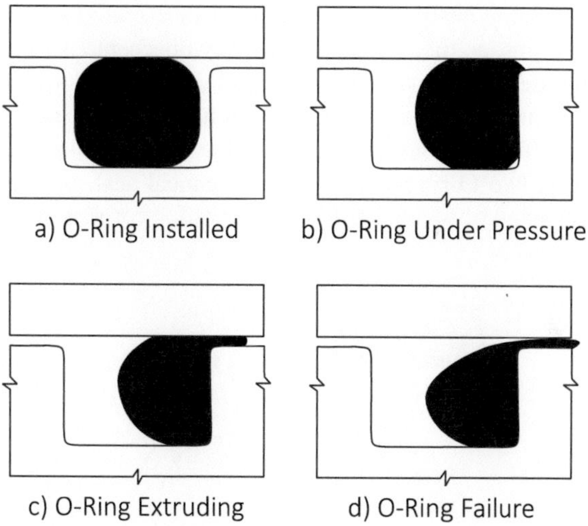

recorded to say they are more concerned with sufficient compression as opposed to swell [5].

The final types of dynamic seal presented here are the simple face seal and the mechanical face seal. Face seals are used in places where one surface is flat and the O-ring in its gland is pressed against the surface (Fig. 13). The relative motion in a simple face seal is transferred through the seal itself, resulting in friction and wear. The design is such that even at its best, there will be a certain amount of leakage. Applying more force may decrease the leak but will increase the friction and wear.

In high pressure or critical applications, where no leakage is acceptable, or in places where frequent seal replacement due to wear is undesirable, a mechanical

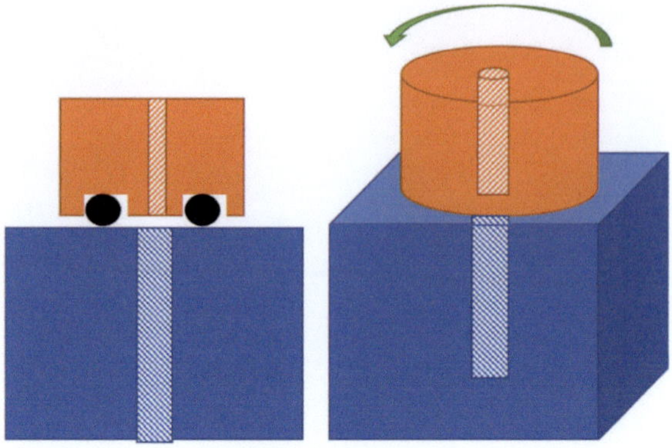

Fig. 13 Diagram of two facing surfaces being sealed with a simple face seal (Figure courtesy of M. Thom)

face seal may be used. This type of face seal requires two O-rings, but in this case, the seals are pressed against surfaces without relative motion to each other (Fig. 14). This type of seal is much less likely to leak over time but is heavier, more complex, and more costly than a traditional face seal. They are frequently used within fuel systems, as fuel leaks are generally unacceptable. For more in-depth discussion on the design, construction, and installation of a mechanical face seal, the reader is directed to text specific to dynamic sealing in general and mechanical face seals specifically. For our discussion, the reader is directed to recognize that there are at least two O-rings used in a mechanical seal, and frequently non-metallic backup rings, all of which require consideration for material compatibility.

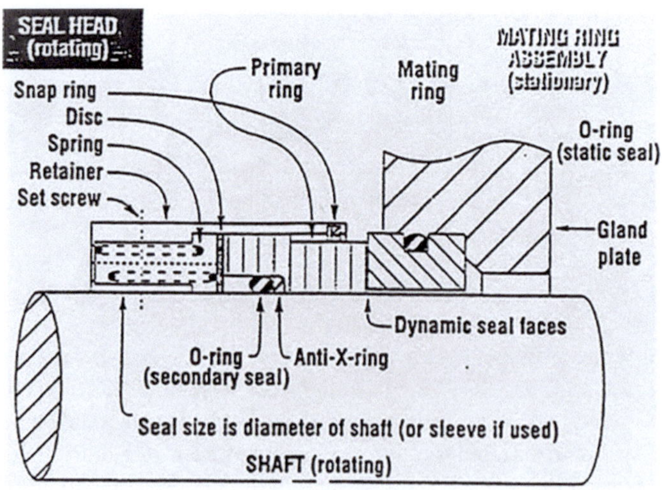

Fig. 14 Diagram of a mechanical face seal [6] (Figure courtesy of the Society of Tribological Engineers)

Backup Rings—In each of the O-ring sealing applications presented, a variety of back up structures may be included. Examples include hard back up rings made of non-metallics such as polytetrafluoroethylene (Teflon™) to provide structural support to O-rings made of softer polymers or using a second O-ring. However, these are only supports. It is only the first O-ring filling the gap which is actually providing sealing. In the case of multiple O-rings, the first O-ring supports 100% of the sealing load. The secondary O-ring provides no load support.

Construction—Throughout this discussion, little mention has been made regarding the construction of these seals. While there are seals used which are metallic such as copper crush gaskets and titanium labyrinth seals, in general seals are constructed of non-metallic materials. Within the aviation fuel system, most seals are manufactured from one of five primary polymer types; nitrile (Buna-N), ethylene propylene diene terpolymer (EPDM), silicone, fluorosilicone, and fluorocarbon. Nitrile, fluorosilicone, and fluorocarbon predominate.

Summary—When considering the material compatibility testing with seals of any type, either testing of the material before forming the seals (usually prepared as flat stock) or testing of the formed seals should be considered. Ideally, both traditional polymer testing using specimens cut from prepared sheets and testing of the final formed seal are recommended. As can be seen from the previous discussion on seals, the development of testing requires considering both traditional material tests, usually to industry standards, and interactions between the fuel and material when in use, generally without industry standards. Furthermore, the development of evaluation criteria may be application specific requiring collaboration with the hardware designers. The discussion above should inform the reader as to the complexity of this test development. The final consideration is the impact the materials may have on the fuel itself, such as thermal degradation because of extracting chemicals from the seal materials.

2.6 *Adhesives*

Adhesives as a class are generally self-explanatory. They are materials which hold together two separate surfaces which are in contact with each other [7]. Within the aviation fuel-wetted systems, a variety of cured and self-curing adhesives are used. From heat resistant, structural adhesives for adhering metal to metal, to cyanoacrylates (SuperGlue™), components are stuck together to construct the aircraft, to prevent threaded hardware from coming loose, and to install foams and seals; adhesives are located throughout the fuel-wetted systems. Adhesives used in US manufacture of aircraft include a variety of formulations in the chemical groups of epoxies, vinyl phenolics, nitriles, acrylics, silicones, cyanoacrylates, and methacrylates.

The selection of the adhesive is determined by the application, the environment to which it is exposed, and the operational strength required from the bond. Generally speaking, the compatibility of the adhesive is related to a loss of structural integrity or dissolution of the bond.

2.7 Composites

Composites are lightweight structural components consisting of resin and fabric interwoven in layers. Fiberglass composites have long been used for duct work, covers, radomes, and other low load bearing parts. The advent of epoxy/graphite and BMI/graphite composites aircraft skins and other high load bearing parts have placed composites as structural members of aircraft fuel tanks. Modern aircraft today uses their wings as integral fuel tanks with composite skins. Early designs of composite fuel tanks always included an epoxy fuel tank coating as a barrier between the composite and the fuel. But data have shown that there is little need for this coating and some recent aircraft have the fuel directly in contact with the structural composites. Most material specifications have been revised to include tests to determine the adhesion or compatibility of sealants and adhesives directly to composites in the presence of jet fuel. These considerations should also be reviewed when evaluating aviation gasolines which are more likely to have a solvency effect on adhesives.

2.8 Foams

Foams inside fuel systems are mainly of two categories: firstly, as insulation for hydraulic lines running inside the fuel area, and secondly as explosion suppression foam. Insulation is generally a closed cell foam with fuel resistant skin bonded over the hydraulic lines to maintain thermal requirements. Most of these foams are fluorosilicone or ether urethanes. The skin is generally the same material formed into an outer layer or barrier. Explosion suppression foams are usually reticulated open cell foams formed into blocks and bonded or placed inside fuel tanks. These are usually ether urethanes. See examples of open and closed cell foams on Fig. 15.

In general, explosion suppression foam is used in wet wings. The foam is installed before the wing skins are put into place, and as such will be in place for potentially long periods of time. Because of the potential for fuel system plugging due to degradation of the open cell fire suppression foams, compatibility testing is critical.

Fig. 15 Examples of closed cell (L) and open cell (R) foam (Photos courtesy of M. Thom)

3 Fuel Composition Considerations

3.1 Effect of Aromatics

The effects caused by the aromatics have been directly observed in fuel exposed non-metallics primarily as swelling. As has been discussed in Sect. 2.5, components such as O-rings are concerned with gland fill and squeeze. Based on the designer's experience with a seal/gland combination in a specific application, an expected amount of swelling is considered. The combination of that expected amount of swell with desired squeeze and fill over time informs the selection of O-ring material, diameter, and cross-section. If that expected swell is never achieved, then the selected seal may be too small in use and result in leakage. If that expected swell is exceeded, then the selected seal may be too large in use, resulting in degradation of the seal or even damage to the gland and hardware.

In the late 1980s and 1990s, the US military made a switch from JP-4 to JP-8. JP-4 was a gasoline type fuel with high aromatic content which caused considerable swelling in the predominately nitrile seals in use at the time. When the military began using JP-8, a kerosene type fuel, the seals were found to shrink, and the systems leaked. An inspection of the O-rings removed from these systems found them to be compressed into a square cross-section, often hardened, and prone to crumbling. Investigations concluded that the seals, which had spent their entire service life in JP-4, had swelled to fill the gland to 100% or more, taken a set, and then shrunk when exposed to JP-8. The now hard, square, and smaller O-rings no longer provided sealing, and the system leaked. Replacement of the old seals with new O-rings that had never been exposed to JP-4 eliminated the problem. But this event resulted in a requirement that a minimum amount of aromatics be present in JP-8/Jet A turbine fuel upon which the designers could reliably expect a certain amount of swell from the seal's polymer.

3.2 *Effect of Sulfur Compounds*

Organic sulfur compounds are naturally found in all petroleum crude oils. The amount of sulfur can range from parts per million to several percent depending on the origin of the crude oil, as discussed in Chap. 2. Sulfur can be present as sulfides, disulfides, sulfonic acids, thiophenes, and mercaptans. During refining, some of the sulfur is converted to SO_x that reacts with atmospheric sodium to cause sulfidation attack of some reactor and supply line materials. Residual sulfur in the finished jet fuel, which can be reactive or unreactive as indicated in Fig. 16, can cause degradation of some supply lines and aircraft materials over time.

All fuel specifications, therefore, limit the amount of sulfur compounds permitted to remain in the fuel to a very small amount. For example, MIL-DTL-83133 limits the total sulfur to 0.30 mass percent and the total mercaptan sulfur to 0.002 mass percent. Mercaptan sulfur is severely limited since it is very reactive to some materials. Because of the desire to use Jet A/A-1 in some military ground diesel vehicles with less than 15 ppm sulfur requirements, sulfur in jet fuel may continue to decrease if there are no overriding safety concerns. In aviation gasoline, sulfur is limited to a maximum of 0.05 mass %.

The materials that are the most severely affected by sulfur compounds are metals that easily form ions in fuel, and sulfur-based polymers such as polysulfides and polythioethers. Metals such as cadmium (Cd) are readily attacked by sulfur in the fuel to form ions in the fuel which in-turn readily attack other materials. For example, the Cd plating on fasteners where cadmium plating is used as a corrosion protection, is worn away by sulfur compounds which form Cd ions in the fuel. In 2013, the author

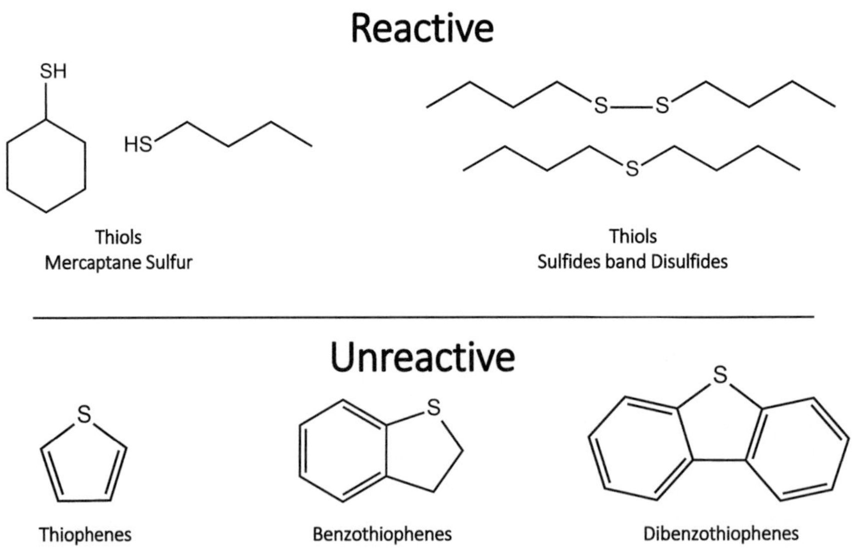

Fig. 16 Residual sulfur groups

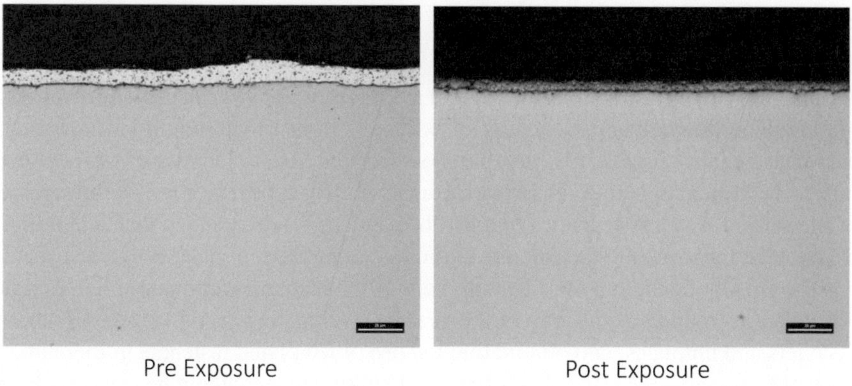

Fig. 17 ×500 photograph of pre and post sulfur exposed cadmium specimens (Photos courtesy of A. Fletcher)

performed an in-house, unpublished study on eight fuels with varying levels of sulfur compounds including 0.000% and 0.002% decanethiol, 0.003% decanethiol, and 0.003% mercaptan that looked at the effect of sulfur containing fuels on cadmium. Cadmium plated fasteners were weighed and submerged in the different fuels at 93.3 °C (200 °F) for 56 days and charted weight loss showed that the mercaptan fuel degraded the Cd coating significantly over time. Figure 17 is a 500 photo of the pre and post exposure. The study also included fuel with US military additives such as antistatic additive, fuel system icing inhibitors (FSIIs), and corrosion inhibitors. The results showed that there was a reaction between the sulfurs and the additives, causing a much greater weight loss than fuel without the additives. It is highly likely that the sulfur interacted with diethylene glycol monomethyl ether (DiEGME) fuel system icing inhibitor (FSII) additive to enhance ion formation.

Polymers that have sulfur in their backbone receive a double hit from sulfur in the fuel. Reactive sulfur can attack the polymer backbone creating chain scission and weakening the material. Sealing compounds such as polysulfides and polythioethers show significantly weaker bond strengths and reduced tensile strengths after mercaptan exposure. In an unpublished study by the author in 2005, a 15% tensile strength reduction in polythioether sealants was seen after seven days submersed in 0.003% mercaptan fuel.

The other negative effect that sulfur backboned polymers experience is a reaction to metal ions in the fuel that cause chalking of sealants. Since sulfur is the main cause of metal ion formation in fuels, its presence in the fuel weakens fuel tank sealants via chalking (the degradation of the polymer from the outermost layer). Many aircraft have had to have their integral fuel tank sealant stripped and reapplied due to chalking.

3.3 Effects of Metal Ions

As described in the last section, ions in fuel are free radicals that can degrade materials via breaking down the cross-linking of polymers. Ions of cadmium, silver, nickel, etc., cause chain scission of which the most famous case is "chalking" of fuel tank polysulfide sealants. Named chalking from the whitish appearance left on the surface of the sealant. Extensive studies in the 1960s and 1970s determined that "chalking" was not the fuel drawing the calcium carbonate filler used in sealants to the surface, but the erosion of the polymer leaving behind the calcium carbonate filler. Action was taken to reduce metal ions in the fuel by overcoating metal surfaces such as fasteners and nutplates, and eliminating the use of Cd plated fasteners in fuel tanks. Sealant formulae were also revised to resist chalking and chalking tests were added sealant specifications. Sealant chalking tests are conducted by subjecting the sealant to jet reference fluids that are spiked with metal ions. AMS 2629 Type II fluids were developed by adding high levels of copper and cadmium ions. A visual check of the immersed sealant is performed after soaking for 120 h at room temperature. Results are determined to be slight, moderate or heavy.

3.4 Effects of Additives

Jet fuel may blend additives into the finished fuels to improve certain properties. Fuels receiving additives are mostly military fuels. In the US, there are three basic types of additives: static dissipation, corrosion prevention, and ice inhibiting. These additives are necessary because unlike commercial aircraft, military aircraft spend much time on the ground with mostly full fuel tanks. Some aircraft see extreme cold at certain airports/bases and some fly at high altitudes for long durations of time thus requiring fuel system icing inhibitor (FSII) additive. Although the volume percent of the additives are low, some additives have a large effect on materials.

One of the worst-case effects is from DiEGME which turns out to be a great paint stripper. A number of military aircraft have seen large issues with fuel tank coating being removed in large chunks. Not only is this a corrosion problem for the metallic structure from which the tank coating has been removed, but the fuel filters are readily clogged, causing the fuel supply to the engine to be reduced. DiEGME has a lower vapor pressure than kerosene-based jet fuels and tends to vaporize into the ullage space. The DiEGME vapor is then absorbed by the fuel tank coating, and over time strips the coating off the substrate.

Many investigations were conducted to determine this root cause of the loss of fuel tank coatings in older aircraft. New test methods and requirements for the fuel tank coating were implemented to reduce the effects of DiEGME on fuel tanks coatings. Also, several studies were conducted to reduce the amount of anti-icing additive in certain aircraft. Figure 18 is a photo of DiEGME additive removing the coating from an aircraft fuel tank.

Fig. 18 Example removal of fuel tank coating by DiEGME (Photo courtesy of the US Department of Defense, open use)

3.5 *Effects of Acids in the Fuel*

Acidic fuels are known to degrade some fuel tank, fuel system, and engine materials. Thus, most fuel specifications, though not ASTM D1655, restrict the total acid number of the fuel to about 0.015 mg KOH/g. A 1999 in-house study performed by the author and unpublished, showed that fuel with acid numbers of about 0.020 mg KOH/g caused long-term degradation of epoxy adhesives, polysulfide sealants, nitrile fuel bladders, nitrile seals, and steel fasteners. Brief uses of high acidic fuel have been allowed, but long-term use has generally been prohibited.

It is well known that thermally stressing fuel greatly increases the acid number. Consideration needs to be given to fuel systems that recirculate the fuel from the engine back into the tanks or use the fuel in heat exchangers. It has been observed that fuel system materials in the area where fuel is recirculated back to the tank from the engine degrade very rapidly compared to the materials in the rest of the tank.

3.6 *Effects of Water/Fuel Interface*

Every jet fuel comes with a certain amount of dissolved water and sometimes free water. As temperatures decrease water accumulates in the bottom of the tanks. Vented tanks can accumulate quite a bit of water as moisture condenses on the walls and finds its way to the water layer at the bottom of the tanks. Some aircraft need routine maintenance to drain the tanks of water. For large body aircraft, pints or quarts of water can be removed during maintenance. Also draining the tanks never gets all the water and pockets of water stay in low lying areas.

Thus, it is prudent to consider the water layer and the water/fuel interface. When the water layer is not considered, materials like ester urethanes can be severely affected. In the 1960s ester urethanes were used as fuel tank sealants. They tested well for fuel compatibility; however, hydrolytic stability was not considered. In aircraft that used the sealant, it was found that over the course of years the sealant would revert into an acidic liquid and not only enable fuel leaks but attack other materials as well. This caused expensive deseal/reseal programs. Another case involved urethane foam used as an explosion suppressant. Open cell, ester-based urethanes were used in some aircraft that tested well on fuel but reverted to a powder in the presence of water. These lessons have taught the importance of considering the long-term hydrolytic effects of any material used in fuel tanks.

The interface of the fuel and the water has also caused some sealants to disbond. A number of adhesion problems occurred with polysulfide sealants in the bottom of aircraft fuel tanks. As a result, it is now required that sealant specification perform peel tests with the samples immersed in a 50% fuel/50% salt water solution. Peel strengths are recorded for the water layer, the interface and the fuel layer. The implementation of this requirement has resulted in very few instances of adhesion problems at the water/fuel interface.

4 Testing Considerations and Methodologies

Testing of non-metallics serves two purposes in aviation and aerospace. The first relates to the design and distribution of the non-metallic material or part. The second relates to the exposure of existing materials to changes in the fuel or its additives. Not only do researchers deal with "what is the purpose of a test," they must also consider "what does the industry need to know." For example, does a newly designed, finished part respond the same as an existing new part? Does a new formulation of an existing non-metallic perform the same as contemporary formulations? Does a change to the manufacturing process still result in material or components that are the same as historical material or parts? Does a new design or a new type of material provide the same or better performance when used in an existing product? Does a new design or new material respond differently as compared to existing designs or materials when new, but still works the same or better over time?

Alternatively, if a change is made to the chemistry of a fuel, or a new additive is considered for the fuel, what happens to all the existing non-metallics used in all aircraft ever built and still flying? Do new materials perform the same as always? Do in-service (old) materials perform the same way? Is their service life at least equivalent? Does the new fuel formula improve the performance of existing non-metallics?

To answer those questions, one must decide on what data are needed. In one case, we are testing new materials for a specific application. It is generally a single new material or component being evaluated for a specific application. In the second scenario we are considering a change that impacts an existing industry where both new materials and in-service materials must be considered. Each of these different philosophies are presented below.

4.1 Testing New Materials/Designs

As has been presented in Sect. 2, there are measurably extensive lists of non-metallic materials used throughout an aviation fuel system. Not only are there multiple categories of materials; elastomers, plastics, inks, coatings, etc., within categories there are a variety of types; silicone, Teflon™, polyurethane, dyes. And within those types there are figuratively endless formulations for each type of material.

In the US and Europe, the ability to track individual parts, literally to the batch number of an individual seal, could be described as astounding. The is due to the requirement to be able to track materials, for example, to comply with an Airworthiness Directive (AD) or in support of an incident investigation. Depending on the criticality of use, parts such as seals and gaskets may not only have batch and lot identification, but also certificates of analysis specific to a lot. These certificates contain a list of properties, the test methods used to measure them, the test results, and where specified, the property test limits to which the material must comply.

These lists of tests may be specified by a material standard from an international standards organization such as ASTM, SAE, or EI, or they may be a company specific standard such as Boeing, Pratt and Whitney, Embraer, or Continental. While the entire certification process is beyond the scope of this discussion, it is important to understand that a certified design requires full documentation of the components used. For the manufacturer procuring the certification, parts used to construct a design are described by their physical properties, i.e., a -20 O-ring, but also by their material standard, i.e., ES-2000. By describing the material by its standard, the designer does not need to have all the, perhaps proprietary, details of the construction in the provided certification package.

The use of a standard now forms the basis for two different functions. The first function is confirmation of a new formula or design. A polymer compounder can develop a new recipe, perhaps using a new type or source of a component in a formula, or a new way of producing a part. The standard to which they wish to market the product will provide the minimum requirements to be able to comply with and sell as a specified standardized material.

The second function of the standard is for quality control and receiving. In this case every batch of material, for example, a polysulfide sealant to AMS-S-8802, is tested to make sure the individual batch complies with the requirements. This minimizes bad batches of sealant ever reaching the market and being used in an aircraft. Furthermore, the receiver of the product can review the Certificate of Analysis to confirm it meets the standard before accepting and using a product, or the receiver can test properties in-house before using to confirm the material "meets spec."

The testing philosophy in the case of a new material is "I have a freshly prepared material sample, and I want to comply with the requirements of an existing specification with a set of requirements under specified and standardized conditions." The new material is the unknown, and the test conditions, including the fluids used to perform the test, are the controls. Every time a batch is prepared, or a part is manufactured, it will be the same within the specified requirements of the standard.

4.2 Testing New Fuels

At the time of this writing, extensive efforts were underway to develop alternative fuels to meet more stringent emissions goals, to accommodate more sustainable sources and processes, and to find ways of reducing fuel use by maximizing propulsion efficiency for both turbine and piston aircraft. These goals have existed since the 1980s and increased in levels of effort in the following decades. One of the most basic challenges was to ensure the aircraft functionality and performance, when changes were made to the fuel formulation, remained at the high levels of safety for which the industry has always strived. Part of that focus was to confirm that the changes made to the fuel did not have a negative impact on the functionality and performance of the non-metallics used throughout the fuel system. This required considering not just one material or component, or even all the components on a single aircraft. This required considering all the materials still in use on every aircraft still flying, across an industry which had grown, consolidated, and lost manufacturers. Even aircraft which no longer had an active company support system were considered.

The testing philosophy in this case also required considering not only every material's performance as a new part, but what would happen to parts already in the field (used). While it was recognized that replacement of components might be required, it was necessary to identify which components had to be replaced, with what, and how quickly after a new fuel blend entered the market. This required considering the test protocols for how to test these used components and to determine what was an acceptable value.

For testing new fuels, the researcher has at their disposal the standardized tests which are currently being used for specifying and monitoring non-metallics. In this case the philosophy is to use the existing tests, keeping all the processes and procedures as written but making the non-metallic item the control, and the fuel the variable, as opposed to QC testing where the fuel is the control, and the non-metallic item is the test article. This has the benefit of a test with familiarity and history, but care must be taken in considering the resultant test values. There is no reason to expect the absolute values from the materials to be the same when the test fluid used is changed. Whether that difference is still acceptable becomes part of the evaluation process.

While this methodology is potentially viable for those materials for which a standardized test exists, a different approach must be considered for materials which do not have a standardized test, or for which there are not even non-standard or company specific tests with which to evaluate the materials.

Finally, test methods must be devised for testing completed, fielded (used) parts. Again, existing test methods may exist for testing new hardware, perhaps even in fuel, but how these tests are to be applied to fielded (used) parts must be considered.

In all cases, an agreement must be reached about how to evaluate the test data once the data have been collected. To date, the most common method is to evaluate the data against existing test limits, where they exist. Ideally a new material will perform the same way in the new fuel as it has done in the past in contemporary fuel. In cases where no limits exist, then an evaluation by comparison is frequently used, where the results for the new fuel with a new part is at least equivalent, if not better than the results from a contemporary fuel. For used parts, the testing is primarily comparative, as used parts are seldom expected to conform to new limits.

4.3 Standardized Laboratory Testing

The benefit to using standardized tests, such as ASTM test methods, is that statistical populations of test data have been collected, and in many cases the accuracy and precision of the results have been determined. The methods for exposing materials have been developed and accepted, and standardized so that when a material is reported to have been exposed for 24 h at 92 °C in ASTM Fuel B, everyone recognizes the test method and can agree that the tensile results from Lab A can be directly compared to the equivalently developed data from Lab B. This does **not** mean the result "is" anything, good or bad, with respect to anything except this standard test. It only means the two values resulted from the same test method.

However, because the design engineer knows the seal made to a specific standard will always be the "same" within the limits of what is specified by the material standard, they no longer need to know, in most cases, what the test value is under every condition it may encounter in use. Testing of the new design confirms that an AMS 83485-210 O-ring works. As long as the O-ring meets the standard, it can be allowed it will continue to work in the design. This works because the manufacturer of the O-ring will always confirm the material by using a standardized laboratory test. If the same value is obtained on this batch as the last one, because it was tested exactly the same way, the material is still the same.

Additionally, because of the use of standardized laboratory tests, it is reasonable to add additional requirements on the desired value or tolerance. For example, when the material is tested per ASTM D1212, under a set of specified conditions, the material must have a tensile strength of a specific value, with an allowed tolerance. It may be a subset of an existing material, but because of the standardized testing, it can be specified. As another example, a designer has a specific environment that is outside of the standard and they are interested in exposure to a different fuel, for example, JP8+100. The standardized laboratory tests provide a basis for testing with which other industry experts have familiarity and which can be duplicated.

4.4 Non-standardized Testing

Standardized testing has been used for most testing, but there are a few non-standard tests that have to be developed to help facilitate compatibility evaluations. These tests are generally not defined by ASTM or other standards and do not have reliability values or collections of data. Generally, these tests are particular to certain laboratories and may include specialized test equipment or jigs. Examples of these tests are the measurement of volume swell via optical techniques and compression stress relaxation to measure sealing force degradation over time.

Compression Stress Relaxation can be performed in-situ or incrementally. In-situ testing continually measures the force of elastomers as it is aged in a fluid at a given temperature. The equipment for this test must include a way to place the elastomer in compression and measure the return force then subject it to a fluid or air and place it in an oven or refrigerator/freezer. A variety of in-situ test equipment is commercially available. Incremental stress relaxation measurements can be made with the aid of special fixtures and load frame measures. Elastomers can be placed into a fixture and measurements of force can be made in an Instron or other load measuring equipment. The fixture is then removed from the load measuring equipment and subjected to fluid/temperature conditioning. Load measurements can be made periodically to produce load reduction over time plots. Figure 19 is a picture of a commercially available in-situ stress relaxometer.

4.5 Basis for Long-Term Effects on Materials

Section 4.3 discussed the use of standardized laboratory tests. These tests had the advantage of familiarity, repeatability, relative ease of execution, and relatively low cost (albeit not necessarily cheap). But as discussed, these tests are designed primarily to design and control production of the non-metallic material or part. Aircraft in use, however, have life spans of decades, and there are parts with service lives of multiple thousands of hours. How then to predict the life of a non-metal material, the service lives of which are typically fractions of the lives of metal components in which they work?

In general, there are two ways to evaluate the useable life of a non-metallic. One is to test it under the conditions it will experience for hundreds and thousands of hours. In the US, paint manufacturers take pride in having sample panels which are exposed to the elements literally for years. While this method has the advantage of being able to directly correlate to actual use, in general, it is not practical to test all non-metal components used in aircraft and distribution systems in this way.

Fig. 19 Optical dilatometry equipment (Photo courtesy of A. Fletcher)

The alternative is to execute testing under accelerated conditions where, for example, one day at a specified accelerated condition has been experimentally determined to equate to one year in the field. The most common method of accelerating exposure is using heat. In the case of most non-metallics discussed here, industry experience is robust enough that standardized accelerated testing has been developed which is generally accepted as predictive. While not directly able to correlate 30 days at 100 °C (212 °F) to 30 months in use, for example, the designers have reasonable confidence that an acceptable result, however that may be defined, with the standard accelerated tests, has a reasonable likelihood of acceptable performance over a desired operating life. It is not precise and not always successful but is generally sufficient to support a first order approximation. Individual applications develop corporate experience with which it becomes possible to correlate accelerated results with field results over time, improving the predictions from the laboratory testing.

Ultimately, it is neither practical nor cost effective to test long-term life results directly, and accelerated testing can address this shortcoming. But by elevating the temperature, one is technically accelerating the degradation of the non-metallic because of the heat exposure and then correlating it to degradation of the material because of aging. While they are frequently similar, they are not necessarily the same. These results may be further confounded when the additional variable of fluid (chemical) degradation is included in the testing. The result is that while accelerated

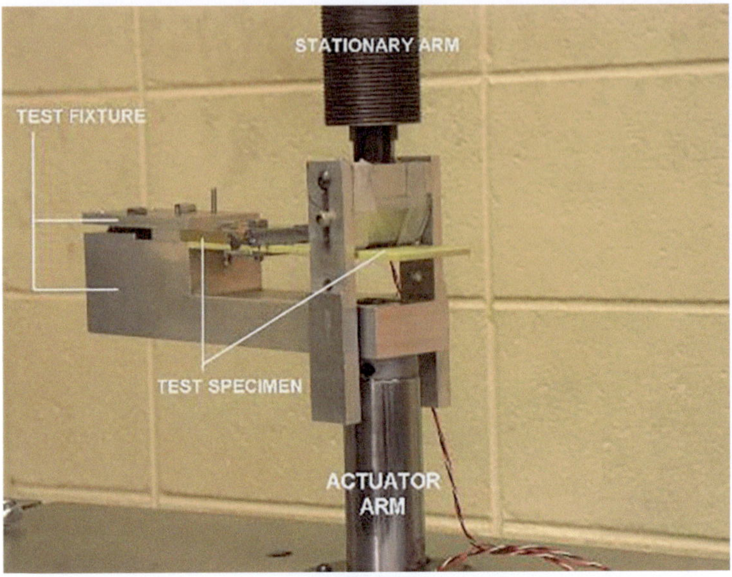

Fig. 20 Fatigue testing of fuel tank sealant (Photo courtesy of A. Fletcher)

testing can provide information regarding life span, the interpretations must be made with the understanding the results may be different in actual use.

4.6 Detail Testing for Vulnerable Materials

If standard tests reveal that there may be a vulnerability of a certain material, it may be worth performing detailed performance tests for the application of the material in the given fuel system. For example, if compression set of nitrile O-rings shows that the seals may fail before routine maintenance is scheduled, detailed tests such as compressive stress relaxation or static/dynamic seal rig testing may be performed to better ascertain the service life of the seal. Another example is that a sealant material that has questionable tensile/elongation data in the fuel, may be further tested within a fatigue rig to determine if the sealant will fail prematurely in the fuel. Figure 20 is a photo of fatigue testing being performed on a fuel tank sealant.

4.7 Data Analysis

Since physical property testing does not always simulate the exact use of the material in the fuel, there is always the need to understand and analyze the data for the aircraft and fuel system for which it is being proposed. For example, tensile/elongation data are easy to produce with reliability, but do not really simulate the function of the materials in the fuel system. Take the example of O-rings, having a significant change in tensile or elongation which in turn indicates that there is an interaction of the fuel with the sealing compound. However, it cannot be directly related to the service life of the O-ring. So, when we review physical property data, there is always the need to ask how the data relate to the function of that material in the fuel system.

4.8 Pertinent Standards

While it is not the purpose of this text to provide an inclusive library of test standards, Table 3 does provide the reader with a short list of typical test methods used with non-metallic materials.

4.8.1 Cost Versus Data Quantity and Quality

One of the challenges to the aviation/aerospace design engineer relates to data. What constitutes enough data? If one looks at any statistics textbook, one recognizes that the more individual data points in a population, the closer one gets to the "actual value." It is beyond the scope of this text to address the intricacies of statistics and interested readers should consider any number of available texts and internet sites for further education on determining population sizes and performing statistical analyses on the data. This discussion will be built on the foundational assumption of the more data, the better the ability to draw accurate conclusions, and the better the precision. This precision occurs at two levels, one within the test itself; how many specimens are necessary to have reasonable assurance that the result is precise? The second is the accuracy of the test itself; how many times do you have to run a new test to know what its precision and accuracy is?

One of the advantages of standardized tests is that the test has been run so many times, there is a reasonable expectation of the precision and accuracy of the test itself. ASTM standards often provide r^2 and R^2 values to give the user a qualitative understanding of how repeatable and reproducible results are when running the method in compliance with the provided methodology. This gives a feeling of confidence in the results, and a level of familiarity. Because repeatability and reproducibility are understood, a small number of specimens can be tested to determine the "value" for a given material.

In general, non-metallic test methods are performed using either three or five specimens in a sample set. Because the precision and accuracy of the test methods have been determined over time on large populations, experience supports such a small sample set. This sample set is actually only one of millions that have been tested within the industry. If one sample set does not perform as expected, there is enough statistical experience on which to base the observation and subsequent

Table 3 List of typical test method used with non-metallic materials

Application	Example std	Title
Coatings/Paints	ASTM D3359	Standard test methods for measuring adhesion by tape test
	ASTM D3363	Standard test method for film hardness by pencil test
Potting materials	KSC-STD-132	Potting and molding electrical cable assembly terminations, standard for
Sealants	AS5127/1	Aerospace standard test methods for aerospace sealants two-component synthetic rubber compounds
Polymers/Elastomers—flat stock	ASTM D297	Rubber products—chemical analysis
	ASTM D395	Test methods for rubber property—compression set
	ASTM D412	Test methods for vulcanized rubber and thermoplastic elastomers—tension
	ASTM D471	Test method for rubber property—effect of liquids
	ASTM D573	Test method for rubber—deterioration in an air oven
	ASTM D624	Test method for tear strength of conventional vulcanized rubber and thermoplastic elastomers
	ASTM D1149	Test methods for rubber deterioration—cracking in an ozone controlled environment
	ASTM D1329	Test method for evaluating rubber property—retraction at lower temperatures (TR test)
	ASTM D2137	Test methods for rubber property—brittleness point of flexible polymers and coated fabrics
Polymers/Elastomers—shaped stock	ASTM D297	Standard test methods for rubber products—chemical analysis
	ASTM D395	Standard test method for rubber property—compression set method B
	ASTM D471	Standard test method for rubber property—effects of liquids
	ASTM D573	Standard test method for rubber—deterioration in an air oven
	ASTM D1414	Standard test methods for rubber O-rings
	ASTM D1329	Standard test method for evaluating rubber property—retraction at lower temperatures (TR)

(continued)

Table 3 (continued)

Application	Example std	Title
	ASTM D2240	Standard test method for rubber property—durometer hardness
	ASTM D7426	Standard test method for assignment of the DSC procedure for determining Tg of a polymer or an elastomeric compound
	ASTM D3677	Standard test methods for rubber identification by infrared spectrophotometry
	ASTM E1131	Standard test method for compositional analysis by thermogravimetry
Adhesives	ASTM D907	Terminology of adhesives
	ASTM D1002	Standard test method for apparent shear strength of single lap-joint adhesively bonded metal specimens by tension loading
	ASTM D5649	Standard test method for torque strength of adhesives used on threaded fasteners
	ASTM D6396	Standard test method for testing of pipe thread sealants on pipe tees
	ASTM D5363	Standard specification for anaerobic single-component adhesives (AN)
	ASTM D5649	Standard test method for torque strength of adhesives used on threaded fasteners
Composites	ASTM D2344	Standard test method for short-beam strength of polymer matrix composite materials and their laminates
	ASTM D2583	Standard test method for indentation hardness of rigid plastics by means of a barcol impressor
	ASTM D3039	Standard test method for tensile properties of polymer matrix composite materials
	ASTM D5379	Standard test method for shear properties of composite materials by the V-notched beam method
	ASTM D5687	Standard guide for preparation of flat composite panels with processing guidelines for specimen preparation
Hoses	ASTM D380	Standard test methods for rubber hose

(continued)

Table 3 (continued)

Application	Example std	Title
Filters	ASTM D1094	Standard test method for water reaction of aviation fuels
	ASTM D1331	Standard test methods for surface and interfacial tension of solutions of paints, solvents, solutions of surface-active agents, and related materials
	ASTM D2276	Standard test method for particulate contaminant in aviation fuel by line sampling

investigation as to why were the results unexpected. The operator? A problem with the instrument? Or a problem with the material? Often several more sample sets will be tested to help determine the cause.

But what if it is a new formulation for which there is not a statistical population with which to compare the results? What if it is an invention for which a specified limit has not yet been generated? What if it is a test method being developed for which there is no known repeatability or reproducibility statement? What if an existing test method is being used in a new way? In most cases, the answer is to run many more sample sets, growing the population until it is large enough to be statistically predictive. While it is costly in time, resources, and money, the only way to improve the accuracy and precision of a result is to run more tests.

The challenge comes to be as the actual cost of an individual test increases, it becomes cost prohibitive to run hundreds of sample sets. When performing the testing for alternative jet fuel as specified in ASTM D4054 and for alternative aviation gasoline as specified in ASTM D7826, there are specified tests which have costs, resource requirements, and access limitations that can mean testing on a population of one. While both understandable and practical due to costs and access, for example, to run a gas turbine icing test, the evaluation must consider the trade-off of cost versus data quantity.

There is a saying, "once is an occurrence, twice is a coincidence, and three times is a trend." Consider the sample set requirement for three to five specimens. When one result in five is statistically different, it may be possible to determine the source and eliminate it as an outlier. With only one result, it is not possible to know anything but the value of the result. Even with two specimens, one is left with the quandary of which value is the "correct" value? Even with three to five specimens, if there is no identifiable reason to exclude a result, the anomalous value must remain. But as the sample set size increases, the amount this one value can skew a result is minimized. So, when the D4054 sample set has a population of one, does an unanticipated, or worse, undesirable result mean the fuel being tested is unacceptable or might there have been an issue with the test? Even if it were possible to repeat the test and now an acceptable result is observed, is this result the outlier or was the unacceptable result the outlier?

The second consideration is the assessment of how good are the data obtained. Is the test method one that by its nature has a lot of variability? Or is it a test method that has such good repeatability and reproducibility that it is generally accepted a single result is both accurate and precise. For example, by the very nature of how a sheet of polysulfide is prepared, the variability of the tensile results obtained in testing with properly prepared Die C tensile bars will be high. The sealant sheet will have voids and air bubbles due to its construction, and while they may be minimized, they remain a source of variability. On the other hand, the variability in the results from a properly prepared fluorosilicone test piece is expected to be measurably lower. While the test method ASTM D1414 has a precision statement for the test method itself, the reported precision in the collected data (i.e., error bars) is related to the material which is being tested.

On the other hand, the test method itself could have large variability. The adhesion test for rating surface coatings, ASTM D3359 is dependent on the skill of the operator to make uniform scores and to apply the pressure tape repeatably for each test. In this case widely varying results may not be different results, or even variations in the actual coating adhesion, but are a function of the skill required to run the test.

The conclusion is that one must balance the cost of running the test, especially multiple times, against the generation of a larger population, thus improving the quality of the data being reviewed. If the test method has a history of being reliable and reproducible, then a smaller data set is likely to be acceptable. If the test method is new, or the typical value for a material under specified test conditions has not been determined, then the risk of an erroneous interpretation goes up as the sample size goes down.

4.9 Standardized Test Methods Versus Use Conditions

In the preceding sections, emphasis has been placed on the value of standardized testing. As has been presented, the use of a standardized test comes with information built in, potentially a determined precision and accuracy statement, and general industry familiarity. When a material's physical properties are presented as data collected using standardized tests, the user can have a certain level of confidence in the value being accurate. But that accuracy is specific to the test being run. When a plastic has been tested using ASTM D543 *Standard Practices for Evaluating the Resistance of Plastics to Chemical Reagents*, the results collected are those specific to the protocols used in the method. The behavior of the material outside of that specified environment is extrapolated based on industry experience.

It could be asked, is it entirely predictive to have a static piece of plastic immersed in a jar of test fluid at one temperature and then tested for hardness changes if the actual use of the part is under a load and it has only incidental fuel contact? It would seem more predictive to devise a test rig to hold the part under the anticipated load and develop a method of intermittent exposure by the actual fuel. This method has the benefit of more closely replicating the actual use conditions. It has the disadvantage of having no experience base for how well it simulates the use conditions, how variable the results are, and if it even is capable of determining differences between sample sets.

With the recent emphasis on alternative fuels and fuel additives development, this question has become more topical. Given the history of aviation, it is entirely possible to have non-metallic materials which have changed very little since their original inclusion in an aviation fuel system design.

One of the observations made during the development of JP8+100 was the presence of parts which had not been tested with fuel in decades, and some of which had never been tested against contemporary turbine fuel. The result was that there were data which were not what was expected when tested in the baseline kerosene, much less the target JP8+100. Because of the familiarity with the standardized tests, it was unlikely that the results were incorrect. Given that many of the materials tested were standardized materials, it was possible to confirm that the materials being tested were still meeting specifications. It was more likely that the target limits chosen to evaluate against were incorrect. It was possible to conclude that the limits selected for the testing did not represent the design, or that the material limits were developed from the specific test fluids of the standardized tests, not for the actual jet fuel. The use of the standardized test methods and standardized materials facilitated defending a comparative analysis between baseline jet fuel and JP8+100, even where the two results did not meet the target values.

More recently, the investigations into alternative aviation gasoline formulations identified an even larger gap in time between the original design of aircraft fuel systems and today. Often there were no target values for a given non-metallic material or component. They were never tested originally, and currently only tested for compliance to a production formula or standard. To evaluate new fuel formulations, the tests, the exposures, and the target values all needed to be developed. On the one hand, it was attractive to use standardized test methods and protocols. They were familiar, defined, and had precision statements. The sources for selecting the methods were standard industry practice and the methods in the material standards. While the methods were familiar and had industry acceptance, there was no information on how the results related to the actual materials in the aircraft fuel systems. And the test methods used for the material standards were designed to control the exposure variables, including the test fluid, to confirm the batch of material was correct. The target values used in the material standards had no relationship to actual use or to aviation fuels. A legitimate question was asked, just because the non-metal would not meet the QA/QC target value, did it necessarily mean it was not acceptable for use in an aircraft? Additionally, could extrapolations from the test conditions to other operating conditions be made with any confidence?

The alternative to using existing "fuel in a jar" tests was the development of tests simulating actual use. While there would be no target value, it could be assumed that a comparative result in the test between a baseline aviation gasoline and the alternative fuel would support acceptance of similarity. Developing such a test included defining the test conditions, in most cases related to time of exposure and temperature of exposure, and defining the test article. It was recognized that a finished part may behave differently than a slab stock sheet with a standardized test specimen. It could be argued, when evaluating the use of the component, the results of slab stock testing were meaningless. Unless you were using a standardized test to get a target result, the slab stock might not be predictive. There was an argument against this path. Nothing was known about the predictive ability of the simulated use tests. There were no precision statements. They had no historical population, so statistics had limited value. There was no known correlation between the test results and actual use, so the predictive ability was limited to the conditions at which the test was run, and extrapolation was risky.

The choice of test method comes down to balancing the familiarity and precision of a standardized laboratory test, with a possible lack of predictive target values and not simulating actual use, against an unknown test with no precision information and no test history, but that better reflects how the material is actually used. This evaluation must also include a consideration of the cost versus quantity assessment discussed in Sect. 4.6.

5 Conclusions

When considering the testing of non-metallic materials with fuel, it is important to recognize the complexity of the task. The top five conclusions regarding the challenges in testing and evaluating the data from non-metallic testing presented are:

1. Recognition of compatibility being related to both the functionality of the non-metallic material when exposed to the fuel, and the impact on the health of the fuel due to extraction of chemicals from the non-metal.
2. The development of a list of test materials is complex and expansive. The complexity is a result of the number of components involved across the entire piston and turbine aircraft population covering nearly a century of designs. Not only is the list of types of components long, but there are also a variety of non-metal types (fluorosilicone, fluorocarbon, polyimide, etc.) used to produce those components. Within these types of materials, there is a near infinite list of individual formulations specific to each type.
3. Testing non-metals requires consideration of testing standardized material sample forms to standardized test methods versus testing of specific finished components to standardized tests versus testing specific finished components to non-standardized tests reflecting actual use.

    - Not all materials to be considered are continuously exposed to the fuel, i.e., fuel-wetted. There are critical materials which may be exposed by incidental contact with the fuel either by accidental content (i.e., splashing) or by periodic contact.

4. Evaluation of the resulting data requires consideration of what evaluation criteria are applicable, and what the absolute values may mean versus the use of the material.

   - Recognition that values developed for the certification of a material may or may not be applicable when the material and test method become the controls, and the exposure fluid becomes the variable.

5. Consideration of cost versus sample size, and recognizing the limitations of a small sample population, up to and including a population of one.

Acknowledgements The authors thank Micki Overberg and Moshe Rabaev for their reviews and the editorial team for their time, comments and contributions to the production of this chapter.

References

1. Parker SP (1984) Red. McGaw Hill Dictionary of Chemistry
2. EI/JIG 1530 Quality assurance requirements for the manufacture, storage and distribution of aviation fuels to Airports. http://www.jigonline.com; https://publishing.energyinst.org/featured/1530
3. Parker O-ring handbook (2007). www.parkerorings.com
4. Van Rensselar J (2017) Introduction to static seals. Tribol Lubr Technol 73(2)
5. Baere Aerospace Consulting, Inc., US Department of Defense, US Army Research Center (TARDEC), & US Air Force National Aviation Fuel Research Center (NAFRC) (2006) Developing sealing guidance for fuel-wetted components in next generation US Warfighters. University of Dayton Research Institute (UDRI)
6. McGuire N (2017) Introduction to dynamic seals. Tribol Lubr Technol 73(7):42
7. Guralnik DB, Friend JH (eds.) (1953) Webster's new world dictionary of the American language: College edition. World Publishing Company

Chapter 7
Aviation Fuel Stability

Steven Zabarnick, Robert E. Morris, and Josefa Ben-Asher

Abstract The term "jet fuel stability" covers a variety of fuel properties that result from the fuel's chemical composition upon production, as well as changes to these properties upon fuel degradation in the supply chain and upon use in aircraft engines. Stability is ordinarily separated into storage stability, the ability of a fuel to resist degradation during supply chain transport and storage, and thermal stability, the ability of the fuel to resist degradation during exposure to relatively high temperatures. This chapter summarizes the fuel degradation processes which occur in these two regimes, details the test methods employed, and offers guidance in preventing and mitigating the effects of these degradation mechanisms.

1 Types of Jet Fuel Stability

Jet fuel stability refers to the ability of the fuel to retain its required properties and composition prior to its use as a combustion source in the aircraft engine. Here we differentiate between two types of fuel stability: storage stability and thermal stability. Storage stability is the ability of a fuel to resist degradation during supply chain transport and storage. Storage stability conditions occur at relatively low temperatures (i.e., ambient temperatures up to about 60 °C) for relatively long periods of time (i.e., weeks to years). Fuel degradation during storage can result in the formation of detrimental species, such as hydroperoxides and acids, as well as precipitates, which can plug filters and close tolerance passages. Thermal stability is the ability of the fuel to resist degradation during exposure to relatively high temperatures (usually

S. Zabarnick (✉)
University of Dayton Research Institute, Dayton, OH, USA
e-mail: Steven.Zabarnick.Retiree@udri.udayton.edu

R. E. Morris
Nova Research, Inc, Alexandria, VA, USA
e-mail: rmorris@novaresearchonline.com

J. Ben-Asher
The Israeli Institute of Energy and Environment, Tel Aviv-Jaffa, Israel

M. Rabaev et al. (eds.), *Aviation Fuel Problems and Solutions: From Refinery to Wingtip*, https://doi.org/10.1007/978-3-032-05514-9_7

over the range of 120–300 °C) over relatively short thermal exposure times (i.e., seconds to minutes). Fuel is heated during flow through aircraft fuel systems, where it can be used to cool multiple system components, such as fuel/oil heat exchangers. Exposure to the highest temperatures occurs in the combustor fuel nozzles, prior to injection and atomization in the combustor. Fuel degradation during thermal exposure can result in significant fuel oxidation resulting in detrimental oxidation products, such as particulates and surface deposits. Particulates can plug close tolerance nozzle passages which can interfere with the nozzle spray pattern, while surface deposits can inhibit heat transfer, cause nozzle plugging, and interfere with proper operation of valves.

2 Storage Stability

Storage stability refers to the resistance of a fuel to undergo chemical processes that can lead to a deterioration in properties or performance, during storage. Civilian and commercial jet fuel systems are designed for high volume and high turnover, with short-term storage stability requirements in days or weeks. Due to the need for strategic fuel storage, military fuels are required to remain stable for up to five years of ambient storage. For jet fuels, storage stability issues are relatively rare but they can exhibit poor stability in certain situations.

2.1 Theories of Sediment Formation Mechanisms

Due to the refractory nature of fuel insolubles, the exact chemical mechanisms through which fuels produce insoluble gums and sediments are not well known. One theory involves the presence of sediment precursors, i.e., compounds that are either high molecular weight or those that react to form larger conjugated products. Field Ionization Mass Spectrometric (FIMS) analyses that have indicated the presence of monomers, dimers, trimers, etc. (Fig. 1) were interpreted as evidence of this mechanism. As these molecules grow in size, they become more polar and hence precipitate from the nonpolar fuel. The finding that the heteroatomic content of fuel sediments is typically several orders of magnitude greater than in the bulk fuel has been interpreted as evidence of this mechanism. Figure 2 shows a morphological study that revealed the structure of fuel insolubles as spherical particles.

This has suggested a theory of sedimentation based on the polymer science concept of excluded volumes. In this theory, as molecule chains grow in size, they are less solvated by the fuel and contract until they reach a minimum excluded volume, whereupon precipitation from solution occurs. It is likely that both polarity and molecular size play a role in the deposition of particulates and gums from fuels in storage.

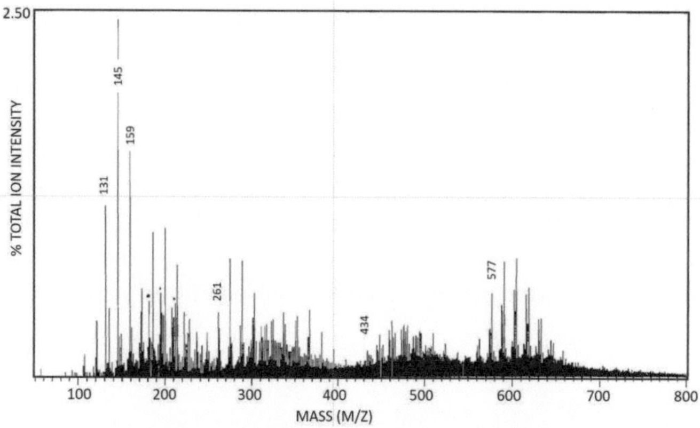

Fig. 1 FIMS analysis of fuel sediment showing repeating molecular sizes

Fig. 2 Electron micrograph of 1.8 μm spherical particles in sediment from Jet A1 after one month storage under air pressure at 185 °C

2.2 *External Influences*

Storage tank materials can exert an influence on long-term storage stability. Steel is the predominant tank material, but others include polymers and concrete. Salt caverns and rock caverns are generally used for strategic storage of crude oils. Vertical cylindrical steel tanks are typically located above ground, while steel tanks located underground are usually positioned horizontally. Above ground tanks are subject to atmospheric and solar heating, where the fuel near the vessel walls is warmed. As the fuel is warmed it becomes less dense and circulates to the surface where it can acquire oxygen and moisture from the headspace. As the tank cools, the oxygenated fuel with entrained water descends into the bulk fuel. In addition, diurnal cooling at sunset will cause air to be drawn into the tank, introducing moisture. Entrained moisture can induce tank corrosion and tends to concentrate polar additives in the water bottoms. Any vented tank withdraws air as fuel is drawn from the tank, introducing oxygen and moisture. These external influences can increase the likelihood of the formation of gums and sediments from unstable fuels, as well as generation of particulates from rust.

2.3 Aspects of Poor Storage Stability

Reactivity Fuels that contain reactive constituents can react to form insoluble gums and sediments, and potentially degrade the performance. These processes include autoxidation, which is a free-radical chain reaction that leads to insoluble products and thermally reactive oxygenated compounds. Under some conditions, fuels with higher olefin contents can undergo acid-catalyzed and oxidative addition reactions that lead to insoluble products. While acid-catalyzed addition reactions may not be prevalent in jet fuels, there are some co-oxidation reactions between free radicals and unsaturated compounds that can occur.

Compatibility Occasionally, when two stable fuels are comingled in the fuel storage system, the resulting blend is found to be unstable, forming unacceptable quantities of insoluble particulates or gums. This is typically referred to as "fuel incompatibility" and has been attributed to either polarity changes which cause deposition of asphaltenes or chemical synergism. Synergistic interactions typically have been found to be a consequence of interactions between key reactive organonitrogen and organosulfur species.

Contamination The introduction of polar high molecular weight constituents from untreated light cycle oils that contain polycyclic aromatics can lead to poor storage stability due to chemical interactions with such electron-rich organic molecules. Non-polar high molecular weight contaminants, e.g., lubricants and greases, can also manifest as insolubles and/or gums. Contamination with catalytic transition metals, most notably copper, can cause a fuel to undergo autoxidative degradation, although jet fuels are not typically subject to autoxidation during storage, but almost always fail requirements for thermal stability in the presence of trace levels of copper, as discussed below.

Additive Issues Water that has accumulated in tank bottoms extracts Diethylene Glycol Monomethyl Ether (DiEGME) anti-icing additive from the fuel. In some cases, when fuel components and polyacrylate particulates dislodged from water monitors were introduced into this DiEGME-enriched water phase, a gelatinous

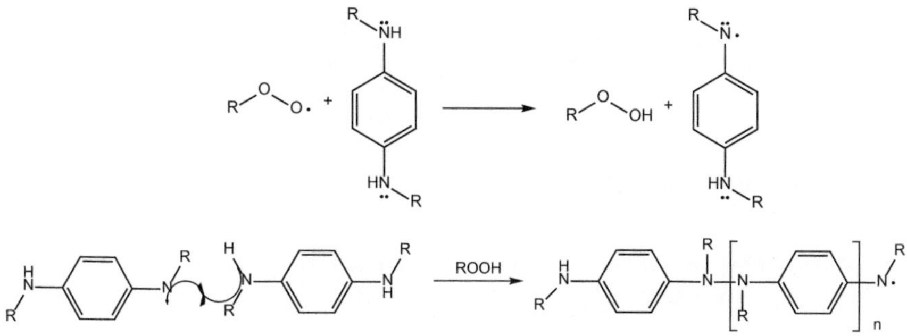

Fig. 3 Oxidative polymerization pathway for phenylenediamine antioxidant additives

non-Newtonian gel (a.k.a. "Apple Jelly") formed. This problem has largely been eliminated by removal of superabsorbent polymer fuel water monitors.

Phenylenediamine stabilizer additives are very effective antioxidants, but are no longer allowed in military jet or diesel fuels because they can undergo oxidative polymerization to form a polyphenylenediamine precipitate as shown in Fig. 3.

2.4 Accelerated Aging and Stability Tests

During prolonged storage, chemical and physical processes can occur. The chemical processes include the association of fuel with atmospheric oxygen (formation of peroxides) and the subsequent formation of soluble and insoluble gums. These processes are related to the storage temperature, the presence of atmospheric oxygen, and catalytic influence of trace metals.

In order to predict fuel behavior in storage, accelerated aging tests were developed to estimate the tendency of stored fuel to form gums, resins, and deposits by exposing it to extreme conditions (e.g., oxygen, pressure and heat) during the test. These include:

ASTM D873 [1] Standard Test Method for Oxidation Stability of Aviation Fuels (Potential Residue Method). Jet fuel is exposed to extreme conditions: pure oxygen, 100 °C, 7 atm. for 16 h. Extent of gums and resins formed during the test are evaluated.

ASTM D4625 [2] Standard Test Method for Middle Distillate Fuel Storage Stability at 43 °C (110 °F). The fuel is subjected to oxidation in mild conditions in air at the temperature of 43 °C. Samples are analyzed for an increased content of resinous compounds, structural changes in the resins (IR spectrum), and change of the color of the test sample.

ASTM D3241/IP 323 [3] Standard Test Method for Thermal Oxidation Stability of Aviation Turbine Fuels. Method for evaluation of thermal oxidation stability of jet fuel. Though the test is part of jet fuel specifications one can look at it as an accelerated aging test. During the test, the fuel is exposed to extreme conditions. Jet fuel with marginal results is not suitable for long-term storage.

2.5 Additional Stability Tests

ASTM D5304 [4] Test Method for Assessing Middle Distillate Fuel Storage Stability by Oxygen Overpressure.

ASTM D3703 [5] Test Method for Hydroperoxide Number of Aviation Turbine Fuels, Gasoline and Diesel Fuels. Formation of hydroperoxides is the first stage of jet fuel oxidation during storage. Therefore, evaluation of storage stability of jet fuel must include evaluation of hydroperoxide number of the sample. The presence of hydroperoxides at levels above 0.5 ppm active oxygen can be an indication of the onset of destructive processes. It is important that the sample not be exposed to light at any time during sampling or transportation to laboratory.

ASTM D381 [6] Method for Determination of the Existent Gum Content of Aviation Fuels. Existent gum content of more than 3 mg/100 ml may be an indicative of inherent instability and a tendency to initiate rapid oxidation processes. Jet fuels with existent gum less than 1 mg/100 ml may be considered for long-term storage.

ASTM D1319 [7] Method for Hydrocarbon Types in Liquid Petroleum Products. This test method covers the determination of hydrocarbon types of total aromatics, total olefins, and total saturates in jet fuels. Although the total olefin content is not explicitly restricted by ASTM D1655 [8] or UK Def Stan 91-091 [9] specifications, it should be no more than 1% v/v for stocks intended for long-term storage to avoid potential stability issues.

ASTM D156 [10] Standard Test Method for Saybolt Color of Petroleum Products. In relevant jet fuel specifications, there is a requirement to report the value of Saybolt Color. No maximum or minimum limits are defined. Experience with stored fuels indicates that during storage Saybolt Color tends to decrease. Jet fuels with Saybolt Color less than 13 probably have a prior storage history and therefore should not be designated for prolonged storage.

ASTM D5453 [11] Standard Test Method for Determination of Total Sulfur by Ultraviolet Fluorescence. Some jet fuels meeting military specifications have been observed in recent years to develop undesirable color upon standing. While these jet fuels met all applicable specifications initially, yellow color developed upon storage for extended periods of time. This change in color was associated with the presence of nitrogen compounds in the fuel. Some studies have shown that high sulfur fuels exhibit poor storage stability. Moreover, a synergistic effect between sulfur and nitrogen compounds has been observed, resulting in reduced stability as exhibited by excessive reactivity and formation of resinous agglomerates. Although sulfur content does not change during storage, stored jet fuel with sulfur contents higher than 0.15% mass and Saybolt Color less than 13 should not be used.

2.6 *Contamination Testing*

Three types of contaminants can accumulate in aviation fuel during long-term storage: solid particulates, water and microbiological debris. Normally tanks used for prolonged storage shall be designed to minimize ingress of water and particulates. Unfortunately, this is not always possible in the absence of available tankage. When the fuel is stored in tanks equipped with a floating roof, external dirt and water can enter. In addition, polymerization processes of unstable compounds can form species that initially dissolve in fuel but eventually separate becoming particulates. Aviation turbine fuel cleanliness requirements differ in various specifications: ASTM D1655 specifies fuel free of undissolved water, sediment or suspended matter. Def Stan 91-091 limits the content of particulate contamination at point of manufacture to max 1 mg/l by Millipore method (ASTM D5452/IP423 [12] and other methods), but does not set a requirement for particulates content downstream. U.S. Air Force military specification MIL-DTL-83133 [13] requires a maximum concentration of contaminants of 1 mg/l along the entire supply chain. It is advisable to adhere to this requirement of maximum 1 mg/l when filling a tank for long-term storage. It is not sufficient to fill a tank with a clean product; the amount of contaminants in the stored fuel must be monitored periodically to ensure that the fuel is still on-spec and fit for use. The first examination should be performed several days after filling the tank in order to allow the contaminants to settle. Subsequently testing is required periodically during storage to monitor potential deterioration or contamination. Tests should be performed within an interval from 6 months to 1 year according to the written procedures of the storing organization.

2.6.1 Particulates Contamination Test Methods

Particulates contamination monitoring can be performed by the following test methods:

- Visual inspection
- Gravimetric/Colorimetric test (Millipore)
- Filtration time
- Particle counting.

Visual appearance check (ASTM D4176 [14]) All significant international specifications for aviation fuel (ASTM D1655, Def Stan 91-091, MIL-DTL-83133, GOST 10227, TS-1 [15]) begin with the requirement that fuel be free from solid matter and undissolved water at ambient fuel temperature. Therefore, the easiest test to evaluate contamination of stored fuel is ASTM D4176. A representative fuel sample is transferred into a graduated glass cylinder and inspected for free water and particulates. If dirt cannot be visually detected, the fuel passes. This test provides a rapid pass/fail method for contamination evaluation in aviation fuel. Its limitation is its subjectivity, because the human eye can detect particles only larger than 30–40 μm.

Gravimetric/Colorimetric tests (Millipore) (ASTM D5452/IP 423) A gallon of aviation fuel is fed through a pair of matched weight membranes with a pore size of 0.8 μm. The membranes are dried in standard conditions and weighed. The weight difference between the two membranes is the measure of the particulate content of the sample. The color on the membrane can be used for evaluation of the type of contaminant (rose brownish—rust, yellow brownish—sand, black—soot). The obvious advantage of this method is the simplicity of execution and numerical results (though some claim that the precision of the test is low). A disadvantage is that particles smaller than 0.8 μm (micron) pass through the membrane and are not weighed. In reality, a situation can arise in which the result of this test is low while filters in the field become clogged.

Filtration Time MIL-DTL-83133 Appendix C This method is a variation of the gravimetric method. In this test, a gallon of fuel is passed through a 0.8 μm membrane and the time required for passage is measured. Fuel cleanliness is reported as filtration time. If the time is more than 15 min, the fuel volume is recorded and a comment is added "greater than 15 min". The advantages of the method are: simple to perform, sensitive to the presence of water and particles smaller than 0.8 μm. The disadvantage: no possibility to follow contamination trend or origin.

Particle Counting (D7619 [16]**, IP 564** [17]**, IP 565** [18]**, and IP 577** [19]) This method features dispersed particles determination in the range from 4 to 30 μm, up to a maximum of 40,000 cumulative counts per ml. The advantage of the method: the particles can be classified according to their size and quantified in each range. Only a small amount of sample is needed for testing. However, presenting the information in ISO CODE makes it very difficult to assess the results. The results do not present the nature of the contaminant making it difficult to evaluate the problem.

2.6.2 Water Accumulation and Microbial Contamination During Storage

During long-term storage, water vapor in air condenses due to temperature changes. Since water is denser than fuel, it will settle to the tank bottom. This potential for free water accumulation at the bottom of the tank and its contact with fuel, introduces a further risk of contamination—microbial growth. To minimize microbiological contamination of the stored fuel regular draining of the free water is strongly recommended. In operational tanks, water is drained on a daily basis or at least once a week. In long-term storage tanks, it is recommended to drain at least once a month. To verify periodic drainage, microbiological testing should be performed yearly or at six-month intervals according to the tank design in which the product is stored. Microbiological contamination in aviation jet fuel has been traditionally measured by laboratory method IP 385. Since then rapid methods (kits) have been developed. The analysis does not require professional knowledge nor laboratory facilities. This book deals extensively with the subject in Chap. 12.

2.7 Testing Frequency

Organizations engaged in long-term storage shall have a documented process to ensure good quality of aviation fuel product during storage. Some organizations recommend a bi-annual inspection of parameters indicated in Annex B2 EI/JIG standard 1530 [20]. Others inspect the stored fuel on an annual basis for requirements of the specification that the fuel was originally certified to and all existing amendments/revisions to confirm the product meets the current requirements. Others perform annual tests that include, not just specification requirements, but also accelerated tests and trace metals content evaluation. In any event, the time interval between inspections should not exceed 12 months.

2.8 Stock Renovation Considerations

A decision to renovate a stock should be based not only on the jet fuel current chemical quality and the anticipated long-term stability but also on a variety of additional factors outlined in this subsection. For evaluation of current and future chemical quality, some stockpiling organizations use human expert advice. Others rely on software for automated assessment and recommendation with regards to the potential stability of the stored jet fuel. Some organizations use both methods: automated assessment in parallel with human expert recommendation. In case of conflict, human expert recommendation prevails. The automated assessment is used to draw the attention of the human expert to possible problems and sharpen their observations.

Additional factors that should be taken into consideration: What is the storage period so far? Is it possible to detect a steady decrease in the quality of the stored fuel? What is the remaining storage period based on the approved procedures of the stockpiling organization (e.g., no longer than five years in case of jet fuel)? Does the organization prefer to store fuel for as long as possible for economic reasons and sell only for strategic reasons, or when a lucrative deal can be identified? Does the fuel have an accumulation of contaminants or microbiological debris as a result of a leaking roof or inappropriate drainage procedures? How soon is the cleaning date of the storage tank (designated by the relevant regulations)?

3 Thermal Stability

At high temperatures, aviation turbine fuels can oxidize and produce insoluble deposits that are detrimental to aircraft propulsion systems. Deposits formed from jet fuels of poor thermal stability pose operational problems. These problems, which date back to the mid-1950s, include clogging of the injector nozzles, a decrease in the efficiency of the heat exchangers, and malfunction of the fuel metering controls. Although current commercial and military jet fuels normally exceed specification thermal stability requirements, intermittent incidents of operation with relatively low thermal stability fuels have continued to the present. This problem is further exacerbated by the higher temperatures of newer production engines—a consequence of improving efficiency and increasing power demands. The chemical and physical mechanisms that lead to formation and agglomeration of insoluble thermally induced reaction products have been envisioned to consist of a series of processes as depicted in Fig. 4.

3.1 Testing Methods

International jet fuel specifications require that jet fuels pass a minimum thermal stability requirement. This is normally evaluated using the ASTM D3241 JFTOT method [22]. As the JFTOT is designed for fuel qualification, short test times and low required fuel volumes are desirable. As such, it employs significantly accelerated conditions, i.e., relatively high temperatures, which do not necessarily represent the conditions to which fuel is subjected in aircraft engines. To more closely evaluate real engine conditions, the industry has developed a number of research methods that can provide thermal stability evaluations that may more closely represent engine conditions. The JFTOT system and some of these research methods are detailed in subsequent sections.

3.2 Jet Fuel Thermal Stability Tester (JFTOT) Specification Method (ASTM D3241)

3.2.1 Historical Development

Early turbojet and turboprop engines did not impose unusually high temperatures on the fuel. However, the situation changed with the introduction of high-pressure turbojets, which significantly increased the heat load on the fuel. In some engine designs, the nozzles and fuel manifolds were in contact with high temperature discharge air, which caused nozzle plugging. After it was reported that static stability tests did not correlate with fuel performance in these situations, a full-scale dynamic test rig,

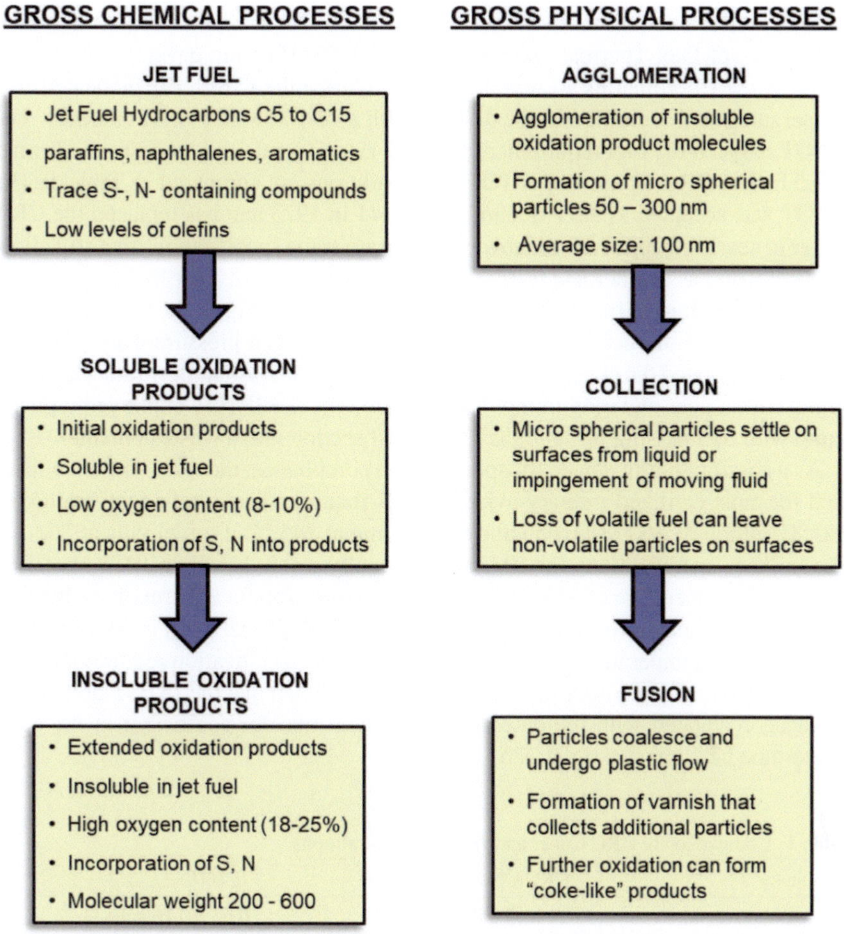

Fig. 4 Chemical and physical processes that lead to the thermal deposition of insoluble gums and sediments in jet fuels [21]

the ASTM-CRC Fuel Coker was developed and standardized as ASTM D1660 [23]. In the Coker, fuel was pumped over a preheater tube and through a heated 20 μm filter. The fuel temperature at the exit of the preheater and the filter temperature were controlled. For Jet A, JP-5 and JP-4 fuels, the Coker was operated at a fuel out temperature of 149 °C, a filter temperature of 204 °C at a flow rate of 2.7 kg/hr (~3.4 L/hr) for a test duration of 5 h. The visual appearance and color of the preheater tube deposits and filter pressure drop were used to assess the thermal stability of the fuel.

Because the fuel temperature, rather than the heated surface temperature, was controlled in the Coker, it was found that this did not adequately predict heat exchanger performance. In addition, the five gallons of fuel required per test caused some shipping problems and allowed only one test to be conducted in a day. To address these deficiencies, two scaled-down CRC Coker replacements, the Alcor and Erdco Jet Fuel Thermal Oxidation Testers (JFTOT) were developed and evaluated by the Coordinating Research Council. Unlike the Coker, the preheater tube temperature was controlled in the JFTOT with a fuel flow rate of 1.5 ml/min. The JFTOT reduced the fuel requirement from 19 L to 600 mL and the test time from 5 to 2.5 h. The CRC Coker and JFTOT test conditions are compared in Table 1. The JFTOT was accepted as ASTM standard D3241 in 1973 and has replaced the CRC Coker in nearly all applications, however there are some specifications that still allow ASTM D1660 as an alternate test.

In early versions of the JFTOT, shown schematically in Fig. 5, the fuel was passed in a closed loop through the heated section with nitrogen pressure over a floating piston that isolated the stressed from unstressed fuel. A constant flow rate was maintained with a gear metering pump. Later versions of the JFTOT employed a piston pump with fuel passing through the heated test section into a waste container.

As a pass/fail rating device for specification compliance, the test conditions specified for most civil and military aviation fuels require a heater tube temperature of 260 °C, although ASTM D1655 allows retesting at 245 °C if a fuel fails at 260 °C. The military specifications for JP-7 and thermally stable kerosene JPTS specify a heater tube temperature of 335 °C. The filter pressure drop for a thermally stable fuel is 3.3 kPa (25 mm Hg) for all JFTOT conditions. JFTOT breakpoint is defined as the maximum test temperature at which a fuel passes the specification requirements for heater tube deposit and filter pressure drop. It was shown that the JFTOT breakpoint could be correlated with the deterioration of heat transfer coefficient in the shell single-tube heat transfer rig at 225 °C.

Table 1 Comparison of CRC Coker and JFTOT test conditions

Parameter	CRC Coker	JFTOT
Fuel volume	19 L (5 gal.)	0.6 L (0.16 gal.)
Fuel flow rate	60 mL/min (6 lb/hr)	3 mL/min
Test time	5 h	2.5 h
Temperature control	Fuel-out	Heater tube
Heating method	Cartridge heaters in preheater tube and filter housing	Electrical resistance of heater tube
Pump location	Before test section	After test section
System pressure	1.0 MPa (150 psig)	3.45 MPa (500 psig)
Heater tube length	330 mm (13 in.)	60 mm (2.375 in.)
Residence time in heated section	10 s	13 s
Filter	25 μm stainless steel	17 μm stainless steel
Flow regime	Laminar	Laminar
Rating criteria	Tube deposits and filter pressure drop	Tube deposits and filter pressure drop

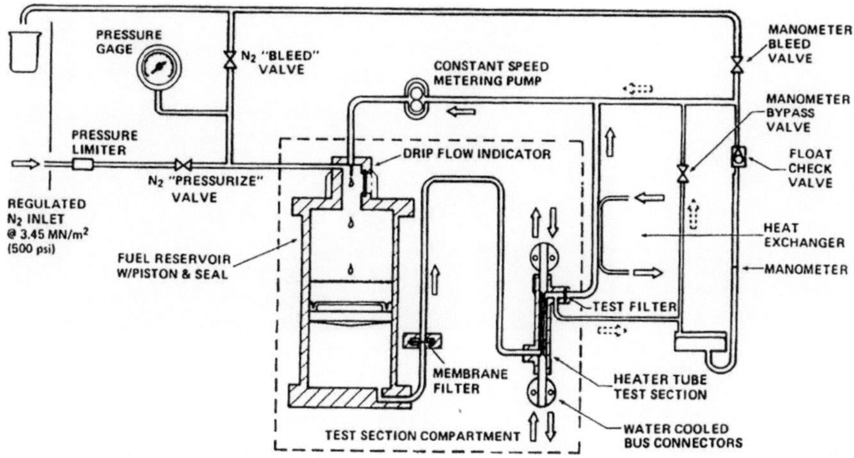

Fig. 5 Gas drive JFTOT apparatus

The heater tubes in both the Coker and JFTOT are composed of 6161 T6 aluminum and the heater tube metallurgy has been found to influence the extent of thermal deposition. A comparison of 304 stainless steel and 6161 aluminum heater tube deposits by carbon burn-off indicated roughly twice the amounts of deposits on stainless steel tubes. Moreover, surface chemistry studies had shown that magnesium, present at nominally 1% in the aluminum alloy, can migrate to the surface of the heated JFTOT tube and had been measured to be as high as 25 atomic percent on the surface. The presence of magnesium on the surface was found to exert an inhibiting effect on tube deposition. Subsequent modification of the material specification for aluminum JFTOT heater tubes eliminated the impact of magnesium on the tube surface.

3.2.2 JFTOT Tube Rating Methods

Visual Tube Rating The process of evaluating the amount of deposition developed on the heater tubes is referred to as tube rating. This was initially performed with the Coker and later the JFTOT, by the visual tube rating (VTR) method, introduced in ASTM D1660 in the 1950s and still included in ASTM D3241. The VTR is based on an operator comparing the color and intensity of tube deposits with an ASTM standard in a device that illuminates the tube and standard (Fig. 6). The subjective nature of the measurement and poor precision between operators of the VTR results was shown to result in random errors that exceeded the differences between the readings. These deficiencies led to various attempts to standardize tube deposit rating and make the measurements more quantitative.

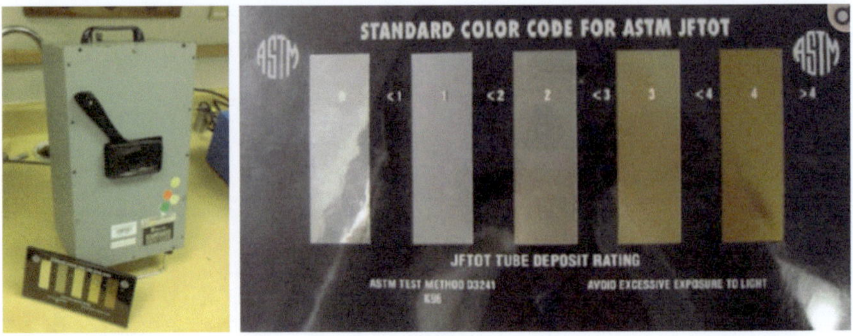

Fig. 6 Visual tube rater (VTR)

Tube Deposit Rater (TDR) The tube deposit rater was included in the ASTM method in 1972 as a less subjective measurement of heater tube deposits. The TDR was based on the measurement of attenuated reflected light from a heater tube deposit. The heater tube in the TDR apparatus (Fig. 7) is spun to get an average value radially around the tube, and the maximum TDR reading is obtained by examining the stationary tube. The TDR photocell is manually moved down the length of the tube and the maximum TDR rating is reported. Often deposits with thicknesses that are multiples of the wavelength of the incident light can produce interference fringes that are reported as "peacock" deposits and classified as a failure. Auger spectroscopy and optical interference measurements have shown that the thicknesses of these peacock deposits are generally on the order of ¼ to ½ the wavelength of the light and can correspond to a VTR rating of 3 or greater, thus confirming the classification by VTR and TDR as a failure. While less subject to operator judgment, the TDR reading could be influenced by the optical properties of the deposit. In addition, the TDR would occasionally provide a low rating to tubes that had obvious heavy deposits. Efforts to explain the differences between how the TDR rated the tubes and the VTR were unsuccessful, and the TDR was removed from ASTM D3241 in 1995.

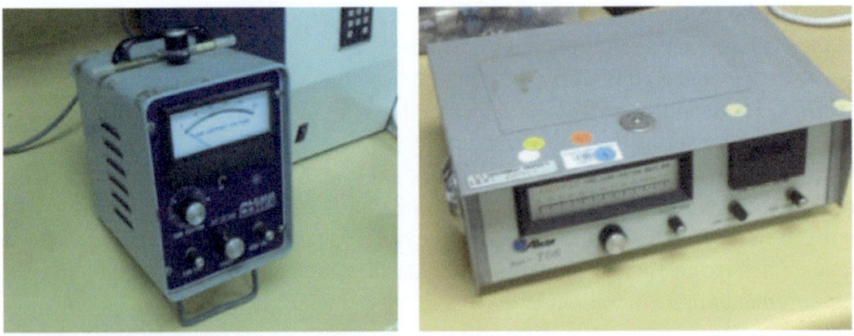

Fig. 7 Tube deposit rater (TDR) versions

Interferometric Tube Rater (ITR) In this method, monochromatic light having a wavelength of 680 nm is directed onto the tube in a direction perpendicular to the surface through a fiber optic assembly that contains both the source and detector optics. The intensity of reflected light represents the quantity of light that emerges through the deposit coating after reflecting off the metal tube surface. Since there is a large difference in the refractive indices between the air and the deposit, the reflected light wave undergoes a phase change. As a consequence, the emergent wave interferes constructively or destructively with the incident wave, depending on the thickness of the deposit. As the detector is scanned across a deposit, the light intensity changes periodically as the deposit thickness passes through multiples of the wavelength of the incident light. At greater deposit thicknesses, absorption of the light by the deposit dominates, limiting measurement to thicknesses below approximately 2 μm. The thicknesses at each tube location are averaged, and the thickness profiles thus determined are integrated to obtain the deposit volume. ITR instruments, such as the one shown in Fig. 8, are commercially available, and the method was included in ASTM D3241 and D1655 in 2014.

Ellipsometric Tube Rater (ETR) Ellipsometry is a well-established method in the semiconductor industry for measuring the thickness of thin films and has been successfully applied to the measurement of JFTOT tube deposit thickness. It is a non-destructive measurement of the change in polarization of laser light after it is reflected from a surface, where the polarization change is proportional to the deposit thickness. The polarization is characterized by an amplitude ratio (Ψ) and phase difference (Δ). In practice, the change in the polarization of light reflected from the JFTOT tube is measured by taking the amplitude ratio of two perpendicularly polarized laser light beams. When the data are fitted to a model of the layer, the refractive index, absorption coefficient and thickness of the deposit can be calculated. Since this measurement requires that the light pass through the deposit and be reflected back to the surface, the measurement will be limited to relatively thin, semi-opaque films up to 250 nm thick. This limitation is mediated by choosing an incident light wavelength that minimizes the absorption. By integrating the thickness measurements taken from the heater tube, the deposit volume is calculated. In addition, the maximum thickness is also reported. Early JFTOT-focused ellipsometers were large bulky instruments,

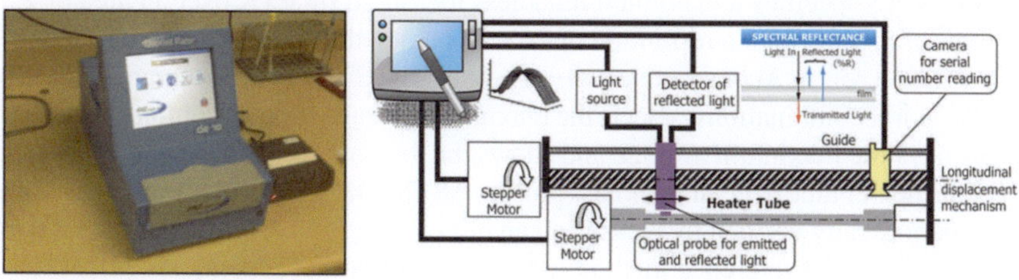

Fig. 8 Interferometric tube rater (ITR)

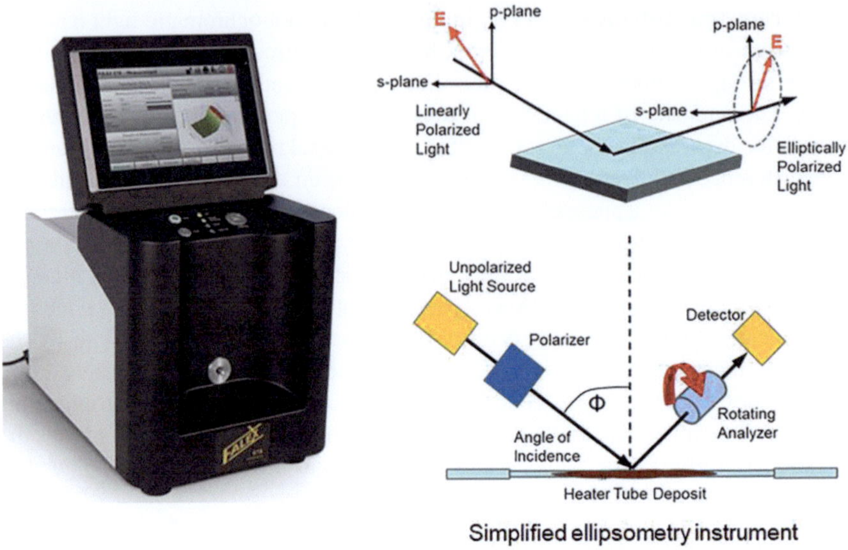

Fig. 9 Ellipsometric tube rater (ETR)

but current instruments, as shown in Fig. 9, are smaller and more compact. Ellipsometric tube ratings (ETR) were included in ASTM D3241 and D1655 in 2014 as an approved method to determine deposit volumes and maximum thickness.

Summary of Current ASTM D3241 JFTOT Heater Tube Measurement Specifications

- Annex A1 specifies that the VTR deposit code be less than 3.
- Annex A2 specifies the maximum thickness measured by ITR shall be less than 85 nm, and deposit volume is reported to the nearest 0.00001 mm^3. The ITR instrument acquires measurements at 50 points along the length from 5 to 55 mm and radially at 24 points. Baseline calibration is conducted with a new unused heater tube and ASTM D3241 specifies that calibration will be periodically verified with an ITR reference heater tube.
- Annex A3 specifies a maximum thickness measured by ETR over an area of 2.5 mm^2 of 85 nm. With a resolution of 24 x 50 points, the 2.5 mm^2 area represents 6 points in a row or 3 consecutive pairs of points. The ETR device software identifies all 6-point combinations, selects the maximum average 6-point combination, then displays the maximum average and shows the location of the maximum on a plot of the heater tube deposit thickness profile. ASTM D3241 requires that calibration be periodically verified by measuring an ellipsometer reference tube. The ETR is specified as the reference method in the U.S. military specifications MIL-DTL-5624 [24] for JP-4/JP-5 and MIL-DTL-83133 for JP-8.

Other Tube Rating Methods There have many other methods proposed to provide more quantitative measurements of JFTOT heater tube deposits. Carbon burn-off has been successfully used to obtain quantitative measurements of tube deposit weights, but this was limited to stainless steel tubes, which were not specified for use in the JFTOT. The dielectric breakdown was also investigated but had limited precision with thin deposits and was affected by metal incorporated into thermal deposits from copper-contaminated fuel. Other methods were also explored to acquire real-time deposition measurements in the JFTOT to develop kinetics and possibly a single-test breakpoint measurement. This included correlations of JFTOT breakpoints with laser-induced fluorescence (LIF), the use of a JFTOT heater tube holder with a quartz window and a scanning fiber optic interferometer, and an interferometric method using a series of fiber-optic light pipes fitted around the length of a specially designed JFTOT tube holder.

3.3 *Thermal Stability Research Methods*

The ASTM D3241 is the only thermal stability device included in jet fuel specifications. Its design and operation are directed toward a pass/fail evaluation of fuel thermal stability for commercial and military usage. Until the recent development of tube metrology methods (such as interferometry and ellipsometry), measurement of tube deposition employed relatively crude visual ratings. In aircraft engines, fuel deposition occurs over thousands of hours of engine operation, with many thousands of gallons of fuel used. D3241 was designed to be relatively rapid (2.5 h) and only require small fuel volumes (600 ml), and as a result, the test temperatures (near 260 °C) are accelerated (i.e., increased) from those used under most actual gas turbine engine conditions. As a result of these limitations, a number of research methods have been developed to better understand fuel oxidation and deposition as it occurs in aircraft engines. Here we present a brief and non-exhaustive review of some of these methods.

3.3.1 Quartz Crystal Microbalance (QCM)

The QCM system (ASTM D7739 [25]) was originally developed for jet fuel deposition measurements by Sandia National Laboratories and further refined by the University of Dayton Research Institute (UDRI). Its primary advantage is its extremely sensitive methodology for measuring surface deposits, in-situ and in real time. A drawing of the system is shown in Fig. 10. The original Sandia design incorporated a 5 MHz quartz crystal suspended in jet fuel contained in a 100 ml Parr reactor. The crystal was attached to an oscillator circuit for monitoring of the oscillator frequency, which decreased as fuel deposits attached to the crystal. The Sandia system was operated with an overpressure of air or oxygen. UDRI refinements to the QCM system included addition of a pressure transducer and oxygen sensor, and operation at lower pressures to limit the oxygen availability. Lower oxygen levels more closely match the ~70 ppm of fuel dissolved oxygen encountered during engine operation. The system

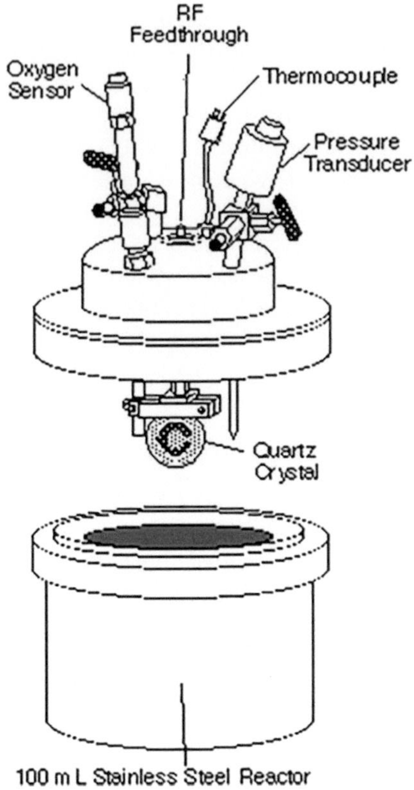

Fig. 10 Drawing of the quartz crystal microbalance reactor system

uses only 60 mL of fuel and typically employs a 15-h test time at 140 °C, although test time and temperature can be varied. Disadvantages of the QCM system are its batch reactor nature and relatively long test time, although shorter test times can be employed. The in-situ and simultaneous measurements of deposition, oxygen, and pressure provide the unique ability to monitor fuel oxidation rates, determine additive methodology (i.e., antioxidant character, detergent/dispersant character, etc.), measure deposit level relative to amount of oxygen consumed, etc. Figure 11 shows example QCM plots of two fuels, one a fast oxidizer and low depositor, the other a slow oxidizer and high depositor. Data such as this have helped determine the role of fuel constituents, such as heteroatomics on jet fuel oxidation and deposition.

3.3.2 High Reynolds Number Thermal Stability (HiReTS) Instrument

The HiReTS (ASTM D6811 [26] or EI 482 [27]) was a commercially available (Stanhope-Seta) system which was designed to incorporate many of the advantages of the JFTOT (e.g., short test times, low fuel volumes) but with high Reynolds number flow typical of fuel nozzles, along with a quantification methodology for deposition.

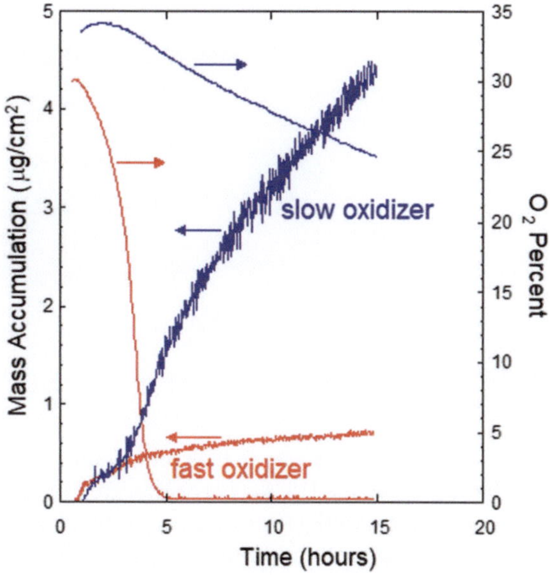

Fig. 11 QCM plots of deposition and oxygen for two fuels with differing oxidation/deposition characteristics

It was developed to provide a turbulent flow deposition test in response to concerns over the ability of the laminar flow in the JFTOT to realistically simulate turbulent flow regions in jet engine fuel systems. The system employs fuel flow through an electrically- heated capillary tube to enhance Reynolds number at a flow rate of 35 mL/min, an inlet Reynold's number of 7500 and a residence time of about 0.01 s. The test is conducted to maintain a constant outlet temperature, typically 290 °C with a test duration of 2 h. Deposition is quantified during the test by a movable pyrometer at nine positions 2.5 mm apart over the last 21 mm of the heated test section. Deposits formed on the internal surfaces of the tube insulate the tube wall from the convective cooling due to the fuel flow, resulting in a decrease in temperature which is related to the amount of deposition. As the test progresses, the change in this temperature map is quantified to produce a metric for thermal stability. The system computes a "HiReTS Number," which is a sum of the absolute value of the temperature differences from a clean tube along the tube during the test run. As such, low HiReTS numbers indicate favorable relative thermal stability. Figure 12 shows an example of the progression of the temperature map during a HiReTS test. The profile is flat at the beginning of the test (front), but develops a hump with time as the deposit within the capillary is developed. Fuels with higher thermal stability will exhibit less change with time.

3.3.3 Flowing Heated Tube Rigs

Most thermal stability research rigs employ fuel flowing through heat tubes (heat exchangers) with post-analysis of the internal tube deposit by carbon analysis via combustion to CO_2 (i.e., carbon burn-off) and spectroscopic IR analysis. The various

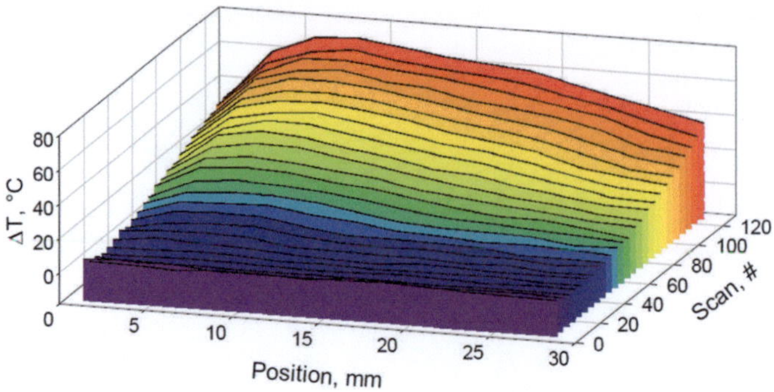

Fig. 12 Changes in heated test section external temperature during a HiReTs test

flavors of these devices vary in their temperature, flow, and heat transfer characteristics. For example, the US Air Force Research Laboratory (AFRL) Near-isothermal Flowing Test Rig (NIFTR) uses a copper block heat exchanger and very low flow rates (as low as 0.1 ml/min) with temperatures near 185 °C. These flow, heat transfer, and temperature conditions result in an almost isothermal reaction environment (over >90% of the tube length) which is ideal for chemical kinetic studies of fuel autoxidation. In contrast, the AFRL ECAT rig (ECAT has no particular meaning in English) uses higher flow rates and with reduced heat transfer effectiveness, resulting in a highly non-isothermal environment but the ability to run at higher temperatures, including the pyrolytic regime in which carbon–carbon bonds are readily broken [28].

The largest thermal stability rigs are fuel system simulators which attempt to simulate the individual components of aircraft fuel systems including wing and body tanks, fuel/oil heat exchangers, nozzles, etc. These simulators provide more realistic conditions which closely approximate those of aircraft fuel systems, but use much larger quantities of fuel (often many thousands of gallons) and longer test times (up to many weeks). Simulators can be designed to "fly missions," with varying fuel flow, temperatures, and pressures to simulate engine throttle variation during aircraft missions that can be repeated hundreds of times. This complexity, large fuel volumes, and long test times enables very close simulation of the conditions in aircraft fuel systems without significant acceleration of any important parameters. Of course, the downside is in cost and time. Fortunately, the smaller rig systems can be employed to complement the larger simulators, with initial exploratory, proof-of-concept work being done in the small systems, while the simulators are used to verify the small-scale results at more realistic conditions only after the exploratory work is concluded.

3.4 Thermal Stability Mitigation Strategies

3.4.1 Refinery Processing

One of the goals during refining of jet fuel is production of fuel with adequate thermal stability to pass the ASTM D3241 JFTOT requirement in the fuel specifications. As mentioned previously, jet fuel thermal stability is negatively impacted by the presence of fuel heteroatomics and dissolved metals. Thus, processes that remove or limit the levels of these species may be useful in producing fuel with improved thermal stability. So called "straight run" fuels are produced using limited or no significant processing to remove these species. As a result, straight run fuels are more likely to have thermal stability issues.

In recent decades, the introduction of low sulfur diesel specifications has increased the use of hydroprocessing during petroleum refining. Hydroprocessing can be an effective way of removing fuel heteroatomics via high-pressure, high temperature reaction of hydrogen with these fuel species, resulting in conversion of fuel nitrogen to NH_3, fuel oxygen to H_2O, and fuel sulfur to H_2S. These light species are then readily separated from the jet fuel fraction. The severity of hydroprocessing greatly affects the heteroatom level of the resulting fuel, with severe hydroprocessing capable of reducing heteroatomics to very low levels (<10 ppmw). These severely hydroprocessed fuels typically exhibit very good thermal stability. Less severe hydroprocessing can also improve thermal stability but this is dependent on the composition of the original fuel.

Merox™ treatment is a refinery process used to lower the mercaptan sulfur level of fuels, to minimize corrosion and odor issues. Merox™ is a catalytic process that converts mercaptans to less corrosive and odiferous disulfides. Thus, the Merox™ process does not actually remove fuel sulfur, but merely converts the form of the sulfur. Mercaptans and disulfides are species that have both been shown to degrade fuel thermal stability and thus, Merox™ processing is not likely to have a significant positive impact on fuel thermal stability in general.

As all petroleum-based jet fuel is produced via distillation processing, it is interesting to consider the effect of distillation characteristics on thermal stability. Narrowing the distillation cut, and in particular reducing the heavy end of the fuel during production, can reduce the concentration of some problematic species, such as metal salts. This will include copper species, which are known to negatively impact thermal stability at copper levels as low as 10–20 ppbw.

In recent years, the industry has seen an increasing number of thermal stability failures during jet fuel transport and storage, especially during transport in multi-product pipelines. Sometimes, the fuel passes thermal stability at the point of manufacture and at the pipeline receipt, but is found to be degraded at downstream storage locations. These failures are still not well understood, but may be attributed to mixing of two or more fuels within the transport and storage system, resulting in a fuel blend that is less thermally stable than any of the constituent fuels. For example, it has been shown that while particular S, N, and O containing species do not by themselves cause fuel deposits, adding combinations of these species can result in very high deposition levels.

The fuel industry has considered various methods to mitigate the thermal stability failures of these fuels, including clay treatment and downgrading and/or re-refining of the fuel. Clay treatment has been recently investigated for its ability to improve thermal stability. Clay treatment is normally used to reduce the level of fuel surfactants which could interfere with filter/coalescer separators which remove water from fuel. As clay is also known to adsorb other polar fuel components, it has been studied for mitigation of these thermal stability failures. Results show that clean, unused clay is effective at removing polar heteroatomics and improving thermal stability, but only for very low fuel/clay volume ratios. The clay is rapidly deactivated in its ability to improve thermal stability and thus clay treatment will not be effective at treating large volumes of fuel.

3.4.2 Fuel Additives

Chapter 5 has already provided an introduction to the main additives found in jet fuels. The use of additives for improving the thermal stability of jet fuel has been studied for many years. Additive types that have been investigated include antioxidants, metal deactivators (such as N,N-disalicylidene-1,2-propane diamine), detergents, dispersants, hydroperoxide decomposers, and various combinations of these additives. Current commercial jet fuel specifications, such as ASTM D1655 and UK Def Stan 91-091, only list metal deactivator additives (MDAs) and antioxidants as being approved for improving thermal stability. Thus, the other fuel additives cannot be used for the purpose of thermal stability improvement in commercial fuels.

MDA should only be used when the fuel is known to contain significant metals content, such as >20 ppbw of copper, but D1655 does allow MDA use for fuels which pass the JFTOT at 245 °C even if metal contamination is unproven. Antioxidants, such as hindered phenols, are currently approved to prevent the formation of oxidation deposits in aircraft engines and to improve the oxidation stability of fuel during storage.

Additive combinations have been used for many years in military applications in the U.S. Air Force specifications for JPTS (MIL-DTL-25524 [29]) and JP-8 (MIL-DTL-83133). JPTS requires the use of the JFA-5 package which consists of low concentrations of antioxidant, metal deactivator additive (MDA), and detergent. JP-8+100 incorporates much higher, and more effective levels, of hindered phenol antioxidant, MDA, and a combination detergent/dispersant. The so-called "+100 package" has been shown to reduce deposit formation up to ten-fold by synergistic interactions between the additive components, in slowing oxidation (by the antioxidant and MDA combination) and preventing agglomeration and surface adherence of polar precursors and particles (by the detergent/dispersant combination).

3.4.3 Mitigation of Metal Contamination via Metal Deactivator Additives

The use of a copper complexing agent to counteract the catalytic action of dissolved copper in gasoline was first proposed in the 1930s. This approach was based on the addition of a hydroxyaromatic Schiff base which has a significant affinity for these dissolved copper ions. The complexing agent forms a tetradentate coordination complex which is fuel soluble and capable of counteracting the catalytic activity of copper by restricting electron transfer with the copper atom. The compounds currently approved as the active ingredient as a metal deactivator additives are N,N′-disalicylidene-1,2-propanediamine and N,N′-disalicylidene-cyclohexyldiamine. It has been shown that, while the presence of MDA will generally suppress the catalytic influence of dissolved copper, it is not effective in reducing thermal deposition when added to copper-contaminated fuels after long-term storage. This illustrates the impact of copper to accelerate the formation of copper-free soluble reaction products and leads to the recommendation that early addition of a metal deactivator is necessary if the fuel is likely to come into contact with sources of copper, particularly for fuels that will be stored for extended periods of time.

The effectiveness of a variety of metal chelators in gasoline was examined by oxygen absorption measurements in sealed reactors. Various structural factors in the chelants, such as ring size and resonance stabilization determined how well these compounds suppressed oxygen consumption in gasoline autoxidation. It was also shown in that study that disalicylidene chelants were among the most effective, combining the desirable features of ring size, electron donation, and coordination number for copper.

Thermal oxidation stability of jet fuels is generally assessed with JFTOT procedure, in accordance with ASTM D3241. It is generally found that the presence of MDA will significantly reduce deposits on JFTOT heater tubes over the temperature range of 260–310 °C. On this basis, it was hypothesized that MDA was somehow inhibiting thermal deposition on heated metal surfaces through the formation of a contiguous passivating monolayer of the additive. However, in single-tube heat exchanger tests conducted at temperatures as high as 350 °C under turbulent flow and in full-scale nozzle testing under similar conditions, the effectiveness of the metal deactivator to suppress thermal deposition appeared to be only temporary. These findings have prompted many to question if the standard laboratory JFTOT test is valid in the presence of the metal deactivating additive, since these induction periods are not observed in the JFTOT. The presence of MDA on metal surfaces after exposure to copper-free jet fuels containing the additive was confirmed with secondary ion mass spectrometry and laser induced surface analysis techniques. However, in a series of thermally programmed desorption (TPD) experiments it was shown that chelant was physisorbed and thus only weakly chemically bound to the 304 stainless steel surface. The data also revealed concurrent growth rates of both chemisorbed and physisorbed state in localized regions, thus disproving the monolayer passivation mechanism. It is nevertheless accepted that the presence of a metal deactivator additive in a fuel will in generally result in significant reductions in JFTOT heater tube deposition, but it

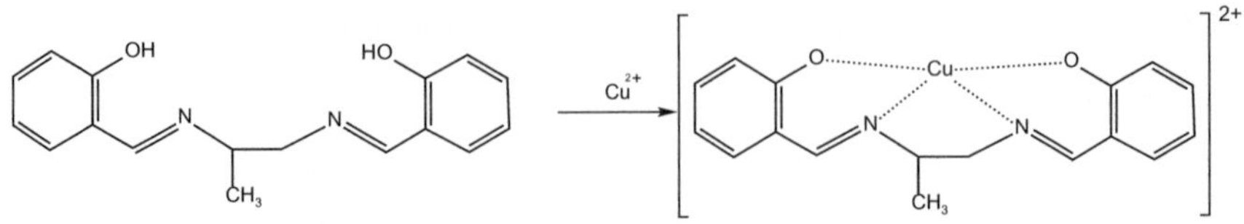

Fig. 13 Tetradentate chelation of divalent copper by N,N′-disalicylidene-1,2-propanediamine

remains a topic of some disagreement whether or not the JFTOT accurately predicts the thermal stability of jet fuels that contain MDA in jet aircraft fuel systems and engines.

Metal deactivator additives are added to fuels to counteract the catalytic activity of copper held in solution by complexation with natural fuel constituents. These additives are derivatives of disalicylidene Schiff bases that form stable tetradentate copper coordination complexes, as shown in Fig. 13.

The geometry of the Cu-MDA complex constrains the copper d-orbitals into a strained square planar configuration, instead of their preferred tetrahedral configuration. This configuration change renders any d-orbital electron transfer reactions energetically unfavorable, thus effectively eliminating any catalytic activity. When a soluble organo-copper compound comes into contact with a heated surface that is above its decomposition temperature, it can decompose on contact, leaving copper on the surface. Surface analyses of stainless steel JFTOT heater tubes have revealed that the Cu-MDA complex thermally decomposed to deposit copper in those areas of the tubes where the surface temperatures exceeded 300 °C. It was also shown in that study that the impact of these copper-enriched surfaces on fuel oxidation in the JFTOT is most evident when there is still available dissolved oxygen. Under standard test conditions in the JFTOT, the fuel is deoxygenated by the time it reaches the region of the heated test section where the decomposition temperature of the Cu-MDA complex was reached. Thus, in the JFTOT, copper deposits from Cu-MDA decomposition exerted no further impact on fuel oxidation. While it is likely that such copper-enriched surfaces could be developed in heated turbulent flow systems, the availability of dissolved oxygen would still be a necessary prerequisite for the catalytic action of copper on wetted surfaces in aircraft fuel systems and engines.

Multi-pass testing has been conducted with copper-contaminated JP-5 thermally stressed in a heated stainless-steel tube at 177 °C. With dissolved copper concentrations ranging from 268 to 300 μg/kg, the formation rate of carbonaceous deposits within the heated zone was 115 μg/hr. This deposition rate was not significantly reduced by the presence of MDA. However, MDA exerted a profound effect on the distribution of copper in the fuel. Samples of fuel effluent taken after each pass through the heated test section showed that the distribution of copper was being shifted from the fuel to the adherent deposits within the heated test section with each successive pass, with virtually all the copper being depleted from the fuel after five passes at 177 °C. However, in the presence of either MDA or the Betz + 100 additive, virtually all the copper remained in the fuel, either as fuel—soluble complexes, or as dispersed insoluble particulates smaller than 0.45 μm. These findings suggested that with bulk fuel temperatures up to 175 °C, the active ingredient in the metal deactivating additive was sufficient to retain all the dissolved copper in solution, either as suspended particulates or as the soluble metal complex. The findings from thermal desorption mass spectrometric studies of the thermal stability of copper-MDA complexes indicated that these complexes were thermally stable at surface temperatures of up to 350 °C, so this should not be an issue in recirculating jet engine fuel systems. Although the surface temperatures within the fuel injectors can reach Cu-MDA decomposition temperatures, the residence times in the injectors is very likely too short for copper deposition to be an issue in the presence of these additives. MDA has been shown to be beneficial in preventing or suppressing thermal deposition from copper-contaminated fuel in fuel nozzles. Moreover, there seemed to be some measurable benefits from the use of MDA in jet engine injector nozzles when used with copper-free fuels. MDA is a component of the U.S. Air Force JP-8+100 additive package.

References

1. ASTM D873. Standard test method for oxidation stability of aviation fuels (potential residue method). ASTM International, 100 Barr Harbor Drive, PO Box C700, West Conshohocken, PA 19428-2959, United States
2. ASTM D4625. Standard test method for middle distillate fuel storage stability at 43 °C (110 °F). ASTM International, 100 Barr Harbor Drive, PO Box C700, West Conshohocken, PA 19428-2959, United States
3. ASTM D3241/IP 323. Standard test method for thermal oxidation stability of aviation turbine fuels. ASTM International, 100 Barr Harbor Drive, PO Box C700, West Conshohocken, PA 19428-2959, United States
4. ASTM D5304. Test method for assessing middle distillate fuel storage stability by oxygen overpressure. ASTM International, 100 Barr Harbor Drive, PO Box C700, West Conshohocken, PA 19428-2959, United States
5. ASTM D3703. Test method for hydroperoxide number of aviation turbine fuels, gasoline and diesel fuels. ASTM International, 100 Barr Harbor Drive, PO Box C700, West Conshohocken, PA 19428-2959, United States
6. ASTM D381. Standard test method for gum content in fuels by jet evaporation. ASTM International, 100 Barr Harbor Drive, PO Box C700, West Conshohocken, PA 19428-2959, United States
7. ASTM D1319. Standard test method for hydrocarbon types in liquid petroleum products by fluorescent indicator adsorption. ASTM International, 100 Barr Harbor Drive, PO Box C700, West Conshohocken, PA 19428-2959, United States
8. ASTM D1655. Standard specification for aviation turbine fuels. ASTM International, 100 Barr Harbor Drive, PO Box C700, West Conshohocken, PA 19428-2959, United States
9. Defence Standard 91-091. Turbine fuel, aviation kerosine type, Jet A-1 NATO Code: F-35, Joint Service Designation: AVTUR, Issued by the United Kingdom (U.K.). Ministry of Defence. Glasglow, UK. https://www.dstan.mod.uk/
10. ASTM D156. Standard test method for saybolt color of petroleum products (saybolt chromometer method). ASTM International, 100 Barr Harbor Drive, PO Box C700, West Conshohocken, PA 19428-2959, United States

11. ASTM D5453. Standard test method for determination of total sulfur in light hydrocarbons, spark ignition engine fuel, diesel engine fuel, and engine oil by ultraviolet fluorescence. ASTM International, 100 Barr Harbor Drive, PO Box C700, West Conshohocken, PA 19428-2959, United States
12. ASTM D5452/IP 423. Test Method for particulate contamination in aviation fuels by laboratory filtration. ASTM International, 100 Barr Harbor Drive, PO Box C700, West Conshohocken, PA 19428-2959, United States
13. MIL-DTL-83133. Detail specification: turbine fuel, aviation, kerosene type, JP-8 (NATO F-34), NATO F-35, and JP-8+100 (NATO F-37)
14. ASTM D4176. Test method for free water and particulate contamination in distillate fuels (visual inspection procedures). ASTM International, 100 Barr Harbor Drive, PO Box C700, West Conshohocken, PA 19428-2959, United States
15. GOST 10227-1986 (revised 1995; 2013 version cancelled). Jet fuels. Specification, maintained by the Euro Asian Council for Standardization, Metrology and Certification (EASC). Covers TS-1 and RT grades
16. ASTM D7619. Test method for sizing and counting particles in light and middle distillate fuels, by automatic particle counter. ASTM International, 100 Barr Harbor Drive, PO Box C700, West Conshohocken, PA 19428-2959, United States
17. IP 564. Determination of the level of cleanliness of aviation turbine fuel—laboratory automatic particle counter method. Energy Institute, London
18. IP 565. Determination of the level of cleanliness of aviation turbine fuel—portable automatic particle counter method. Energy Institute, London
19. IP 577. Determination of the level of cleanliness of aviation turbine fuel—automatic particle counter method using light extinction. West Conshohocken, PA 19428-2959, USA: American Society for Testing and Materials International. Energy Institute, London
20. EI/JIG 1530. Quality assurance requirements for the manufacture, storage and distribution of aviation fuels to Airports. http://www.jigonline.com; https://publishing.energyinst.org/featured/1530
21. Hazlett RN (1991) Thermal oxidation stability of aviation turbine fuels. ASTM
22. ASTM DD3241. Test method for thermal oxidation stability of aviation turbine fuels (JFFOT). ASTM International, 100 Barr Harbor Drive, PO Box C700, West Conshohocken, PA 19428-2959, United States
23. ASTM D1660. Standard test method for thermal stability of aviation turbine fuels (suspended). Method of test for thermal stability of aviation turbine fuels (withdrawn 1992). ASTM International, 100 Barr Harbor Drive, PO Box C700, West Conshohocken, PA 19428-2959, United States
24. MIL-DTL-5624. Detail specification: turbine fuel, aviation, kerosene type, Grades JP-4 and JP-5
25. ASTM D7739. Standard practice for thermal oxidative stability measurement via quartz crystal microbalance. ASTM International, 100 Barr Harbor Drive, PO Box C700, West Conshohocken, PA 19428-2959, United States
26. ASTM D6811. Standard test method for measurement of thermal stability of aviation turbine fuels under turbulent flow conditions (HiReTS method) (withdrawn 2013). ASTM International, 100 Barr Harbor Drive, PO Box C700, West Conshohocken, PA 19428-2959, United States
27. EI 482. Determination of the thermal oxidation stability of aviation turbine fuels—HiReTS method. Energy Institute, London. https://knowledge.energyinst.org/search/record?id=9906. Accessed Jan 5, 2025
28. Kuprowicz NJ, Zabarnick S, West ZJ, Ervin JS (2007) Use of measured species class concentrations with chemical kinetic modeling for the prediction of autoxidation and deposition of jet fuels. Energy Fuels 21(2):530–544. https://doi.org/10.1021/ef060391o
29. MIL-DTL-25524. Detail specification: turbine fuel, aviation, thermally stable

Chapter 8
Supply Chain Risk Management (SCRM) of Fuel for Aviation

Richard Bruce Knapp, Barbara Mooney, Curtis T. Stough, James M. Eberhardt, Renrick Atkins, Rebecca Purcell, Lori Taylor, Kelley Jackson-Cherry, Cheng Yen, Yan Guo, Makeena Stephens, Daniel Baniszewski, Keith Mills, Mark Mizrahi, Michael Heidbreder, and Patricia Wilkins

Abstract Supply Chain Risk Management (SCRM) within the petroleum industry often represents a complex logistical challenge. Risks include aspects such as planning, identification, analysis, response planning, and monitoring. The petroleum supply chain involves diverse processes, from sourcing to delivery, with a focus on collaboration due to its segmented nature. Risk factors can be found at different supply chain segments, including source of supply, refinery, distribution, and storage.

R. B. Knapp · B. Mooney · C. T. Stough · J. M. Eberhardt · R. Atkins · R. Purcell · L. Taylor · K. Jackson-Cherry · C. Yen · Y. Guo · M. Stephens · D. Baniszewski · K. Mills · M. Mizrahi · M. Heidbreder · P. Wilkins (✉)
Defense Logistics Agency Energy, Fairfax, VA, USA
e-mail: Patricia.Wilkins@dla.mil

R. B. Knapp
e-mail: Richard.Knapp@dla.mil

B. Mooney
e-mail: Barbara.Mooney@dla.mil

C. T. Stough
e-mail: Curtis.Stough@dla.mil

J. M. Eberhardt
e-mail: James.Eberhardt@dla.mil

R. Atkins
e-mail: Renrick.Atkins@dla.mil

R. Purcell
e-mail: Rebecca.M.Purcell@dla.mil

L. Taylor
e-mail: Lori.Taylor@dla.mil

K. Jackson-Cherry
e-mail: Kelley.Jackson-Cherry@dla.mil

C. Yen
e-mail: Cheng.Yen@dla.mil

M. Rabaev et al. (eds.), *Aviation Fuel Problems and Solutions: From Refinery to Wingtip*, https://doi.org/10.1007/978-3-032-05514-9_8

Internal and external risk factors are identified and addressed through audits, assessing operational breakdowns, personnel changes, cybersecurity, environmental compliance, and external disruptions. Mitigation planning is crucial for successful supply chain execution, including the need for an integrated end-to-end approach. For risk mitigation in the aviation fuel supply chain, it is important to understand the relationship between different supply chain elements and the critical need for risk management. The various modes of shipment (ocean tankers, pipelines, road and rail tankers, drums, and ISO tank containers) each have their own potential issues and mitigation strategies. Quality technical supply chain risks are addressed by adhering to standards, testing, and certification. Competition for raw materials in fuel additives is of importance, as there are challenges and strategies for managing supply chain disruptions. The limited industrial base for specialty fuels and severe weather impacts on fuel additives are also examined, along with the significant threat of cross-contamination. Correct application of these SCRM principles is important for maintaining reliable fuel supply in the aviation industry.

1 Introduction

Risk may be defined in financial terms as the likelihood that an outcome or investment's actual gains will differ from an expected outcome or return. To effectively understand and manage risk, organizations normally develop and implement various levels of risk management tools. There are a wide variety of tools that the industry leverages; however, they share numerous commonalities in addressing the Supply Chain Risk Management process. Supply Chain Risk Management (SCRM) is the process of identifying, assessing, and mitigating the risks of an organization's supply chain. Implementing global Supply Chain Risk Management strategies can help an enterprise operate more efficiently, reduce costs, and enhance customer service. Within petroleum supply chains, organizations often focus on developing and implementing guiding standards for testing and evaluating critical performance factors

Y. Guo
e-mail: Yan.Guo@dla.mil

M. Stephens
e-mail: Makeena.Stephens@dla.mil

D. Baniszewski
e-mail: Daniel.Baniszewski@dla.mil

K. Mills
e-mail: Keith.Mills@dla.mil

M. Mizrahi
e-mail: Mark.Mizrahi@dla.mil

M. Heidbreder
e-mail: Michael.Heidbreder@dla.mil

such as testing for fuel quality to avoid contamination, developing an environmental remediation and protecting plans, establishing a formal public affairs office to maintain good relations, as well as instituting a series of tools to track and manage fuel as it is acquired, shipped, stored, and delivered. The petroleum supply chain is a complicated system, and its management refers to how organizations manage the flow, including many processes that are involved in transforming raw materials the organization consumes into finished products or services the organization offers. It includes the planning and management of the activities around sourcing, procurement, and conversion, as well as logistics management functions.

One of the main reasons companies implement global supply chain management strategies is to increase their competitive advantage through market reach. However, the benefits of extending supply chains can increase the risk of quality, safety, business continuity, reputation, and more.

Every company, regardless of industry, is exposed to risk and, in some manner, performs an assessment or evaluation of internal and external risk. Internal and external risk factors are identified and addressed through internal and external audits and include the following criteria specific to the bulk petroleum supply chain: Internal Factor Evaluation (IFE) and External Factor Evaluation (EFE).

2 The Risk Management Process

The management process of risk may include five main elements: planning and execution, identification, analysis, response planning, monitoring, and controlling.

2.1 Planning and Execution

Strategic and operational planning are critical elements in identifying risk because they define the operating parameters of the organization. From a strategic perspective, an organization needs to understand where it sits in the sphere of influence with respect to the market. This can be determined by a multitude of factors such as size, reputation, financial strength, capabilities, and competition. From an operational perspective, a company needs to fully understand its ability to execute, which includes processes, personnel, finances, communication, and management oversight/internal controls. Additionally, planning and execution revolve around understanding the entire process from the source of supply to delivery.

2.2 Identification

Identification covers three segments: source of supply, intermediate storage management and final storage/delivery. The petroleum products supply chain risk environment is a combination of internal and external factors, unlike many other supply chains. Fundamentally, it comes down to the ability to execute a fuel delivery to the customer but with a greater emphasis on collaboration. This is partly due to the nature of the supply chain in that no one entity owns the entire function. For example, the military sector may be heavily reliant on the use of bulk jet fuel supply, just like the airline industry; however, neither entity owns any piece of the supply chain included in the extraction of crude oil or refining. Yet, there is a reliance on successful extraction of crude oil and refining to produce large volumes of specification compliant jet fuel for flight operations. The beginning of the supply chain is at the point where products are shipped/received and possibly stored for future use. In some cases, these entities have pipelines, ocean vessels, railways, etc., to receive the product, but none of them own the large-scale infrastructure for fuel production so they rely on agreements with suppliers. These agreements are critical risk factors because they ultimately trigger the suppliers' demand signal to produce and deliver specific types of petroleum products such as jet fuel. The ability to monitor petroleum supplies around the globe is critical across the entire supply chain. Without monitoring, companies would have tremendous challenges with producing, shipping, storing, and reallocating petroleum to ensure the supply chain is stable and robust to meet current and future demands.

2.3 Analysis

Analysis describes tools, techniques, approach, application, and management in risk analysis. The bulk petroleum supply chain is an industry lead in logistics execution and has set numerous benchmarks in automation, technology, and advanced analytics. While appearing simple in design, the supply chain is extremely resilient if the analytics to support operations are present. For example, if a pipeline is out of service due to an unforeseen event such as a mechanical failure, or natural disaster, the modeling and tracking tools are available. Some supply chain models allow for real time visibility over fuel assets and product movements (land and sea based). The petroleum supply chain requires constant monitoring and live feeds of real-time data to enable on the spot informed decision making to reduce the impact of petroleum supply chain disruptions.

2.4 Response Planning

Response planning is a product of understanding the interdependencies and key relationships within the petroleum supply chain. Having real-time situational awareness over assets and the various nodes/modes of movement for products is essential to the function of the supply chain. A critical element of managing risk is knowing when and where a part of the supply chain is impacted, as well as understanding what needs to be done and when. Often, this is performed through a node analysis to identify and determine capabilities for movement, storage, and distribution of petroleum products.

Knowing what product is available in real-time allows companies to identify what is and is not an acceptable operational risk. For example, in case of running an airport in the continental US where there is a robust network of supply options in a particular area, it is possible to arrange agreements with suppliers to provide coverage in the event of a major disruption at one or more to the "normal" nodes of distribution. By understanding this capability to meet customers' needs, there is an option for effectively mitigate/reduce the risk factors for operating the airport. However, if there is a single source of supply for the location it is important to elevate the risk and identify countermeasures to reduce the risk to an acceptable level. This may be a function of finding an alternative delivery node, such as a railway or truck distribution plan. Other options might be the construction of a secondary pipeline or additional storage terminal capacity. The need to monitor risk is a direct factor in performing the IFE and EFE audits on a recurring basis (often quarterly or as needed based upon contingencies that arise over time). An example of an internal/external audit can be tied to how a company identifies, understands, plans for, and reduces risk for fuel additive supply. Often petroleum companies do not produce or distribute fuel additives such as corrosion inhibitor, static dissipator, or fuel system icing inhibitors. If a customer requires these additives specific steps must be taken to isolate the product, inject/mix the additives to the required specification and arrange delivery. Since additives are normally provided by an external service provided, an assessment of the industrial capability to provide the quantity of additives at the right time in the right location is also required. If only a limited supply is available an external risk audit will be required to identify the levels of production/transportation, and storage for additives to ensure on specification product can be provided to customers. This type of risk assessment often includes market analysis of products (i.e., number of suppliers, available quantities, locations and transportation options for the products, etc.). Once a company understands these economic variables, they can perform an internal risk audit to determine the best strategy which may be to pre-position additives along the supply chain at intermediate storage nodes, or at the point of delivery. The internal audit will look at company strengths and capabilities to determine how much risk can be mitigated as acceptable to maintain operations. The first item normally identified is a financial review of the available resources to determine the viability of staging additives at various points along the supply chain. The company may also identify the amount of additive that can be pre-purchased and for how long it can be stored (shelf life). By performing the external and internal audits, the company can now leverage internal capabilities to developed specific additive supplier capabilities to meet demand. Additionally, the company can determine the correct amount of risk they are willing/able to take regarding the availability of additives along the supply chain. Normally, these decisions are closely monitored with performance management tools using metrics to determine effectiveness of the company's risk management strategy, and adjustments can be made as trends develop over time.

2.5 Monitoring and Controlling (Review, Reassessment)

Fully understanding risk and effectively managing/mitigating risk are essential factors in running a successful business. Just having the risk factors identified is not enough. Constant control must be maintained by monitoring and managing risks as the business environment changes. What works today might not fit tomorrow's needs.

3 Industrial Application of SCRM to Assess Critical Supply Chain Segments

Source of Supply (SOS) for petroleum products such as jet fuel is defined as the point of origin where crude oil is being collected and prepared for refining. This includes areas such as oil rigs, drilling platforms, fracking platforms, etc. Regardless of the type of platform, the SOS is a critical element in the supply chain that is subject to both internal and external risk ranging from geological, socio-political, and environmental, to financial, military, terrorism, or cyber-attacks. Each of these operational platforms requires an assessment of the potential for disruption, the likeliness of occurrence, overall impact, and mitigation strategies/matrices to expedite response times and reduce the overall long-term impact on operations. While critical, the SOS is often not impacted or targeted beyond socio-political, geographic, or weather-related means. Mitigation strategies are often implemented using "known" solutions such as providing physical security around fixed platforms, tethering afloat drilling rigs, and ensuring Occupational Safety and Health Administration (OSHA) standards are being followed.

Refinery is defined as the point in which raw oil is processed into refined product such as gasoline, diesel fuel, kerosene, jet fuel, lubricating oils, and distillates. Refinery operations are normally focused on risks related to the safety of operations to ensure no mishaps occur. Mitigation strategies normally include regular safety inspections, equipment testing, maintenance, and overhaul, as well as standard product quality control measures.

Distribution is defined as the nodes/modes of transportation and intermediate storage of refined products. Distribution is often the most targeted and most challenging segment of the supply chain. The risk associated with distribution is due to the nature of distribution modes/nodes such as tanker, railcar, tank truck, pipeline, and barge, which is often difficult to monitor, protect, and maintain. While in transit, fuel is vulnerable to a litany of potential disruptions such as weather, geological, socio-political, military, cyber, and physical terrorist attacks. Mitigation strategies normally include procedures for vessel acceptance and loading, tamper-proof seals and tagging modes of transportation with radio-frequency identification (RFID) type tags to track progression through the distribution system. However, some events in the cyber area have proved to be exceptionally vulnerable, for example, the risk of pipelines shutdown. With vulnerability comes another mitigation strategy for distribution, which often includes a node management approach in which an alternative distribution network or capability is established, regularly tested, and often used as a supplemental option to mitigate risk. This builds resiliency into the supply chain.

Storage is defined as the point at which refined product is often kept prior to the final point of sale (POS). Storage can be identified as the intermediate where the product's destination is unknown or serves as a pre-positioned stock that will later be distributed as required. Additionally, some customers required specific additives to be included in the product, such as Fuel System Icing Inhibitor (FSII), Static Dissipator Additive (SDA), and Corrosion Inhibitor (CI) for aircraft fuels, or additives packages with cleansing agents for gasoline distribution locations. Refer to Chap. 5 for expanding on risks and problems in the field of additives for aviation fuels. Regardless of the final point of sale, storage has the second highest risk factor associated with it due to its sedentary nature. Normal risk may include weather-related contamination or damage, product cross-over (an incorrect product entering the storage tank), engineering failure or even seismic activity. For this reason, physical security and strict operational procedures are often the first line of defense to protect storage operations. Other risk factors include regular inspections, fire suppression systems, overfill/tank volumetric tracking indicators, and emergency warning systems.

Often the POS includes some sort of storage facility with an additional mode of delivery, such as a gasoline station drawing product directly from the storage tanks, or an aviation fuel site where the product is drawn directly from operational tanks and delivered to the skin of the aircraft. Risk is often minimal for these types of locations mainly because the majority of the product is moved underground through hidden pipelines/pumps and is filtered and tested for contaminants constantly. Mitigation strategies normally include physical safety inspections, extensive product quality control measures, and regular product accountability/reconciliation to identify issues such as potential leaks.

4 Understanding Internal and External Supply Chain Risk Management

Internal supply chain risk(s) are those areas that can be controlled and influenced significantly by the supplying organization, such as infrastructure, financial, personnel, operational, and production risks. These risk factors are most often broken out into two categories often referred to as either supply and demand sides, or customer and supplier sides operations. Regardless of which category the risk factors fall into, the assessment, management, and mitigation strategies are normally the same given the organization goals and objectives. Examples of internal risk factors that are commonly assessed by petroleum organizations include many of the following:

- Internal operational breakdowns (communication, financial, IT and control system, etc.)
- Changes in key management, personnel, business processes and loss of critical technical expertise
- Lack of planning for disruptions due to contingencies
- Lack of effective cybersecurity policies and controls
- Noncompliance with environmental regulations or labor laws
- Inability to meet demand based upon poor demand planning or forecasting.

External risk factors may include:

- Contingency, unexpected surges in demand, logistics node disruption (political or military conflict, natural disaster, including earthquakes, weather related events, etc.)
- Interruptions at critical supply chain points from the source of supply up to the point of sale
- Socio-political, financial, legal, and other related economic restrictions or factors
- Environmental and regulatory compliance.

Since the early 1980s companies have been recognizing risk and developing or adopting various risk management tools to assist them with Supply Chain Risk Management. Regardless of which approach a company chooses to meet its SCRM needs, the basic tenets of internal and external risk identification and significance will exist. The factors are identified in the first column that are aligned with strategic goals and objectives which are often called critical success factors. The significance of any impact is identified as the weight (note all weights combined cannot exceed 100%). This is followed by a rating derived by determining the likelihood of impact to the organization such as a refinery going out of service for an unplanned extended period. The rated score is a simple multiplying effect of taking the risk weight times the rating likelihood of occurrence to determine where the company should prioritize its assets to manage or mitigate that particular risk element.

The results of these internal and external audits drive the organization's risk management strategy and often feed the overarching strategy to alleviate concerns and promote healthy growth such as strengthening cyber security policies, operations, visibility, and staffing.

An organization could also fall victim to supply chain financial risk if something happens that threatens the financial health of a supplier, such as bankruptcy. In addition, a company might suffer reputational risk if a supplier engages in unethical behavior, such as bribery, child labor, or anything that could reflect badly on the company's brand.

The mitigation of these risks is crucial, and many businesses manage this problem by diversifying their suppliers. Such action helps to minimize disruptions, meet demand, improve flexibility/resiliency, and ensure effective quality control over the final product.

5 Common Impacts of Poor Planning and Execution of SCRM Practices

Companies actively engaged in petroleum supply chain operations may focus on day-to-day operations with no formal plan to deal with any potential disruption. While applicable to successful short-term goals and objectives this myopic focus often increases the likelihood and significance of the impact of other risk factors on the supply chain that result in major disruptions, inefficiencies, and demise of the companies' abilities to execute in both the short and long-term. Examples of major organizational supply chain impacts that result when internal risk factors are not observed include the following:

- Internal operational breakdowns (communication, financial, IT system failures, etc.)
- Changes in personnel or business processes leading to ineffective and untimely decision-making
- Lack of planning resulting in supply disruptions and an inability to react
- Lack of effective cybersecurity policies and controls resulting in compromised systems
- Non-compliance with environmental regulations or labor laws leading to legal entanglements
- Inability to meet demand based upon improper demand planning or forecasting.

6 The Importance of Mitigation Planning

Mitigation planning is one of the critical success factors in supply chain execution. Regardless of the operational focus or elements represented by an organization in the bulk petroleum supply chain, there needs to be an integrated end to end approach to assessment and management of operations. Each element of the supply chain is dependent on the other performing in an optimized and integrated manner.

7 Risk Mitigation Impact

Understanding and managing risk factors assure smooth execution of operational goals and objectives to keep products moving through each supply chain element. Any disruption in a phase of the supply chain directly impacts all other elements. For example, in the event an intermediate storage location/node is shut down for non-compliance with environmental standards, it automatically triggers a response from suppliers at the refinery to either reduce operations or redirect the product to another node in the supply chain. While numerous nodes exist for moving products such as rail, tankers, and pipelines, there are numerous isolating elements to be considered such as product specificity, location of demand for a specific product. If not properly managed at each step in the supply chain, small inefficiencies can quickly accumulate into major impacts or supply chain failures.

8 Mode of Shipment (Potential Issues and Mitigation)

The mode of shipment of aviation fuel may have a significant impact on its quality. This section discusses the potential issues and mitigation strategies for each of the four main modes of shipment: ocean tankers, coastal/inland waterway vessels/barges, pipelines, road tankers and rail tank cars, and drums and intermediate bulk containers (IBC) and ISO tank containers.

8.1 Ocean Tankers, Coastal/Inland Waterway Vessel/Barges

Water-borne movements of aviation fuel may involve quick coastal or inland waterway deliveries or extended ocean voyages. Cargoes can be dedicated to one product, or multiple products, the latter significantly increasing supply chain risk. Previous cargoes, compatibility, compartment cleaning, and the ability to take representative samples are key considerations for maintaining quality.

Examples:

- Incomplete or incorrect cleaning results in the loaded product not meeting requirements
- Review of previous cargo fails to identify a product that contaminates loaded product even after cleaning protocols (e.g., crude; linseed oil; molasses)

- Restrictive access prevents representative sampling and so fails to detect particulate contamination or stratified/non-homogenous products.

Given the extensive use of water-borne transport for petroleum products, the industry has sought to develop robust guidelines under Hydrocarbon Management, HM50, Guidelines for the Cleaning of Tanks and Lines for Marine Tank Vessels Carrying Petroleum and Refined Products [1].

8.2 Pipeline

Pipeline systems provide efficient movement of petroleum. Ideally, a system is dedicated to aviation fuels; however, multi-product systems may move multiple grades and types of products. System design is important to confirm compatible metals and materials are used, and that capabilities like filtration are correctly established.

Multi-product pipeline systems create risks that adjacent products in an interface moving through the pipe may contaminate each other. Management of quality predetermines an interface cut (mid, head, or tail) to direct product to a receipt line or receipt tank to protect the quality of one product. The interface cut may be based on the quantity or a product property such as density which is monitored during pumping. Maintaining pressure on the pipeline can minimize the area of the interface and resultant contamination. Multi-product pipelines moving both aviation fuel and fuels with bio-content, dyes, and some additives require a risk assessment. Not all pipelines will allow all the desired materials to be shipped if interface management is too demanding. Pipeline operator procedures address sources of contamination including involved products, manifold separation, dead-legs of static product, valve separation, and booster pumps.

Examples:

- Incorrect interface cut contaminated aviation with commercial diesel, resulting in failing thermal oxidation stability and potential contamination with biofuel and FAME (fatty acid methyl ester).
- Pipeline drag reducing agent (DRA) in diesel precipitated out of solution when in contact with ethanol gasoline blends and contaminated later movement of aviation fuel.
- Dyed product contaminates aviation fuel.
- Higher sulfur product contaminates ultra-low sulfur product during movement.
- Dead-legs were not removed or drained regularly, and static product or accumulated water created off-specification product issues.

To help mitigate these risks pipeline companies tend to set higher aviation fuel product quality requirements than standard specifications. In addition, formal procedures are available to assess other additives which may come into contact with aviation fuel such as EI 1535 Minimum Criteria to Determine Acceptability of Additives for use in Multi-product Pipelines Co-transporting Jet Fuel.

8.3 Road Tankers and Rail Tank Cars

Road and rail tanks feature the construction materials aluminum, stainless steel, or carbon steel (which requires a compliant epoxy coating). Before a new tank can be brought into service for aviation fuel a soak test must be conducted. Equipment dedicated to aviation is ideal, however, multi-product use can be addressed with evaluation of previous cargoes and appropriate cleaning. Bottom loading is preferred due to the safety risks associated with top loading. Inspection for cleanliness and absence of water can be performed by checking low point drains (only required for first cargo for dedicated transport; required every load when non-selective couplings are used). Identification of water or heavy sediment results in the load being rejected and notification to appropriate management. Marking of grades on equipment helps assure the correct product is handled, for example, as in EI 1542 Identification Markings for Dedicated Aviation Fuel Manufacturing and Distribution Facilities, Airport Storage and Mobile Fueling Equipment [2].

Previous cargoes, compatibility, compartment cleaning, and the ability to take representative samples are key considerations for maintaining quality.

Examples:

- Transloading of diesel to aviation fuel introduces FAME above limits.
- Samples taken prior to settling time result in failed visual or particulate contaminant tests.

8.4 Drum and Intermediate Bulk Container (IBC) and ISO Tank Container

Documented procedures are available for the selection and use of drum, intermediate and ISO tank containers. These cover internal and external inspection (and rejection criteria), flushing/cleaning/waste disposal, storage, and release of product, plus cite statutory requirements for filling/loading. For example, steel drums are required for use with aviation fuel and IBCs must be stainless steel or aluminum. Sealing materials at seams/bottom/top of drums plus at bung openings must also be compatible with aviation fuel (galvanized and plastic IBCs are prohibited). Filling systems are segregated to eliminate cross-contamination, and require filtration identified in EI 1581 [3] or EI 1590 [4] (Avgas only). Color coding and marking of drums helps reduce the possibility for incorrect product loading or use. Drums are not normally reused for aviation fuel, but protocols exist for careful reuse. When drums and IBCs are removed from aviation service, markings must be completely removed. Storage of drums will be in well-ventilated buildings as indoors or following outdoor protocols. After an allotted period in storage, representative sample(s) of the batch are tested with a new CoA issued to document quality.

Challenges for markings and proper storage create ongoing issues.

Examples:

- Drums stored outdoors exposed to direct sunlight may lead to deterioration of key jet fuel properties, including thermal oxidation stability.
- A laboratory analysis might confirm an additive quality using a retention sample stored in the laboratory, instead of a representative sample taken from a drum or IBC.
- An aviation fuel drum might be re-used to store a non-aviation product while retaining the original labeling, constituting a major flight safety risk.

9 Quality Technical Supply Chains Risks and Challenges

The use of proper specifications and standards is key to assuring the quality of aviation fuel from point of manufacture through the supply chain to point of delivery. Quality assurance applies actions for certification of product at the point of manufacture, then additional testing and recertification through the supply chain to ensure properties remain within established limits. The use of incorrect or outdated standards or test methods creates questionable data supporting the quality of the product. Allowing the use of test reports past established timelines also introduces risk. Re-testing and re-certification must absorb and execute emergent changes to industry guidance that occurred since the previous data were generated. Conversely, incorrect application of higher test standards can create reported test failures not supported by requirements, and results in delays to shipments or unnecessary disposal of product.

Incomplete/improper testing or failure to maintain records creates a traceability shortfall, complicating the resolution of quality issues. Traceability becomes challenging when dealing with the eventual mixing of the original batch with the product through the supply chain creating new batches. Management of fungible products with the commingling of batches during transportation also complicates traceability. The inherent risks make actions earlier in the supply chain even more important to maintain quality from refinery to customer or skin of the aircraft.

9.1 Fuel Additive Raw Material Competition

Fuel additives are chemical compounds added to petroleum products to enhance the performance and efficiency of fuel systems, see more in Chap. 5. The additive supply chain may involve multiple companies such as raw material suppliers, the source of supply, chemical additive manufacturers, fuel refineries, and distributors. The competition for raw materials by chemical additive manufacturers has impacted the fuel additive industry. Mono-unsaturated fatty acids are one example. The mono acid industry has begun redirecting its resources to the more lucrative cosmetic industry, the ever-growing area of renewable fuels, and for utilization in diesel fuels. Some of the limitations are due to sanctions or other geopolitical conflicts. The limiting of these resources has begun to directly affect the supply chain management of some military specification fuels. When raw materials are constrained, there is a risk of supply chain disruption and uncertainty that can affect the industry's operations and profitability. Good practice for fuel additive suppliers would be to assess the risk of not receiving raw materials. Such an event would cause a major disruption in the supply chain and highlight the value of identifying ways to mitigate the risk such as evaluating alternative sources of supply. Contract management also plays a vital role in limiting supply chain risk. Suppliers may also wish to negotiate contracts that include volume commitments, price adjustments and length of term for assured delivery. Refineries and storage terminals should evaluate the risk of not receiving additives and develop mitigation strategies to ensure the final product is delivered on time and on specification. For example, putting in place arrangements to manage additive inventory to guarantee supplies are sufficient to meet the demand. This involves working closely with fuel additive suppliers to establish minimum stock levels and conducting regular assessments of demand patterns.

Tall Oil Fatty Acid (TOFA), a subgroup of mono-unsaturated fatty acids, is one of the main ingredients in producing Lubricity Improver Additives (LIA) utilized in various fuels. Due to the redirection of resources the suppliers of these LIAs have seen supply chain delays causing further disruptions at refineries and storage terminals. As TOFA becomes scarce due to high demand, suppliers are seeing the price of raw materials increase and supply chain are experiencing delays in the receipt of feedstock. LIA suppliers have developed some risk management solutions to combat these supply chain issues, while other manufacturers have decided to no longer produce such additives, further increasing the demand while decreasing the number of suppliers. Some LIA additive producers are looking for alternative raw materials for TOFA such as Esters to use as a substitute, but this presents challenges as well. Securing the raw materials from alternative sources of supply and evaluating the additives through qualification testing to be included as an approved product can take a year or more to complete. This is an important aspect for aviation fuel where the D4054 [5] approval process involves a wide range of analysis, as discussed in Chap. 5.

Other manufacturers are looking to synthetic lubricity additive to supply the market when the TOFA supply is heavily strained. An outcome of such events is LIA suppliers requesting additional lead times in contracts to combat supply chain delays. While additive stockpiling might be considered, holding an inventory of LIA, for example, involves having the optimal storage conditions for the additive and having personnel with a technical understanding of the storage and handling. LIA is a necessary additive for the military fuel specifications and stakeholders have started looking at alternatives that do not use TOFA as a key component. These are being evaluated for potential effectiveness and compatibility in military fuels.

9.2 Limited Industrial Base for Specialty Fuels

There is significant interest in ensuring the availability of high-quality military specification fuels. However, the fuel industry has begun seeing supply chain issues, particularly in military specification fuel production. The move to alternative fuels is leading to changes in refinery facilities as a new supply chain is developed. Refineries that have specialized in producing military specification fuel are now investing in new equipment and processes to produce alternative fuels compliant with environmental mandates. The new tax incentives in California are further limiting military specification fuel options. The demand of military specification fuel, in particular, Aviation Turbine fuel JP-5 ($\geq$+60.0 °C flash point) is relatively small compared to commercial fuel, leading to lower sales, volumes, and revenue. There are over 700 fuel manufacturers worldwide but 2024, there are only 19 with the capability to produce JP-5. With limited demand, some suppliers find it not profitable to produce and sell fuel that meets military specification standards, as it may not justify the significant investment required. To mitigate these risks, it's imperative users of these fuels perform the necessary work to incorporate alternative production pathways becoming more widely used throughout the industry into their relevant specifications.

9.3 Severe Weather Impact on Fuel Additives

Severe weather disruptions have long been an issue for the petroleum industry, especially those refineries located in certain geographical areas, such as the Gulf Coast of the United States. In 2021, there were two severe weather events that impacted fuel additives, specifically fuel system icing inhibitor (FSII).

Fuel additives are more closely linked to the chemical industry than the petroleum refining industry. Of all the fuel additives, fuel system icing inhibitor (FSII) is the most different. FSII is the industrial chemical Methyl Carbitol or Diethylene Glycol Monomethyl Ether (DiEGME) which may be injected into jet fuel at low concentration ($\leq$0.15% v/v) to mitigate trace free water freezing in fuel systems. FSII was not directly developed as a fuel additive and has other pre- and post-commercial applications outside the fuel supply chain. To complicate matters further, there is a limited supplier base for FSII because of the hazards associated with manufacture. This supply chain was pushed to the limit in 2021 when two uncharacteristic natural disasters impacted the Gulf Coast of the United States which has a large presence of chemical manufacturers.

In February 2021, a historic deep freeze hit Houston and totally disrupted the petrochemical industry. The uncharacteristically cold weather damaged process equipment used to manufacture feedstocks for the production of FSII. In August 2021, Hurricane Ida caused significant infrastructure damage to the Gulf Coast, again interrupting the ability of chemical producers to manufacture and ship products. Two disruptive weather events occurring in rapid succession revealed many weaknesses in the supply chain and risk elements that are unique to the chemical industry when compared to the refining industry. One of these is the narrow supplier base with potentially only a few chemical companies making a commodity product at any one time. Each supplier also tends to have one or two manufacturing facilities that are dedicated to producing a small subset of products. If these manufacturers experience a loss of manufacturing capability at their designated facility, there is no backup site and the product on the market becomes scarce. During the 2021, event every contractor in the United States that injects FSII was vying for the same small pool of FSII due to storm damage.

The nature of FSII also poses some unique challenges as it chemically attracts water in the supply chain. While the fuel industry has some experience with this phenomenon, these weather disasters required technical support decisions that were outside the realm of known practice. Much of the product is sold packaged in drums but it is also transported in tank trucks. Locating suitable transport that was appropriate for the product was a major challenge. Unlike fuel haulers, chemical haulers may take a wide variety of different chemicals in a trailer. There is no set of reliable prior cargoes that are coming in and there are some prior loads that present problems for the maintenance of product quality. In those cases, the truck must be rejected, and a new truck located.

Another specific challenge faced in these circumstances was the shipment of partial cargo loads. The limited availability of the product necessitated the shipment of partially loaded trucks to key locations to avoid disruption of operations. Shipments of products are typically full trucks for two reasons. The first reason is somewhat obvious as a full truck minimizes the impact of transportation costs. The second reason ties back to the affinity of the product to attract water from wherever it can be found. Partial truckloads do not minimize headspace and therefore humidity in the trailer could possibly impact the product during transportation. The impact of water in the distribution system is still unknown but experience determined the product was unlikely to degrade past the point of usability during transportation. There is limited testing capability for FSII outside of the manufacturer so there is still work to be done to develop a working technical expertise to guide quality decisions.

Fuel additives other than FSII are closely tied to the fuel industry and have a good working relationship with the fuel community. FSII is primarily an industrial chemical that happens to be used in fuel and there is not the same sort of technical partnership between manufacturers and users. Understanding how site-specific conditions impact the product is key for the user to understand their own risk posture. Site specific conditions that can impact product performance include local weather, product turnover rates and even tank size. The manufacturers are engaged into making the product but not necessarily supporting the product in use. Each customer application can be very different making full technical support difficult.

9.4 Cross-Contamination

Aviation fuel contamination poses a significant threat to air travel safety. The distribution of jet fuel using various means of transport is one of the principal risks for a contamination event to occur. To avoid this, strict cleaning and handling instructions are put in place to prevent jet fuel from coming into contact with a multitude of contaminants. Many jet fuel handling/cleaning guides identify contaminants of concern and how to mitigate quality issues. In particular, fatty acid methyl esters (FAME), light/heavy petroleum products such as gasoline/naphtha, low/high sulfur diesel, heating oils, and crude oils lend themselves to easily contaminate aviation fuel when using a multi-product distribution system.

FAME is the byproduct of the transesterification of oils and fats into a finished product used as fuel commonly known as biodiesel. FAME usage has increased substantially across the globe as nations aim to lower carbon footprint and produce fuel from renewable sources. Many of these countries have laws mandating the use of FAME at various levels. In the 2020s, the typical FAME levels in on-road diesel fuels are given below:

- Australia—7%
- Brazil—12–15%
- Canada—5%
- European Union—7%
- Indonesia—35%
- United States of America—5%.

Not listed is the FAME content in heating oils, which is expected to increase over the coming years similar to on-road diesel.

Aviation and diesel fuels typically share parts of the distribution infrastructure (pipelines, barges, and tank trucks) which increases the likelihood of contamination. The increase in FAME across multiple diesel products presents a problem when transporting aviation fuels since the trace contamination limit is 50 mg/kg. These issues arise when a load of aviation fuel follows a load of diesel fuel containing FAME. If not properly cleaned, FAME can adsorb to the walls of a tank or pipelines and contaminate the aviation fuel resulting in off-specification product. As well as being a surfactant it may change the balance of fine contamination through pipelines and cause a sudden change in fine particle and other contamination level in the bulk stream. FAME is also hygroscopic, meaning it can readily absorb water from its environment. This tendency to absorb water can have a detrimental impact on aviation fuels. Excess water can affect the water separability and cold flow properties, causing emulsions to form, and may also introduce water freeze issues and other problems affecting flight. In addition to this, FAME promotes fungal and microbial growth as it is a nutrient source for microorganisms. Fuel stability becomes a problem when there is FAME contamination. This instability stems from the tendency for FAME to reduce the oxidation stability of fuels in long term storage, promoting the formation of particulates, acids, gums, and other deposits that can foul fuel lines and injectors.

Other than FAME, a common source of aviation fuel contamination is from incompatible cargoes used in multiproduct transport systems. This includes gasoline, ultra-low/high sulfur diesel fuels, intermediate/residual fuel oils (IFOs/RFOs), crude oils, and chemicals. Many of these products may contain highly reactive chemical compounds, metals, or molecular species that affect jet fuel quality. Like FAME, depending on the level of contamination, these products can affect the cold flow properties by increasing the freezing point, retaining higher amounts of water and affecting the fuel volatility by decreasing the flash point (gasoline) or increasing the distillation temperatures (diesel and heavy petroleum products). Intermediate fuel oils and crudes contain metals that can induce polymerization reactions in the fuel which lead to deposits and gum formation affecting thermal stability. Even at trace levels, some of these metals have been shown to negatively affect fuel quality. Unfortunately, petroleum-based products are not the only source of contamination. Pure chemicals such as industrial polymers, surfactants, chemical additives, and precursors to renewable fuels such as vegetable oils can have significant consequences to aviation fuel quality. These products mimic many of the deleterious effects on aviation fuel quality as FAME and other petroleum products.

There are ways to stem contamination events from leftover fuel and chemical residues. Publications such as HM50, EI/JIG 1530 [6], MIL-STD-3004-1 [7] and many more cover cleaning requirements and best practices for fuel handling. HM50 and EI 1530 are commercial quality assurance guides that provide information on the proper storage, cleaning, and handling of commercial aviation products. MIL-STD-3004-1 provides similar information, but specifically includes military petroleum products and in some cases include strict guidelines for prior cargoes carried before aviation fuel can be loaded. While costly, dedicated pipelines, tank trucks or waterborne craft are useful to ensure the quality product reaches its destination. It is imperative that the buyer and supplier agree on the correct method of transportation (dedicated versus non-dedicated) and follow proper fuel handling procedures along with maintaining a robust quality assurance program to prevent costly fuel issues.

10 Common Quality Supply Chain Fuel Issues—Instability

In the aviation industry, fuel instability is a significant concern within the supply chain as unstable fuel can cause harmful deposits to form in fuel lines, filters, and engine components, leading to engine malfunctions or failures. Fuel instability can also reduce fuel efficiency and engine performance, increase emissions, disrupt operational readiness, and cause aircraft safety risks.

Fuel instability has many factors that have not been fully elucidated. Multiple methodologies and technical approaches have been developed to research fuel composition, chemical reactions, oxidation pathways, and molecular precursors to fuel instability. Decades of research and scientific inquiry into the chemical processes that occur in turbine fuel at elevated temperatures, which leads to coking and insoluble formation, have been limited by the sensitivity and selectivity of instrumentation. Chapter 7 goes in-depth on the topic of fuel storage and stability.

In some instances, fuel can meet specification limits at the point of manufacture but degrade in the fuel distribution system. In such cases there may be remediation options down-stream, for example, use of clay filtration or use of metal deactivator additive, but identification of the root cause may be challenging. Historically, investigations primarily utilized specification tests to analyze the fuel and, while beneficial in determining if the fuel met specification requirements, these are insufficient to identify the root cause of the instability. Chapter 14 describes a range of more advanced analytical methods and approaches for investigation of root cause which can be beneficial.

In some situations, the root cause of thermal stability failures can be determined for example: a high concentration of catalytic metals (i.e., copper), contamination by high molecular weight compounds (i.e., lubricants, pipeline drag reducers, etc.), or other chemical contaminants. However, in some cases the root cause is undetermined, due to the nature of the contaminant, for example, traces of a surfactant impacting water separation. With respect to fuel thermal-oxidative properties, it is hypothesized that the bulk hydrocarbon distribution has little to no impact, but the trace heteroatomic compounds that are present are responsible for the thermal stability failures.

11 Summary

In summary, SCRM is a combination of factors essential to ensure the success of the supply chain operations. SCRM consists of several overarching principles to manage petroleum products from point of manufacture to the point of use.

Among the many tools to identify, assess, and analyze risk there are additional techniques to help ascertain the appropriate levels of impact based on a plethora of factors depending on the segment of the supply chain being studied. Tools such as regression analysis can be applied to the standard risk evaluation tools as an aid to provide more specific levels of clarity when developing mitigation strategies. These add on tools are easily applied and allow organizations to better determine the likelihood and significance of a risk occurrence when additional variables such as timing, delivery mode/node, and speed are factored in.

The key to risk is understanding not only what applies directly to operational interest, but how the interrelated segments of the supply chain can be leveraged to help mitigate internal risk factors and those external factors within the supply chain itself, thus driving resiliency and stability.

Acknowledgements We would also like to thank Alisdair Clark, Moshe Rabaev, and Liel Mazar for the chapter's professional and editorial review.

References

1. Energy Institute (2024) HM 50: guidelines for the preparation of tanks and lines for marine tank vessels carrying petroleum and refined products, 6th edn. Energy Institute. ISBN 978 1 78725 411 4
2. Energy Institute (2024) EI 1542: identification markings for dedicated aviation fuel manufacturing and distribution facilities, airport storage and mobile fuelling equipment, 10th edn. Energy Institute. ISBN 978 1 78725 454 1
3. Energy Institute (2024) EI Specification 1581: specifications and laboratory qualification procedures for aviation fuel filter/water separator, 7th edn. Energy Institute. ISBN 978 1 78725 437 4
4. Energy Institute (2014) EI 1590: specifications and qualification procedures for aviation fuel microfilters, 3rd edn. Energy Institute. ISBN 9780852937075
5. ASTM D4054. Standard practice for qualification and approval of new aviation turbine fuels and fuel additives. ASTM International, West Conshohocken, PA. https://doi.org/10.1520/D4054
6. Energy Institute (2019) EI/JIG standard 1530. Quality assurance requirements for the manufacture, storage, and distribution of aviation turbine fuel to airports
7. MIL-STD-3004-1. Quality assurance for bulk fuels, lubricants, and related products

Chapter 9
Common Practice in Aviation Fuel Handling Equipment

Pierre Poitras, Jim Gammon, Daniel E. Kadlecek, and Amy Carico

Abstract Ensuring flight safety starts with proper handling of aviation fuel. This chapter explores key practices for maintaining equipment, managing fueling operations, and avoiding mistakes like misfueling and contamination. It highlights the importance of delivering the correct grade and quantity of fuel, with a focus on over-wing fueling challenges. Practical measures such as using grade-selective nozzles, clear wing decals, and verification steps are discussed to reduce the risk of errors. The chapter also addresses the increasing problem of additive misfueling, particularly confusion between Diesel Exhaust Fluid (DEF) and Fuel System Icing Inhibitor (FSII). It explains the consequences of such mistakes and emphasizes the importance of regular testing and calibration to detect issues before they escalate. Maintenance and storage of fueling equipment are equally critical. From using dust caps to prevent contamination to identifying wear and tear early, the role of fuelers in maintaining fuel quality is emphasized. Proper storage of sampling buckets and attention to hydrant systems are also covered as essential parts of the process. The chapter compares fueling operations at large airports, where equipment is maintained rigorously, to remote locations, where longer storage times and periodic maintenance increase risks. Regardless of the setting, the chapter underscores the need for proactive maintenance and strict adherence to industry standards to ensure the reliability of fuel handling systems and, ultimately, flight safety.

P. Poitras (✉)
Fuel Plus Consulting Inc., Chelsea, Canada
e-mail: pier.poitras@gmail.com

J. Gammon
Gammon Technical Products, Manasquan, NJ, USA
e-mail: jim.gammon@gammontech.com

D. E. Kadlecek · A. Carico
ExxonMobil Technology and Engineering, Baytown, TX, USA
e-mail: daniel.e.kadlecek@exxonmobil.com

M. Rabaev et al. (eds.), *Aviation Fuel Problems and Solutions: From Refinery to Wingtip*, https://doi.org/10.1007/978-3-032-05514-9_9

1 Introduction

The manufacture and distribution of fuel, from the refinery to the aircraft wingtip, is a complex process. While earlier chapters have detailed potential challenges and strategies to address them, once fuel is loaded onto an aircraft, the ability to detect and manage fuel-related issues becomes significantly limited. The airport, therefore, represents the final line of defense in ensuring fuel safety for flight. For this reason, procedures for handling fuel at airports or small fueling stations must be rigorous, and operators must maintain vigilance.

Industry documents such as EI 1540 [1], EI 1550 [2], JIG 1/2/4 [3–5] , ATA-103 [6], ICAO 9977 [7], EN 12312-5 [8], and others outline standards and procedures for managing airport fuel quality and handling equipment. This chapter serves as a practical supplement, offering real-world examples and context that highlight the reasons behind these standards and procedures. Additionally, a review of common problems and their solutions provides readers with valuable insights into critical airport fuel handling practices and establishes a solid foundation for troubleshooting fuel quality issues.

2 Airport Fuel Receipt

Fuel receipt at an airport fuel facility can occur via various supply routes. It is crucial for operators to thoroughly understand the specifics of their facility's fuel supply chain to ensure that receipt procedures and testing are appropriate for managing the fuel quality risks inherent in the receipt process.

2.1 Dedicated Versus Non-dedicated Fuel Receipt

Dedicated supply systems handle only one grade of fuel throughout the delivery system. Examples include jet fuel-dedicated bridgers loaded and unloaded via grade-selective couplings or pipelines transporting a single grade of fuel. In such systems, the risk of cross-product contamination—such as jet fuel being mixed with diesel or motor gasoline—is minimized. Consequently, receipt procedures may be less stringent, and quality control testing of the fuel may be limited. This risk-based approach relies on comprehensive product integrity throughout the supply chain (see Chap. 8).

Conversely, non-dedicated supply systems handle multiple products (e.g., non-dedicated bridgers or multi-product pipelines), increasing the potential for cross-product contamination. These systems require more rigorous receipt procedures and thorough quality control testing. Standards and procedures referenced earlier are essential for identifying contamination risks in such systems. Operators must understand the potential signs of contamination to prevent unfit fuel from entering the airport facility. This section will highlight examples of contamination during product receipt and outline actions operators should take to address these issues.

2.1.1 Dedicated Supply System

Regardless of the supply route, fuel must always be sampled and tested upon receipt at an airport fuel facility. At a minimum, testing in dedicated systems should include checks for appearance, density, and electrical conductivity (if a static dissipating additive is used).

Fuel appearance must be clear, bright, and free from undissolved water or particulate matter. Additionally, it must not deviate from prior test results. Visual inspection is a key method for identifying contamination. For example, jet fuel with a faint pink hue may indicate cross-contamination with dyed diesel, potentially caused by misaligned delivery tanks at a supplying terminal. Similarly, fuel that does not appear bright or "sparkle" may contain haze or water. While filter/water separators can remove undissolved water, stabilizing agents like surfactants may hinder effective water separation, indicating a broader contamination issue that should not be ignored.

When testing density or conductivity, results must meet specification requirements and remain consistent with previously reported values. Significant deviations may signal contamination. For instance, during the receipt of jet fuel from a dedicated bridger (tanker type tractor trailer delivery truck), a measured density of 0.7927 kg/L (left picture in Fig. 1) differed from the 0.7850 kg/L value on the release note (right picture in Fig. 1). This exceeded the allowable variation of 0.003 kg/L under most standards. Investigation revealed the truck had previously delivered diesel fuel and was not properly cleaned before returning to jet fuel service. Consequently, the contaminated fuel was rejected, and the truck returned to the supplying terminal. This example underscores the importance of adhering to grade-change protocols, even in dedicated systems.

Sampling procedures are equally important. For deliveries involving multiple compartments (e.g., multi-compartment bridgers, railcars, or barge/marine vessel tanks), it is standard practice to treat each compartment as a separate delivery. This approach prevented a significant contamination event at an airport when a railcar containing 100 kL of motor gasoline was mistakenly included in a multi-railcar jet fuel shipment. Density checks during receipt testing flagged the error, as found on the Certificate of Quality (COQ), preventing the gasoline from being offloaded into a one-million-liter jet fuel tank.

Representative sampling is essential to confirm product integrity. Multi-compartment road bridgers, as shown in Fig. 2, often use a single connecting manifold. When sampling these compartments, the manifold must be drained to ensure the sample reflects the compartment's content, not residual fuel from the manifold line.

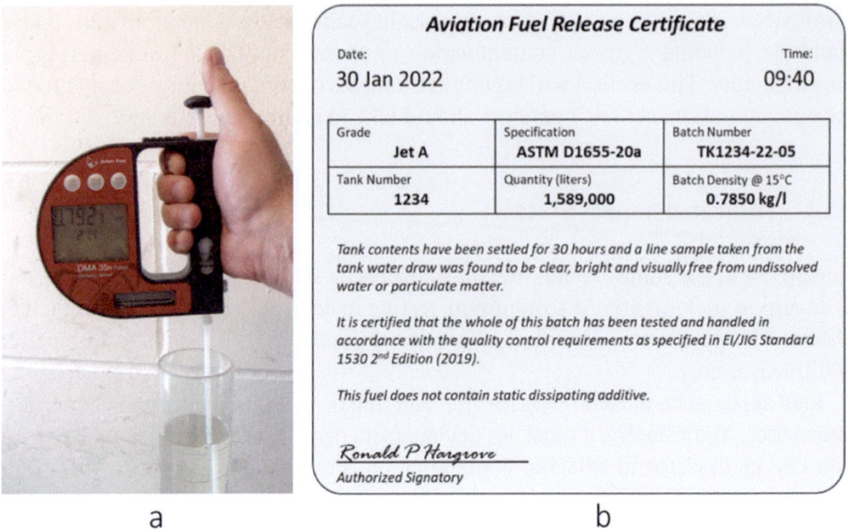

Aviation Fuel Release Certificate

Date: 30 Jan 2022

Time: 09:40

Grade	Specification	Batch Number
Jet A	ASTM D1655-20a	TK1234-22-05
Tank Number	**Quantity (liters)**	**Batch Density @ 15°C**
1234	1,589,000	0.7850 kg/l

Tank contents have been settled for 30 hours and a line sample taken from the tank water draw was found to be clear, bright and visually free from undissolved water or particulate matter.

It is certified that the whole of this batch has been tested and handled in accordance with the quality control requirements as specified in EI/JIG Standard 1530 2nd Edition (2019).

This fuel does not contain static dissipating additive.

Ronald P Hargrove

Authorized Signatory

Fig. 1 Density testing—(left) fuel density measurement during receipt and (right) release certificate from the supplying terminal. Note that instrument complying to ASTM D7777/IP559 is safer to use that hydrometers and thermometers

Fig. 2 A multi-compartment bridger with each compartment being sampled individually

2.1.2 Non-dedicated Supply Systems

As fuel distribution systems grow increasingly complex, greater emphasis is placed on product quality documentation. This documentation provides critical information about the fuel batch being received. In non-dedicated systems, the risk of cross-product contamination is higher if grades are not adequately segregated or protected during receipt. Consequently, additional testing is required when receiving fuel from such systems. This may involve either a Recertification Test (RT) or a full Certificate of Analysis (COA).

Upon receipt, representative fuel samples should be taken to ensure the test results accurately reflect the entire batch of fuel in the receiving tank. These results must then be compared to the prior certified batch results, such as a refinery COQ or a previous COA, to identify potential cross-product contamination. Test certificates also provide essential traceability of the product, complementing the bill of lading information at the time of custody transfer or airport delivery.

Any off-specification results or significant variations in key properties warrant further investigation, similar to processes for dedicated supply systems but across a broader range of quality control parameters. Additionally, close attention must be paid to declared additives to prevent overdosing if re-additization is necessary. For example, Metal Deactivator Additive (MDA) is used to recover jet fuel thermal stability but has a maximum allowable concentration. Traceability of additive content through test certificates ensures these limits are not exceeded. Misuse or overdosing of MDA can obscure underlying issues with poor thermal stability in the fuel.

Figure 3 illustrates a Recertification Test report indicating the addition of MDA before testing. This suggests that the fuel initially had poor thermal stability. However, the report lacks supporting information on the pre-MDA test results, raising questions about the justification for MDA usage and whether the batch being tested had been treated with it. Note that MDA is typically used to counter-react metallic species such as copper, zinc and yellow-type alloys affecting the fuel thermal stability. MDA concentration must be declared on COQ when used at point of manufacture.

2.1.3 Other Receipt Modes

A less common delivery mode that poses quality risks without proper testing is small fuel deliveries via isotainers or drums. Sampling is essential, whether upon receipt or before use. Representative sampling of multiple containers ensures that the fuel batch meets specifications and that the containers were correctly filled with the appropriate grade during manufacturing. This is especially important at airports serving both turbine- and piston-engine aircraft to prevent misfueling.

Note that some light aircraft with retrofit diesel engines may outwardly resemble Avgas aircraft but require jet fuel. These aircraft should be clearly marked "Jet Fuel Only" near the fill connection, but markings may be obscured by wing mats, requiring vigilance by operators.

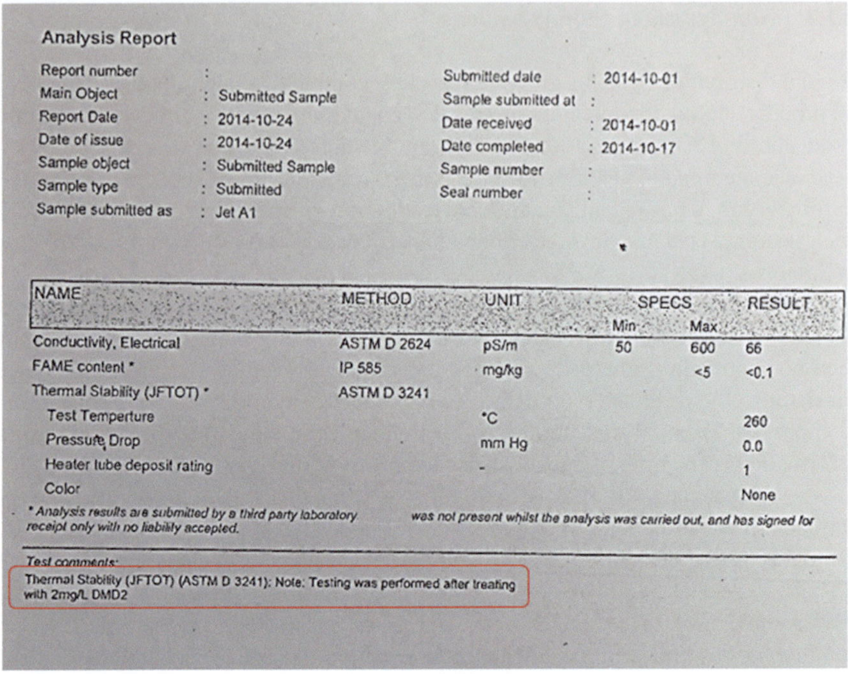

Analysis Report

Report number :
Main Object : Submitted Sample
Report Dale : 2014-10-24
Date of issue : 2014-10-24
Sample object : Submitted Sample
Sample type : Submitted
Sample submitted as : Jet A1

Submitted dale : 2014-10-01
Sample submitted at :
Date received : 2014-10-01
Date completed : 2014-10-17
Sample number :
Seal number :

NAME	METHOD	UNIT	SPECS Min	SPECS Max	RESULT
Conductivity, Electrical	ASTM D 2624	pS/m	50	600	66
FAME content *	IP 585	mg/kg		<5	<0.1
Thermal Stability (JFTOT) *	ASTM D 3241				
Test Temperture		°C			260
Pressure Drop		mm Hg			0.0
Heater tube deposit rating		-			1
Color					None

*Analysis results are submitted by a third party laboratory was not present whilst the analysis was carried out, and has signed for receipt only with no liability accepted.

Test comments:
Thermal Stability (JFTOT) (ASTM D 3241): Note: Testing was performed after treating with 2mg/L DMD2

Fig. 3 A portion of a recertification test report (highlighted in red) showing thermal stability testing after the addition of MDA, but without reporting the pre-MDA test results

Similar to airport locations where fuel is stored for extended periods, drums at remote sites or fuel caches require strict quality control, proper storage, and careful handling. Drums must be inspected immediately upon receipt, ensuring they are properly sealed and free from leaks. Drums with broken seals or leaks should be set aside and investigated for potential mishandling or sabotage. If contamination concerns are resolved, these drums should still be emptied and filtered into intermediate tanks as a precaution.

Mark drums or affix labels with the following information:

- Fuel grade
- Filling date
- Batch number
- Safety data and transportation details (as appropriate).

Ideally, markings should be lithographed onto the drum, as temporary labels may degrade when exposed to the elements.

Drums should be stored in a cool, ventilated area and protected from environmental elements to maintain fuel quality and safety. Proper storage conditions involve positioning the drums either upright or on their sides. When stored on their sides, the bungs should be placed horizontally, at the 3 o'clock and 9 o'clock positions, to ensure that the seals remain immersed in the fuel and prevent air contamination (see Fig. 4). If drums are stored upright, it is essential to angle them slightly to avoid water pooling on the bungs or, alternatively, to use protective covers.

To prevent direct contact with the ground, which could lead to contamination and degradation, drums must be stored on dunnage with adequate bracing. Outdoor storage requires placing drums on their sides to avoid water pooling on drumheads, which can result in water contamination. For concrete storage areas, a sloped floor design is highly recommended, allowing controlled drainage and preventing any potential fuel leaks from spreading (see Fig. 5). Safeguarding the physical environment and integrity of the product is essential.

For outdoor storage, it is advisable to group drums by lot, limiting each group to no more than 50 drums. This approach helps streamline inventory management and minimize potential product degradation and corrosion. Additionally, older fuel batches should be used first to maintain product quality and reduce the risk of damage to drum markings.

Fig. 4 Fuel cache in isolated areas

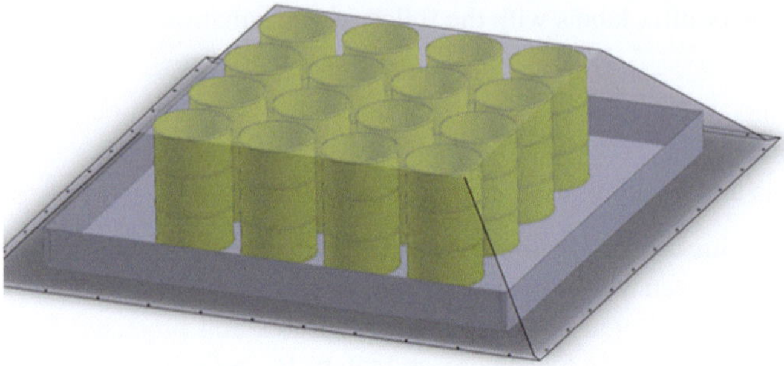

Fig. 5 Diagram of secondary containment for storing fuel drums and preventing environmental fuel leaks

In emergencies, never store drums directly on the ground; always ensure proper drainage, particularly on sloped concrete floors. Drums positioned vertically outdoors should be closely monitored, as water seepage through the bungs can lead to product contamination and potential safety risks. Proper care in handling and maintaining the integrity of storage setups is essential to safeguard both fuel quality and the environment.

2.1.4 Receipt of Fuel Containing Additives

When fuel containing additives (e.g., Static Dissipating Additive (SDA), Fuel System Icing Inhibitor (FSII), Metal Deactivator Additive (MDA), etc.) is supplied, several quality control checks are necessary, such as verifying fuel conductivity or FSII concentration. If fuel is received from multiple supply locations, it is possible to receive both additized and unadditized fuel.

For instance, a location that did not typically receive or use fuel with SDA had a supplier deliver a batch of fuel containing SDA. The airport lacked the equipment to measure fuel conductivity and had to rely on an airport in another country to provide a conductivity meter. Once the fuel was confirmed to be on-specification upon receipt, it was mixed with unadditized fuel in a common receipt tank, as it was a dedicated supply location. During this process, it was crucial to continuously monitor fuel conductivity while filling the tank to prevent over-dilution of the SDA-containing batch and ensure that the final tank contents still met the minimum fuel conductivity requirements.

Note: In locations or countries where the use of static dissipating additives is not mandated, fuel conductivity is typically not monitored. However, the fueling infrastructure is designed to ensure adequate electrostatic charge relaxation. Fuel conductivity is mainly for safe ground handling as oppose to aircraft fuel system requirement.

3 Fuel Quality Control in Airport Storage

Fuel quality control in airport storage serves as the "second line of defense" and is important because it is often the last opportunity to identify contaminated fuel before it enters the airport fueling system. While numerous standards govern on-airport fuel quality control, some unusual scenarios and potential issues are highlighted below.

3.1 Tank Settling

Proper tank settling is a fundamental step in aviation fuel handling, essential for ensuring fuel cleanliness and protecting engine performance. Once fuel is transferred into airport storage tanks, it is necessary to allow a settling period to enable suspended particulate matter and free water to separate and accumulate at the bottom. Industry guidelines recommend a minimum settling time of one hour per foot (30 cm) or three hours per meter of fuel height [1], though extended settling times are often advantageous for better contaminant separation.

Neglecting adequate settling times can compromise fuel quality, leading to potential damage to aircraft fuel systems. Particulate matter that remains in the fuel can erode pumps, clog injectors, and impair engine performance, while undetected water contamination may freeze at high altitudes, causing blockages in fuel lines. Regular drainage from tank sumps is essential to remove these contaminants, ensuring compliance with aviation fuel quality standards.

Settling also helps identify more serious issues, such as microbiological contamination. For example, if clear water typically drains from a tank but is replaced by cloudy water with a foul odor, it may signal microbial growth. Microbial contamination, often caused by water accumulation at the fuel-water interface, can produce acidic byproducts that corrode tank linings and generate sludge, further degrading fuel quality.

Another factor impacting tank settling is "tank breathing," where tanks draw in air due to fuel depletion or external air pressures. Airborne contaminants, such as dust from nearby construction sites, fungus spores or pollen during the flowering period, or organic matter like yeast from adjacent bakeries, can enter the tanks, compromising fuel purity. Notably, incidents of yeast contamination have been reported, where microbial activity produced gas bubbles visible at roof vents, indicating ongoing fermentation within the tank.

Conversely, overly fine mesh screens on tank vents intended to block contaminants can become clogged with sand, bird nests, or insect debris. This obstruction restricts airflow, creating pressure imbalances that may damage the tank structure and reduce its operational efficiency. Tanks in remote locations, subject to infrequent maintenance, are particularly vulnerable to these challenges (see Fig. 6).

By implementing proper settling practices, coupled with regular sump drainage and proactive monitoring, operators can safeguard fuel quality and prevent costly

Fig. 6 Fuel stored in Arctic condition

equipment failures. Additionally, maintaining appropriate vent designs and isolating storage tanks from environmental contaminants can mitigate risks and enhance the longevity of fuel systems.

3.2 On-Airport Filter Management

Airports are equipped with filters to remove residual particulate matter or free water that does not settle in storage tanks. Filters are governed by EI 1550 (*Handbook on Equipment Used for the Maintenance and Delivery of Clean Aviation Fuel*) [2], but a common issue at airports and upstream supply locations is the interconnection of pressure relief and air eliminator lines.

Often, for simplicity, these lines are connected and piped to a separate product recovery tank. While less of a concern for airports with a single grade of fuel, it can become problematic when multiple grades are stored. Interconnected lines may allow leakage between vessels due to failing check valves or varying operating pressures, resulting in cross-contamination of fuel grades.

3.3 Additives Used On-Airport

Only a few additives are typically introduced into fuel at airports. Static dissipative additive (SDA) may be added to fuel tanks upon receipt, while fuel system icing inhibitor (FSII) is often introduced during aircraft refueling. Although SDA is usually added upstream before the fuel reaches the airport and is therefore not commonly stored at airports, FSII is routinely injected at smaller airports serving the general aviation market.

As with fuel, maintaining the purity and proper application of additives is critically important. Specific procedures are usually required to ensure additives are used correctly. Proper labeling and storage are essential, especially when additives are handled at the point of use or during aircraft refueling. Additives should always be stored in a dry, sheltered area and, if possible, positioned horizontally to prevent water ingress through drum bungs.

FSII, for instance, is highly sensitive to water contamination. Drums or reservoirs containing FSII must have a desiccant dryer to prevent particulate contamination and control humidity. FSII is hygroscopic, which makes it highly effective at inhibiting ice formation in aircraft fuel tanks. However, if water contamination exceeds 0.5%, FSII's ability to dissolve in fuel is compromised. Undissolved FSII tends to settle at the bottom of fuel systems, where it can chemically degrade polymeric and metallic materials, such as aluminum.

Figure 7 illustrates two cases of FSII contamination. In Fig. 7a, the additive, which was injected in concentrated form during fueling, dissolved the polyurethane baffle material in a helicopter fuel tank. Figure 7b shows a red layer of excess FSII and water removed from the bottom of an helicopter auxiliary tank. Both examples led to significant fuel system issues, such as filter bypass and poor engine performance.

In another example, an operator improperly added FSII to a jet fuel storage tank. Although the correct amount of additive was calculated, the operator mixed it with fuel in a bucket and poured the mixture into the tank. FSII's poor solubility in fuel caused it to settle at the bottom of the tank, damaging the epoxy lining. As a result, the additive failed to disperse into the fuel, leaving the aircraft operators with insufficient FSII protection.

In recent years, there have been incidents involving the accidental use of unapproved additives, such as Diesel Exhaust Fluid (DEF), in aircraft fuel systems (see Chap. 15). DEF is an emissions-reduction additive used in fueling vehicles and is not intended for use in aircraft. These errors were often caused by confusion between similar additive names, such as DEF and DICE (an FSII tradename), to poorly labeled containers or using the same container to transfer the two chemicals from the shipping container to the service reservoir. DEF contamination can have catastrophic effects because it primarily consists of 67% water and 33% urea. When DEF is added to jet fuel, the urea precipitates, leading to blocked fuel filters, contamination of fuel system components, and potential turbine engine stalling. Figure 8 shows the impact of DEF contamination, with (a) a fuel filter clogged by urea and (b) urea particles passing through a mesh filter.

a b

Fig. 7 Two cases of FSII contamination illustration—**a** polyurethane baffle material dissolved by excess FSII in a helicopter fuel tank and **b** excess FSII/water phase removed from an aircraft tank

a b

Fig. 8 Impact of DEF contamination—**a** the aircraft fuel filter is contaminated with DEF (Urea), and **b** the mesh filter shows urea contamination in aircraft fuel systems

3.4 *Fuel Drum Quality Control*

Where possible, drummed fuel should be inspected annually, within 12 months of the filling date, to verify continued compliance with its specification. Samples sent to the laboratory must be composite samples, consisting of equal parts taken from the top, middle, and bottom of the drum, with a minimum total volume of 4 L. Test results and retest dates must be documented.

It is recommended that one in forty (40) drums per batch per location is sampled and tested. When multiple fuel batches are sampled, each batch should be tested separately in accordance with ASTM D4057 [9].

If the test results indicate no significant changes (i.e., results remain within fuel specification limits), the fuel may be stored for an additional 12 months. Drums containing fuel that no longer meets specifications should be clearly marked, and any original labels or markings should be removed or obscured. Aviation fuel failing to meet specifications must no longer be used in aircraft but may be repurposed for other applications if deemed fit.

The date of the last inspection should be clearly marked on the drum in the following format: RE-INSPECTED (Month and Year).

3.4.1 Drawing Fuel Samples from Drums

Before drawing a sample from a drum, place the drum in a tilted position on a support block for at least 10 min to allow the contents to settle. Carefully clean the bungs and the top surface of the drum.

Remove the bungs slowly to release any excessive pressure gradually.

Use either a small, dedicated transfer pump with a fuel-compatible hose in good condition or a manually operated stainless-steel sampling tube (commonly known as a "drum thief") to draw a 1-L sample. This sample is used to perform a quality control check before transferring fuel into an aircraft, ensuring fuel quality is confirmed. Avoid sampling from the very bottom of the drum to prevent drawing any potential contaminants.

Fuel stored in drums is particularly prone to contamination by rust particles and water.

Note: Unopened drums stored under proper conditions should not contain contaminants.

Regardless of whether a hand-operated or electric pump is used to transfer fuel from the drum to the aircraft, the fuel must pass through an approved filter separator or an equivalent filter of similar performance before entering the aircraft's fuel tank.

4 Airport Fuel Hydrant Systems

Airport fuel hydrant systems provide an efficient and safe method to dispense jet fuel to aircraft. These systems pump fuel from a storage facility through filters into an underground piping network that delivers fuel to aircraft stands via hydrant servicers or carts. A key advantage of these systems is the ability to supply unlimited fuel to an aircraft in a single refueling, unlike the limited capacity of tank-based refuelers.

However, hydrant systems present potential fuel quality concerns related to contamination. While cross-product contamination is minimal within the airport perimeter, environmental contaminants such as microbes, dirt, and water can be significant challenges. Operators can mitigate these risks by:

1. Preventing contaminants from entering the system, and
2. Maintaining sufficient fuel velocity in the hydrant lines to move contaminants to designated low points for removal.

Constructing clean hydrant systems and using effective fuel filtration help prevent contaminants from entering the system. To remove contaminants already in the system, maintaining a fuel velocity of at least 1 m/s is critical. This ensures settled particles or water are flushed to low points where they can be effectively removed.

4.1 Hydrant System Design and Construction

Hydrant systems vary in complexity but generally include a main header line and various spur lines connected to aircraft stands. While designing for future airport expansion, excessive capacity can lead to low-velocity operation, making it difficult to flush contaminants effectively. Fuel velocity below 1 m/s allows contaminants to settle, increasing the risk of free water accumulation promoting microbiological growth and corrosion caused by sulfate-reducing bacteria in anaerobic conditions. To prevent this, systems should be designed to maintain velocities exceeding 1 m/s during regular operation.

During construction, maintaining the cleanliness of piping and valves is essential for future fuel quality. Pipe plugs should prevent dirt and water ingress, and only actively worked-on pipes should be placed in hydrant trenches to avoid flooding from rainwater. As piping and valves are welded, debris should be removed continually to avoid future contamination. Cleaning during construction and prior to commissioning is significantly easier than post-installation. Figure 9 illustrates a hydrant installation where inadequate safeguards during construction resulted in significant contamination and delays.

Fig. 9 Improper storage (**a**) and installation (**b**) of hydrant piping, allowing significant particulate matter and water ingress

4.2 Hydrant System Operation

Proper operation of hydrant systems ensures clean, dry fuel delivery to aircraft. While filtration prevents most particulates and free water from entering, condensation may occur due to fuel cooling in underground pipes. Therefore, hydrant systems should include numerous low points equipped with drains to collect and remove free water and particulates.

Most operational standards recommend draining these low points weekly to prevent contaminants from passing into aircraft fueling equipment. Prolonged water accumulation can promote microbiological growth. During draining, the line volume should be flushed with sufficient velocity to push settled contaminants to a flushing vehicle. This vehicle should include a sampling point for visual confirmation of cleanliness.

If low-point flushing is insufficient, the entire hydrant system may need to be flushed. For simple systems, this might involve flushing into a vehicle tank at the most remote hydrant pit, acknowledging the safety risks of using fixed-volume tanks. Complex systems should circulate fuel back to storage via filtration to separate contaminants. Severe contamination may require "pigging," depending on system design. Operators must notify airlines of such activities, and fueling operations should pause until the system is deemed clean.

Inactive hydrant sections require monthly cycling to move contaminants to low points. Maintaining logs of hydrant usage helps determine appropriate flushing frequencies and prevents stagnation.

Hydrant pit boxes and valve chambers are also potential entry points for contamination. Subsided pit boxes may collect rainwater, which, if not removed, could lead to system contamination. Regular inspection of valve chambers is critical to prevent submersion of hydrant valves. Additionally, dirt and water on pit valves can compromise fueling operations. Dirty valves hinder coupler connections and risk leaks during fueling. Figure 10 shows significant contamination in a hydrant pit box, highlighting the importance of cleaning pit valve surfaces before connecting to fueling vehicles. Failure to do so risks contamination of the hydrant servicer and, ultimately, the aircraft.

Fig. 10 A hydrant pit box showing significant contamination

4.3 Hydrant System Maintenance

Hydrant systems are complex pieces of equipment that require regular maintenance to ensure optimal performance and safety. To this end, process control systems are often employed to automate tasks like turning pumps on and off and cycling them periodically to prevent operational issues. Additionally, monitoring fuel throughput and filter conditions is crucial to avoid overloading specific components.

A critical aspect of hydrant system maintenance is ensuring the integrity of the pipeline. This can be achieved through the use of hydrant tightness control systems, which continuously monitor for leaks. Alternatively, manual inspections can be conducted to assess pipeline condition. It's important to note that leaks in a hydrant system pose a dual threat. Not only can they lead to environmental contamination of the surrounding soil, but they can also act as aspirators. In such cases, the venturi effect can draw contaminants, such as particulate matter or water, into the hydrant system through any openings in the pipeline. This is particularly concerning for systems operating at low pressure but high velocity [10].

For these reasons, it is critical to routinely monitor hydrant system integrity. Regular inspections and maintenance should be conducted to ensure that all components are functioning correctly and that there are no leaks. Technologies such as acoustic monitoring, pressure testing, and visual inspections can be employed to detect and address any issues promptly. Leak testing utilizing fuel additives such as Tracer A® may be used as permitted in most governing fuel specifications, but may also be restricted by local environmental requirements. Recommended Practice EI 1560 contains additional troubleshooting information related to unusual operating conditions or observations related to hydrant systems and their maintenance.

5 Fueling Equipment

Fueling equipment refers to the systems used to deliver fuel to aircraft. Maintenance and monitoring practices for fueling equipment cover a wide range of tasks, including:

- Monitoring and limiting filter vessel differential pressure
- Controlling fueling pressure and flow rates
- Inspecting hoses for wear, leaks, and age
- Checking nozzle strainers, nozzle condition, and electrical bonding wire, clips and system integrity
- Ensuring deadman controls and emergency stops function properly
- For refueling trucks with tanks, managing high-level controls during tank loading.

5.1 Equipment Maintenance Issues

Fuel quality is often compromised by inadequate maintenance of refueling vehicles. One common issue involves contaminants entering the fuel tank due to poor inspections. For example, regular inspection of the tops of refueler tanks is crucial to prevent water pooling, especially rainwater. Refueler tanks are equipped with "rollover protection" to safeguard the manway and vent during rollovers. However, if drainage pathways on the tank top are blocked, rainwater and debris can accumulate. This pooling can lead to water and dirt entering the tank through leaks, damaged gaskets, or when the power take-off (PTO) is engaged, causing the tank vent to open (Fig. 11).

Operators often implement improvised solutions to maintenance challenges without realizing their potential impact on fuel quality. For instance, Fig. 12 shows a filter vessel where grease was applied to secure an O-ring and create a seal on a horizontal vessel. However, grease is a known aviation fuel contaminant that can affect thermal stability, water separation, and gum content in fuel. Similarly, Fig. 13 illustrates a fueling nozzle contaminated with lithium grease, demonstrating that grease should never be used on fuel-wetted components.

a b

Fig. 11 Refueling truck tank top views with rain downspout

Fig. 12 Horizontal filter vessel with greased-contaminated seal

Fig. 13 Fueling nozzle contaminated with lithium grease

Regular inspections of nozzle strainers are critical. The presence of foreign material in nozzle strainers indicates underlying issues. For example:

- Metal filings can result from meter failure or pipe welding
- Black, rubber-like particles suggest hose interior deterioration

Fig. 14 Debris found in filter nozzle strainer

- Fibers are particularly concerning, as they often indicate a damaged or burst filter element (Fig. 14).

Refuelers for aviation gasoline (Avgas) and jet fuel differ significantly in operation. Avgas refueling typically uses handheld overwing nozzles without automatic shutoff features, and these should never be left running unattended. Jet fuel trucks commonly utilize underwing refueling with single-point nozzles (also known as pressure refueling). Flow is controlled by a manual "deadman" device, ensuring that fuel only flows when actively operated by the refueling personnel.

Care must be taken to avoid cross-contamination between Avgas and jet fuel. Jet-fueled aircraft and helicopters sometimes require the smaller spouts used for Avgas, increasing the risk of error. Special attention is needed to prevent jet fuel from being loaded into Avgas aircraft, as their engines cannot burn it. Exceptions exist, such as light aircraft with diesel engines, but refueling personnel must exercise extreme caution due to the visual similarities between these aircraft and Avgas-powered planes.

6 Fuel Delivery to Aircraft

Although fueling itself is not equipment, the process of delivering fuel to aircraft represents the culmination of all prior equipment, testing, and procedures. Individual airlines are responsible for developing their own refueling protocols, but there are many established best practices shared across the aviation industry. Fueling vendors must adhere to the specific requirements outlined in each airline's fueling manual, but they can also refer to industry-level guidance documents such as the IATA Guidance Material on Standard Into-Plane Fueling Procedures and JIG 1 Aviation Fuel Quality Control and Operating Standards for Into-Plane Fueling Services.

While numerous factors must be considered when fueling aircraft, this section focuses on common issues related to fuel quality. Although it is preferable to address and mitigate fuel quality challenges upstream of the refueling process, opportunities for contamination or quality issues can still arise during aircraft refueling.

6.1 Awareness of Operations

Ensuring fuel quality during aircraft refueling is fundamentally dependent on maintaining operational awareness. Despite advancements in automated systems and high-tech equipment, there is no substitute for a fueler who remains attentive and prepared to respond to any irregularities that may arise. Vigilant oversight across the airport is essential for effective fuel quality management, particularly during refueling operations, where proactive monitoring can prevent serious issues.

One of the core practices that support awareness and fuel quality is the routine sumping (draining) of filter vessels in fueling equipment (see Fig. 15). This task serves a dual purpose: It removes free water and particulate matter that has accumulated within the system and acts as an early diagnostic tool for identifying potential problems. A sudden and significant increase in free water or particulate matter during sumping can be indicative of upstream issues such as filtration failures or deficiencies in hydrant system maintenance or construction. Operators who are well-acquainted with the usual appearance and volume of sumped fuel are in a better position to recognize irregularities and take appropriate action.

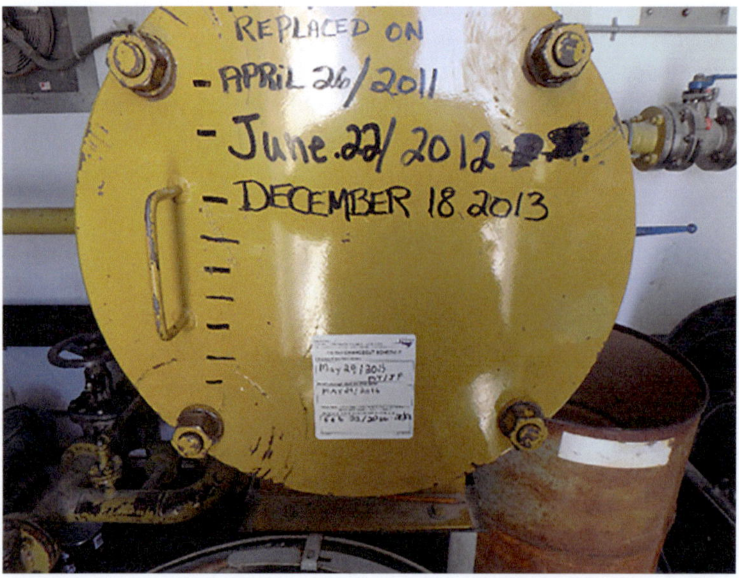

Fig. 15 Improper usage and maintenance of filter vessel

It is essential to perform sumping under pressure while fuel is flowing, as this ensures that contaminants are removed efficiently and minimizes the risk of air entering the system. Sumping filter vessels while the fuel is static can introduce air, especially if air eliminators are missing, defective, or have been compromised (e.g., painted over). Such conditions can create hazardous situations, impacting fuel system performance and, ultimately, flight safety. Industry best practices emphasize that sumping should only be performed while fuel is actively circulating.

A common pitfall is failing to maintain a high enough velocity during sumping to properly clear contaminants. For example, opening the valve partially and allowing fuel to trickle out may create a false impression of acceptable fuel quality because the lower velocity does not effectively remove accumulated particulates and water. This inadequate sumping can lead to the buildup of contaminants in the fuel system, posing significant safety risks.

The consequences of inadequate fuel quality management during operations can be severe. This practice is particularly significant given past incidents, such as the November 2010 event involving the Airbus A380-842 (VH-OQA), where an uncontained engine rotor failure (UERF) led to substantial structural and system damage due to oil feed failures within the engine [11]. While this incident stemmed from manufacturing defects that allowed oil to ignite and rupture the engine, it highlights the broader importance of strict operational and procedural adherence in preventing failures that could lead to serious outcomes. The investigation by the Australian Transport Safety Bureau (ATSB) revealed how non-conforming parts, like thin-walled oil feed stub pipes, could cause critical issues; similarly, improper fuel system management could result in severe risks. This incident serves as a stark reminder that comprehensive awareness and diligence are essential throughout all stages of operation to prevent potential issues from escalating into catastrophic failures.

6.2 Monitoring Differential Pressure (dP)

Effective differential pressure (dP) monitoring is another cornerstone of fuel quality management. Differential pressure measures the difference between the inlet and outlet pressures of the filter vessel and typically fluctuates as the flow rate changes during operation. Operators must be familiar with their equipment's expected dP values, including the baseline when new filter elements are installed (see Fig. 16). Monitoring for deviations in dP throughout fueling operations is essential. dP readings shall be corrected with variable flow rates. Filter vessels should be operated as close as possible to its maximum nominal flow rate for proper filtration.

Gradual increases in dP over time are normal and expected. However, sudden spikes or drops in dP often indicate underlying issues requiring immediate investigation. For instance, a significant increase in dP might suggest the presence of large quantities of particulates or free water in the fuel supply. If unaddressed, such conditions could lead to filter ruptures, allowing off-specification fuel to reach the aircraft. Installing dP limiting devices, such as dP switches (typically mounted on

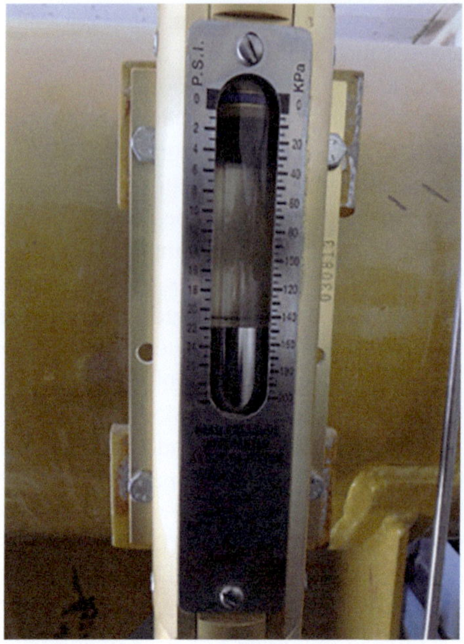

Fig. 16 Excessive DP reading across a filter vessel

the dP gauge), can automatically shut down flow when maximum operating dP is exceeded. However, these devices are not yet universally adopted.

A notable example illustrating the importance of dP monitoring is the Cathay Pacific Airways Flight CPA780 incident on April 13, 2010 [12]. During descent into Hong Kong, both engines of the Airbus A330 suffered fuel contamination caused by clogged fuel control units. This contamination resulted from fuel system anomalies that might have been detected earlier with proper dP monitoring practices. Despite the compromised engine performance, the flight crew skillfully executed a safe emergency landing.

Conversely, a dramatic drop in dP can also signal a serious problem. While a gradual decrease in dP near the end of fueling is expected due to back pressure from the aircraft, a sudden and severe drop may indicate structural failure in one or more filter elements. Fuelers must remain attentive to these changes in real time and act promptly when abnormalities occur. The dP gauge readings must be recorded and monitored over time of the filter elements service period.

6.3 Correct Grade/Type/Quantity of Fuel Delivered

Misfueling remains a common and potentially catastrophic issue in fueling operations. Ensuring the correct fuel grade (e.g., Avgas vs. Jet A1), proper additive packages, and accurate quantities is critical. Misfueling is one of the most significant risks during fueling operations, particularly during overwing fueling. A key mitigation against misfueling is proper labeling of equipment to identify fuel products and grades. EI 1542 provides critical information on the proper labeling of aviation fueling equipment.

The risk of misfueling is highest during overwing fueling. EI 1597 [13] provides best practices, such as using wing decals to indicate the proper fuel grade, grade-selective nozzles, and order verification procedures. However, wing decals can become faded or damaged and are often obscured by wing mats during fueling. Additionally, not all aircraft are designed to accept grade-selective nozzles. Larger, oblong nozzles (commonly referred to as "duckbill") are used for jet fuel, while smaller-diameter nozzles are for Avgas. Newer aircraft typically feature grade-selective orifices compatible with these nozzles. However, older aircraft may not accommodate the larger jet fuel nozzle, requiring the duckbill to be removed to use a smaller nozzle. As alternate Avgas grades become available in the general aviation sector as replacement to 100LL grade, it is essential that storage tanks, refueling vehicles and hoses are properly identified to reflect the fuel grade being provided. In addition to signage, to ensure flight safety, all fueling transactions must be acknowledged by the pilot and the fuel supplier while being duly documented.

This practice has led to incidents where the duckbill nozzle was not reinstalled, resulting in confusion about the fuel type being delivered. For example, a smaller nozzle intended for Avgas was mistakenly used to deliver jet fuel. To mitigate such risks, fuelers should follow stringent procedures for removing and refitting larger nozzles and confirm the fuel grade before loading. Ultimately, a vigilant, conscientious fueler remains the most critical safeguard against misfueling.

Recently, the industry has seen a rise in misfueling incidents involving additives, particularly the confusion between Diesel Exhaust Fluid (DEF) and Fuel System Icing Inhibitor (FSII). As discussed earlier, inadvertent loading of DEF into an FSII tank can have severe consequences. Fuelers commonly verify FSII volumes by measuring the refractive index of a mixture of FSII and water. Unfortunately, DEF has a nearly identical refractive index, making errors difficult to detect. The industry has developed new testing methods to identify DEF in FSII, and regular testing of FSII reservoirs is now a recommended practice.

Additive injection systems should also undergo regular calibration, either at full or partial flow, following the manufacturer's instructions.

6.4 Maintenance and Storage of Fueling Equipment

Proper maintenance of fueling equipment, discussed earlier, is essential. Fuelers play a crucial role in identifying maintenance needs and ensuring actions taken during fueling do not compromise equipment integrity.

For instance, using dust caps on nozzles prevents contamination of nozzle surfaces when equipment is not in active use. Fuelers can also identify early signs of wear on components like bonding cables or hoses, enabling proactive maintenance before

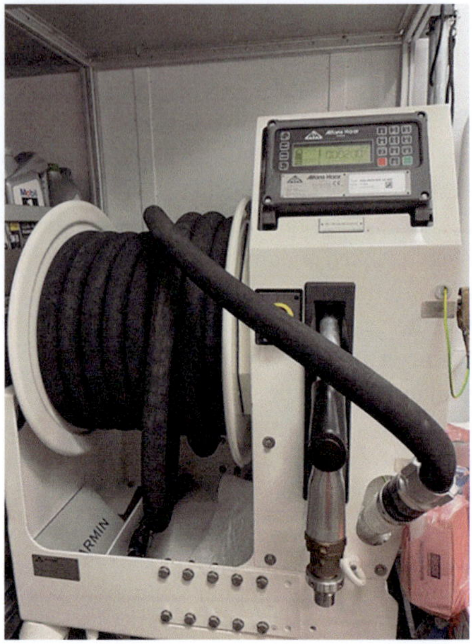

Fig. 17 Hose reel in a superyacht fueling chamber

failure occurs. Similarly, fuelers are well-positioned to spot issues with hydrant systems, such as missing hydrant pit valve covers or water accumulation, and report these to system operators for prompt resolution.

White buckets, commonly used for fuel sampling and visual evaluation, must be stored properly to avoid contamination. Improperly, stored buckets can accumulate contaminants that distort the visual appearance of fuel samples, potentially masking fuel quality issues.

Fueling hoses are critical to deliver on-specification fuel to the aircraft. Hoses manufacturing shall comply to ISO 1825 or EI 1529 latest edition, see Fig. 17. Hoses shall be placed in service within two years of manufacture date.

At time of installation, new hoses should be subjected to a soaking and flushing cycle to ensure that debris and chemical free hoses are used. Several industry standards such as ATA 103, EI 1740, JIG 1, 2 and 4 shall be consulted for maximum use of refueling hoses. Interesting to note that Offshore rigs facilities and Superyachts with helicopter capability are mandated by CAP 437 [14] to pressure test their fueling hoses every two years for safety reasons. Short of conducting this pressure test, hoses should be changed to avoid environmental disaster at sea.

7 Conclusion

As demonstrated in this chapter, fuel quality is closely tied to the equipment used for storing, handling, and delivering aviation fuels. Regular maintenance, timely repairs, constantly looking for changes, and adherence to industry standards are key for safe and efficient fueling operations.

At major airports, fuel is typically stored and dispensed using well-maintained equipment with short residence times, often just a few hours before delivery to an aircraft, minimizing contamination risks. Conversely, fuel stored at remote locations often remains for extended periods and is maintained less frequently due to limited accessibility. These sites pose a greater risk to fuel quality.

Regardless of the facility's size or location, maintaining fueling equipment to industry standards is essential to ensure fuel quality and flight safety.

Acknowledgements The author would like to thank Madi Mohtadi for her helpful comments and additional insights during the preparation of this chapter.

References

1. EI 1540 (2023) Design, construction, commissioning, maintenance and testing of aviation fuelling facilities, 6th edn. Energy Institute, London. https://publishing.energyinst.org/topics/aviation/aviation-fuel-handling/ei-1540-design,-construction,-operation-and-maintenance-of-aviation-fuelling-facilities
2. EI 1550 (2019) Handbook on equipment used for the maintenance and delivery of clean aviation fuel, 3rd edn. Energy Institute, London. https://publishing.energyinst.org/featured/ei-1550-handbook-on-equipment-used-for-the-maintenance-and-delivery-of-clean-aviation-fuel
3. JIG 1 Aviation fuel quality control & operating standards for into-plane fuelling services. Joint Inspection Group (JIG). http://www.jigonline.com
4. JIG 2 Aviation fuel quality control & operating standards for airport depots & hydrants. Joint Inspection Group (JIG). http://www.jigonline.com
5. JIG 4 Aviation fuel quality control and operating standards for smaller airports. Joint Inspection Group (JIG). http://www.jigonline.com
6. ATA-103. Standard for jet fuel quality control at airports. Air Transport Association of America, Inc. (ATA) d/b/a Airlines for America, 1275 Pennsylvania Ave. NW, Suite 1300, Washington, D.C. 20004. http://www.airlines.org
7. ICAO (2012) Manual on civil aviation jet fuel supply (1st edn., Doc 9977). https://store.icao.int/en/manual-on-civil-aviation-jet-fuel-supply-doc-9977;
8. BS EN 12312-5 (2021) Aircraft ground support equipment. Specific requirements—aircraft fuelling equipment. The British Standards Institution
9. ASTM D4057. Practice for manual sampling of petroleum and petroleum products. ASTM International, West Conshohocken, PA, https://doi.org/10.1520/D4057
10. Reader-Harris M (2015) Orifice plates and venturi tubes. Springer International Publishing, Basel, Switzerland
11. Australian Transport safety bureau in-flight uncontained engine failure overhead Batam Island, Indonesia 4 November 2010 VH-OQA Airbus A380-842. Australian Transport Safety Bureau, 4 Nov. 2010

12. Civil Aviation Department (2013) Report on the accident to airbus A330-342 B-HLL Operated by Cathay Pacific Airways Limited at Hong Kong International Airport, Hong Kong on 13 April 2010 Civil Aviation Department
13. EI 1597 (2017) Procedures for overwing fuelling to ensure delivery of the correct fuel grade to an aircraft, 2nd edn. Energy Institute, London. https://publishing.energyinst.org/topics/aviation/aviation-fuel-handling/ei-recommended-practice-1597-procedures-for-overwing-fuelling-to-ensure-delivery-of-the-correct-fuel-grade-to-an-aircraft
14. CAP 437. Standards for offshore helicopter landing areas. https://www.caa.co.uk/our-work/publications/documents/content/cap-437/

Chapter 10
Aviation Fuel Filtration Systems

John Leonard, Kimberly Morrison, and Samantha Glenn

Abstract This chapter provides an overview of aviation fuel filtration systems. Filtration systems are designed primarily to remove particulate contamination from fuel products as they are transported from refinery fuel tanks, ultimately to aircraft. Filtration systems also strip dispersed and free water from fuels. EI maintains EI 1550, handbook on equipment used for the maintenance and delivery of clean aviation fuel. The handbook includes all published and approved technology and provides additional details that are not covered in this chapter. This chapter will describe different types of filtration equipment, filter media, and the most common problems that can degrade filter performance.

1 Introduction

1.1 Why Filter Aviation Fuel

When aviation fuels leave the refinery, they meet ASTM or other internationally recognized specification requirements. ASTM Specifications D910 [1], D6227 [2], and D1655 [3] detail the properties of leaded and unleaded aviation gasoline (AvGas) and aviation turbine fuels (Jet A and Jet A-1), respectively (see Chap. 1). As the fuel moves from the refinery to aircraft fuel tanks, it can be exposed to a variety of contaminants (dirt, rust, water, surfactants, etc.). When filters are functioning as designed, the rate at which filters need to be replaced depends on the filter's

J. Leonard (✉) · S. Glenn
Facet, Tulsa, OK, USA
e-mail: John.leonard@filtrationgroup.com

S. Glenn
e-mail: sglenn@filtrationgroup.com

K. Morrison
Chevron, Houston, TX, USA
e-mail: kmorrison@chevron.com

M. Rabaev et al. (eds.), *Aviation Fuel Problems and Solutions: From Refinery to Wingtip*, https://doi.org/10.1007/978-3-032-05514-9_10

capacity, the filtered fuel's contaminant load (i.e., number of particles per liter, water concentration, or the combination of the two), and the volume filtered.

1.2 Contamination Sources and Impacts

The three primary categories of aviation fuel contaminants are water, particulates, and microbes. Unless they are removed from the fuel before it is delivered to aircraft fuel tanks, these contaminants can create various problems that can have catastrophic consequences. For example, free or dispersed water in jet fuel is likely to freeze at the −40 to −70 °C outside temperatures characteristic of cruising altitudes (9000 m to 12,000 m). The ice formed at altitude can cause fuel line plugging and consequent loss of fuel flow to engines (see Chap. 15). As discussed in Chap. 12, water promotes microbial growth, which can degrade fuel tank surfaces and coatings, foul sensors, and plug filters. Similarly, particulates suspended in jet fuel can plug filters and fuel lines on aircraft. Surfactants and polar molecules, present naturally or as a result of additives, at the refinery or at major fuel storage depots can disarm clay treaters (see Sect. 3.2) and coalescers (see Sect. 3.3). Filtration removes non-dissolved contaminants, thereby keeping the fuel clean and dry. Jet fuel is typically filtered at each stage of the distribution system from the refinery bulk storage tank to the aircraft fuel hydrant (see Fig. 1).

Fuel leaving the refinery tank farm can be as warm as 40 °C. Water solubility in Jet A fuel is approximately ~100 mg L^{-1} at 40 °C and decreases as the fuel cools (solubility ≈20 mg L^{-1} at 0 °C). Consequently, dissolved water in fuel at the refinery is likely to fall out as the fuel cools during transport. All tanks breathe. As fuel is drawn from a tank, outside air enters through its vent to fill the vacuum created by the removed fuel. Moisture in air that enters tanks through vents also contributes to free water in the fuel. Fuel transported by water in tanker ships or barges is likely to

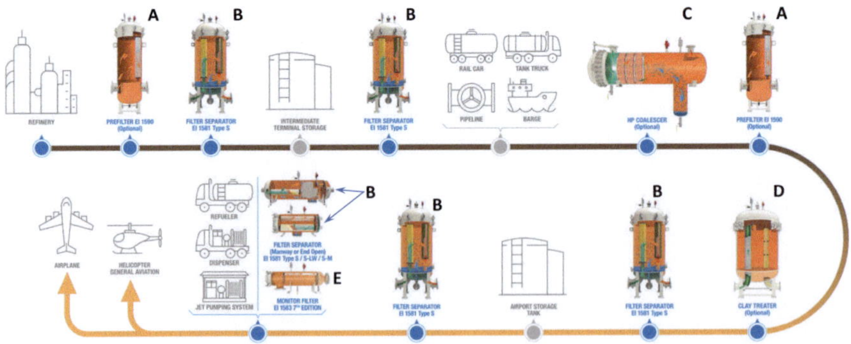

Fig. 1 Typical fuel flow from refinery to aircraft: A—EI 1590 [4] compliant prefilter; B—EI 1581 [5] filter/water separator; C—HP coalescer; D—clay treater; E—filter monitor (being phased out)

be water saturated. High water concentrations can shorten clay treater performance life and overwhelm separators. In clay treaters, water can create channels and disarm the clay by adsorbing it to sites that would otherwise adsorb surfactants and other polar contaminant molecules.

The primary sources of particulate contamination include rust and scale, product carryover from ships' cargos, airborne dust, refinery drier salts, sea salt, and wear debris released due to pump failures. Rust and scale are composed of iron oxides and iron hydrates. Rust (iron-based minerals) is typically the most abundant particulate in aviation fuels. The second most common particulate is sand, as evidenced by the presence of silicate (SiO_2). Elementary sulfur (S) contamination is symptomatic of microbial activity (biodeterioration—see Chap. 12). Aluminum and calcium are present in the clay used in clay treaters. Their presence in fuel indicates that an upstream clay treater has failed. Fuel contaminated with seawater is likely to have elevated sodium and chloride concentrations. High concentrations of particulate matter can shorten particulate filter life and disarm clay treaters. When fuel is stored for prolonged periods, oxidation reactions can promote the formation of organic particles.

Microbial contamination is discussed in considerable detail in Chap. 12. Here, we will simply note that uncontrolled microbial contamination can plug filters and disarm separators and clay treaters.

Although uncommon events can cause filter failure, it is normal for the pressure differential (ΔP—in some texts it is written as "DP") across filters to increase as particulate contaminants accumulate on and in filter media. Depending on the filter type, its nominal porosity (see Sect. 3.1.2.1), and its design capacity, the volume of fuel filtered before ΔP indicates that a filter that needs to be replaced is generally predictable. As will be discussed below, acceptable flow rates and contaminant removal percentages vary based on filtration unit design (i.e., filter medium and unit capacity). Thus, each type of filtration system has an expected performance life in the volume of fuel filtered at specified flow rates and maximum ΔP. Some high-capacity filters can process more than 2000 m^3 (500,000 gal) of fuel before they need to be replaced. When the maximum ΔP occurs before the expected volume of fuel has been filtered, the condition is identified as premature filter failure. Atypically high water, particulate or microbiological loads will cause premature filter failure. This chapter describes the recommended filtration practices for contaminant removal and discusses the unique factors (i.e., those other than water, particulates, or microorganisms) that can contribute to premature filter failure.

Consensus documents such as Energy Institute EI 1550 [6] and individual filtration manufacturers' operating manuals provide detailed recommendations for filtration unit selection and operation. This chapter provides an overview of the filtration systems used at each stage of the fuel distribution system from refinery to aircraft. It will then provide a more detailed description of each type of filtration system and the problems most likely to affect its performance. Common failure mechanisms are discussed, and root cause analysis is represented at a high level.

1.3 Aviation Fuel Cleanliness

Before we discuss filtration, we will clarify the meaning of clean, dry, on specification fuel (CDF; some stakeholders use the phrase "clean, clear, and bright"—CCB)—fuel that is fit for use. The three aforementioned ASTM specifications do not list upper control limits for water, sediment, or suspended solids among product specifications. In their respective workmanship sections, they each indicate that fuel should be "free of undissolved water, sediment, and suspended matter" [3]. The Energy Institute (EI) defines *clean and dry* as containing <0.26 mg L^{-1} solids, <15 μL L^{-1} free, and entrained water (not dissolved) and <10 fibers L^{-1} [7].

1.4 Filtration Limitations

Filtration can remove particles and non-dissolved water but cannot affect the fuel specification properties. As will be discussed below, filters remove a percentage of the contamination suspended in fuel passing through them. Filters are rated based on their nominal pore size (NPS) and ratio of particle loads entering and leaving the filter—i.e., the beta-ratio (β-ratio). Ultrafilter NPS is in the 0.01 μm range. Fuel prefilters are typically 100 μm NPS. The NPS of filters used for fuel polishing is usually 1–5 μm (see Sect. 3.1.2). A filter's β-ratio is defined by Eq. (1):

$$\beta = \frac{n_u > x}{n_d > x} \tag{1}$$

where n_u is the number of particles entering the filter (upstream), n_d is the number of particles downstream of the filter, and x is the NPS rating. This means that a 10 μm NPS filter will have a greater β-ratio for particles ≥50 μm diameter (dia) than it will for those that are 10 μm dia. Filters are also rated based on their flow rate capacity—i.e., the m^3 min^{-1} (or gallons per minute—gpm) they are designed to handle. Filtration units should never be used at flow rates greater than those for which they are rated.

2 Commonly Used Filtration Technologies from Refinery to Wingtip

Chapters 8 and 9 provide detailed discussions of supply chain risks and fuel transport, respectively. Chapter 11 discusses the best practices for collecting samples to be tested fuel for contamination at critical points in the distribution system. This chapter reviews the fuel distribution system in the context of primary points where filtration is used to minimize contamination as aviation fuel is transported from the refinery to the wingtip and within the aircraft. Fuel distribution systems vary globally. In some regions, fuel is delivered directly from refinery tank farms to airports—either through pipelines or transport tank trucks. In the U.S., the distribution system is quite complex in terms of fuel transportation modes and the number of terminal tank farms along the distribution path. As illustrated in Fig. 1, once aviation fuel leaves a refinery tank, it can be transported by pipeline, barge, tanker vessel, railroad tank car, tank truck, or some combination of these modes. Fuel from multiple refineries is comingled at one or more depots (tank farms)—becoming a fungible product once it is commingled. Fuel specifications such as ASTM Specifications D910 [1], D6227 [2], and D1655 [3] apply at points of sale.

Fuel being delivered to aircraft is expected to meet the relevant specifications. To help ensure that contamination does not prevent delivered fuel from meeting specifications, consensus standards such as EI/JIG 1530 [7] prescribe the types of filtration equipment to be used at each stage of the distribution system. Fuel leaving refinery tank farms is typically filtered through a 60 mesh strainer (not shown) to remove large particles and passed through a filter/water separator that meets EI 1581 [5] requirements (Fig. 1A). Before delivery to airport depot tanks, fuel may optionally pass through a Model HP coalescer (Fig. 1B) and then in series through a Model M prefilter (not shown). Model F clay treater (Fig. 1D), and another filter/water separator. Figure 1 includes filter monitors (Fig. 1E), although these units are being phased out of fuel distribution systems and they are not supposed to be in use any more. The fuel passes through filter/water separators as it is transferred to fuel trucks or hydrant vehicles and as it is delivered from hydrants to aircraft. Fueling hoses are fitted with strainers not coarser then 60 mesh at the hose to aircraft connection (not shown).

2.1 *Refinery Output*

To help ensure that fuel entering the distribution system from refineries is free of substantial contamination, devices are commonly installed to remove larger particles (i.e., $\geq$40 μm). Typically, refinery discharge lines will be equipped with a strainer (see Sect. 3.1), a prefilter (see Sect. 3.2), or both. If the refinery's drier unit tends to release particles, a high-capacity clay bed treater might also be installed. Strainers are commonly rated to 60–100 μm and are installed upstream of prefilters or microfilters. Fuel entering the midstream transport should be clear and bright and carry $\leq$1 mg L^{-1} particulate matter.

2.2 *Intermediate Depots (Terminals)*

Aviation fuel is filtered before it enters and discharged from depot tanks. Typically, a Type S filter/water separator—either alone or in series with a microfilter—removes

Table 1 Rust particle and water droplet settling rates in Jet A fuel as a function of size (diameter) [9]

Size (μm dia)	Settling rate (m h^{-1})	
	Rust particles	Water droplets
60	–	1
40	–	0.6
20	1.3	0.2
17	1	–
15	0.8	–
10	0.5	0.1
5	0.1	–
4	–	0
2	0	–

water and particulates from fuel as it enters depot tanks. Depending on fuel turnover rates, depot tanks can function as separators. The rate of particle settling is determined by Stokes' Law [8]. The rate is related to a particle's size (radius), specific gravity (SG), and the specific gravity of the fluid through which it is settling. Anderson and Clark [9] provided tables listing average settling rates for rust particles and water droplets in jet fuel (Table 1).

Rust particles smaller than 2 μm dia and water droplets smaller than 4 μm can remain in suspension indefinitely because of Brownian motion. When possible, fuel should be permitted to settle in receiving tanks for at least 24 h (i.e., at least 1 h per m of fuel column height) after receiving the product. This minimizes the water and sediment likely to be in the fuel drawn from the tank. In well-designed tank systems, the bottom layer is recirculated through a bypass loop (kidney) filter/water separator, and fuel is discharged through a floating suction line with an inlet near the top of the tank (see Chaps. 9 and 11). As it is discharged from the depot, fuel is again filtered through a 40 μm prefilter. If the water concentration in the fuel is near saturation, the best practice is to install a coalescer (Sect. 3.3.1). Fuel leaving depots should be clear and bright and have a particulate load ([PL]) of $\leq$1 mg L^{-1}.

2.3 Airport Tank Farms and Hydrants

Fuel is filtered before it is discharged into airport bulk storage tanks. I some cases, a bag-type clay treater operated upstream of a 10 μm or 40 μm prefilter. The fuel is then filtered again upstream of the hydrant. One or more Type S or S-LW filter/water separator filters remove water and particulates to $\leq$30 μL L^{-1} and <1 mg L^{-1}, respectively. Commonly, a 2 μm NPS or 5 μm NPS microfilter is placed upstream of the filter/water separator. Although 1 μm NPS and 3 μm NPS microfilters are available, they are used less commonly. Fueling equipment (pump carts, pumping stations, and tank trucks) have strainers and microfilters installed just upstream of their fueling hoses. Historically, water monitors (see Fig. 1 and Sect. 3.4) were operated upstream of the other post tank filtration units. However, it was determined that the polyacrylamide gel in the monitors could be eluted by the fuel flowing through. Sodium polyacrylate can form a gel and foul aircraft fuel system components.

A fuel quality survey performed in 1999 [10] tested fuel for particulate loads at 18 airports. The authors found that Jet A fuel delivered to the airport tank farms had particulate loads ([PL] ranging from <0.00 mg L^{-1} to 0.34 mg L^{-1} ($\overline{[PL]}$= 0.12 mg L^{-1}, where $\overline{[PL]}$ is the average [PL]). Although $\overline{[PL]}$ were all <1 mg L^{-1}, individual values exceeded that limit. The survey results also indicated that >80% of rust particles in jet fuel were <4 μm. The authors concluded that elevated [PL] reflected inadequate filtration system condition monitoring and maintenance. A separate survey concluded that all stakeholders responsible for fuel quality control must do their part to ensure that the fuel delivered to aircraft is clean and dry.

3 Filtration Equipment

3.1 Prefiltration

Prefiltration entails the use of relatively inexpensive filters to remove dispersed water and large particles from fuel to reduce the loads on downstream filters. The choice of filtration equipment is dictated by industry standards (e.g., [6, 7]), operating conditions (e.g., daily volumes passing through filters and flow rates), and the fuel's expected cleanliness (based on historical data). If the fuel is either extremely dirty, wet, or both (for example, fuel coming out of an old pipeline, barge, or tanker), the prefiltration array might include a 60 μm mesh filter to remove large solids, a *dehydrator* for water removal (aka Hay Pack or "HP"), and a microfilter to remove solid particulates that pass through the mesh filter. If it's just moderately dirty (pipeline or tanker truck), then a microfilter alone can suffice.

3.1.1 Dehydrator (HP Coalescer)

Design

A dehydrator (also called an HP coalescer) is the simplest form of aviation fuel filtration. It employs bonded and unbonded glass fibers, metallic wool of various types, or treated wood fibers (excelsior) to catch water slugs, filter out large particulates, and coalesce large water droplets. As depicted in Fig. 2, fuel enters the vessel and passes through tortuous path filter elements that are designed to coalesce water droplets so that they become sufficiently large to swiftly drop out of the fuel on the downstream side of the elements. Water and particulates settle into a trap, and dewatered fuel leaves the unit.

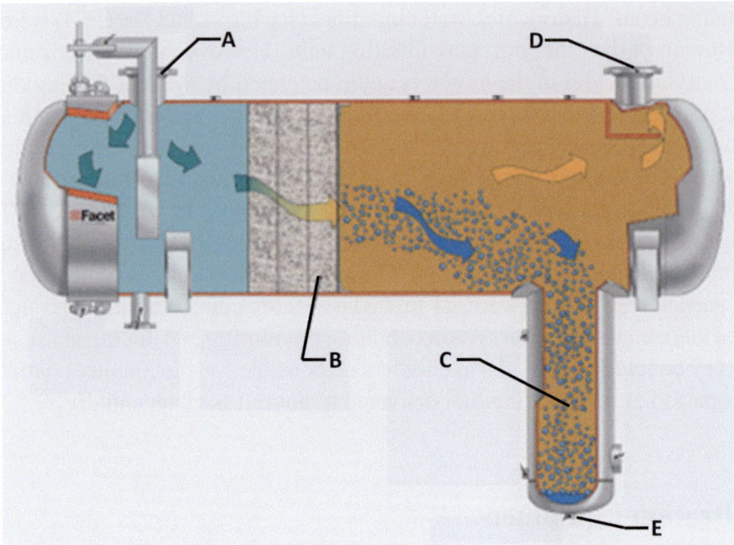

Fig. 2 Dehydrator—A: fuel inlet; B: coalescing medium; C: sump; D: fuel outlet; E: water drain valve

A well designed and functioning dehydrator can reduce dispersed water from a concentration of [H_2O] >0.5% in the fuel entering the unit to [H_2O] in the 50 mg L^{-1} to 100 mg L^{-1} range in the discharged fuel.

Dehydrator vessel construction (see Fig. 2) should comply with ASME Code Section VIII—Division 1 (Rules for Construction of Pressure Vessels) [11].

The internal coating should meet the requirements of the latest edition of EI 1596 Design and Construction of Aviation Fuel Filter Vessels [12]. Flow rates through this vessel type range from 22 to 5400 L min^{-1} (6–1425 gpm).

Dehydrator Failure Mechanisms and Their Prevention

High water concentration (i.e., >100 mg L^{-1}) in the fuel exiting the unit is the primary indicator of dehydrator failure. Most commonly, dehydrator failure occurs when non-specification replacement parts are used, or torque specifications are not followed.

Replaceable components do not meet specifications—seals can be incomplete if either non-specification filter elements or gaskets are used. Poor filter element seals can allow fuel to bypass elements. When this happens, water separation will be incomplete. Non-specification gaskets can be manufactured from incompatible elastomers. Dehydrator unit leaks are likely to occur. Only specified filter elements and gaskets should be used.

If filter element holders or twist-on filter elements are not tightened to specified torques, seals are likely to be incomplete. Unless the filter element seals are complete, fuel can bypass the element. The dimensions of non-specified elements are not necessarily identical to those of specified elements. Consequently, it can be impossible to obtain complete seals when using non-specified elements. Ensure that each time filter elements are replaced, the wingnuts on the element retainers are tightened to the specified torque. Similarly, if excessive torque is applied, bolt threads can be stripped, seals distorted, or both. The result will be fuel bypassing the filter medium.

The other operator-related failure mechanisms include neglecting to drain the water sump with sufficient frequency and operating the unit at greater than capacity flow rates. Both actions prevent appropriate water removal. Sumps are typically equipped with water level sensors that control drain valves. Fouled sensors or corrosion on valves' moving surfaces can prevent normal operation. When this occurs, sensors and valves can be cleaned. If heavily fouled or corroded, they might need to be replaced. Filtration units have specified initial pressure differentials (ΔP) across the medium. If the initial ΔP is less than the expected value, if ΔP does not increase as the volume of fuel filtered increases, or the water concentration in the exiting fuel's $[H_2O] > 100\ mg\ L^{-1}$, inspect the entire filter array for correct fit. Ensure that all gaskets are seated properly and that none of the retainer bolts are stripped.

Uncontrolled microbial contamination can also degrade filter element performance. Microbial growth on filter elements can impair their ability to coalesce water. Bioburdens insufficient to affect the ΔP across the filter can change the electrical charge properties of filter fibers. This can prevent water droplet size growth and allow water droplets $<10\ \mu m$ diameter to remain dispersed in the fuel. If the water concentration in the discharged fuel is increasing but ΔP is not, then microbiological contamination is a likely cause. The remedy is to disinfect the dehydrator's internal surfaces and replace the elements (see Chap. 12).

3.1.2 Microfilter

Design

A microfilter (Fig. 3) removes finely ridged solid contaminants (primarily iron oxide/rust and dirt) from the fuel stream. It serves as an initial filter that protects the finer, more expensive filtration downstream, including the filter/water separator (FWS).

This vessel's orientation can be either vertical (Fig. 3a) or horizontal (Fig. 3b). The choice between filter orientation is primarily based on space availability. Vertical housings have a smaller footprint. Moreover, water is easier to remove through the bottom drain. Horizontal housings provide easier access to filter elements. Although housings are typically angled to facilitate water draining, it can be difficult to drain horizontal housings thoroughly. That said, it is typically easier to perform maintenance on horizontal vessels. The vessel's primary construction complies with ASME

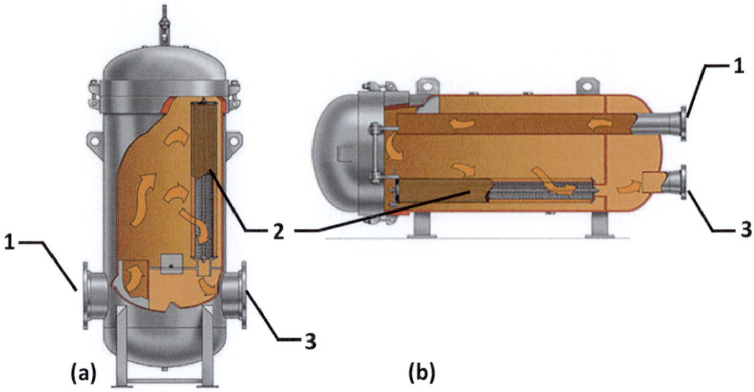

Fig. 3 Microfilters—**a** Vertical vessel; **b** horizontal vessel; 1: fuel inlet; 2: filter element; 3: fuel outlet

Code Section VIII-Division 1. Its design should meet the requirements of the current edition of EI 1596 [12].

The filter elements themselves should be qualified for the EI design and performance Standard 1590 [4]. Typical choices of NPS ranges are 1, 2, 3, and 5 µm. Filter element performance life is inversely proportional to NPS—elements with smaller NPS must be replaced more frequently than those with larger NPS. From both practical and financial standpoints, it is recommended that operators choose the porosity that gives the FWS downstream an extended service life. Depending on the typical particle loads of the incoming fuel, it can be advantageous to place two or more housings with successively smaller NPS in series. Microfilter elements are much less expensive and hold more solid contaminants than coalescer elements found in fuel-water separators (FWS).

The ΔP across newly installed filters should be between 3.4 and 28 kPa (0.5 and 4 psi). Filters should be replaced as ΔP approaches 103 kPa (15 psi). As noted previously, the volume of fuel that can be filtered before ΔP approaches 103 kPa depends on the incoming fuel's cleanliness. Under normal operating conditions, a microfilter unit should be able to process >1800 m^3 (>500,000 gal) between filter changes.

Micro-filter elements should be changed out when (1) ΔP reaches 103 kPa (15 psi); (2) if there's a sudden increase in ΔP (e.g., an inflection point in Fig. 4); or (3) after 3 years of use, whichever comes first.

Filter media will fail when ΔP becomes too great. This failure point is typically provided by filter manufacturers. Once ΔP exceeds the failure point (peak of the curve in Fig. 4), the filter loses its integrity. When this happens, ΔP decreases rapidly as unfiltered fuel flows through the failed element. When changing the elements, the vessel's interior should be thoroughly wiped down, and the epoxy coating should be checked for integrity. If flaws are found, check with the vessel manufacturer for repair options.

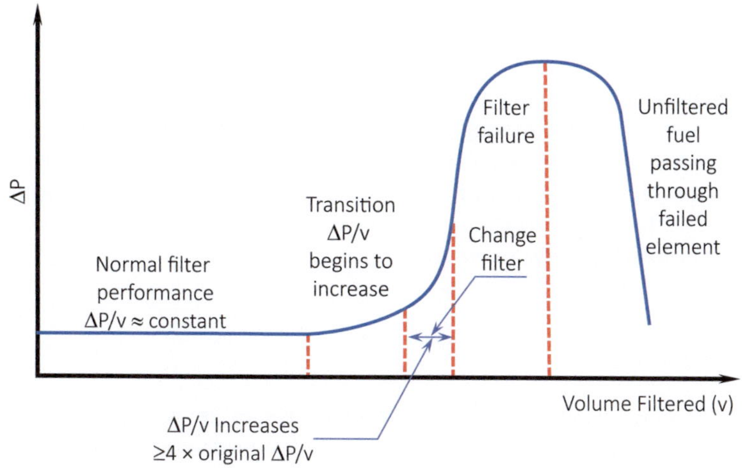

Fig. 4 Pressure differential (ΔP) as a function of fuel volume filtered. Filter elements should be replaced when the $\Delta P/v$ changes a rate that is $\geq 4\times$ the original rate

Failure Mechanisms and Their Prevention

As with dehydrator units, the most common failure mode is filter bypass caused by using non-specification filter elements, filter housing gasket seals, and improperly torqued filter element retainer nuts. This failure mechanism can be avoided by using only specified elements and following the manufacturer's instructions.

Excessive flow rates or upstream pressure can cause microtears to develop in filters. As a result, fuel passing through these tears will not be filtered. Ensuring that flow rates and upstream pressure are below the unit's design limit will prevent filter failure by this mechanism.

Unlike inert particles, microbes trapped by filter media can reproduce in and on the elements (Chap. 12). As microorganisms reproduce within filter media, they can form communities on the downstream faces of filter elements and thereby increase the bioburden in the effluent fuel. Moreover, as discussed in Chap. 12, microorganisms can produce extracellular polymeric substance (EPS—*slime*). Slime accumulation within filter elements will accelerate the rate at which filter failure occurs. The combined process of microbiological growth and slime production on and in filter elements is a primary cause of particulate filter failure. When it occurs, the housing should be disinfected, and the filter elements replaced (see Chap. 12).

If fuel with water concentration >100 mg L^{-1} is passed through particulate filter elements, it can cause fiber swelling and a consequent rapid increase in ΔP. When this occurs, the filter elements must be replaced. The risk of water-stress on particulate filters can be reduced by placing a dehydrator (Sect. 3.1.1) or fuel/water separator unit (Sect. 3.3) upstream of particulate filters.

3.2 Clay Treater

Clay treaters are used within refineries and downstream of refineries in an as-needed basis in the Americas due to fungible systems. Specifically, when MSEP or thermal stability is lost due to contamination. Additionally, jet fuels in the Americas are produced and dispensed to conform to D1655 [3]. Any additive addition is normally conducted at the wing, downstream of any clay treating that would remove static dissipator or metal deactivator additives.

3.2.1 Design

Technically, clay treaters are not actually filters, as they are not intended to remove physical contaminants. Instead, clay treaters are designed to remove surfactants and other polar molecules (for example, corrosion inhibitors) by a process known as adsorption (adhesion to the surface area of the clay).

Surfactants are compounds that can interact with fuel or water and lower the interface tension between two liquids or between a solid and a liquid. Surfactants can also interfere with the filtration media's surface chemistry or properties. For our application, this would mean lowering the interfacial tension between aviation fuel and water. When this happens, it becomes more difficult to remove the free water from fuel by coalescing. When surfactants adsorb onto filter media, filters such as coalescers can become disarmed and no longer function as intended for water separation.

Surfactants most commonly arise during refining, see more in Chap. 5—Additives. Additionally, surfactants adsorbed from other fuel products onto tanker vessel walls (pipelines, ships, barges, railroad cars, and trucks) can contaminate aviation fuel. Residues of chemicals used to clean fuel system internal surfaces (tank shells, filter/water separator housings, etc.) can also contribute to surfactant contamination in fuel. Because fuel distribution systems are mostly fungible—carrying multiple products from multiple refineries—the risk of cross-contamination has increased. Drag reducer additives might be added to products transported in fungible or multi-product pipelines. For products transported through pipelines, cross contamination risk is mitigated but not totally eliminated by directing the leading edge of each product tender into a transmit tank at depots. Acids and salts in ship and barge tanks can react with the product to produce sodium naphthenates and sulfonates.

Clay treater vessels (Fig. 5) are always vertical, built to ASME Code Section VIII-Division 1 [11] and the internal coating complies with the standard given by EI 1596 [12].

The elements are usually either cans (canisters or cartridges) or bags holding approximately 7.26 kg (~16 lb) of *Attapulgus clay* and have an 18 cm (~7 in) diameter. The cans can be mounted in three different ways: (1) Singularly, (2) 3-high removable pedestal, or (3) removable bundle assembly (Fig. 6 a and b).

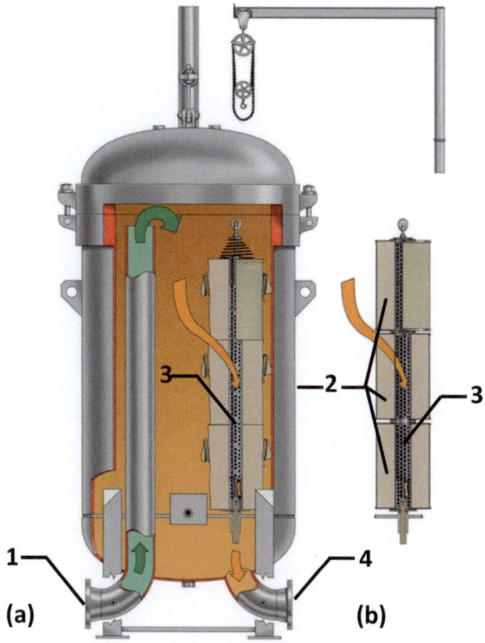

Fig. 5 Clay treater—**a** Vessel; **b** stacked clay canisters; 1: fuel inlet; 2: clay canisters; 3: lay canister core; 4: fuel outlet

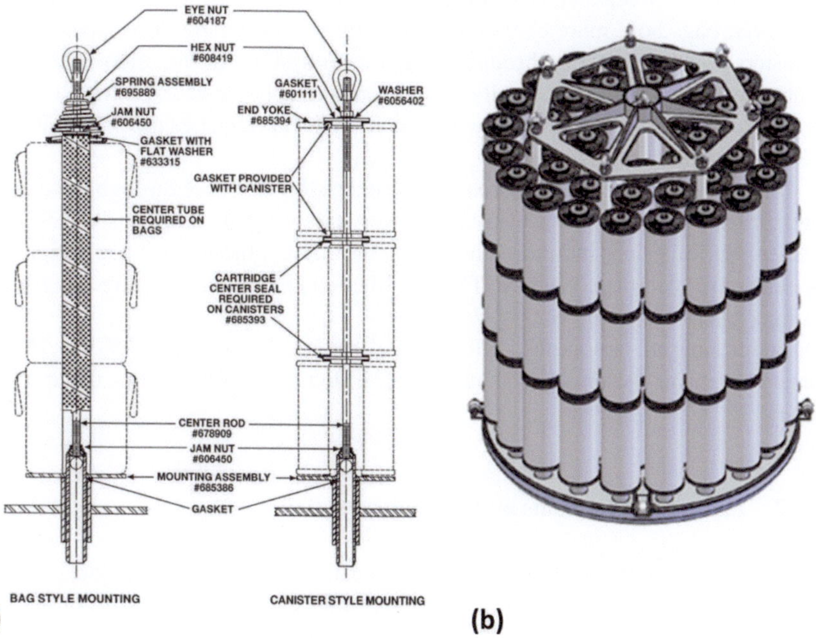

Fig. 6 Clay element mounting—**a** Bag and canister clay container options; **b** 108 element mounting configuration with 36 three-element pedestals

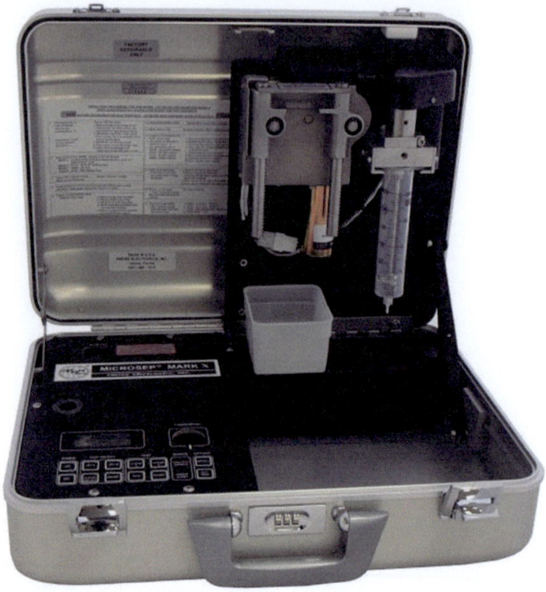

Fig. 7 ASTM D3948 compliant micro-separometer

The seal between clay canisters is typically better than that between bags. However, bags hold more clay and are generally considerably less expensive than canisters. For fuel flowing through a clay treater system for the removal of surfactants, the slower, the better. Ideally, the flow should not exceed 26 L min^{-1} (~7 gpm) per bag or can. High-capacity clay treaters installed at refineries are sometimes clay beds (clay-filled vessels).

Due to the clay's extreme porosity and fuel's slow flow rate through the clay, the pressure differential across clay treaters is very rarely significant. Consequently, clay treater performance is monitored by testing the effluent fuel's water separation characteristics in accordance with ASTM Test Method D3948 [13]. Testing is performed using a Micro-separometer (Fig. 7).

Under normal operations, effluent fuel will have an MSEP/WISM (Micro-separometer Analysis/Water Separation Index Modified) unitless rating in the 90–100 range.

3.2.2 Failure Mechanisms and Their Prevention

The primary failure mechanism for clay treaters is water contamination. Under normal operating conditions, the surface area of the clay used in treaters is ~4,000 m^2 kg^{-1} (~13 acres lb^{-1}). Polar molecules adsorb onto the clay crystals and are thereby stripped from the fuel as it flows through the medium. Water destroys the clay's crystalline structure—dramatically reducing porosity. The effect is a rapid increase in ΔP (normally <172 kPa–<25 psi—at 19 L min^{-1} to 27 L min^{-1}—5 to 7 gpm).

If there's a sudden jump in ΔP, immediately cease operation and check for water contamination. Because water competes with surfactants for adsorption sites, water contamination will also cause MSEP/WISM ratings to decrease. Water can also cause channeling through the clay. When this occurs, the ΔP will not change. Additionally, the water separability ratings of the effluent fuel will be unchanged from that of the influent fuel. Bags/cans that have become contaminated with water must be replaced. As with the micro-filter, when changing the elements, the vessel's interior should be thoroughly wiped down and the epoxy coating checked for integrity. If flaws are found, check with the vessel manufacturer for repair options. It is recommended that, even if the MSEP ratings are in an acceptable range, clay should be changed out every 2–3 years.

Less commonly, microfilter failure will permit particulate matter to reach the clay treater. Typical clay treater porosity is in the 163 μm (90 mesh) to 595 μm (30 mesh) range. Particles bypassing microfilters can decrease clay filter porosity. This will cause ΔP to increase but will not affect MSEP/WISM ratings. When clay media has become contaminated with particulates, it must be replaced. Additionally, the corrective measures stipulated for microfilters (Sect. 3.1.2) should be taken.

As with microfilters, poor seals between stacked filter elements or between elements and the filter housing can allow fuel to bypass the clay. When this occurs, MSEP/WISM ratings will decrease, but ΔP will not change. To prevent bypass issues, ensure that all seals are complete.

Clay treater cartridge manufacturers provide charts showing the expected relationships between fluid viscosity, ΔP, and flow rate. If the ΔP across newly installed clay cartridges exceeds the expected value by >5%, the most likely cause is that the unit is undersized for the flow rate. This problem must be addressed either by reducing the flow rate or replacing the unit. If the ΔP across newly installed clay filters is >5% below the expected value, the likely cause is improperly installed and secured bags or canisters.

3.3 Filter/Water Separators

As summarized in Sect. 2, the distribution system by which fuel is transported from refinery tanks to airport depot tanks varies based on location. Because system components are vented, opportunities for intrusion of dust particles and water exist at each stage of the fuel distribution system, regardless of how simple or complex the system might be. Additionally, the shear forces of fuel flowing through lines can erode fine particles off of system surfaces. To prevent particle and water accumulation in the fuel as it moves from the refinery to the airport, two-stage filter/water separators (FWS) are used at each stage in the distribution systems.

3.3.1 Design

Filter/Water Separators are available as vertical (Fig. 8) or horizontal (Fig. 9) units. Regardless of the vessel's orientation, its construction must comply with both the current edition of ASME Code Section VIII—Division 1 [11] and EI 1596 [12]. Per EI 1596, the maximum permissible flow rate for a single vessel is 19600 L min^{-1} (2500 gpm).

Regardless of the FWS's orientation, fuel entering the unit initially flows through the coalescer cartridge's core and through the coalescer medium to enter the unit's chamber. The fuel then flows from the chamber, through the separator cartridge's medium and through the separator's core to the unit's outlet. Both vertical and horizontal FWS have design variations that include single or multiple coalescer and separator elements.

Dispersed water droplet size increases as the fuel flows through the coalescer element. After leaving the element, the water droplet size is sufficient to quickly settle to the sump at the device's base. The separator element prevents dispersed water from passing, thereby promoting water droplets settling from the fuel in the unit's chamber. Typically, FWS is equipped with water detectors and an automatic drain valve to discharge bulk water before it reaches a level at which it can contact the filter elements.

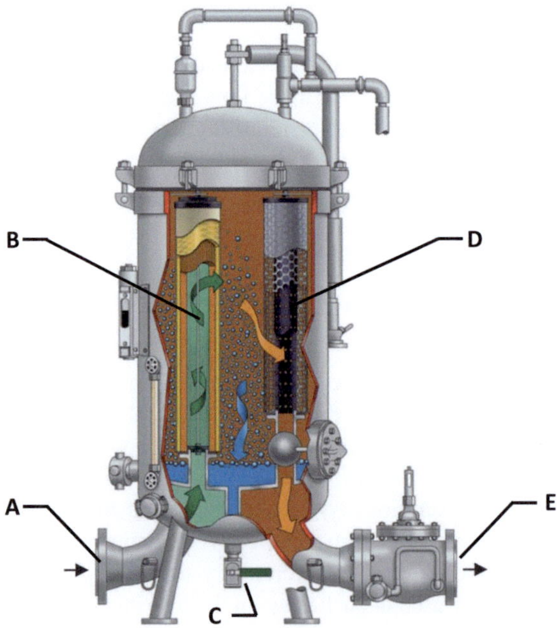

Fig. 8 Vertical filter/water separator—(A) Fuel inlet; (B) coalescer element; (C) water drain valve; (D) separator element; (E) fuel outlet

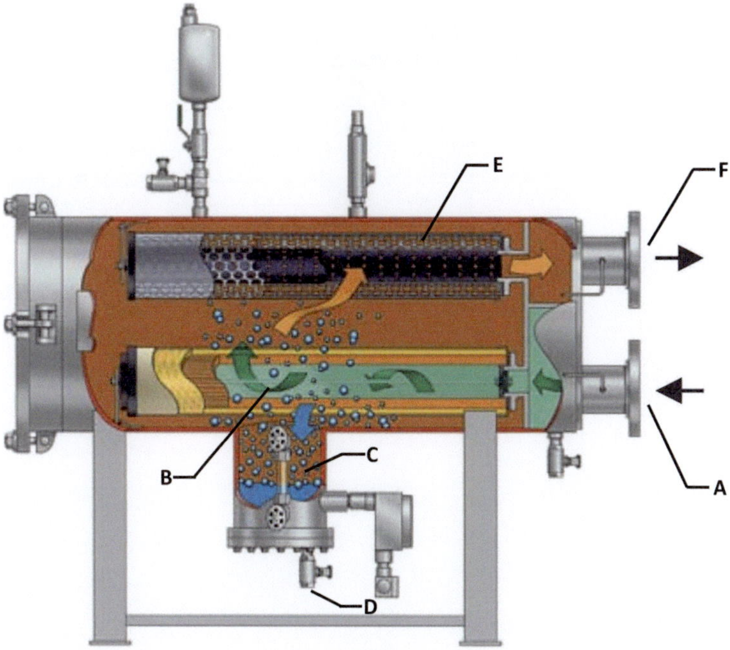

Fig. 9 Horizontal filter/water separator—(A) Fuel inlet; (B) coalescer element (cartridge); (C) sump; (D) water drain valve; (E) separator element (cartridge); (F) fuel outlet

Coalescer

The design and performance of the coalescer must meet the current standards of EI 1581 [5]. Coalescer elements are fabricated from three components and a structural core. The three coalescer components include an internal pleated media section, followed by coalescing/droplet growth section(s) and finally an outer cotton sock (Fig. 10). The flow direction outward is from the inner core.

The pleated section catches particulates (NPS rating range from <1 to 25 μm). The coalescing medium functions similarly to dehydrator media (Sect. 3.1.1.1). Water droplets grow to a critical size as they pass through the coalescing and droplet grow media so that they readily fall out of the fuel after it exits the coalescer. Coalescer elements operate optimally at $\Delta P = 103$ kPa (15 psi) and fail at $\Delta P = 517$ kPa (75 psi).

When the ΔP of the FWS starts to rise, the primary cause is the accumulation of contaminants in this pleated section. As stated earlier, the micro-filter is in place to slow or eliminate this entirely.

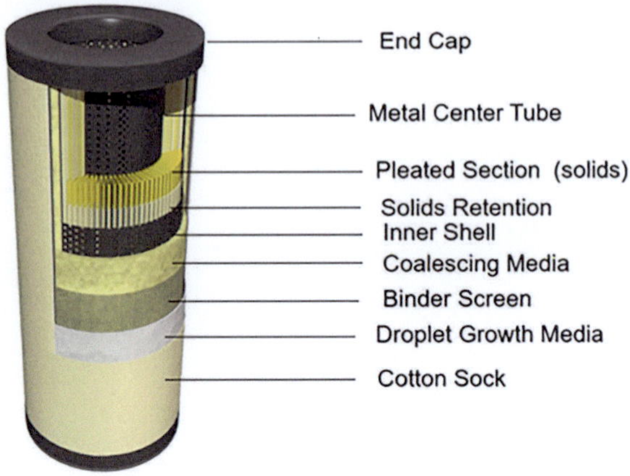

Fig. 10 Filter/water separator—cutaway view showing the multiple layers through which fuel flows outward from the element's core

Fig. 11 Filter/water separator, separator elements—**a** Elements with non-PTFE synthetic water barrier; **b** elements with PTFE water barrier

Separator

The separator is a hydrophobically (water repelling) treated screen. Fuel flows from outside the filter cartridge toward the inner core. For airport applications, separator media may be either polytetrafluoroethylene (PTFE—Fig. 11b) or other hydrophobic, synthetic material (Fig. 11a).

Their design and performance must also meet the current standards of EI 1581 [5] and EI 1582[14].

The primary differences between separator materials are:

(A) PTFE is more robust (will stand up to cleaning procedures better than the synthetic)
(B) Synthetic performs better—50 μm porosity versus 70 μm for PTFE. It also has a lower price point.

Types, Fuel Categories, and Elements

Specification EI 1581 EI 1581 designates three categories of fuel: C, M and M100:

(1) Civilian (C): Jet A or Jet A-1.
(2) Military (M): Jet A with corrosion inhibitor, anti-icing agent and anti-static agent (JP8/F24).
(3) Military (M100): Category M fuel that has been treated with a +100 thermal stability additive (JP8+100).

Filter elements filter must match the fuel type. For example, Category C elements may **only** be used on Category C/Jet A/Jet A-1 fuel. Category M elements may be used on both Category C and Category M Fuel. Category M100 elements may only be used on M100 fuel. Category C fuel with only the anti-icing additive is still considered Category M for the purposes of filter element selection.

Additionally, EI 1581 lists four types of filter/water separators (S, S-LW, and S-M), based on the filter's location along the supply chain and the likely particulate loads and free-water content:

(1) Type S: Receipt and issue from fuel tank farms—filtration points where substantial amounts of free water and particulate loads are likely.
(2) Type S-LW: Into-plane—where minimal water or particulate loads are expected.
(3) Type S-M: Into-plane (fullers, carts, and hydrant dispensers)—where minimal water or particulate loads are expected. S-M elements combine the functionality of older S-LD and S-LW elements.

Figure 12 illustrates the fuel types and categories. Additionally, all elements in an FWS must be made by the same manufacturer.

3.3.2 Routine Maintenance

At a differential pressure of 103 kPa (15 psig) or one year in service (whichever comes first), the vessel should be shut down, the elements removed, and the interior wiped down and inspected (repaired, if necessary). Filter manufacturers require that the coalescers be replaced annually regardless of whether or not $\Delta P \geq 103$ kPa (15 psig). However, there are exceptions (for example, see ATA 103 [15]). Most manufacturers provide guidance and instructions for cleaning and reusing separator elements.

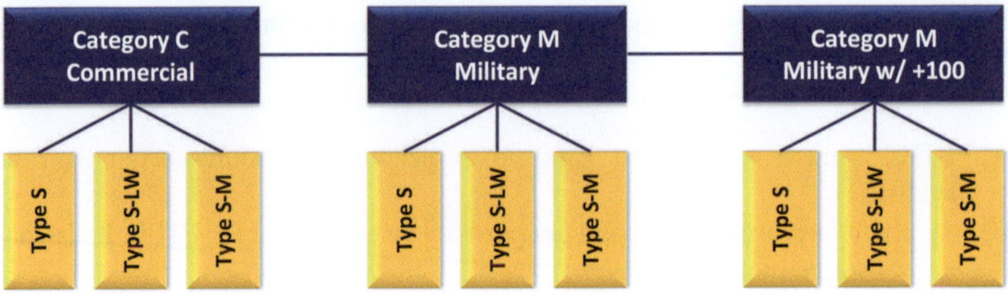

Fig. 12 Types and classes of filter/separator elements (*Source* EI 1581 [5])

3.3.3 Element Shelf Life

Elements should be stored in a dry environment in their original packaging. Storage in climate-controlled rooms is not necessary, but continuous storage at temperatures >45 °C (>113 °F) should be avoided. Storage at elevated temperatures can cause the paper and gaskets to become brittle. Any cartridges which show signs of exposure to moisture should be discarded. Most elements, if stored correctly, have a shelf life of 5 years. Optimally, paper separators and water absorbent monitors are put into service within 2 years of the date of manufacture.

3.3.4 Recommended Filtration Vessel Accessories

Several accessories can be fabricated onto FWS housings. Together, they improve FWS performance and reduce the risk of filter failure. Some of the most commonly installed accessories are:

(1) Provision for automatic air elimination
(2) Direct reading differential pressure gauge
(3) Manual sump drain
(4) Upstream and downstream sampling connections (an outlet sample connection on one vessel can be considered the inlet sample connection on a vessel immediately downstream)
(5) Pressure/thermal relief valves
(6) Placard indicating month and year of last filter change
(7) An operational data plate in accordance with EI 1596 [12]
(8) An automatic water defense system with provision for operational tests (filter/water separators only).

Industry standards such as EI 1550, EI 1596, and local regulatory bodies (for example, ATA and JIG) provide guidance regarding requirements for different accessories.

3.3.5 Filter/Water Separator Failure Mechanisms and Their Prevention

Common FWS failure mechanisms include coalescer inactivation, separator blanking, and element bypass. As discussed in Sect. 1.2, microbial growth on coalescer fibers can make normally nonpolar sites on the fiber matrix polar. This inhibits the mechanism by which water droplets are knocked out of suspension. If the fuel in the vessel downstream of the coalescer elements remains hazy, investigate for the presence of microbial growth within the coalescer medium. Similarly, microbial growth and EPS (biofilm) accumulation on separator element surfaces, can face off (i.e., plug) the surface pores and prevent fuel flow through the elements. Microbiological contamination can degrade coalescer performance without affecting ΔP, but ΔP will increase rapidly when separator element fouling occurs. If microbiological contamination is suspected, follow the manufacturer's recommended protocol for cleaning and disinfecting the vessel. After cleaning and disinfecting the vessel, replace all coalescer and separator elements.

Bypass issues have been discussed in previous sections. The most common cause of bypass is improper element installation. Only specified elements should be used. Additionally, it is essential to follow installation torque specifications for twist-on elements or retainer rod nuts. If the fuel flowing from an FWS is hazy, confirm that all seals are intact and that elements are properly secured so that fuel cannot bypass seals. Any precipitous ΔP decrease indicates that one or more elements or seals have become damaged. Also, if the concentration of free water in the fuel flows from an FWS $\geq 15\ \mu L\ L^{-1}$ (ppm), either the coalescer, separator, or both element types are malfunctioning. In both cases, the corrective action is to replace all elements.

As fuel transits the distribution system, it can accumulate electrostatic charge. If the FWS if not properly bonded, static discharge can destroy the coalescer and separator elements (Fig. 13). Static discharge can be prevented by ensuring that FWS housings are bonded.

3.4 *Into-Aircraft Filtration*

3.4.1 Equipment and Design

Aviation Fuel Monitor (Monitor)

Monitors (Fig. 14) were introduced in the late 1980s. After several incidents, investigations, and studies on the migration of monitor media (specifically, sodium polyacrylate) and its effect on turbine engine throttle controllers, in November of 2017, the Energy Institute announced that EI 1583 [15] would be withdrawn on December 31, 2020. The last date for monitor use was July 2023. As of this writing, monitors should no longer be used. They are mentioned here because of their historical use and to ensure that stakeholders responsible for filtration systems no longer have monitors in operation at their facilities.

Fig. 13 FWS elements are destroyed by static discharge

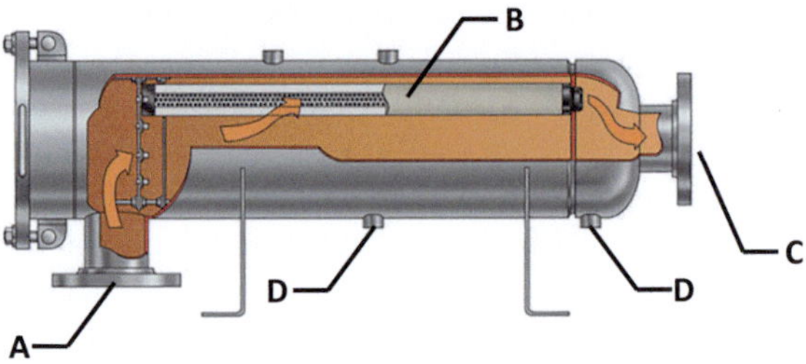

Fig. 14 EI 1583-compliant monitor vessel—(A) Fuel inlet; (B) monitor element; (C) fuel outlet; (D) water drains

Filter/Water Separators

Figure 1 shows that the primary filtration technology used as fuel flows from hydrant into aircraft is the FWS (Sect. 3.3). There are two types of FWS for into-plane fueling: S-LW and S-M [7, 14]. The primary difference between S-LW and S-M is the coalescer's diameter and length. The element length, obviously, has a direct effect on vessel size.

- **S-LW**: Coalescer diameters are all 15 cm (6 in) with lengths ranging from 40 cm (16 in) to 140 cm (56 in). Vessel lengths and diameters ranging up to 200 cm (80 in) and 50 cm (20 in), respectively. In other words, both the vessel length and diameter can vary.
- **S-M**: For this type the coalescer diameters are all 5 cm (2 in) with the length fixed at 76 cm (30 in). So, the vessel length is also fixed (107 cm approximately 42 in). As the flow rate increases, so does the number of coalescers and the diameter of the vessel.

Water Barrier Filters

Water Barrier Filters (WBF) compliant with EI 1588 [16] are designed as drop-in replacements for monitor elements in vessels designed originally for 5 cm (2 in) EI 1583 [17] monitors. Elements have a nylon-polytetrafluoroethylene, hydrophobic barrier layer that effectively blocks particulates and water from entering the aircraft (similar to Gore-Tex®)[1] but permits fuel to pass through. Performance characteristics are comparable to monitors': NPS between 0.5 μm and 5 μm and dispersed water removal to <5 μL L^{-3}. Unlike monitors, WBF are not susceptible to surfactant disarmament.

When using this technology, the lack of an integral water sump in the vessel makes it important to continually monitor for water build-up. Excessive water in the vessel can lead to high velocity and ΔP.

In accordance with JIG Bulletin No. 147 of 05 April 2023 [18], Airlines for America (A4A) and the Joint Inspection Group (JIG) have conditionally approved WBF for use as monitor replacements.

As with other filter media, the primary failure mechanisms include incomplete seals and element damage that allow fuel to bypass elements. Both are signaled by a drop in ΔP or an initial ΔP <3.4 kPa (0.5 psi). Ensure that elements are installed correctly and damage-free. Heavy particulate contamination loads or dispersed water concentrations >50 μL L^{-3} can plug WBF. If there is a risk of contaminated fuel flowing from the hydrant or tanker, an FWS should be installed between the hydrant and WBF.

Dirt Defense/Electronic Water Sensor (DD/EWS)

The dirt defense element (EI 1599 [19]) is an element that drops directly into vessels designed originally for 2″ EI 1583 [16] monitor technology. The dirt defense element prevents any solid contaminant from getting into the aircraft but allows free and entrained water to pass.

The electronic water sensor (EI 1598 [20]) is situated immediately downstream of the filter vessel and shuts down the flow of fuel to the aircraft when it senses free water and shuts down when refueling operation reaches criteria dictated by the operating standard. The JIG standard is set to 30 μL L^{-1} for $\geq$10 s or 50 μL L^{-1} for $\geq$5 s. The ATA 103 [15] standard is set to 20 μL L^{-1} for $\geq$10 s or 50 μL L^{-1} for $\geq$5 s. Both are at the time of writing and are subject to revision. Always refer to the most current version of the operating standards for the current limits.

The DD/EWS system signals that there is a water concentration in the post-unit fuel that exceeds the aforementioned limits and automatically stops refueling. This system might not be suitable for installations at airports where excessive dispersed water concentrations are a recurring issue. To continue the aircraft fueling, a system that removes free water must be used once the EWS shuts down fueling.

---

[1] Gore-Tex is a registered trademark of W. L. Gore & Associates, Inc., Newark, DE, USA.

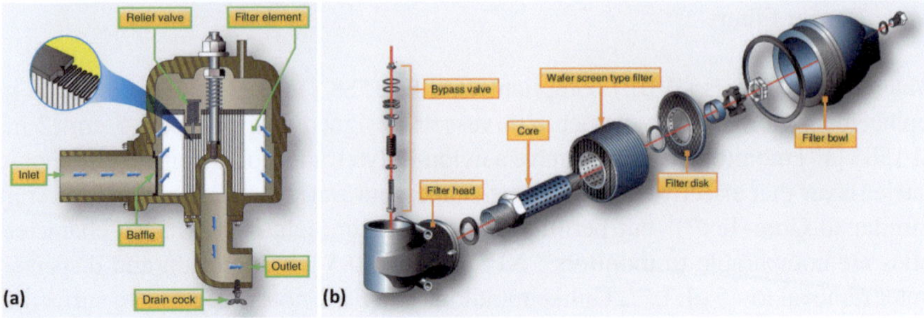

Fig. 15 Aircraft fuel filters—**a** screen filter unit; **b** pleated filter unit

3.5 In-Aircraft Filtration

3.5.1 Design

As noted in the introduction, all tanks breath. Consequently, fuel in aircraft tanks can accumulate water and particulate contamination. Chap. 12 and the IATA Guidance Material on Microbiological Contamination in Aircraft Fuel Tanks [21] discuss microbiological contamination that can develop in aircraft tanks and plug aircraft fuel filters. Onboard fuel filters are designed to protect engine components from fuel contaminants that accumulate after the fuel has been delivered to the aircraft. Each airframe model has an aircraft maintenance manual (AMM) in which the manufacturer specifies all maintenance tasks, including filter inspection, condition monitoring, and replacement.

Depending on the complexity of a given aircraft's fuel system, an aircraft can have one or more onboard fuel filters (Fig. 15a and b).

Typically, there is a filter located at the outlet of each fuel tank. If the aircraft has fuel booster pumps, the filter is installed just upstream of the pump. This element is designed to block particles $\geq$100 μm dia. A wire mesh filter (Fig. 16(2)) is likely to be located just downstream of the low-pressure fuel pump located at the engine compartment's firewall (Fig. 16(1)). A water absorbent, particulate filter (Fig. 16(5)), is designed to remove particles in the 10 μm to 200 μm range downstream of the high-pressure fuel pump (Fig. 16(5)). Both of these final filter units have pressure relief valves to permit fuel flow to the engine when the filter element is plugged.

3.5.2 Failure Mechanisms and Their Prevention

Onboard filtration is particularly critical because flow restrictions to aircraft engines can result in the engines shutting down. When this occurs in-flight the results can be catastrophic. The use of specified filters that are properly installed minimizes the risk of failures due to human error.

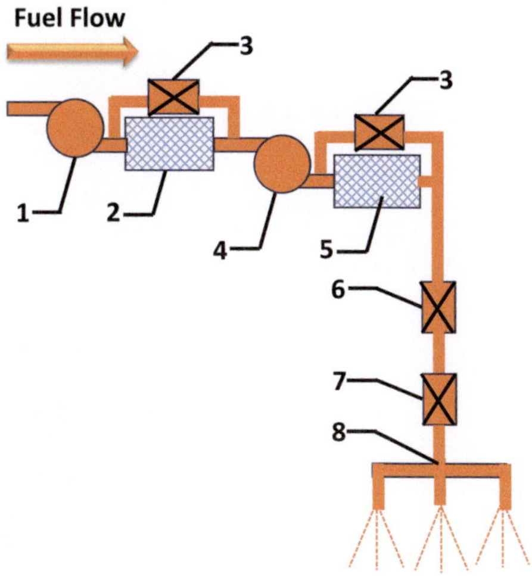

Fig. 16 Aircraft engine fuel filters—(1) low pressure fuel pump; (2) screen-type fuel filter; (3) fuel filter bypass valves; (4) high pressure fuel pump; (5) pleated medium fuel filter; (6) fuel metering valve; (7) fuel shutoff valve; (8) fuel nozzles

Electronic gauging on the aircraft's control panel signals when filter bypass valves have been actuated. As with other filters, the two most common causes of aircraft fuel filter plugging are particulates and polymeric residues. Polymeric residues can be EPS or non-biological polymers. One example of the latter is a gel-like polymeric contamination caused by the reaction of polyacrylamide gels (PAG) from aviation fuel monitors with the anti-icing agent diethylene glycol methyl ether (DiEGME), water, and other fuel contaminants. Under certain conditions, PAG from monitors dissolves in the DiEGME and forms a viscous substance called "Apple Jelly" (Fig. 17). This was a primary reason for the ban on water monitors.

As noted in Sect. 3.4.1.1, the risk of apple jelly formation was eliminated by discontinuing the use of aviation fuel monitors.

More recently, filter plugging was attributed to a polymeric residue that formed on fuel filters that had been treated with an isothiazolone-one based fuel treatment microbicide (Chap. 12). An investigation determined that the operators whose aircraft experienced this problem had not followed either the microbicide or aircraft manufacturer's use instructions. Subsequently, the microbicide manufacturer eliminated aviation fuel treatment as an approved end-use for the product.

Both examples of non-biological polymeric substances plugging fuel filters illustrate the importance of not using unauthorized fuel additives and using authorized additives strictly in accordance with the manufacturer's instructions. They also illustrate the importance of following AMM guidance on filter inspections and replacement. Ensuring that aircraft fuel system contamination control practices, as delimited in AMMs, are followed minimizes but does not entirely eliminate the risk of fuel filter plugging. Monitoring ΔP across filters provides the most reliable means of determining when filters require replacement. Routine fuel tank draining and inspection are essential for minimizing the risk of premature filter plugging. Similarly, best practice fuel stewardship to ensure that aircraft only receive fuel that meets workmanship criteria also reduces filter plugging risk.

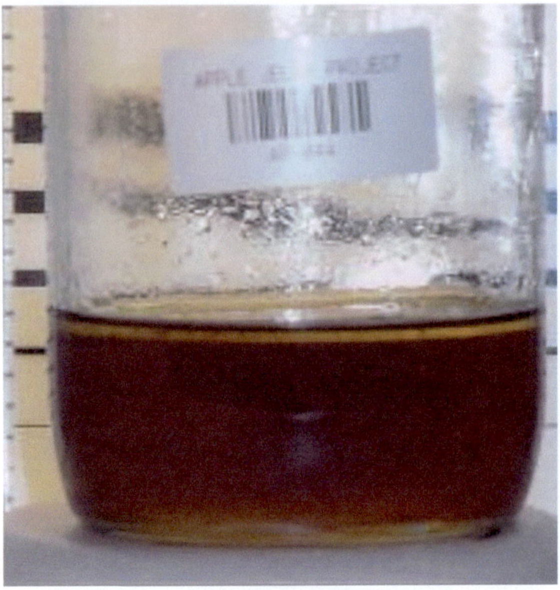

Fig. 17 Jet A contaminated with apple jelly (*Source* DTIC ADA472258: Investigation of apple jelly contaminant in military jet fuel: defense technical information center: free download, borrow, and streaming: internet archive)

4 Filtration System Condition Monitoring

4.1 Setting Expectations

Filtration units are designed to remove contaminants from fuel as they flow through the filtration medium or media. The volume of fuel that can be filtered before elements require replacement will depend on the cleanliness of the fuel entering the unit. There are times when it is more cost effective to install several units in series. If the incoming fuel is expected to have [PL] >2 mg L^{-1}, the particle size distribution should be determined [22]. As appropriate, place two or more filtration units in series, with each downstream unit having a smaller NPS than the preceding one (for example, one set up for a filtration vessel in series might include a 40 μm prefilter, 10 μm microfilter, and 2 μm microfilter). One or more of these units might be an FWS. If clay treaters are placed downstream of additive injection points, they will strip the additive out of the fuel. Consequently, when clay treaters are used, they should be located upstream of additive injectors.

Filter life depends on the particulate loads reaching the elements and the total surface area of the elements in the vessel. For example, 150 mm (6 in) dia Í 1422 mm (56 in) long pleated elements provide a greater microfilter performance life than the narrower diameter or shorter elements. Similarly, 32-element vessels will provide longer service than 16-element vessels. As a rule, filter life also increases with NPS, although this increased life might come at the cost of effective particle removal.

Decreased performance life is typically a symptom of degraded fuel workmanship (i.e., increased contamination loads) rather than filter medium malfunction.

4.2 Approving Filter Media

EI 1550 provides detailed guidance on criteria for performing bench tests to assess filter medium performance. As noted above, the types of tests performed are dictated by the operator's performance requirements. Typically, bench tests are performed using single filter element rigs. Particulate removal performance should be assessed based on expected [PL] and particle size distributions. If the fuel's particulate load includes known minerals, their concentrations should be measured before and after filtration. Depending on the filter's chemical composition and construction, a given medium can remove certain minerals more effectively than others. Conversely, if the fuel contains additives such as corrosion inhibitor, lubricity additive, or static dissipator additive, it is essential to ensure that the filter element does not strip these from the fuel. Filter elements from different manufacturers can be nominally the same but have differences sufficient to affect performance. For example, an element that is 2 mm shorter than the specified element will allow fuel to bypass the element. Elements with identical ratings can be fabricated from different materials—particularly the polymeric fibers from which woven or non-woven media are created. Different polymer chemistries are likely to interact differently with additives, etc.

4.3 Selecting Vessels

Ensure that the vessels meet the specifications and capacities discussed in Sect. 3. Filtration units must not be operated at flow rates greater than those specified. Additionally, flow rate and upstream pressure variations can reduce filter life and degrade the fittings that secure filter elements in place. All filter vessels should have a readily accessible bottom drain and two pressure gauges (one at the inlet and one at the outlet) to determine ΔP.

4.4 Condition Monitoring and Maintenance

4.4.1 Water

The best practice is to drain water daily using the vessel's bottom drain. Drain water into a clear glass bottle and observe for masses (flocs) of microbial biomass, rust particles, and other debris. To assure complete evacuation of the sump, the system should be pressurized immediately prior to the water draw. Record the volume of water solids drained to facilitate early detection of atypically wet or dirty fuel.

4.4.2 Pressure Differential (ΔP)

Record ΔP and flow rates daily. ATA 103 [15] advises that pressure gauges should be accurate to ±13 kPa (2 psi) and that microfilter elements be changed when ΔP ≥103 kPa (15 psi). FWS elements should be replaced when ΔP ≥151 kPa (22 psi). ATA 103 also recommends changing filter elements if the day-to-day corrected ΔP ≥34 kPa (5 psi—pressure data should be corrected for actual flow rates).

4.4.3 Clay Treaters

Increased aluminum (Al) and calcium (Ca) concentrations in fuel samples downstream of clay treaters indicate that clay is carried along with the fuel as it passes through the medium. When a clay treater is functioning properly:

- Effluent fuel conductivity is less than influent fuel conductivity.
- Water separability by ASTM Test Methods D3948 [13] or D8073 [23] for the effluent fuel should be greater than for the influent fuel and be >99%.
- The interfacial tension of effluent fuel should be the same as that of pure fuel of the same grade.
- ΔP at rated flow rate should be ≤103 kPa (15 psi).

If the fuel that has passed through the clay treater fails one or more of these tests, replace the bags or canisters.

4.4.4 Visual Inspections

- Filter elements—when replacing filter elements, inspect the used elements for signs of microbial contamination, heavy rust loads, or unusual residue accumulation. Send any affected elements to a qualified testing lab for further analysis.
- Vessel—when replacing elements, inspect the vessel's seals and internal surfaces for signs of damage or deterioration. Sludge and sediment accumulation on the bottom of the vessel should be negligible.

4.5 Collaboration with Filter System Manufacturers

Filtration unit manufacturers have a wealth of experience and test data for the elements and vessels they produce. Whenever systems are not performing as specified, they are your best first source of subject matter expert support.

5 Conclusion

The equipment discussed in this chapter is continually under review by JIG, A4A and EI resulting in regular updates and modifications to both equipment and procedures that address changes in the field. EI maintains EI 1550 to include all published and approved technology and covers more details on this chapter. Make sure your operation is always up to date. Also, operating parameters (fuel source, weather, etc.) vary from location to location. Make sure the equipment being used is appropriate for your specific location. If you're unsure, consult a subject specialist.

Acknowledgements The authors would like to thank Frederick Passman for his significant professional guidance and review, to John Buffin, Moshe Rabaev and Liel Mazar for their helpful comments and additional insights during the preparation of this chapter.

References

1. ASTM D910-24 (2024) Standard specification for leaded aviation gasolines. ASTM International, West Conshohocken, PA. https://www.astm.org/d0910-24.html
2. ASTM D6227–24 (2024) Standard specification for unleaded aviation gasoline containing a non-hydrocarbon component. ASTM International, West Conshohocken, PA. https://www.astm.org/d6227-24.html
3. ASTM D1655-23a (2023) Standard specification for aviation turbine fuels. ASTM International, West Conshohocken, PA. https://www.astm.org/d1655-23a.html
4. EI 1590 (2014) Specifications and qualification procedures for aviation fuel microfilters, 3rd edn. Energy Institute, London. https://publishing.energyinst.org/topics/aviation/aviation-fuel-handling/ei-1590-specifications-and-qualification-procedures-for-aviation-fuel-microfilters
5. EI 1581 (2016) Specifications and laboratory qualification procedures for aviation fuel filter/water separators. Energy Institute, London. https://publishing.energyinst.org/topics/aviation/aviation-fuel-handling/ei-specification-1581-specifications-and-laboratory-qualification-procedures-for-aviation-fuel-filterwater-separators
6. EI 1550 (2019) Handbook on equipment used for the maintenance and delivery of clean aviation fuel, 3rd edn. Energy Institute, London. https://publishing.energyinst.org/featured/ei-1550-handbook-on-equipment-used-for-the-maintenance-and-delivery-of-clean-aviation-fuel
7. EI/JIG 1530 (2019) Quality assurance requirements for the manufacture, storage and distribution of aviation fuel to airports (A4), 2nd edn. Energy Institute, London. https://publishing.energyinst.org/featured/1530
8. Stokes GG (1850) On the effect of internal friction of fluids on the motion of pendulums. Transactions of the Cambridge Philosophical Society, 9, part ii, pp 8–106. https://babel.hathitrust.org/cgi/pt?id=mdp.39015012112531&view=1up&seq=208

9. Anderson SD, Clark AQ (2016) Fuel cleanliness—the sources and fate of contaminants from refinery to aircraft. In Morris RE (ed) Proceedings of the 14th international symposium on the stability and handling of liquid fuels, 04–08 October 2015, Charleston, South Carolina, USA. https://iash.conferencespot.org/2015/pdf/2015a003/2015fl008
10. Hughes VB, Rugin PD (2001) A survey of solid contamination types and levels found in a range of airport fuel loading systems. In Giles HN (ed) Proceedings of the 7th international conference on the stability and handling of liquid fuels, 24–29 September, 2000, Graz, Austria. https://iash.conferencespot.org/2000/pdf/2000a038/2000fl038
11. ASME (2023) BPVC section VIII-rules for construction of pressure vessels, division 1. The American Society of Mechanical Engineers, New York. https://www.asme.org/codes-standards/find-codes-standards/bpvc-viii-1-bpvc-section-viii-rules-construction-pressure-vessels-division-1/2023/print-book
12. EI 1596 (2019) Design and construction of aviation fuel filter vessels, 3rd Ed. https://publishing.energyinst.org/topics/aviation/aviation-fuel-handling/ei-specification-1596-design-and-construction-of-aviation-fuel-filter-vessels
13. ASTM D3948-22 (2022).Standard test method for determining water separation characteristics of aviation turbine fuels by portable separometer. ASTM International, West Conshohocken, PA, https://www.astm.org/d3948-22.html
14. EI 1582 (2011) Specification for similarity for EI 1581 aviation jet fuel filter/separators. Energy Institute, London. https://publishing.energyinst.org/topics/aviation/aviation-fuel-handling/ei-1582-specification-for-similarity-for-ei-1581-aviation-jet-fuel-filterseparators
15. ATA 103 (2023) Spec 103: standard for jet fuel quality control at airports. Airlines for America. https://publications.airlines.org/CommerceProductDetail.aspx?Product=322
16. EI 1588 (2022) Laboratory tests and minimum performance levels for aviation fuel water barrier filters, 2nd edn. Energy institute, London. https://publishing.energyinst.org/topics/aviation/aviation-fuel-handling/1588
17. EI 1583 (2020) Laboratory tests and minimum performance levels for aviation fuel filter monitors (obsolete). Energy Institute, London. https://publishing.energyinst.org/topics/aviation/aviation-fuel-handling/ei-1583-laboratory-tests-and-minimum-performance-levels-for-aviation-fuel-filter-monitors
18. JIG Bulletin 147 (2023) Adoption of WBF technology for into-plane filtration. Joint Inspection Group, Cambridgeshire. https://www.jig.org/documents/bulletin-147/
19. EI 1599 (2017) Laboratory tests and minimum performance levels for aviation fuel dirt defence filters, 2nd edn. Energy Institute, London. https://publishing.energyinst.org/topics/aviation/aviation-fuel-handling/ei-specification-1599-laboratory-tests-and-minimum-performance-levels-for-aviation-fuel-dirt-defence-filters
20. EI 1598 (2012) Design, functional requirements and laboratory testing protocols for electronic sensors to monitor free water and/or particulate matter in aviation fuel, 2nd edn. Energy institute, London. https://publishing.energyinst.org/topics/aviation/aviation-fuel-handling/ei-1598-design,-functional-requirements-and-laboratory-testing-protocols-for-electronic-sensors-to-monitor-free-water-andor-particulate-matter-in-aviation-fuel1
21. IATA (2023) Guidance material on microbiological contamination in aircraft fuel tanks, 6th edn. International Air Transport Association, Montreal. https://store.iata.org/IEC_ProductDetails?id=9680-06&_gl=1*1vhjlj1*_gcl_au*MjA4MzgwMjI1OS4xNzIwODAwODU0*_ga*MjkwNzQyNDY2LjE3MjA4MDA4NTQ.*_ga_PLLG1EY0X0*MTcyMDgwMDg1My4xLjAuMTcyMDgwMDg1My42MC4wLjA.&_ga=2.73741138.1116726301.1720800854-290742466.1720800854
22. ASTM D7619-22 (2022) Standard test method for sizing and counting particles in light and middle distillate fuels, by automatic particle counter. ASTM International, West Conshohocken, PA. https://www.astm.org/d7619-22.html
23. ASTM D8023-22 (2022) Standard test method for determination of water separation characteristics of aviation turbine fuel by small scale water separation instrument. ASTM International, West Conshohocken, PA. https://www.astm.org/d8073-22.html

Chapter 11
Insights into Aviation Fuel Sampling Processes

Moshe Rabaev, Liel Mazar, Yalfal Siyum, Haim Judah, and Frederick J. Passman

Abstract To ensure flight safety, it is crucial to supply engines with the appropriate quality of fuel consistently. The fuel quality is verified through proper fuel sampling methods for field and laboratory testing. This chapter outlines official standards and reviews relevant literature on fuel sampling techniques, including tools and containers designed for various sampling points. The impact of local defects—such as poorly positioned sampling valves, sloping piping sections, or tilted vessels—on sampling accuracy is also examined. Further, the chapter explains methods for obtaining representative and indicative samples from fuel lines, tanks, and filters to diagnose specific issues. Special attention is given to the unique challenges associated with sampling from aircraft fuel tanks. The chapter also addresses incidents requiring sample collection for forensic analysis, detailing solutions for areas with limited access and sampling from ground-based and aircraft filters. Lastly, the chapter provides an in-depth discussion on sampling for microbiological testing, including the associated chain-of-custody documentation required for such tests.

M. Rabaev (✉)
Independent Consultant, Moshe Rabaev Consulting, Beer Sheva, Israel
e-mail: rabaevm@gmail.com

Blechner Center for Industrial Catalysis and Process Development, Department of Chemical Engineering, Ben-Gurion University of the Negev, Beer Sheva, Israel

L. Mazar
Independent Consultant, Jerusalem, Israel
e-mail: liel.mazar@cservices.co.il

Y. Siyum · H. Judah
Israeli Air Force, Tel Aviv-Yafo, Israel
e-mail: yalfalsi@idf.il

H. Judah
e-mail: Haim.judah@idf.il

F. J. Passman
Biodeterioration Control Associates Inc, Princeton, NJ, USA
e-mail: fredp@biodeterioration-control.com

M. Rabaev et al. (eds.), *Aviation Fuel Problems and Solutions: From Refinery to Wingtip*, https://doi.org/10.1007/978-3-032-05514-9_11

1 Introduction

For aviation to be safe, it is essential to have a constant supply of high-quality fuel, which is ensured by using appropriate sampling for testing. The reliability of test data depends on the quality of the samples collected. This includes collection from suitable locations, use of appropriate containers, and correct handling between the time samples are collected and the time they are tested. In broad terms, there are two categories of samples: representative samples and diagnostic samples. A representative sample is a portion extracted from the total volume of fuel expected to contain the constituents in the same proportions present in the total volume. That is to say, the fuel property-test results of replicated, representative samples collected from the same fuel volume are expected to be within the specified reproducibility limits of the parameters tested. The properties of each sample should be statistically indistinguishable from those of other representative samples from the same total volume. Representative (average) samples, as defined in ASTM D4057 [1], are typically collected for routine fuel quality tests. In contrast, a diagnostic sample, as defined in ASTM MNL1-9TH-EB [2], will most likely contain detectable concentrations (or quantities) of the contaminants. For example, as discussed in Sect. 10, when microbiological testing is to be performed, diagnostic samples are collected from the system's low-point drains, filter housings, and elsewhere where microbial contamination accumulates and is most likely to be detected.

The primary considerations when planning fuel sampling operations are the purpose of the sample, and the types of tests to be performed. Many aspects determine the choice of sample container (size and material), sampling location, required tools, and the maintenance conditions of the sampled system. For example, non-fragile sample containers are preferable for sampling operational aircraft due to the risk of inducing foreign object damage (FOD). If a diagnostic sample is required to assess the contamination severity of fuel in a stagnant and non-operational pipeline, it is best to take samples from the lowest points. In a pipeline, while the fuel is not flowing, sometimes valuable data can be obtained from the liquid collected from the first drain, and sometimes after massive draining.

The information presented in this chapter provides insights regarding issues to be considered before sampling. It is not intended to be either prescriptive or to replace legal and contractual requirements, nor does it attempt to address all possible scenarios or problems. In particular, it does not supersede operational procedures specified by ASTM International, the Energy Institute (EI), American Petroleum Institute (API), Joint Inspection Group (JIG), Airlines for America (A4A), International Air Transport Association (IATA), IATA Fuel Quality Pool (IFQP), airlines, and other publications. The material presented here is intended to assist in troubleshooting and understanding the common difficulties and challenges encountered in sampling ground and aircraft fuel systems. The procedures described in this chapter are accurate at the time of writing, and it is understood that when dealing with fuel sampling it is important to be familiar with the applicable official standards, aircraft maintenance manual (AMM), and regulatory requirements. As a practical activity

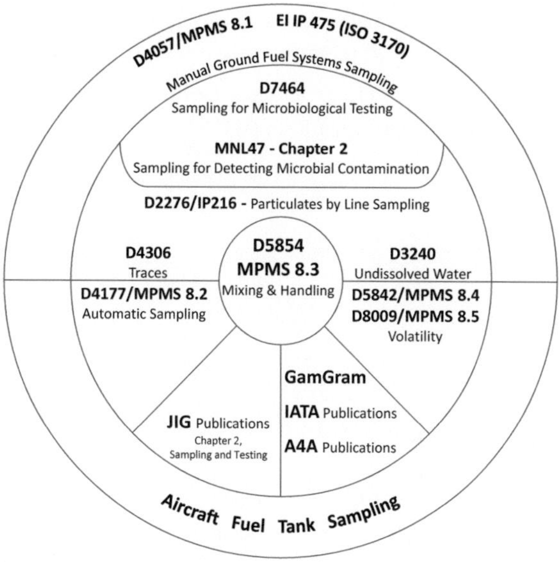

Fig. 1 Visual overview of several standards and practices in aviation fuel systems sampling *(Figure Courtesy of M. Rabaev)*

often undertaken in a hazardous environment, it is also essential to follow good safety practices when sampling fuel, with appropriate personal protective equipment, site knowledge and guidance.

2 Overview of Sampling Standards and Terms

Figure 1 visually depicts the diverse standards and practices documents relevant to sampling within aviation fuel systems. It reveals no universally adopted fuel sampling procedure that pertains to all situations. Instead, individual stakeholders in the fuel distribution chain and aircraft manufacturers have developed specific sampling methods suited to their respective needs.

Descriptions of the standards can be found in Table 1.

3 Sampling Points in a Typical Fuel Supply Chain

An aviation fuel supply chain can be complex and feature many storage and transport options as illustrated generically in Fig. 2.

Throughout the supply chain's different stages there are lots of potential sampling points. Some of them are mandatory, some commercially important, and others are part of good practice or optional. To illustrate this diversity of options, one should

Table 1 Selected aviation fuel sampling standards and other guidance documents

Publisher	Unique identifier	Title	Summary
ASTM	D4057 [1]	Standard practice for manual sampling of petroleum and petroleum products	Procedures and equipment for manually obtaining samples of liquid petroleum and petroleum products, crude oils, and intermediate products from the sample point and their transfer into the primary container. Includes procedures for sampling of free water and other heavy components associated with petroleum and petroleum products. Additional specific information is provided about sample container selection, preparation, and sample handling
API	MPMS[1] Ch 8.1 [3]		
Energy Institute	IP 475 [4] (ISO 3170)	Petroleum liquids—manual sampling	Manual methods to be used for obtaining samples of liquid or semi-liquid hydrocarbons, tank residues and deposits from fixed tanks, railcars, road vehicles, ships and barges, drums and cans, or from liquids being pumped in pipelines Applies to the sampling of petroleum products, crude oils, and intermediate products, which are stored in tanks at or near atmospheric pressure, or transferred by pipelines, and are handled as liquids at temperatures from near ambient up to 200 °C
ASTM	D7464 [5]	Standard practice for manual sampling of liquid fuels, associated materials and fuel system components for microbiological testing	Aspects of sample device preparation and sample handling to prevent sample contamination with micro-organisms other than those originally present in the sample. This practice also covers sample handling considerations that may impact sample perishability

(continued)

[1] MPMS—Manual of Petroleum Measurement Standards.

Table 1 (continued)

Publisher	Unique identifier	Title	Summary
ASTM	MNL 47 Chapter 2 [6]	Sampling methods for detecting microbial contamination in fuel tanks and systems	This ASTM manual brings together the various test procedures for diagnosing contamination in fuels and fuel systems. It also suggests the means for detection and control of microbial contamination. The manual complements ASTM D6469 "Standard Guide for Microbial Contamination in Fuels and Fuel Systems" and also explains in layman's terms how equipment and tests work [7]
ASTM	D2276 [8]	Standard test method for particulate contaminant in aviation fuel by line sampling	A test for the determination of particulate contaminants in aviation turbine fuel using a field monitor. This test is primarily for detection of dirt/ solid contaminants, but it could also detect other contaminants, which cause membrane discoloration, or plugging of the pores and clogging of the membrane. In rare cases contaminants may be white or transparent and thus not clearly visible in this test
IP	216		
ASTM	D4306 [9]	Standard practice for aviation fuel sample containers for tests affected by trace contamination	The most suitable types of containers for handling aviation fuel samples for the determination of critical properties affected by trace contamination
ASTM	D5854 [10]	Standard practice for mixing and handling of liquid samples of petroleum and petroleum products	The handling, mixing, and conditioning procedures required to ensure that a representative sample of the liquid petroleum or petroleum product is delivered from the primary sample container/receiver into the analytical test apparatus or into intermediate containers
API	MPMS Chapter 8.3		
ASTM	D3240 [11]	Standard test method for undissolved water in aviation turbine fuels	The test method covers the measurement of undissolved water in aviation turbine fuels in flowing fuel streams without exposing the fuel sample to the atmosphere or to a sample container

(continued)

Table 1 (continued)

Publisher	Unique identifier	Title	Summary
ASTM	D4177 [12]	Standard practice for automatic sampling of petroleum and petroleum products	General procedures and equipment for automatically obtaining samples of liquid petroleum and petroleum products, crude oils, and intermediate products from the sample point and their transfer into the primary container. Additional specific information is provided about sample container selection, preparation, and sample handling
API	MPMS Chapter 8.2		
ASTM	D5842 [13]	Standard practice for sampling and handling of fuels for volatility measurement	Procedures and equipment for obtaining, mixing, and handling representative samples of volatile fuels for testing compliance with the standards for volatility-related measurements applicable to light fuels
API	MPMS Chapter 8.4		
ASTM	D8009 [14]	Standard practice for manual piston cylinder sampling for volatile crude oils, condensates, and liquid petroleum products	This practice includes the equipment and procedures for obtaining a representative sample of "live" or high vapor-pressure crude oils, condensates, and/or liquid petroleum products from low-pressure sample points, where there is insufficient sample point pressure to use a Floating Piston Cylinder (FPC)
API	MPMS Chapter 8.5		
JIG	JIG 1 [15]	Aviation fuel quality control and operating standards for into-plane fueling services	Joint Inspection Group JIG 1 is the Standard for into-plane fueling operations for major hubs and regional airports
JIG	JIG 2 [16]	Aviation fuel quality control and operating standards for airport depots and hydrants	Joint Inspection Group JIG 2 is the Standard for on-airport storage and hydrant operations for major hubs and regional airports
JIG	JIG 4 [17]	Aviation fuel quality control and operating standards for smaller airports	Joint Inspection Group JIG 4 is the Standard for fueling operations at smaller airports (typically <10 million liters per year, with road or rail supply only)

(continued)

Table 1 (continued)

Publisher	Unique identifier	Title	Summary
EI/JIG	1530 [18]	Quality assurance requirements for the manufacture, storage and distribution of aviation fuel to airports	Mandatory provisions and good practice recommendations for maintaining aviation fuel (jet fuel and aviation gasoline (avgas) quality in refineries and in storage, distribution and transport systems including those delivering to airports. The standard covers facility design and construction; product manufacture; batching; sampling; testing; release; storage and handling; receipt and discharge and quality assurance requirements
Gammon technical products	–	GamGrams [19]	GamGrams are short educational articles, based on years of company experience, designed to help with various aviation fuel handling operations
IATA	–	Aviation turbine fuels specifications guidance material [20]	A document developed by the International Civil Jet Fuel Specification Liaison Group (ISLG) to describe the characteristics of jet fuels and specification requirements. Key recommendations for the handling of fuel are provided together with training guidance for airlines, fuel service providers and original equipment manufacturers (OEMs). In addition, it provides insights into fuels composition, physical and chemical properties, and addresses the issues of additives and cleanliness in fuel handling
A4A	ATA-103 [21]	Standards for jet fuel quality control at airports	Guidance for safe storage and distribution of jet fuel at airports as currently practiced in the commercial aviation industry, including forms that can be used to record performance of tests and inspections. This is the airline standard primarily used in the USA

consider the refinery, low drains, above and underground sampling points on dedicated, multi-product, and fungible pipelines; as well as fuel farms, storage, delivery systems, and transport vehicles such as bowser trucks, railway tankers, barges, ships, aircraft carriers, and aircraft fuel tanks and systems. Additionally, there are various civilian and military applications, temporary airports, etc. A good practice is to

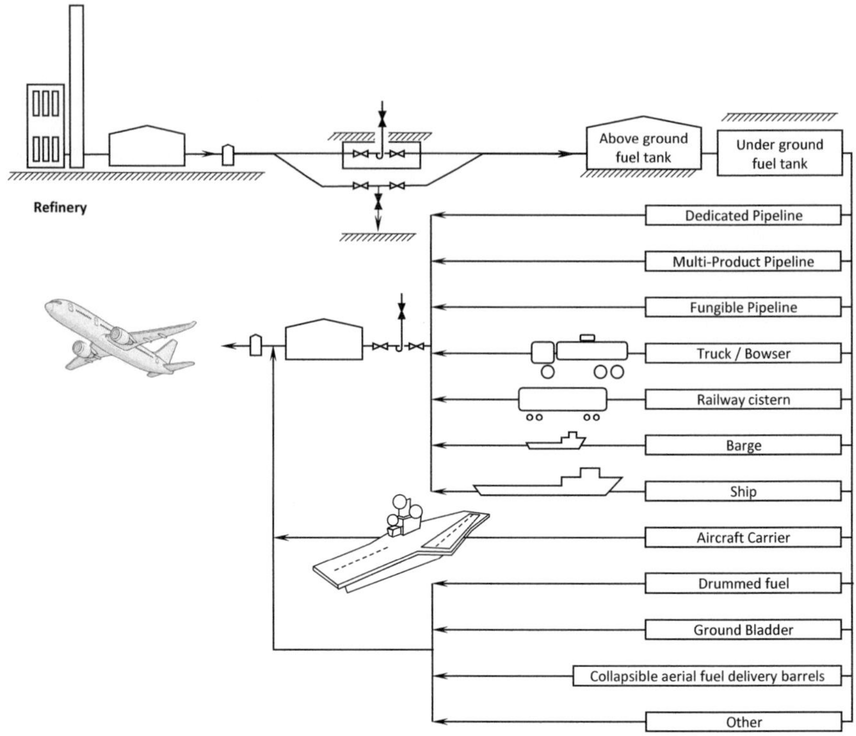

Fig. 2 Simplified typical fuel supply chain from refinery to wingtip (Figure Courtesy of M. Rabaev)

sample fuel for cleanness when fuel is filtered as it moves. These filters usually have inlet and outlet sampling connections in addition to sump (low point) drain valves. Although this is not the primary purpose, the drain valves may sometimes be used to help monitor fuel quality.

One of the most crucial sampling areas is the aircraft's fuel tanks and systems. Aircraft tanks have low point drains for sampling and sometimes filling ports on the top of the tank or inspection ports to allow access. For detailed procedures, refer to the Aircraft Maintenance Manual.

4 Sampling Container Considerations

Some consensus methods listed in Table 1 specify sample container requirements. For example, although most standard test methods specify sample storage in opaque containers, ASTM Test Method D4176 [22] which details visual inspection procedures such as the "Clear and Bright Test", requires a transparent and colorless jar of at least 75 mm (3 inches) in diameter to effectively detect haze, color change, or fine particulates in the fuel. When sampling from aircraft, selecting a suitable sample container is crucial, based on the sampling point and environmental conditions. For instance, in the case mentioned above of the "Clear and Bright Test", a glass jar is the preferred option even though small drops of free water can stick to the jar's walls. However, if the jar happens to break, glass shards can become a personal hazard and increase FOD risk in the aircraft area. As a second example, for ASTM Test Methods D7463 [23] and D7687 [24] (Adenosine Triphosphate (ATP)), HDPE containers are prescribed to minimize the risk of activity loss because traces of water-containing microorganisms can adhere to hydrophilic surfaces. Although it is difficult to visually examine droplets and particles through the walls of a translucent container, the presence of water can be detected by using complementary samples or sampling from the translucent container itself using a syringe, a pipette, or a separatory funnel. HDPE can be used for bottles in rapid-response kits with reagents or separatory funnels. These tools are recommended for processes that have a high sensitivity to water or dissolved compounds in the water. Any change in temperature will directly change the amount of water that may be dissolved in the fuel. As fuel warms, it can absorb more dissolved water, and as the fuel cools, dissolved water can condense into haze or droplets. The following sections will discuss the appropriate sampling tools for different fuel system parts.

A fuel sample can be affected by the sample container, the sampling tools used, and the condition of the sampling environment. Some of these parameters (and it is by no means an exhaustive list) would be:

- Container material and construction.
- Container cleanliness.
- Composition of the container cap material and the cap seal, if caps are used.
- Cleanliness of sampling apparatus like direct core sampler; weighted beaker or bottle; "Thief" or Bacon Bomb for sampling of a fuel reservoir (usually consists of a rope, a cable, or a chain attached to a valved container); a container outside of the tank, with a vacuum pump and a suction hose that is extended into the tank.
- Cleanliness and position of the "thief" rope, cable, or chain. (Note: Static electricity is a hazard when taking samples from a tank; care must be taken to avoid conditions where sparks may occur, which can cause fuel vapor ignition).
- The condition and cleanliness of the fuel sample extraction points in the sampled system. There might be a need to flush the interior of the piping before sampling.
- The flow rate from the fuel sample extraction points in the sampled system.
- Sample type (inline vs. tank) and the design of the sampling system or point.
- The fuel system design of the sampled system.
- Proper flushing of sampling connections when a representative sample is needed.
- Environmental conditions when sampling (rain or dust may compromise sample integrity).

As detailed in Table 1, fuel sampling standards, including those outlined in publications such as ASTM D4306, JIG, IFQP, and ATA-103, thoroughly address the suitability of various containers for aviation fuel samples. These standards provide critical guidelines for selecting appropriate containers that ensure sample integrity and accuracy. Table 2 lists some common ways fuel samples are compromised by incorrect use of sampling containers and sampling tools or specific sampling environment factors and provides various methods to resolve these issues. If it is crucial to determine the undissolved water content in the fuel, the test method outlined in ASTM D3240 might be helpful. This method measures undissolved water in aviation turbine fuels while the fuel still flows without exposing the sample to the atmosphere or a sample container. If there is suspicion of contamination by volatile components, it might be appropriate to use ASTM D5842 or D8009.

The issues listed above in Table 2 affect the quality of the sampled fuel, and if adequate measures are not taken, the fuel sample will not be representative of the bulk fuel. A quality certificate of the fuel sample includes a long list of tests which can be affected by the parameters noted in Table 2.

5 Considerations for Pipeline Sampling

Sampling aviation fuel from pipelines presents particular characteristics, problems, and solutions. The sampling point (usually a tap) should be positioned to enable fuel extraction according to official sampling procedures (e.g., ASTM D4057).

When a representative sample from a pipeline is required for a routine quality test, collecting the sample while the fuel flows at its typical rate is necessary. In contrast, when a diagnostic sample is needed to test for contamination levels in an inactive line containing fuel, in some cases, the sample should be taken while the fuel is static, preferably from the pipeline's lowest point. Samples from these two scenarios are expected to differ significantly. In an inactive pipeline that contains fuel, there will be enough time for dirt, water, and contamination, such as fungi, bacteria, and even sulfate reducing bacteria (SRB), to accumulate in the system's low points.

When taking fuel samples from a flowing system, it is crucial to properly understand safety issues regarding working with internal pressures, static electricity, tools that may create sparks, protective equipment, grounding, and the containment of adverse events or spills. This section focuses on the sampling process itself and not the extensive safety measures which must also be considered. Table 3 details some operational parameters requiring special attention when taking samples from a flowing system.

Sampling fuel pipelines requires familiarity with the sampled pipeline route's geography, topography, and detailed design. Pipelines are complex systems influenced by various factors, some of which are routes passing through hot or cold climates, above- or underground; the design of trenches and the methods of segregation along the route; forks; if the pipelines are dedicated or multiproduct; and special points such as especially long, steep slopes, and renovated segments. It is also necessary to be familiar with the pipeline components such as fuel metering devices, shock absorbance, filters, etc.

Familiarity with a pipeline's complexity and characteristics is crucial, as illustrated by the following example. In the context of a normal and regular operation, with a low

Table 2 Common sample contamination scenarios and effects

Parameter description	Possible scenario	Consequences/Significance	Possible solutions
Cleanliness of sampling container (and funnel, if used)	Presence of water in the sample container or funnel prior to sampling	Extraneous water and materials dissolved in the water represent a source of false results. The water and/or the dissolved materials can react with the chemical components of the fuel and give false values for the fuel system for such parameters as sample opacity, microbial growth, and dissolved ions	Using a clean and dry container and funnel Flushing the container with the fuel to be sampled from the sample point (at least 3 times is considered optimal). This also flushes the sample point, and in cases where forensic sampling is required, the flushed fuel might also be collected to prevent any loss of evidence during the flushing procedure Using a dedicated and convenient container, in accordance with ASTM D4306, authorized beforehand for use in sampling aviation fuel (for investigative purposes or other unique situations)
	Presence of particle contamination	Particles present in the sampling tool may lead to an inaccurate assessment of the fuel system's cleanliness	
	Presence of foreign materials	Residues from previous samples can adhere to the inner walls of a container without being visible. Over time, these residues can undergo chemical changes due to exposure to light, heat, microorganisms, and other factors. This can be difficult to detect and may result in significantly different test outcomes, such as for jet fuel thermal stability, gum, water separation, microbiological testing, and others	

(continued)

Table 2 (continued)

Parameter description	Possible scenario	Consequences/Significance	Possible solutions
Sampling container material	Water adherence to the inner walls of a glass container	Hindered or complete inability to visually discern the presence of water in the sample	Use of multiple containers can be appropriate depending on individual situations: • Visual appearance test: Transparent Glass Container • Water contamination tests: HDPE container • Chemical tests: Amber glass container or epoxy-lined metal canister Around the aircraft: a container, metal, glass with externally protective polymer-coating or glass protected by a metal/wood case for transport and storage (See Fig. 3, Fig. 4, Fig. 11b or Fig. 16a) Using a dedicated and convenient container, in accordance with ASTM D4306, authorized beforehand for use in sampling aviation fuel (for investigative purposes or other unique situations)
	Opaque or translucent container	Inability to appraise the appearance of the sample Reducing light exposure	
	Transparent container	Sensitivity to light May affect thermal stability, conductivity, gum, visual appearance tests, etc	
	Fragile container	Possible source of FOD	

(continued)

Table 2 (continued)

Parameter description	Possible scenario	Consequences/Significance	Possible solutions
Cap material and condition	Reacts on contact with fuel	Certain fuel types, especially synthetic, may dissolve some of the cap or cap seal material	Use caps and seals made of materials that are resistant to the specific fuel sampled Testing the cap and cap seal material, especially if there is a seal in the cap Cork caps are not recommended for use in any situation
	Absorbs foreign contaminants	Materials such as cork absorb and release chemicals into the sample and can affect MSEP[a], thermal stability, gum tests, etc Emphasis on fuel samples that contain synthetic components and have a very high/low presence of aromatic or other components	
	Releases foreign material	Metals, polymers, plasticizers, pigments, etc. Emphasis on fuel samples that contain synthetic components and have a very high/low presence of aromatic or other components	
	Impaired tightness	Polymeric caps for vapor-pressure testing containers may be damaged by aging, interaction with fuels, or other physical factors	

(continued)

Table 2 (continued)

Parameter description	Possible scenario	Consequences/Significance	Possible solutions
Faucet (Tap) condition (Sample point)	Externally unclean	Inclusion of foreign materials in the sample if the opening of the container or the funnel encounters the sides of the faucet	Regular periodic cleaning of the faucet as part of maintenance procedures, especially before sampling
	Internally unclean	Entry of foreign materials and particles, specifically particles and products of fuel aging If the faucet is rarely used, the inner area of the opening, can be a breeding ground for small animals, micro-organisms, insects, etc	Flushing with a large amount of fuel under maximum allowable pressure (according to technical and safety parameters). Special care should be taken for microbial or forensic sampling so evidence will not be lost by the flushing
Flow rate from the faucet (Sample point)	Loss of findings in initial draining	The initial drain quantity is critical. The visual appearance, odor, the presence of particles and water in the initial drain, all give supplementary information concerning the current state of the system. On occasion, these findings can also be expected to appear in the inner parts of the fuel system such as filter vessels, large tanks, and low points in long fuel lines Expertise and familiarity with the sampling points are necessary to differentiate between fluid that is stagnant in the fuel drainpipe and the fuel from the fuel system	Using a transparent container or a white bucket for the initial drain. Meticulous inspection and recording of findings: water—quantity, appearance; particles; color of the fuel; micro-organisms Sometimes (commonly in storage tanks in refineries) a clear and colorless sample of water may be observed with no fuel. This amount of water is the result of condensation collecting in the drain system

(continued)

Table 2 (continued)

Parameter description	Possible scenario	Consequences/Significance	Possible solutions
	Fuel flow stoppage	Sometimes, due to design constraints, the location of the faucet (sample point) is sub-optimal. This can manifest itself as a vacuum at the sampling point, despite the flow and pressure in the fuel line. Mainly due to the venturi effect, installation of sampling points near turns and pipe fittings (i.e., valves, counters, etc.) at certain flow rates can create a vacuum even in an open faucet	Changing the characteristics of the fuel system so that fuel flow is not hampered. Elimination of valves after the sampling point, increase/decrease of fuel flow, etc. Sample probes that extend into the flow stream are preferred to simple sample points These operations require in-depth understanding of the system and must be done with the appropriate authorization
"Thief"cleanliness	Sample contamination	Contamination from residue of previous samples. Particle contamination	Thorough cleaning of the "thief" by flushing it with sampled fuel before sample collection
"Thief"material	Thermal stability failures	Choice of material for thief container is important. Copper/brass thieves can lead to samples with sufficient copper being dissolved to create thermal stability failures, with higher total acid fuels being more at risk of this showing as an apparent problem when it's a sampling error. Also, then pouring sample through a brass/copper funnel compounds the error	Avoiding copper/brass thieves for aviation fuel sampling

(continued)

Table 2 (continued)

Parameter description	Possible scenario	Consequences/Significance	Possible solutions
"Thief's" rope cleanliness	Sample contamination	Contamination from residue of previous samples. Particle contamination	The "thief's" rope, especially if the rope is thick, can absorb fuel or collect particles and contaminants from the floor or from its case. The rope must contain no unnecessary knots, loose strands or other unsmooth areas that may become a source of contamination (See Fig. 5) Cables or chains shall be made of material appropriate to reduce spark and static electricity hazards (see ASTM D4057 [1])

[a]Microseparometer test for jet fuel water separation properties

Fig. 3 Whole (left) and cut-away (right) view of a 5-L steel epoxy-lined UN-A1 aviation fuel sample drum (Figure courtesy of A. Clark)

Fig. 4 Transparent glass bottles protected against light and physical damage by wooden case (Figure courtesy of M. Rabaev)

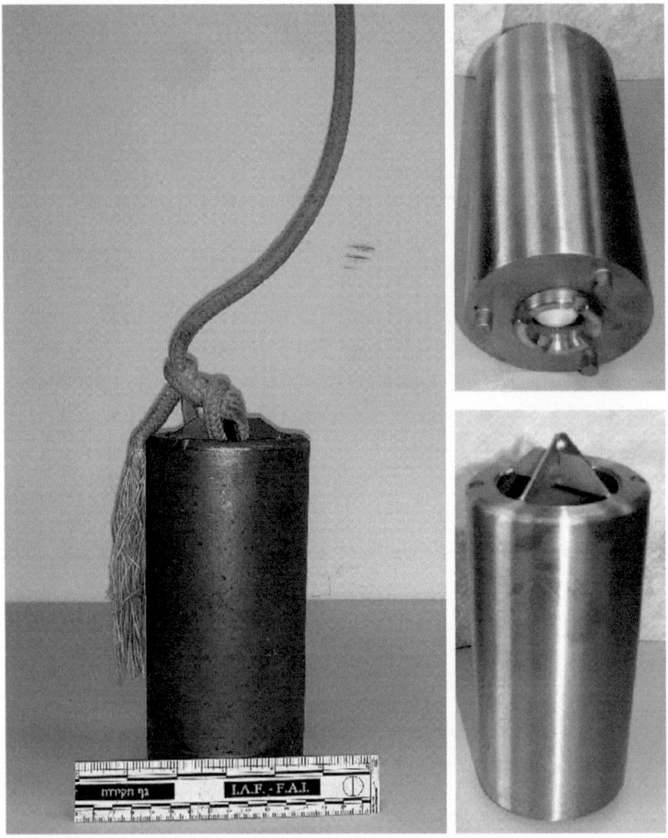

Fig. 5 Fuel Sampler ("Thief") with an unacceptably dirty rope, specifically showing an overcomplicated knot and excess rope, allowing large amounts of dirt and residues of previous samples into the sample. The bottom of the sampler is equipped with a ball, allowing the collection of samples from different levels (Figure Courtesy of M. Rabaev)

frequency of fuel deliveries, it is highly possible that a line that is tens or hundreds of kilometers in length contains several different batches of fuel. Consequently, there may also be different contamination levels at different low points along the line. Thus, getting a representative sampling process through the whole pipeline is challenging. Another example is illustrated in Fig. 6. Let us assume a cross-country pipeline that contains several months of stagnant fuel. At this hypothetical low point, the upstream side of the segregation valve will contain contaminants from the upstream fuel, and the downstream side will contain contaminants from the downstream fuel after the fuel flow has been stopped and the valve is closed.

If the system is still stagnant and a representative sample is required for laboratory or visual tests, it should only be collected after any water and sediment have been drained from the line. Draining should be performed after the flow has been stopped and sufficient time has elapsed to allow water and sediment to settle out of the product.

Table 3 Some operational parameters requiring special attention when collecting samples

Parameter	Possible problem	Consequences/Significance	Solutions
Sampling at lower or higher flow rates than usual	Unusually low or high particulate levels	Misjudgment of the true state of the system	Stabilizing the flow rate of the system at the desirable usual flow rates
Sampling from a partially full line	Layers of water and particles that are usually in equilibrium with the fuel on the internal surfaces of the pipe may be dislodged and migrate downstream	Partial or complete draining of a line poses risks requiring special care. This situation is usually encountered when an active line is drained for maintenance or due to some malfunction. When the line is re-serviced, the fast-flowing fuel generates a flow-front with especially high turbulence	Slow filling of the entire volume of the pipeline Filters and other components may become overloaded with water and dirt. Removing them or replacing with mesh strainers to catch excess dirt on commissioning until the system has stabilized should be considered, including a temporary tank to contain commissioning product
Sampling point location	The design of the fuel system may place sampling points at an incorrect distance from pipe components, on forks or on bypasses	As described in GamGram 6 [19], the location of the sampling point may affect sample properties. Samples taken from improperly located sample points may not be representative of the state of the fuel in the system	Reviewing sample point locations according to industry publications

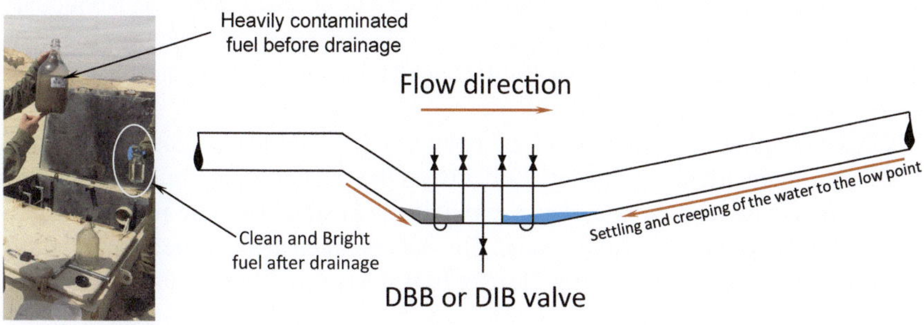

Fig. 6 Pipeline segment includes a trench, a double block, and bleed (DBB) valve, or a double isolation and bleed (DIB) valve (Figure Courtesy of M. Rabaev)

If the fuel line includes a piping branch for the sample tap, the entire volume of the piping must be flushed out at least once (relates to above- and underground pipeline areas). Only after obtaining a clear and bright sample should the fuel be analyzed. It is advisable to conduct and document visual inspections of the fuel at the beginning, middle, and end of the draining process to reveal certain characteristics of the line near the sampling point. Analyzing the findings from different sampling points along the line for water and sediment loads, microbial bioburdens, and fuel quality results (both within the specification and beyond) could be a significant tool supporting decisions concerning the maintenance of the line, the quality of fuel reserves, and the effectivity of regular and special maintenance operations.

In fluid mechanics, the fluid velocity within a pipe is fastest at the center and slows down near walls. In the study of physics and fluid mechanics, the boundary layer refers to a thin liquid layer forming near a surface as the liquid flows along it. This layer is influenced by the way the liquid interacts with the surface, which enforces a no-slip boundary condition, meaning that the velocity of the liquid at the surface is zero. Further away from the surface, the liquid velocity gradually increases until it reaches the bulk flow velocity. The part of the liquid that has not yet reached the bulk velocity is known as the velocity boundary layer. Contaminants such as dirt and water tend to be attracted to the liquid or pipe in what is called the boundary layer. They tend to cling to the surface, and the thickness of this layer varies with the general flow rate and turbulence in the pipe. The higher the flow rate, the thinner the boundary layer will be. In a system that usually operates at a specific flow rate, a layer of sediment will form according to the thickness of the boundary layer. A sudden rise in the flow rate will make the boundary layer thinner and cause some deposits from the sediment layer to detach, as illustrated in Fig. 7.

For example, when the pipe diameter decreases, the flow velocity increases, and the boundary layer consequently becomes thinner. The dramatic decrease in boundary layer thickness occurs when transitioning from a laminar to a turbulent flow regimen. The operator should know the transition values of the various flow rates that may be present at different times.

While changing the flow rate on the boundary layer on the microscopic scale is significant, there are also substantial effects on the macroscopic scale. If there is a sudden change in the flow rate in a system with a constant flow rate regimen, unexpected contaminant levels might appear. The flow regimen dictates the accumulation, for example, of water. When a tiny droplet of water meets a large quantity of water,

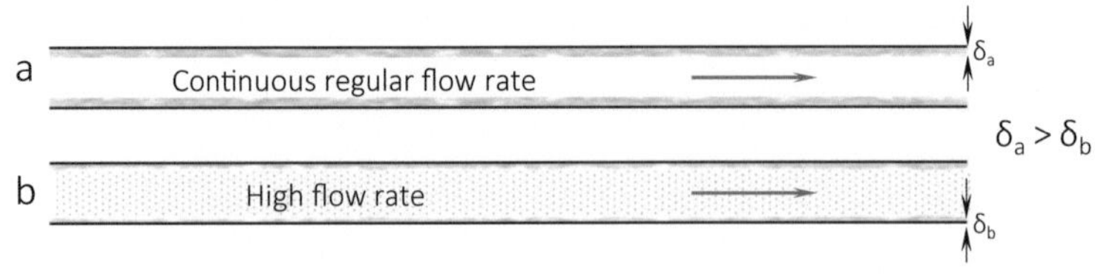

Fig. 7 Boundary layer thickness at different flow rates—**a** flow under typical conditions; **b** flow under high rate conditions relative to the typical flow rate (Figure Courtesy of M. Rabaev)

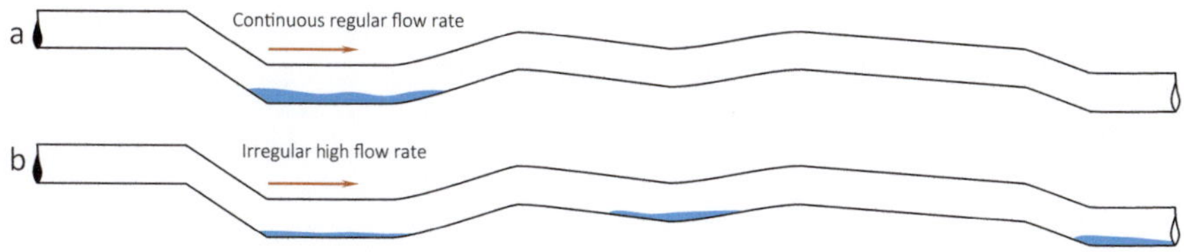

Fig. 8 Schematic Illustration of the influence of switching from—**a** continuous typical flow rate to **b** relatively high flow rate (Figure Courtesy of M. Rabaev)

it will remain in the accumulated water and will not continue downstream. A sudden rise in the flow rates will change the fuel turbulence levels, which could result in the entrapment of the water further downstream (Fig. 8).

In the case of constant flow regimens, it is possible to assume long-term stability of the pipeline system. Long-term cleanliness and dryness levels along the pipeline may be achieved by using a device for pipeline scraper (or "pig"[2]) at frequencies depending on the intensity of use in the system or in accordance with local or international requirements. In addition, as recommended in GamGram 4 [19], flow rates of 6 feet (about two meters) per second will ensure clean and dry pipe boundaries because the turbulence levels create a thin enough boundary layer. Fuel samples from a system with this flow regimen may be expected to represent the content at the stable state of the system.

6 Sampling from Storage Tanks

Tanks come in various shapes and are distinguished by features such as size, roof structure, above or underground location, stationary location or portable, intended for specific purposes and meeting specialized requirements or suitable for numerous products, painted/lined on the inside, equipped with or without a floating suction arm, and many more individual parameters.

Similarly, sampling from a fuel tank is accomplished in many ways: for example, using drainage valves for gravitational drainage, employing hand-powered or electric pumps, or using a "thief." ASTM D4057 is a thorough guide describing sampling tools, methods, limitations, and pros and cons. Proper utilization of this guide assists in making informed decisions about selecting the most suitable and appropriate tool for sampling, considering the specific characteristics of the tank while ensuring safety protocols are met and avoiding the consequences of taking an unreliable sample.

Figure 9 illustrates several concerns when taking a fuel tank sample. If the tank bottom is correctly sloped, the water in the tank will flow toward the drainage valve (Fig. 9a). However, distorted tank bottoms or tilted tanks, with the bottom sloping away from the drain low point (b) or to the sides (c), will cause the water to collect

[2] "pig"—device inserted into pipes and tubes which forces fluids and dirt out of them.

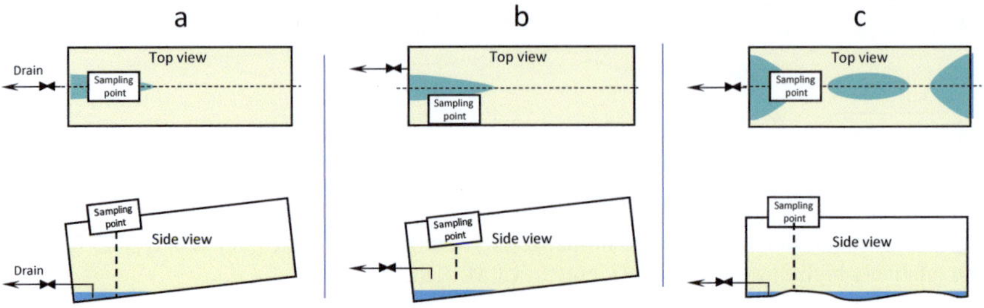

Fig. 9 Some tank slope differences and their effects on the drain and sampling processes—**a** correct slope with sampling point above the low point; **b** tilted; and **c** sagging (low longitudinal center) and hogging (lower ends than at longitudinal center) (Figure Courtesy of M. Rabaev)

at the lowest point away from the drainage pipe. This creates a challenge in properly sampling or completely draining all the water and sediment from the tank.

During sampling, a naive assumption is often made that the tank is perfectly level. However, regardless of whether the tank is a vertical or horizontal cylinder, above or underground, or has any other configuration, there can always be areas with water puddles or layers of water and precipitates that will not be accessible during sampling or drainage. While a perfect solution to these issues may not always be available, a clear understanding of each tank's actual bottom configuration will allow for educated risk management. Figure 10 illustrates a schematic drainage system with limited effectiveness. The apparent configuration of the tank bottom shown below is for illustration purposes only: in reality, it is impossible to see the slope (commonly 2–3 cm per 3 m) without the aid of level-measuring devices. Additionally, the angles of tank floor plate deformations can be greater than the nominal slope of the tank floor. This promotes small-volume puddling (<1L) of bottom-water in the center of each deformed floor plate.

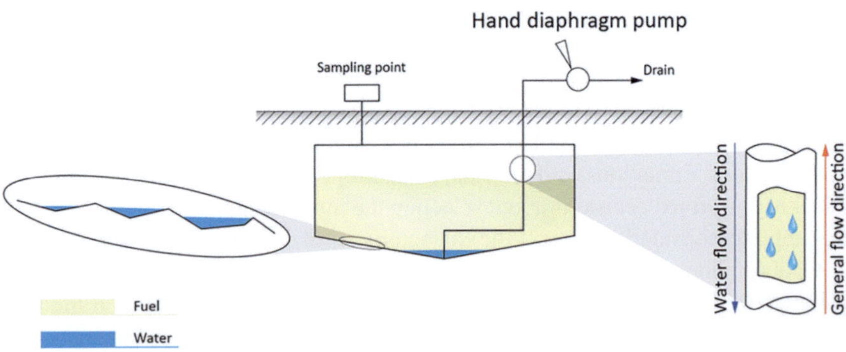

Fig. 10 Distortions of the tank bottom and water flow during the insufficient draining rate process (Figure Courtesy of M. Rabaev)

In situations where drainage (sump sampling) occurs by pumping with a diaphragm (suction) pump from the bottom of an underground tank, it is possible and even highly probable that water will only partly travel up the pipe instead of being completely flushed out. Achieving sufficient velocity is crucial for removing all contaminants, a challenge often encountered during bottom tank drainage. When the pump is used periodically (for example, as with a hand-powered diaphragm pump), if the drainage procedure time and/or velocity is insufficient and a portion of the water is still in the pipe when the pump is stopped during the drainage process, the water in the pipe may return to the tank bottom. Given water's higher density than fuel, it tends to settle back inside the pipe during or after drainage operations. Thus, the drainage rate must be sufficiently high and sustained even after seemingly water-free fuel appears. Even if this is done when drainage is slow with a hand-powered pump, when an automatic pump is used intermittently or with a short drainage time, water may remain and accumulate in the crevices of the bottom of the tank. Obviously, in these situations, the sample taken will not be representative for analysis.

Fuel can form different layers in storage tanks. In some cases, the layers' temperature or density may differ slightly. There may be significant differences in the contamination composition of the fuel regarding contaminants in different fuel layers in the tank. The upper (surface) layer may differ from the fuel lower down in the tank. Water condensation from the ceiling drips down onto the surface of the fluid in the tank and can contaminate the surface layer of the fuel, creating a breeding ground for micro-organisms. This well-known issue is described fully in ASTM MNL 47 Chap. 2 [6].

7 Sampling and Sampling Tools for Aircraft

Sampling from aircraft requires specific certifications for safety in and around aircraft due to the various hazards associated with working in this area. These specialized certifications arise from the need to avoid various dangers entailed in working on aircraft, which derive, among others, from moving parts (elevators, rudders, propellers, rotors, etc.), airflow into the engine intake, emissions from the exhaust, munitions in military aircraft, static electricity, and hot brakes after landing, as well as not causing damage to the aircraft. Special tools for specific aircraft or drain/sample points are commonly needed.

Sampling aircraft fuel is multi-purpose, and it includes periodical water drainage from low points, pilot inspection before a flight, sampling for laboratory analysis of fuel properties, or aseptic sampling to determine the level of microbial contamination in the aircraft's fuel system.

No special sampling tools are necessary beyond a sampling container if the drainage point is simply exposed or covered by an access panel. However, specialized tools are required if the sampling points have press-able openings. In situations with differences in shape, volume, or material of sampling containers, there are also

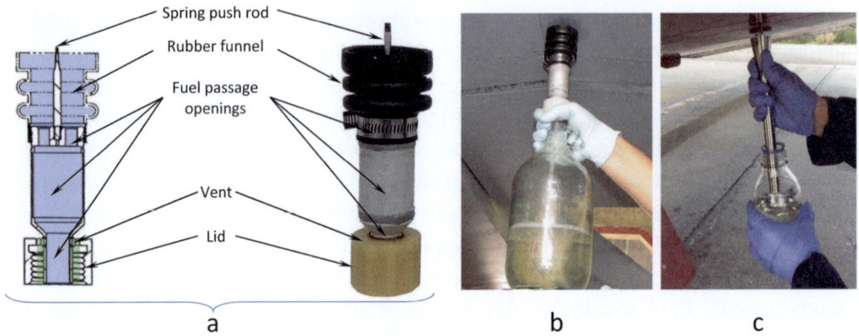

Fig. 11 Aircraft fuel sampling tool for draining/sampling from aircraft's press-able drain points—**a** sampler with integrated rubber funnel; blue area—space for fuel flow, green—vent; **b** sampler with integrated rubber funnel in use; **c** sampler equipped with metal funnel (Figure Courtesy of M. Rabaev)

differences in sampling tools for aircraft, some examples of which are shown in Fig. 11.

Simple draining valves in aircraft do not necessitate specialized tools. However, press-able draining valves, such as spring valves, do require their own dedicated tools. Each of these tools has its own advantages and disadvantages, and the sampler manufacturers will produce tools in accordance with the specifications of the aircraft manufacturer.

Some specialized tools are available from commercial sources and custom manufacturing. Every sampling situation has its own specifications and circumstances, and an in-depth understanding of each particular case is necessary in order to make an informed decision about the suitability of a particular tool for a particular task. Sometimes, a sampling tool feature is disadvantageous in one situation, but advantageous in another.

Several considerations need to be considered when choosing sampling tools for specific tasks:

1. A funnel wide enough to cover the sampling point.
2. A funnel is narrow enough to sample from under panels or a limited area (Figs. 12 and 13).
3. The compatibility of its diameter with the sampling point enables stable flow during sampling.
4. Vent openings enable air expulsion from the sampling container during sampling.
5. The sampling container cap that screws onto the container is made of inert materials.
6. Resistance of the component parts and materials to sterilization under autoclave conditions.

Familiarity with the structure of the aircraft's draining system, including the fuel tank inside structure, is highly recommended. Like other fuel tanks, the bottom is

Fig. 12 Aircraft sampling point. The sampling point is hidden under the aircraft's outer panels. A small diameter sampling tool is needed. **a** sampling point covered under the panel; **b** too wide; inappropriate funnel diameter; **c–e** appropriate funnel diameter (Figure Courtesy of M. Rabaev)

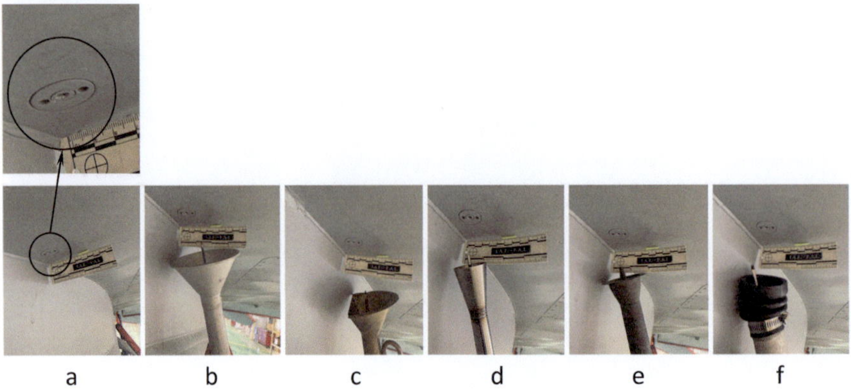

Fig. 13 Aircraft sampling point. The sampling point is close to a vertical surface. A small diameter or cut-edge sampling tool is needed. **a** sampling points close to the vertical surface; **b** too wide; inappropriate funnel diameter; **c** the cut-edge appropriate sampling tool; **d–e** appropriate narrow enough funnel diameter; and **f** the appropriate flexible funnel (Figure Courtesy of M. Rabaev)

not always smooth, leading to water accumulation, precipitating in flexible, rigid, or integral tanks. Aircraft ribs, spars, and other structural components can block water flow and contaminants in the fuel tank to the drainage points. See Fig. 25 in Chap. 12—"Microbiology" for a schematic of drainage points in aircraft wings. The pictures in Fig. 14 show the typical fungal contamination that can occur in areas with impaired fuel flow, such as near the spars or ribs of the wing.

It is sometimes necessary to open the fuel tanks and sample the fuel aseptically using sterile sampling equipment to locate and identify sources of contaminants, especially in cases of fungal contamination.

Fig. 14 Examples of damage from water trapped behind wing spars and ribs (Figure courtesy of M. Rabaev)

8 Sampling from Limited Access Points

Failure analysis, or investigation of specific systems, requires sampling from areas with limited access, such as the bottom and other parts of the aircraft fuel tank, unevenly positioned tanks, filtration systems, and sometimes even fuel tanks of military missiles and other weapon systems. For sampling from these challenging locations, it is possible to use a vacuum system connected to a new, thin leader-hose with a 1/8″ (or other appropriate diameter) made of polypropylene (PP), polytetrafluoroethylene (PTFE), or metal (see the simplified schematic in Fig. 15). The hose can then be more easily directed to the desired sampling area, enabling the suction of fuel into the sampling container. If a metal hose is used, it must be cleaned to the standard of use for pure gas systems because otherwise, the sample could be contaminated by the hose. The hose is connected to the sampling container using a cap made of fuel-resistant material with two holes—one for the hose and the second for a vacuum pipe. It is important to remember that significant safety considerations apply when using such a system, including among others, the risk of static electricity, sparks from metallic parts, shards from glass tools and vessel implosion should the inlet become blocked.

The sampling system should be set up using clean containers, a roll of new hose, a cap with appropriately sized holes, vacuum pumps, and a convenient electrical source or generator for the pump. When using a flexible hose, it may be difficult to lead the hose to the sampling area; in that case, it is possible to use a rigid stick connected to the end of the hose to lead it to the correct place.

For small systems such as a container holding more than one phase of contaminated fuel (see Fig. 16a) or the integral aircraft tank or bladder with an open panel (Fig. 16b), it is possible to use single-application polyethylene pipettes.

Samples taken from these containers under such conditions are intended for non-standard testing and investigation, and adjustments are made to tailor them to the specific test required, see Chap. 14. Laboratory polystyren, or glass pipettes, may be useful in collecting the suspect part of the fuel sample.

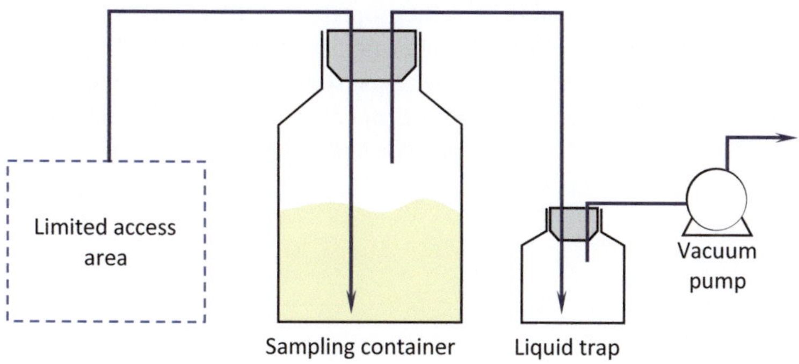

Fig. 15 Schematic illustration of a sampling system for areas with limited accessibility (Figure Courtesy of M. Rabaev)

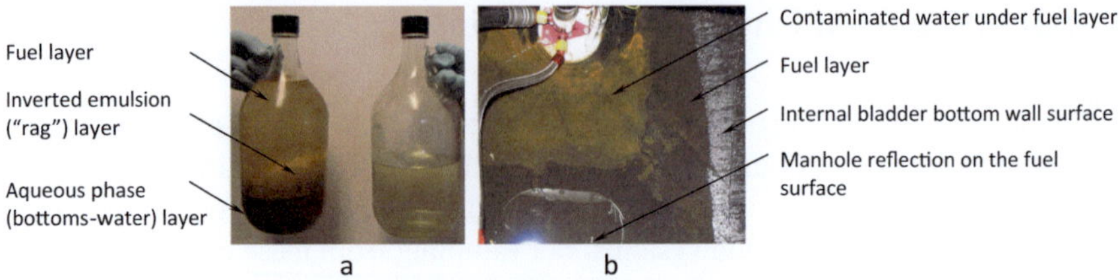

Fig. 16 More than one phase of contaminated fuel—**a** comparison of contaminated fuel beside clear and bright fuel; **b** aircraft's internal tank showing microbial contamination (picture taken from the top manhole and the light brown fluid was sampled using a sterile pipette for identification and characterization) (Figure Courtesy of M. Rabaev)

Polar additives and fuel components may preferentially pass from the organic fuel phase to the water phase. Their concentration in the sample can be substantial (>1 g/kg) as total dissolved solids (TDS), but the impact on their concentration in the fuel is less than the repeatability standard deviation of fuel chemistry test methods. Thus, high TDS in the aqueous phase does not necessarily indicate the presence of contamination. High TDS signals the need for additional testing to assess whether it comprises contaminants, partitioned fuel molecules, or both.

9 Sampling from Filters for Forensic Investigations

By the very nature of their functionality, filtration systems collect impurities, and as a result, they potentially may become a source of contaminants in the aviation fuel supply chain. Naturally, this source of contamination is an important resource for status monitoring, failure investigations, and crash investigations. Finding a rag or rubber glove blocking the fuel filter in a plane does not require any experience or special know-how, but ice crystals can accumulate and block the flow of fuel and then disappear as if they never existed—this requires in-depth knowledge of how the fuel system works, the physical and chemical properties of the water-fuel mixture, heat transfer, fluid dynamics, and various other fields. Chapter 10 and EI 1550 discuss types of filters, problems common to filters, and solutions to those problems in-depth. While this section discusses the sampling process for filters and their unique characteristics, the analytical chemistry methods and processes for interpreting findings from filters, are discussed in Chap. 14.

9.1 Basket Filter

Basket filters are typically installed at the entrance of the fuel pipeline to the facility, to remove larger-sized contaminants resulting from activities such as the opening/

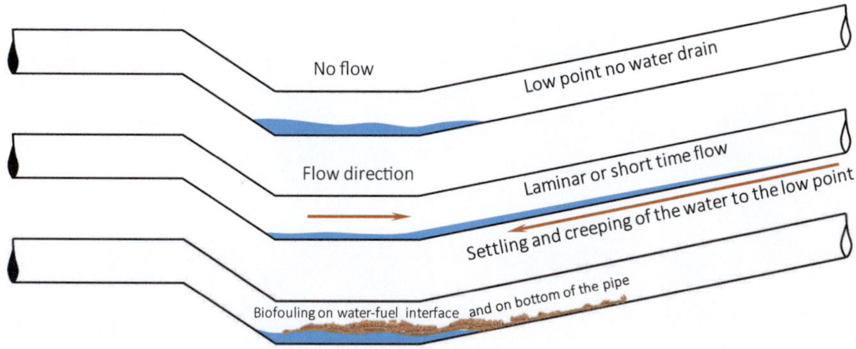

Fig. 17 Characteristic accumulation of contaminants in low points of fuel lines (Figure Courtesy of M. Rabaev)

closing of the system. Sometimes, solid, sludge or microbial contaminants may reach the basket filter as a biofouling film. Fuel lines that feature numerous low points and laminar flow are likely to constitute sources of microbial activity. These conditions usually manifest in fuel lines wider than 8 inches, in sections that are tens of meters long, and below the level of the main part of the fuel line (see Fig. 17).

Under these conditions, contaminants and water will accumulate at the low point, and a lack of drainage points in the section will only worsen the situation. Both laminar and short-term flow will accelerate the accumulation process at the low point. In the case of short-term flow, even if the flow rate is high, all the contaminants and particulates will not clear the low point section before the flow is stopped. Since they are on a slope, they will settle to the bottom of the pipe and begin to flow backward until they accumulate again at the low point. A biofouling film could occur after a sufficient length of time, especially under laminar flow conditions. This film could also travel downstream due to the ability of micro-organisms to grow into colonies, which create water as a nutrient for their metabolic processes. The characteristics and abilities of microorganisms are discussed in Chap. 12. Under these conditions, the contamination can flow and be trapped in a basket filter, as shown in Fig. 18.

Findings from the filter housing and from the basket are areas of great interest regarding technical investigations.

9.2 *Cyclone Separators*

A cyclone separator is usually positioned before the filtration system for long fuel lines in locations where large solids, such as stones, gravel, etc., can occur. The separation of contaminants and particles is achieved by applying centrifugal force created by the partially conical construction of the separator. The fluid enters tangentially to the circumference of the structure at its upper level, which is cylindrical, and

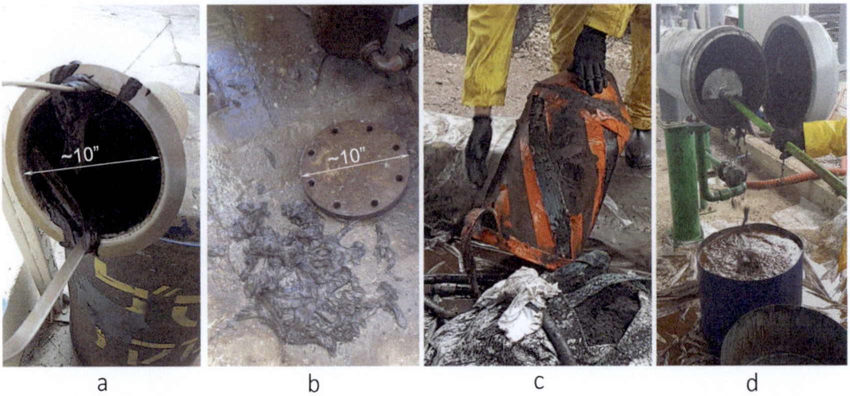

Fig. 18 Contaminants in a basket filter—**a** immediately after opening the filter; **b** basket filter lid and the contamination; **c–d** the residue from cleaning the line using cleaning scrapers (pig) (Figure Courtesy of M. Rabaev)

exits axially from the top of the separator. The fuel's inertial force, tangential to the cylinder's circumference, creates a cyclone with the fuel traveling in a spiral until it reaches the bottom. Due to the centrifugal force and higher pressure of the flow at the walls of the separator, once the fuel reaches the bottom, it is forced upwards axially, leaving behind any heavy particles or contaminants to accumulate at the bottom. A cyclone separator is sampled by way of a drainage valve and may contain solids of varying sizes.

If, through unusual circumstances, mud enters the system, it adheres to the walls of the cyclone separator. After cleaning the separator from foreign material, the material that adhered to the walls may slowly detach from the walls and contaminate the system, damaging gauges, valves, and other attachments to the fuel line. Understanding the status of the cyclone separator is achievable by examining the contaminants sampled from the drainage of the separator. Sticky sludge at the outlet can indicate that there is sludge on the walls of the cyclone separator as well. Sometimes, an experienced technician can determine if there is sludge on the walls of the cyclone separator just by knocking on the housing and recognizing an unusual sound.

Based on the author's field experience, examining the composition of the solids and liquids drained from the cyclone separator is highly recommended. This inspection involves an initial drain into a clear container, which is expected to be neither clear nor clean. As with all separator draining processes, clear and clean samples upon preliminary draining can indicate a problem.

Draining of a large amount of liquid should be done into a large funnel covered by a metal sieve, usually 10–20 mesh. This is done to prevent clumps of contaminants from entering the drainage pipe.

9.3 Aviation Fuel Filters

Filters which form part of the aviation fuel distribution system (e.g., microfilters, filter water separators, clay treaters) are reviewed in EI 1550 and Chap. 10. It is important to note that these filters are discussed in the context of their role within the system. A comprehensive knowledge of the working principles and appropriate selection is required. Installing and changing filters in such industrial applications creates unique challenges requiring skill and knowledge. Despite this, it can be simply stated that the tools needed for fuel sampling are standard and compliant with the tests that may be required to be performed:

1. Clear and Bright test—a clean and transparent container at least 3″ or ~75 mm in diameter.
2. Chemical inspection, as well as particulate testing—a standard specified container.
3. Microbiology testing—a set of disinfected tools, as specified in Sect. 10.

These tests indicate the quality of the fuel, including cleanliness, particles, and microbial contaminants. However, they do not necessarily reflect the condition of the filtration systems. Understanding that passing all tests is possible even without filter elements in place is crucial.

Clay treaters adsorb polar compounds, which include dissolved contaminants, surfactants, pigments, and other polar molecules such as quinones. Clay treaters should not be used if polar fuel additives have been used, such as static dissipator additives, as these may be removed by the clay or suffer a reduction in active concentration. If clay is used, the fuel exiting the treater should be tested to re-determine the additive concentration.

Sometimes, it is necessary to examine the materials (fluids and particles) that have been absorbed by or trapped in filter media. Moreover, there are cases in where it is desirable to examine the materials that were absorbed in the contaminants caught by the filter. Depending on the required testing, it can be performed directly by XRF, FTIR, or other techniques. In some cases, extracting the sample from the filter medium might be necessary. When performing such extractions, use solvents that do not react with the chemical composition of the fuel filter or the contaminants. Commonly, the fuel works well as the extractant.

Chapter 14 discusses analytical chemistry methods for examining samples derived from non-routine situations. Again, it must be emphasized that finding a dirty filter shows that the filter has performed its function. Occasionally, it is necessary to cut the filter open to stretch out the fabric or net for visual, microscopical, and chemical inspection (Fig. 19).

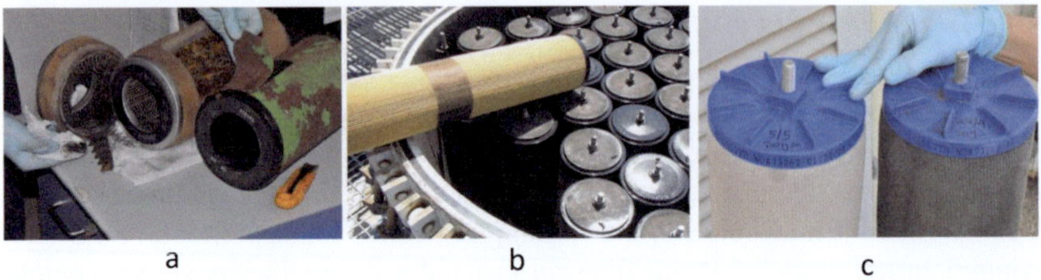

Fig. 19 Catastrophic contamination in fuel filters—**a** failure of a coalescer following the tearing of a PTFE separator due to the contamination in the form of sludge, that originated from the bottom of the tank and was forced through the PTFE separator because of the high flow rate; **b** identification of contamination by visual inspection of filters in a refinery supply point; the new cartridge is for comparison; **c** filters from the downstream section of an airport fuel farm, left used for 24 months under normal conditions; right, after two weeks receiving fuel from the contaminated path, as described on (**b**) (Figure Courtesy of M. Rabaev)

9.4 Aircraft Filters

In those cases where it is necessary to take a fuel sample from an aircraft (other than clear and bright tests), this is typically part of a technical investigation or related to an aircraft incident or accident. Sampling an aircraft's fuel filter requires technical know-how (see IATA's standard on filter investigations [25] and Boeing's article [26]), as well as authorization from the proper organizations. If the filter being inspected has a filter housing, the characteristics of the filter are very similar to those described above. On the other hand, if the filter being inspected is a line filter, disassembly will involve the loss of some of the fluid during the opening of the connections, and therefore, proper precautions must be taken. Suitable precautions include using a standard, clean, wide, and dry funnel suitable for the sampling container and a rag or fabric under the filter to absorb any spilled fluid. The filter will usually be shaped like a corrugated cylinder or cone. These filters have a large surface area and thus can allow for fuel flow even if there are large blockages. Generally, removing the fluids and particles from these filters is an arduous process that requires opening the filter and spreading out the corrugated fabric (see Fig. 20).

Good practice is to handle/store filters between removal and examination, putting them in plastic bags or similar stopping them from drying out.

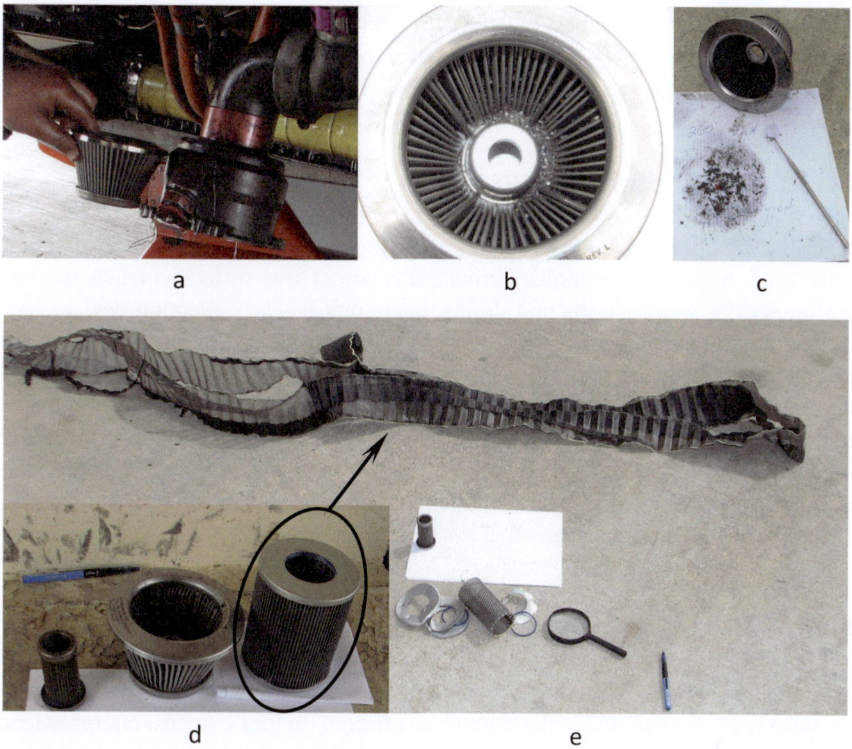

Fig. 20 Disassembled aircraft fuel filter for investigation of the possibility of anomalous particle contamination—**a** examined aircraft filter beside its housing**; b** close up of the inside of the filter, particles are shown; **c** solid particles removed from the filter onto A4 white paper; **d** another filter from the same aircraft; **e** filter that has been cut to uncover the whole surface area, to search for sticky or other contaminant types (Figure Courtesy of M. Rabaev)

10 Sampling for Microbiological Testing

10.1 Sampling Issues Unique to Microbiological Parameters

Microbes are present in aviation fuels and fuel systems as contaminants. The potential hazards associated with uncontrolled microbial contamination in aviation fuel systems are described in Chap. 12. This section focuses on sample collection and handling. Three characteristics of microbial contamination create the need to take special precautions when collecting and handling samples intended for microbiological testing:

- Cross-contamination risk
- Heterogeneity of microbial contamination
- Perishability of samples.

For a detailed guide on best practices for microbial sample collection, see ASTM Standard Practice D7464—"Standard Practice for Manual Sampling of Liquid Fuels, Associated Materials and Fuel System Components for Microbiological Testing" [5].

10.1.1 Cross-Contamination Risk

Microbes are ubiquitous, present in the air as bio-aerosols and on surfaces that may come into contact with samples during collection. These surfaces include, but are not limited to, sample ports and drains, operator's hands, collection devices, and collection containers. To minimize contamination risk, it is important to observe several fundamental industrial hygiene precautions: wearing surgical gloves that must be changed after each sample is collected; using sterile and chemically clean sample containers, preferably not previously used; disinfection of all exposed surfaces (see Sects. 10.4 and 10.5); and minimizing the time during which the sample container is open. Failure to follow these basic guidelines can result in microbial contamination of samples and lead to false positive test results.

10.1.2 Microbial Contamination Heterogeneous Distribution

Most fuel properties are relatively homogeneous—the results will be approximately the same regardless of where the sample is collected. In fuel systems, microbes tend to accumulate in zones that contain free water and on surfaces of the system.

Microbes can be suspended in fuel but are likely to be dormant wherever water activity (a_w) is less than 0.9 (the ratio of the partial vapor-pressure of water in a solution, to the partial vapor-pressure of pure water at the same temperature; see Sect. 1.1.2 in Chap. 12—"Aviation Fuel and Fuel System Microbiology." In aircraft wing tanks, small volumes (from only a few mL to more than a liter) of free water will likely accumulate outboard of each rib (Sect. 7). Consequently, drain samples can be water-free even when the tank's total volume of free water is >1L.

Just as water can accumulate in discrete zones outboard of each rib, biomass (i.e., slime deposits) tends to accumulate irregularly on wing tank vertical surfaces (Fig. 21). Microbiological test methods (Sect. 2.7, Chap. 12) can only detect microbes if they are present in collected samples. The heterogeneous distribution of microbial contamination creates a substantial risk that samples from systems with moderate to heavy contamination might contain negligible bioburdens.

10.1.3 Microbe Parameter Perishability

Microbes are living organisms. Consequently, regardless of whether they are in the system from which samples have been collected or in the sample container, they continue to reproduce, proliferate, metabolize available nutrients, and die. This dynamic means that the correlation between what is detected in a sample and what is

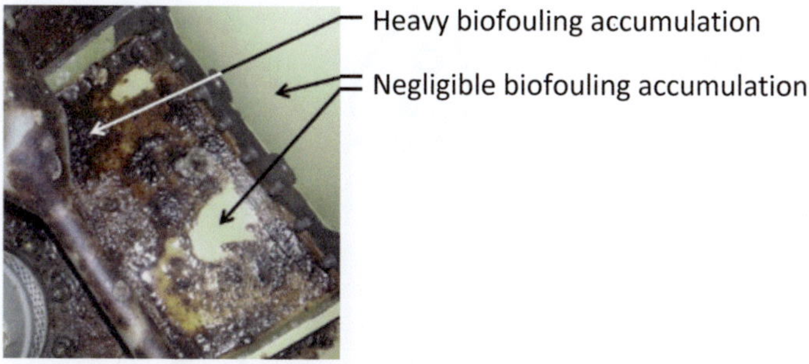

Fig. 21 Aircraft wing section shows the irregular distribution of biomass on the tank surface. Note that all the slime has accumulated on the tank bottom. This is like heterogeneity issues in bulk storage tanks. See Chap. 12 for a more detailed discussion of the distribution of microbial contamination in aviation fuel systems (Courtesy F. Passman)

present in the original system diminishes over time, illustrating the concept of perishability. It is unpredictable whether the microbial population in a sample container will increase, decrease, or stay the same (see Fig. 12, Chap. 12).

Moreover, even if the bioburden seems stable, it is likely that the relative abundances of the different organisms present are likely to shift. Ideally, microbiological testing should commence within four hours after sample collection. When this is impossible, samples kept refrigerated or on ice may still be viable for testing up to 24 h later. Although older samples can be tested, the risk of obtaining erroneous results increases with sample age.

10.2 Sample Collection Devices

The devices used for collecting samples for microbiological testing are the same as those used for conventional sample collection. The various types of sampling devices are reviewed in Sects. 4, 7 and 8. The preferred device for collecting samples from bulk storage tank bottoms is the Bacon Bomb type sampler (Fig. 22). This device makes it possible to capture sludge and sediment that has accumulated on the tank floor, including water and fuel, from the very base of the tank.

Figure 23 shows a bulk storage tank bottom sample that was collected using a Bacon Bomb sampler.

The appropriate sampler should be used to sample different fuel levels. There are many models designed for this purpose; one of the options has a ball on the bottom of the sampler, as shown in Fig. 5.

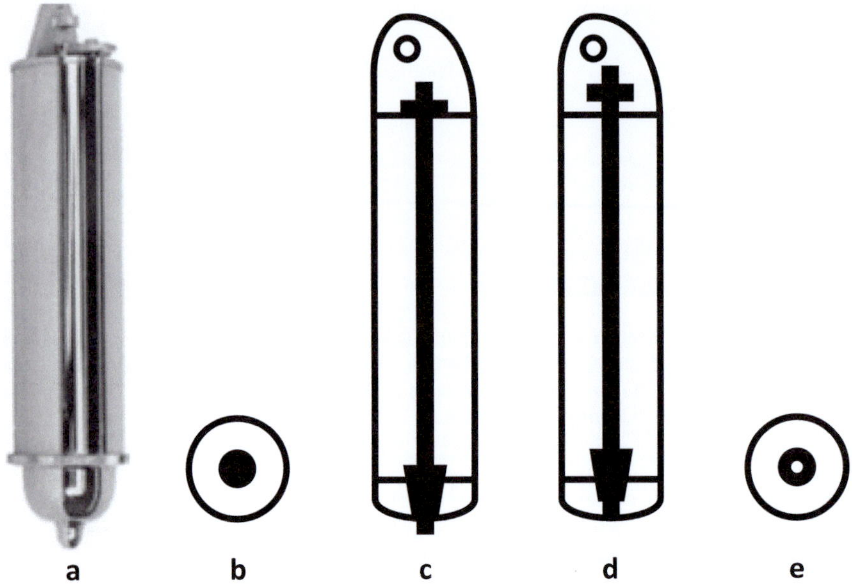

Fig. 22 Bacon bomb sampler—**a** assembled sampler; **b** top view, piston closed; **c** side view, piston closed; **d** side view, piston open; **e** bottom view, piston open. When the bottom of the piston contacts the tank bottom it lifts, allowing fluid, sludge, and sediment to enter. As the sample is lifted off the bottom, the piston closes

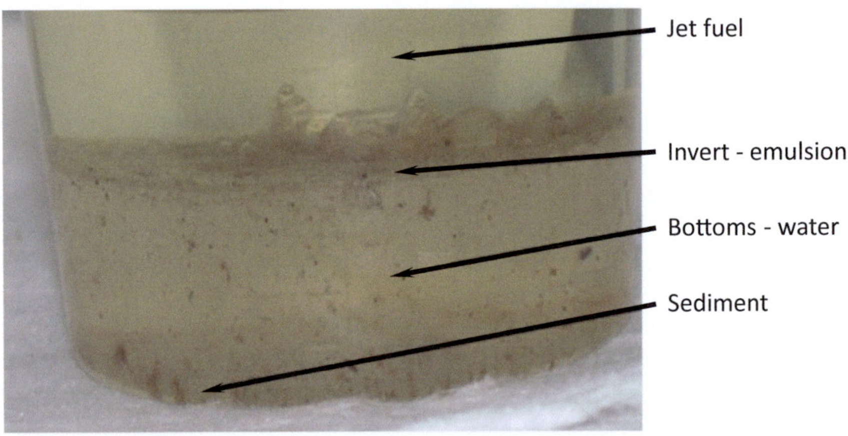

Fig. 23 Bulk fuel tank bottom sample collected using Bacon Bomb sampling device. Note the presence of four layers: fuel, invert emulsion (<2 mm thick), bottoms-water, and sediment (traces) (Figure courtesy of F. Passman)

10.3 Preparing Sampling Devices for Sample Collection

In microbiology testing laboratories, heat is the primary means to disinfect contact surfaces such as inoculating loops and the mouths of test tubes, sample bottles, etc. However, this method is not an option when operating near or when handling combustible and flammable materials like aviation fuels. Nonetheless, it is essential to disinfect sampler contact surfaces before each use. The first step is to remove bulk contaminants—typically residual sludge, slime, and particulate material that has adhered to the surfaces of the sampler from the previous sampling.

Depending on the type of sampling device used, it is often appropriate to first disassemble the device to make it easier to reach all internal and external surfaces. Next, a lint-free cloth or disposable wipe is saturated with alcohol ($\geq$70% alcohol) and used to wipe down all surfaces. All surfaces must be inspected to ensure that all residual material from previous samples has been removed. When sampling devices are heavily fouled, it may be necessary to repeat this process several times. It is crucial to allow sufficient time for the alcohol to evaporate before reassembling the device. After reassembling the device, a final disinfection rinse should be carried out by filling the device approximately 25–30% with alcohol, shaking it vigorously for 30 s, decanting the alcohol, and allowing all residual alcohol to evaporate before collecting a sample. Once disinfected, the sampling device should be handled carefully and used within 5 min to minimize the risk of re-contamination.

10.4 Sampling Best Practice

10.4.1 Aircraft Wing Tank Samples

Wing tank samples are discussed in Sect. 7. Before collecting samples for microbiological testing, it is crucial to clean the area around the sample port. First, the surrounding area must be wiped to eliminate any dirt that might contaminate the sample. Prior to sample collection, the surface must be wiped with an alcohol-saturated cloth (see Sect. 10.3), and the alcohol must be allowed to evaporate fully. The first flow of the sample, which typically includes water and sludge/sediment as well as fuel, should then be collected. If there is a sufficient volume of water to fill the sample container, a second sample should be collected so that fuel-associated water and fuel are available for testing.

10.4.2 Filter Housing, Fuel Cart, Hydrant, and Tanker Truck Samples

Microbial contamination can enter aviation fuel at numerous points between refinery tanks and aircraft. Consequently, the best practice is to periodically sample from each point (Figs. 24 and 25). The general guidance for microbial contamination sampling is the same for all components with drain valves.

The procedure for sampling from drain valves is similar to the one described in Sect. 10.4.1. The external surface of the drain line must be wiped down to remove dirt and debris. An alcohol-saturated cloth should be used to wipe down and disinfect all external and near-outlet internal surfaces. For wing tank drain samples, contrary to best practice for collecting samples intended for fuel specification property testing, it is important to capture first flow samples. If first-flow fills the sample container with water, further samples should be collected until more than 50% of the sample is fuel.

An alternative procedure for collecting samples from filter housings is to use a peristaltic pump to draw samples from the bottom of the housing. Alcohol-rinsed scissors or knife blades should be used to cut a section of unused tubing of a length sufficient to reach from the bottom of the filter housing, pass through the head of

Fig. 24 Filter housing sampling—**a** Jet-A filter housing; **b** filter housing drain; **c** tubing inserted into filter housing for sample collection by a peristaltic pump (Figure courtesy of F. Passman)

Fig. 25 Drain line sampling—**a** hydrant cart drain lines; **b** tank truck drain line. Note that the sample flows through a disinfected funnel from the drain outlet to the sample bottle (Figure courtesy of F. Passman)

(a) (b)

Fig. 26 Bulk tank water drain—**a** drain outlet <0.5 m from tank shell; **b** drain outlet >2 m from tank shell (Figure courtesy of F. Passman)

the peristaltic pump, and into a sample container. The pump can then be operated to transfer the sample from the filter housing to the sample container.

10.4.3 Above Ground Bulk Storage Tank Samples

When sampling from bulk storage tanks, the best practice is to collect both first flow, and post-flush samples from the tank's water drains (Fig. 26).

The first flow captures microbial contamination that has developed within the drain line and is likely to be diagnostic of contamination present in the tank. However, environmental conditions such as diurnal heating and cooling affecting the microbes present in the drain lines of a bulk tank can be substantially different from those within the tank. The likelihood of drain-line bioburdens being different from tank-bottom bioburdens increases as the distance between tank shell penetration and drain-line outlet increases. Consequently, the importance of drawing both pre- and post-flush samples increases with the length of the drain line. The volume flushed should be equal to the volume of the drain line. This is computed as the volume of a cylinder.

10.4.4 Underground Storage Tank Samples

Bottom samples are optimally collected using the Bacon Bomb sampler described in Sect. 4 and 10.4, only if the sample port is directly above the low point. The procedures for collecting microbiological testing samples differ from other samples only in the precautions taken to disinfect the sampler and the funnel used for transferring the sample from the Bacon Bomb to the sample container. To properly disinfect a Bacon Bomb sampler, the device should be disassembled. Any residual material on internal surfaces should initially be rinsed in clean fuel of the same grade to be sampled, or alternatively, alcohol-saturated cloths can be used to clean off all internal surfaces.

Residues are more commonly found around the base of the plunger and the bottom cup. After cleaning and reassembling the device, it should be immersed in alcohol and then fully drained. If this cannot be accomplished, using clean fuel of the same grade to be sampled is generally sufficient. A sample of the rinse fuel should be retained for microbiological testing as a basis for comparison with the sampled fuel.

10.5 Sample Container Selection

Most commonly, samples for microbiological testing are dispensed into either 500 mL or 1 L clear glass Boston round bottles 3″ in diameter/75 mm or greater. Some tests, such as ASTM Test Method D7463 [23], specify using HDPE bottles. Previously unused bottles were preferable, but chemically clean, sterile, and used bottles are acceptable. Clear glass allows for visual inspection in accordance with ASTM Test Method D4176 [22] and makes it easier to draw layer-specific specimens from samples with more than one layer (see Fig. 23).

11 Chain of Custody Documentation

Standard chain of custody (COC) documentation is essential for samples that are to be tested by a different person than the one who collected the sample, at a different location, or both. A typical COC form includes:

1. Identity and contact information of personnel collecting the sample.
2. Tank/vehicle/unit identification.
3. Date and time of sample collection.
4. Sample type (i.e., water drain, Bacon Bomb).
5. For drain samples, whether pre- or post-flush.
6. For Bacon Bomb samples, whether the sample is from a single Bacon Bomb drop or a composite of two or more drops. If the latter, the number of drops should be noted.
7. Disposition of sample from the time of collection to arrival at the laboratory (e.g., maintained on ice, stored in a temperature-controlled cooler at 5 °C (±2 °C), etc.).
8. Date and time of receipt at the laboratory.
9. Identity and contact information of the person receiving the sample.
10. Disposition of the sample from the time of arrival at the laboratory to the time testing was initiated.
11. Identity(ies) and contact information of the personnel performing test(s).
12. Date and time testing initiated.
13. Date and time testing completed.
14. Disposition of samples after testing.

Tamper-proof plastic or lead seals are typically used to ensure sample integrity. These also serve to demonstrate the importance and value of samples, which can represent large cargos of fuel or critical root-cause investigations.

12 Conclusion

Ensuring aviation fuel quality through effective sampling is vital for maintaining the safety and reliability of air travel. The diverse challenges associated with sampling from different parts of the fuel supply chain—from refinery to aircraft fuel systems—require a deep understanding of the operational environment and the specific standards governing sampling procedures. This chapter outlines the necessary tools, techniques, and considerations that must be observed to collect accurate and representative samples, whether for routine quality checks or more specialized diagnostic testing. Proper sample handling, container selection, and adherence to established protocols are essential to preventing contamination and ensuring the integrity of the collected samples. Additionally, the need for meticulous documentation, especially in forensic investigations or when handling microbiological samples, cannot be overstated. By following the guidelines and best practices detailed in this chapter, professionals can better navigate the complexities of fuel sampling, ultimately contributing to aviation's overall safety and efficiency in operations.

Acknowledgements We would also like to thank Dor Granek and Dr. Roland Mureinik for their technical and linguistic consultation, Nisim Vasilevsky, Dr. Udi Akiva, Inna Kaprovsky, and Dr. Shimon Pisnoy for their beta readers' contribution.

To Joel Schmitigal, Jim Gammon, and Dr. Alisdair Clark for the chapter's deep and professional scientific and editorial review.

References

1. ASTM D4057. Practice for manual sampling of petroleum and petroleum products. ASTM International, West Conshohocken, PA. https://doi.org/10.1520/D4057
2. Rand SJ, Verstuyft AW (2018) Significance of tests for petroleum products. MNL1-9TH-EB. ASTM International. https://doi.org/10.1520/MNL1-9TH-EB
3. API MPMS CH 8.1 (2022) Manual sampling of petroleum and petroleum products (ASTM D4057), 6th edn. American Petroleum Institute
4. IP 475: Petroleum liquids—manual sampling (ISO 3170: 2004), REF/ISBN: IP475-2934869. Energy Institute, 61 New Cavendish Street, London, United Kingdom, W1G 7AR
5. ASTM D7464. Standard practice for manual sampling of liquid fuels, associated materials and fuel system components for microbiological testing. ASTM International, West Conshohocken, PA. https://doi.org/10.1520/D7464
6. Hill G (2003) Chapter 2-Sampling methods for detecting microbial contamination in fuel tanks and systems. In: Passman F (ed.) Fuel and fuel system microbiology: fundamentals, diagnosis, and contamination control, pp 14–23. ASTM International, West Conshohocken, PA. https://doi.org/10.1520/MNL10444M

7. ASTM D6469. Standard guide for microbial contamination in fuels and fuel systems. ASTM International, West Conshohocken, PA. https://doi.org/10.1520/D6469
8. ASTM D2276. Test method for particulate contaminants in aviation fuel by line sampling. ASTM International, West Conshohocken, PA. https://doi.org/10.1520/D2276
9. ASTM D4306. Standard practice for aviation fuel sample containers for tests affected by trace contamination. ASTM International, West Conshohocken, PA. https://doi.org/10.1520/D4306
10. ASTM D5854. Standard practice for mixing and handling of liquid samples of petroleum and petroleum products. ASTM International, West Conshohocken, PA. https://doi.org/10.1520/D5854
11. ASTM D3240. Standard test method for undissolved water in aviation turbine fuels. ASTM International, West Conshohocken, PA. https://doi.org/10.1520/D53240
12. ASTM D4177. Standard practice for automatic sampling of petroleum and petroleum products. ASTM International, West Conshohocken, PA. https://doi.org/10.1520/D4177
13. ASTM 5842. Standard practice for sampling and handling of fuels for volatility measurement. ASTM International, West Conshohocken, PA. https://doi.org/10.1520/D5842
14. ASTM D8009. Standard practice for manual piston cylinder sampling for volatile crude oils, condensates, and liquid petroleum products. ASTM International, West Conshohocken, PA. https://doi.org/10.1520/D8009
15. JIG 1. Aviation fuel quality control & operating standards for into-plane fuelling services. Joint Inspection Group (JIG). http://www.jigonline.com
16. JIG 2. Aviation fuel quality control & operating standards for airport depots & hydrants. Joint Inspection Group (JIG). http://www.jigonline.com
17. JIG 4. Aviation fuel quality control and operating standards for smaller airports. Joint Inspection Group (JIG). http://www.jigonline.com
18. EI/JIG 1530. Quality assurance requirements for the manufacture, storage and distribution of aviation fuels to Airports. http://www.jigonline.com; https://publishing.energyinst.org/featured/1530
19. The GamGrams. Gammon Technical Products Incorporation Inc. https://gammontech.com/education/gamgrams/
20. IATA aviation turbine fuels specifications guidance material. https://www.iata.org/en/publications/manuals/tech-fuel-servicing-guidelines/
21. ATA-103. Standard for jet fuel quality control at airports. Air Transport Association of America, Inc. (ATA) d/b/a Airlines for America, 1275 Pennsylvania Ave. NW, Suite 1300, Washington, D.C. 20004. http://www.airlines.org
22. ASTM D4176. Test method for free water and particulate contamination in distillate fuels (visual inspection procedures). ASTM International, West Conshohocken, PA. https://doi.org/10.1520/D4176
23. ASTM D7463. Standard test method for adenosine triphosphate (ATP) content of microorganisms in fuel, fuel/water mixtures, and fuel associated water. ASTM International, West Conshohocken, PA. https://doi.org/10.1520/D7463
24. ASTM D7687. Standard test method for measurement of cellular adenosine triphosphate in fuel and fuel-associated water with sample concentration by filtration. ASTM International, West Conshohocken, PA. https://doi.org/10.1520/D7687
25. International Air Transport Association (2022) Guidance material investigation and categorization of engine fuel filter blockages, 2nd edn
26. Jones M (2008) Engine fuel filter contamination. In: Boeing aero magazine. AERO_Q308_article3, 08 May 2022

Chapter 12
Aviation Fuel and Fuel System Microbiology

Frederick J. Passman, Graham Hill, and Gareth Williams

Abstract Uncontrolled microbial contamination can cause or contribute to a number of problems in aviation fuels and fuel systems. The most common problems include fuel system biodeterioration, biofouling, and degradation of fuel properties. This chapter provides an overview of aviation fuel and fuel system microbiology, its detection, and its control.

1 Introduction

1.1 Microbiology Basics

This section will provide a fundamental overview of microbiological concepts that will enable readers to be comfortable with this chapter's material. Readers interested in a deeper understanding of fuel microbiology are encouraged to consult ASTM D6469 [1], Passman [2], International Aviation Transportation Associations' (IATA's) Guidance Material on Microbial Contamination in Aircraft Fuel Tanks, 6th Ed. [3] and the Energy Institute's (EI's) Guidelines for the investigation of the microbial content of liquid fuels and for the implementation of avoidance and remedial strategies [4]. Madigan et al. [5] provide an extensive treatment of microbial ecology.

Fuel microbiomes—the types and relative abundances of microorganisms present—can differ substantially among samples, based on fuel chemistry and environmental conditions at facilities from which samples are collected.

F. J. Passman
Biodeterioration Control Associates, Inc, Princeton, NJ, United States
e-mail: fredp@biodeterioration-control.com

G. Hill · G. Williams (✉)
ECHA Microbiology Limited, Cardiff, UK
e-mail: gwilliams@echamicrobiology.com

G. Hill
e-mail: ghill@echamicrobiology.com

M. Rabaev et al. (eds.), *Aviation Fuel Problems and Solutions: From Refinery to Wingtip*, https://doi.org/10.1007/978-3-032-05514-9_12

1.1.1 Microbes

Microbes are single-celled organisms, including representatives for each of the three biological kingdoms: Bacteria, Archaea, and Eukarya. Individually, microbial cells are too small to be seen by the naked eye. However, microbes tend to form macroscopic masses comprised of billions of cells and extracellular material as will be explained below. Although it is suspected that there are millions of bacterial species, only a hundred or so have been recovered from jet fuels or fuel systems. Archaea species were first detected in the 1970s. To date, only a few Archaea species have been identified and few have been recovered from fuels. The Kingdom Eukarya includes all organisms that have true cell walls. In our discussion of aviation fuel microbiology, the only relevant eukaryotes are fungi. In particular, among the many types of fungi that exist, only molds and yeasts infect aviation fuel. Molds are filamentous and are characterized by dense vegetative and spore-bearing (it is technically more accurate to refer to these structures as *conidia-bearing*), *hyphae* (also referred to as *mycelia* or *filaments*). When suspended in fluids, molds tend to form spherical colonies (Fig. 1a). At fuel-water interfaces, molds tend to form cohesive, irregularly shaped mycelial masses (Fig. 1b). When viewed under a microscope, mold filaments and spores are readily visible (Fig. 1c). Yeasts are most commonly present as single-cells (Fig. 1d). However, some species of yeast can also exist in filamentous form, depending on their physiological state.

Although numerous lists of the types (*taxa*) of microbes that have been detected in aviation fuel systems have been published, the organism names are of little importance to fuel system operators, as the corrective actions (Sect. 4) are the same, regardless of the types of microbes detected. Moreover, the ease with which microbes

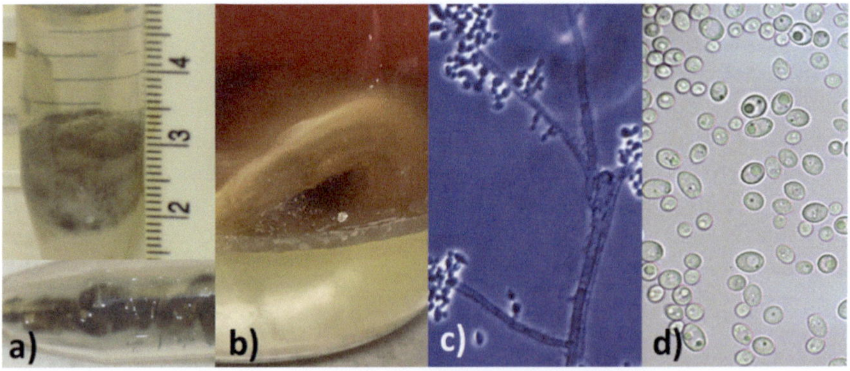

Fig. 1 **a** Mold growth as spherical colonies at fuel-water interface and in bottoms-water; **b** fungal mycelial mat formed at fuel-water interface; **c** mold filaments with spores; **b** and **d** yeast cells. Figure 1a and b images courtesy of Biodeterioration Control Associates, Inc., Princeton, NJ, USA. Figure 1c image courtesy of Conidia Bioscience Ltd, Egham, UK. Figure 1d image courtesy of ECHA Microbiology Ltd, Cardiff, UK

can exchange genetic material among different taxa substantially complicates traditional taxonomic identifications that are based primarily on metabolic and physiological characteristics (i.e., organic molecules microbes can use as nutrients, oxygen requirements, etc.).

1.1.2 Microbial Physiology

Physiology addresses the ways in which organisms function. The most relevant aspects of microbial physiology in fuel systems include proliferation, oxygen requirement, nutrient requirements, and metabolite production.

Water is essential for metabolic activity. Specifically, water activity (a_w)—the ratio of the partial vapor pressure of water in a solution to the partial vapor pressure of pure water at the same temperature. Most known bacteria require an $a_w \geq 0.91$, and fungi require $a_w \geq 0.70$. In jet fuel, a_w can range from 0.4 to 1.0. Note that in fuel, water can be present in dissolved or free (liquid) states; free water may be dispersed as droplets in the fuel or as a discrete phase settled below the fuel or as condensate films on surfaces. When free water is present in a fuel tank or system, it can support microbiological activity. Microbes can survive in an inactive (*dormant*) state indefinitely in fuel that has no free water present.

Proliferation refers to cell reproduction, resulting in the increase in population density (Fig. 2). The time required for a cell to divide to form two new (daughter) cells is a *generation*. Bacterial and fungal generation times are typically measured in minutes or hours. Under optimal growth conditions, some bacterial species have generation times in the 15–30 min range. This means that theoretically, a single cell can proliferate into a population of one billion cells in less than 15 h. This rarely, if ever, happens in nature, but is the basis for culture testing (Sect. 2.7.3). Microbes with generation times greater than four hours are generally considered to be slow growers. It can take these organisms more than five days to reach one billion cells. One billion cells are the approximate number needed to form a visible colony (see Sect. 2.7) in or on nutrient media. The units used to report population densities in fluids are X mL^{-1}, where X is the parameter measured (for example, colony forming units—CFU, the number of gene copies—GC, the concentration of an indicator molecule such as adenosine triphosphate—ATP, protein, or antigen). The parameters that are most commonly used to report population densities will be discussed in Sect. 2.7. Surface sample population densities are reported as either X cm^{-2} or X g^{-1} depending on whether the specimen is from a swabbed or scraped area.

For a given species, the proliferation (growth) rate depends on environmental conditions. The primary environmental factors affecting growth rates include nutrient availability, temperature, pH, salinity, and water.

Temperature affects organisms' growth rates. As illustrated in Fig. 3, an organism's optimal growth temperature, there is a logarithmic relationship between temperature and growth rate until the maximum growth rate is achieved. Growth rates typically plummet rapidly once the temperature exceeds the optimum, as cells' enzymes and

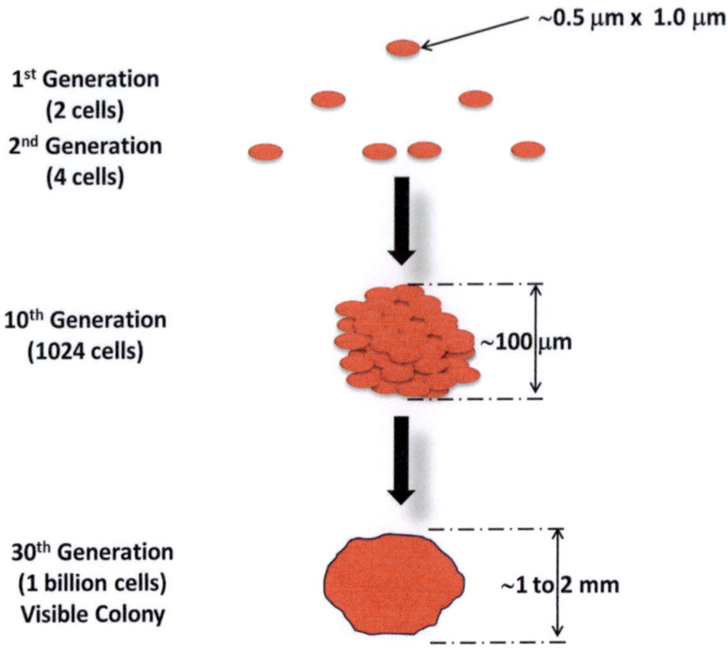

Fig. 2 Bacterial proliferation—increase in the number of cells over time

other biomolecules begin to degrade (i.e., *denature*). Depending on the species, a microbe's optimal growth temperature can range from less than 15 °C to greater than 90 °C. This means that microbes are able to proliferate in fuel systems anywhere on earth. *Obligate psychrophiles* do not proliferate at temperatures greater than 20 °C. *Obligate thermophiles* typically do not proliferate at temperatures less than 45 °C. *Mesophiles*—the microbes most commonly recovered from aviation fuel systems—grow optimally in the 20–40 °C temperature range.

Acidity is measured by *pH*, which is the negative log of the hydrogen ion (H^+) concentration in an aqueous solution. As with temperature, the optimum pH for growth depends on the microbes present. Some microbes—acidophiles—thrive in strongly acidic environments (pH < 2). Others—*neutrophiles*—grow when the pH is between 5 and 8. Alkalinophiles require pH > 8 and grow optimally at pH > 9. In aviation fuel systems, the most commonly detected microbes are neutrophiles. Acidophiles are also common fuel system contaminants, including, for example, fungi which proliferate at acidity below pH 5. As many microbes produce organic acids when they metabolize organic material, they may increase the acidity of their environment, leading to conditions which favor more acidophilic microbes.

Salinity is the salt concentration in a fluid. It can range from < 0.001 mg kg^{-1} (i.e., below the detection limits of available instrumentation) to 260,000 mg kg^{-1} (the saturation limit of salt in water). As with other parameters, some microbes thrive in

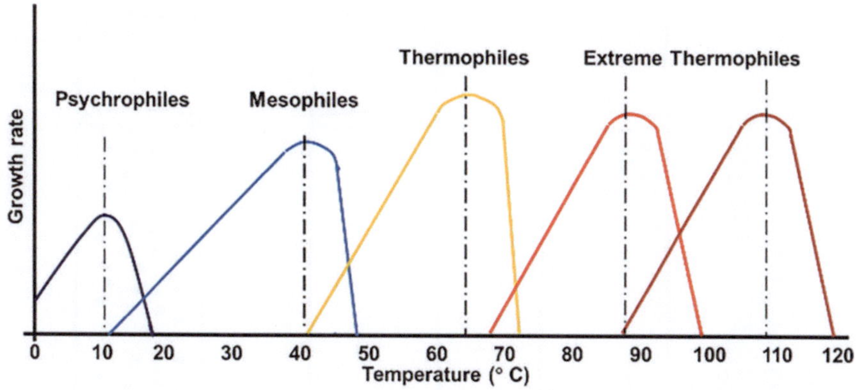

Fig. 3 Relationship between growth rate and temperature. Figure courtesy of the Energy Institute [6]

deionized water, others thrive in brine. Those most commonly recovered from fuel systems prefer salinities in the 100–10,000 mg kg^{-1} range.

Oxygen requirements vary for different microbes. Some microbes require oxygen and others cannot tolerate exposure to oxygen (Fig. 4). Those that require oxygen are called *obligate aerobes*. Those that cannot tolerate oxygen exposure are called *obligate anaerobes*. The oxygen requirement is related to cells' energy metabolism mechanisms. Some obligate anaerobes rely on organic molecules to drive energy metabolism. These organisms are *organotrophic*. Others— *lithotrophic* microbes— use nitrate, nitrite, or sulfate. *Facultative anaerobes* use oxygen when it is available and switch to fermentation to metabolize organic molecules for energy as their environment becomes oxygen depleted.

Metabolism is the umbrella term used for all the biochemical reactions that occur to maintain life within an organism. Metabolism includes *catabolism*—those reactions that breakdown complex organic molecules ingested as food—and *anabolism*— the synthesis reactions that produce new cell material. Catabolic reactions generate energy. Although much of this energy is lost as heat, some is captured through the process of energy metabolism. Energy metabolism stores energy as ATP through a series of cascading reactions in which electrons are transferred from one molecule to another until they are captured by a *terminal electron acceptor*. For the organism types discussed in the previous paragraph, oxygen is the terminal electron acceptor for aerobes; inorganic molecules such as nitrate, nitrite, or sulfate are the terminal electron acceptors for lithotrophic anaerobes; and organic molecules function as terminal electron acceptors for fermenters. The processes of reducing complex organic molecules to carbon dioxide (CO_2) or methane (CH_4) are called *mineralization* and *methanogenesis*, respectively.

Metabolites are molecules produced during metabolism. Some metabolites are reactive intermediates that are used to create new cellular material (enzymes, structural

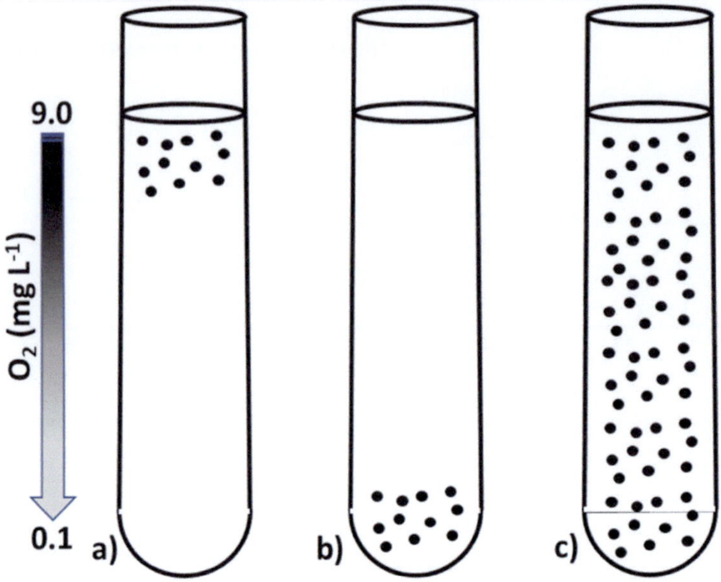

Fig. 4 Microbial oxygen requirements reflected as growth in agar deeps—**a** obligate aerobes grow only near surface where oxygen is plentiful; **b** obligate anaerobes grow only near bottom where there is no oxygen present; **c** facultative anaerobes grow throughout agar, regardless of the amount of oxygen present. Figure courtesy of the Energy Institute [6]

proteins, lipids, genetic material, etc.). Others are excreted as either wastes (for example, one to six carbon, low molecular weight organic acids—LMOA) or as functional molecules (for example, extracellular polymeric substance—EPS—the slimy matrix from which biofilms can be formed). All microbes produce organic acids during catabolism. Microbes such as certain *Acetobacter* species (spp.), *Acinetobacter* spp., *Lactobacillus* spp. and *Pseudomonas* spp. that produce substantially more acid than others (and which have been demonstrated to contribute to microbiologically influenced corrosion—MIC) are classified as *acid producing bacteria* (APB). Other relevant functional groupings implicated in MIC include *iron-related bacteria* (IRB) that either reduce or oxidize iron, nitrifying bacteria (bacteria that oxidize either ammonia—NH_3^+—or nitrite—NO_2^-—during energy metabolism), denitrifying bacteria (microbes that reduce nitrate—NO_3^-—to nitrogen gas—N_2), and sulfate reducing bacteria (SRB). Microbes that are capable of producing prodigious volumes of EPS are classified as *slime forming bacteria*. Note that these are strictly functional classifications. Microbes that are not genetically related can be members of one or more of these groups.

Nutrients are typically classified as either macronutrients or micronutrients (also referred to as trace nutrients). Macronutrients are carbon-based substances that microbes use as building blocks for creating new cell mass, for energy metabolism, or both. *Autotrophs* can use carbon dioxide or carbon monoxide as their primary or exclusive food source. In contrast, *heterotrophs* use organic molecules, and it is this group which is generally the predominant cause of problems due to microbial growth in fuel systems. In fuel systems, the concentrations of one or more of the macronutrients nitrogen, phosphorous, or sulfur can limit the ability of contaminant microbes to grow and proliferate. The variety of macronutrients a specific type of microbe can use is limited. However, the nutritional diversity of microbes as a group is much greater than that of any other type of organism.

In fuel systems, some members of the microbial community can use fuel or fuel additive molecules as food. Microorganisms capable of directly breaking down fuel hydrocarbons are known as fuel hydrocarbon degraders (sometimes referred to as *hydrocarbon utilizers*). Others thrive by feasting on metabolites excreted by hydrocarbon degraders. Consequently, robust fuel-contaminating microbial communities include both hydrocarbon degrading and non-hydrocarbon degrading microbes. In this context, *robust* refers to microbial communities that are metabolically active and capable of surviving external and internal stresses—including microbicidal treatments.

Micronutrients can be mineral elements, such as iron, calcium, sodium, manganese, magnesium, and zinc; or organic molecules (also known as growth factors) that cells are unable to synthesize. Typically, these molecules are needed only in trace quantities. Microbes that need a greater than average number of micronutrients or which can use only a few macronutrients are classified as being fastidious. They can thrive in fuel systems thanks to the production of metabolites by non-fastidious microbes (i.e., microbes that are able to obtain their carbon from a variety of sources, including fuel hydrocarbon and non-hydrocarbon molecules; Fig. 5).

Biofilms are complex microbial communities embedded within EPS or consisting of a matrix of fungal hyphae or a combination of both. Biofilms typically form on surfaces but also at the interface between two liquid phases, such as fuel and water. The chemistry of the EPS matrix can be quite complex. It is comprised of

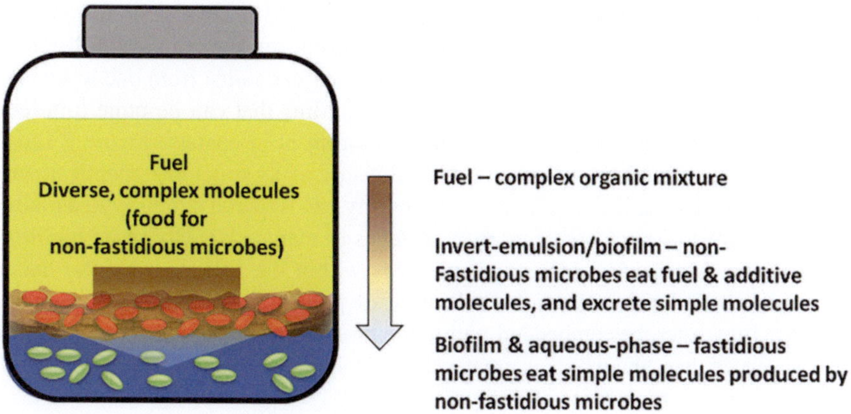

Fig. 5 Relationships between non-fastidious and fastidious microbes in a fuel system

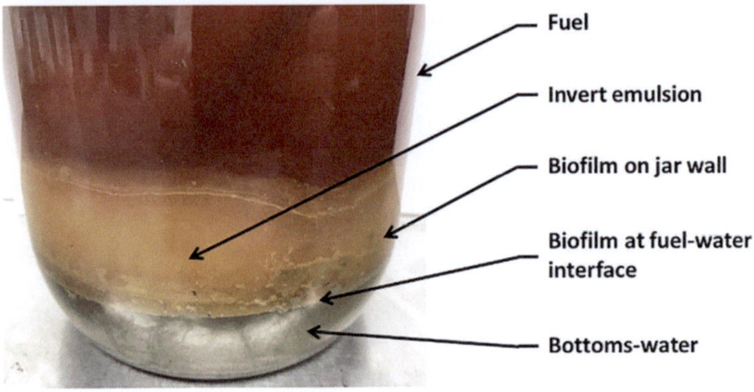

Fig. 6 Biofilm growth at fuel-water interface. Image courtesy of Biodeterioration Control Associates, Inc., Princeton, NJ, USA

water, carbohydrates, proteins, lipopolysaccharides, deoxyribonucleic acid (DNA), and ribonucleic acid (RNA), and generally has a slimy texture. Biofilm thickness can range from 0.2 to >10 mm. Both thickness and texture vary with fluid flow conditions. Biofilms that develop under high-shear, laminar flow conditions (such as those found in fuel transfer lines) tend to be thin and dense. Those that develop under low-shear conditions (as in aircraft wing tanks) tend to be thicker and less dense. Figure 6 shows a biofilm that has developed at the fuel-water interface in a glass jar. This biofilm is membranous and sufficiently tough to remain intact to prevent a pipette from penetrating it to reach the aqueous phase beneath it.

The EPS structure is quite complex. Some zones have dense cell populations while others appear to have no cells at all. Channels within the EPS matrix facilitate nutrient transport to and waste-metabolite transport from biofilm cells. Chemical signals among biofilm cells function analogously to those among human cells in a given tissue—determining each cell's appearance (*morphology*) and function (*physiology*). Genetically identical cells can appear and behave quite differently, depending on their location within the biofilm. Moreover, the number of different types of microbes active within a biofilm community can range from one to several dozen. These factors create conditions within biofilms that can be quite different from those in the bulk fluid with which the biofilm is in contact. Figure 7 illustrates the oxygen gradient within a biofilm. Because of its structural, chemical, and microbiological complexity, a biofilm community can function similarly to a multicellular organism (every organism from sponges to mammals). Moreover, biofilms can function as citadels—protecting microbes within the EPS matrix from harsh environments that can exist in the bulk fluid or preventing microbicidal products from reaching microbes embedded within the biofilm structure. Aristotle is said to have observed that "the whole is greater than the sum of its parts." This is certainly true for biofilm communities.

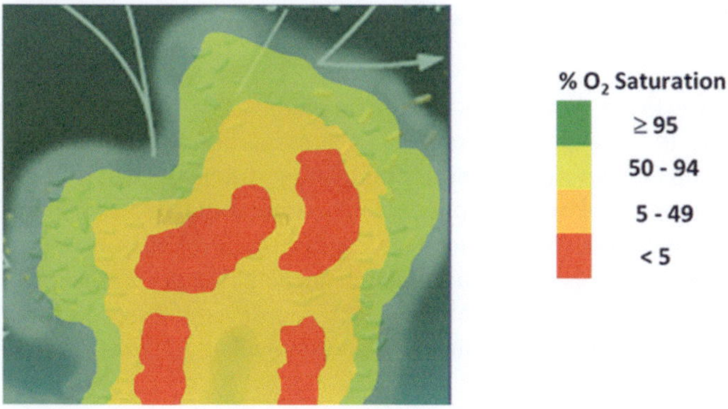

Fig. 7 Oxygen gradient from surface to base of biofilm matrix

Dormancy is a reversible state of minimal metabolism into which microbial cells can enter when their environment is hostile (i.e., wrong atmosphere, insufficient water or nutrients, etc.). Historically, it was believed that only specialized cell structures such as bacterial endospores or fungal spores could survive in a dormant state for prolonged periods (even centuries) under hostile conditions. However, it is now recognized that dormancy is a nearly universal capability among non-spore-forming microbes. For example, fuel with $a_w < 0.9$ can transport dormant cells that are subsequently able to reactivate once sufficient free or dispersed water is available. Like endospores, dormant, non-spore-formers can persist for centuries—becoming metabolically active once environmental conditions are suitable. Dormancy has two important implications in aviation fuel systems. First, it means that microbial contaminants can be transported throughout the fuel distribution system—remaining dormant until they come into contact with free water. Second, because fuel-treatment microbicides are effective only against metabolically active microbes, dormant cells can survive microbicide treatment (see Sect. 4.2.1).

As noted in the discussion of *water*, typically microbes in fuel are dormant. Dormant microbes represent a potential risk rather than a direct or current risk. As long as they remain dormant, microbes are neither proliferating nor producing problematic metabolites. Detection of dormant bacteria, yeasts, or fungal spores in fuel-phase samples is diagnostic of potential growth where or when sufficient water is present to support microbial activity. When metabolically active microbes are recovered from fuel-phase samples (e.g., by detection of ATP see Sect. 2.7.2), it is a strong indication that the fuel contains dispersed, biologically available water (i.e., $a_w \geq 0.91$). Additionally, the detection of dormant microbes (e.g., mold spores) in fuel-phase samples, can be an indication that active microbial growth is occurring in pockets or films of water somewhere in the fuel system. Because fungal spores are hydrophobic, they are often more readily dispersed in fuel. Consequently, dormant microbes can reflect activity in unsampled zones within a fuel system, transport from upstream fuel distribution system components, or both.

1.2 Microbial Growth in Aviation Fuel Systems

Microbial growth can occur at any stage in the supply and distribution of aviation fuel from refinery to the wing; in storage tanks at refineries, terminals, and airports; in cargo tanks, pipelines, refuelers and airport hydrant systems, and in aircraft fuel tanks. Filtration systems used by fuel suppliers can become contaminated with microorganisms. The aviation industry goes to great lengths to control microbiological contamination and has developed guidance documents detailing good practice [3, 7]. This section will discuss where microbial growth can develop in specific aviation fuel systems and the problems and dangers it can pose where microbes proliferate.

1.2.1 Storage Tanks at Refineries, Terminals, and Airport Depots

Microbial growth can occur in fuel tanks wherever even small amounts of free water accumulate. Primarily, this will be in free water that has accumulated on tank bottoms and system low points. Sources of water include residues left behind following tank cleaning and hydrotesting, ingress through seals on floating roof tanks or groundwater seepage into underground storage tanks. Microbial growth can also occur in condensate films that form on tank surfaces from the condensation of dissolved water from fuel due to temperature fluctuations or condensation of humid warm air when it enters a cold tank. It is important to note that only very little water is needed for microbial growth. To give this some perspective, a 1 mm depth water film would "look" like a 300-m-deep lake to a microorganism. Small pockets or films of water may support microbial growth in isolated areas of fuel tanks which are difficult to remove by routine sumping operations. This is particularly true for underground storage tanks or tanks with complex internal structures.

Microorganisms have the potential to grow, reproduce, and produce large amounts of biofilm which can foul surfaces and components in the fuel tanks. Microorganisms can adhere to surfaces and form extensive biofilm. Figure 8 shows microbial biofilm that has grown on the floor of a jet fuel storage tank. It can be seen that a mop has been used in some areas and has exposed the white epoxy coating. Turbulence within fuel tanks, for example during fuel receipts, can dislodge biofilms on the tank bottom and surfaces and suspend them in the bulk fuel phase. This leads to unacceptable levels of particulate matter in the stored fuel, rendering it unfit for use. Typically, fuel displaying an excess of 1 mg L^{-1} of particulate matter or sediment requires corrective actions (see Chap. 10). Microorganisms will eventually tend to settle back to the tank bottom with time. However, the settling times needed for this may be extensive (more than 24 h), and deliveries from the tank may be unacceptably contaminated if insufficient settling time is allowed. The height of the suction point within the tank for fuel deliveries will be critical in determining the extent to which fuel delivered from the tank is contaminated. Even small amounts of contamination in fuel delivered from a contaminated tank can be passed to tanks and infrastructure downstream, "inoculating" these. If water is present downstream, further proliferation can occur, leading to spread of contamination through the fuel distribution chain.

There tends to be seasonal influence on microbial growth in fuel tanks in temperate climates, with a higher risk of microbial growth occurring in the warmer summer months. In climates where there are very high day temperatures and very low night-time temperatures, these temperature fluctuations can promote water condensation and therefore increase the risk of microbial growth. More microbial growth can be expected in hot humid climates. Uneven temperature changes may also cause convection currents within tanks which exacerbate the suspension of microbes in the fuel.

The activity of a group of bacteria called sulfate reducing bacteria (SRB) can result in black/gray discoloration of water and fuel in tank bottoms due to the generation of corrosive hydrogen sulfide. Presence of hydrogen sulfide in fuel can result in the failure of fuel specification tests, for example the copper strip corrosion test (e.g., ASTM D130 [8]), and indicates a corrosion risk in storage tanks and facilities. SRB activity is a well-known cause of MIC in steel fuel tanks. SRB are anaerobic (oxygen hating) and tend to be active within settled water, sludge, and biofilms. The activity of other aerobic (oxygen utilizing) microorganisms can result in localized oxygen-deficient areas which can promote the activity of SRB. SRB feed on organic acids produced by other types of microbes when they degrade fuel. SRB activity can result in extensive and rapid pitting corrosion of steel plates in fuel tanks.

Fig. 8 Microbial growth on the floor of a jet fuel storage tank. Image courtesy of ECHA Microbiology Ltd., Cardiff, UK

1.2.2 Filtration Systems in Aviation Fuel Distribution

Filter water separators (e.g., those designed to requirements of EI 1581 [9]) are used throughout the aviation fuel supply chain and prevent the spread of water and minimize microbial contamination through the supply chain to the end user. However, if poorly maintained or underutilized, these filters may become colonized by microbial growth. Fuel is a continual supply of nutrients for microorganisms that become trapped on the coalescer elements of filter water separators. Water droplets are formed by coalescence within the coalescer elements and further separated from the fuel by the hydrophobic membrane of the separator elements (Chap. 10). When microorganisms are present in the fuel, they can proliferate within the filter element and negatively affect coalescer and separator element performance. When proliferating microbes grow on the outer surface of the coalescer elements, the result is brown spotting referred to as "leopard spotting," where each spot is a colony of microbial growth (Fig. 9). Because microbial growth occurs on the downstream side of coalescer elements, fuel passing through the elements can become contaminated; instead of removing microbial contamination, the filters become a source of contamination. Heavily contaminated coalescer elements can even fail to separate water effectively. The risk of this contamination developing is highest in systems with low flow rates and where filter water separators are intermittently used or underutilized. Water that accumulates in coalescer housings also provides an excellent environment for microbial growth.

Fig. 9 Leopard spotting on the outer surfaces of coalescer elements. Image courtesy of ECHA Microbiology Ltd., Cardiff, UK

1.2.3 Hydrant Systems

Sometimes the design and construction of hydrant systems can encourage microbial growth. For instance, the pipe bore may be wide relative to the fuel throughput if the system was designed to accommodate anticipated levels of higher throughput in the future. In wide bore systems and/or where throughput is relatively low, there will be no natural scouring or flushing effect within the pipes, giving rise to conditions where water and microorganisms can accumulate. Also, hydrant spurs with dead legs will not be flushed effectively and may provide locations where microbial growth can occur. The cleanliness of these systems upon commissioning also needs to be carefully considered, as debris from construction operations and water from hydrotesting can be sources of contamination or nutrients for microbial growth. Although hydrant low points are flushed routinely to remove water and particulate matter, insufficient flow rates or fuel volumes used during the process can leave water behind and become a source of nutrients and oxygen for microorganisms to proliferate.

1.2.4 Aircraft Fuel Tanks

Free water forms in aircraft fuel tanks from condensation of dissolved water in fuel as it cools when the aircraft ascends. Water can also come from condensation of humid warm air when it enters a cold tank through vents as the aircraft descends. Active microbial growth can occur wherever free water accumulates in the tank; this will typically be in pockets on the tank floor (particularly trapped behind spurs and frames), on upper surfaces of internal tank structures, and also on any surface where water condensation forms. Temperature is a particularly important factor that impacts the growth of microbes in aircraft fuel tanks. Microbial growth will occur predominantly during ground time because fuel temperatures in flight are generally too cold for microbial growth, although it is not sufficiently cold to kill many microbial cells. Aircraft on short haul operations are more prone to see growth because fuel does not have time to become as cold for extensive periods. When aircraft are parked for long periods or are underutilized, microbial growth can be a serious concern. For example, the COVID-19 pandemic, which brought global air transportation to a near standstill, resulted in mass parking of aircraft and numerous reports of microbial growth. Microbial growth is more prevalent in parked aircraft for several reasons.

- Temperatures will be warmer, and thus more favorable for microbial growth, all the time.
- Day and night temperature fluctuations will result in condensation.
- Water scavenging systems might not be operational or as effective when aircrafts are parked.
- There will be no turbulence in the fuel tanks to reduce the occurrence of biofilm formation by contaminating microorganisms.

As stated above, large amounts of biofilm can foul surfaces and components in the tanks (Fig. 10). If biofilms are disturbed and become suspended in the fuel,

Fig. 10 Microbial growth in aircraft wing tanks—**a** Microbial biofilm suspended in fuel residues; and **b** microbial biofilm attached to surfaces (Images courtesy of ECHA Microbiology Ltd., Cardiff, UK)

for example during flight or tank filling, they can clog aircraft engine filters. This problem will usually be encountered as full thrust is engaged during aircraft take-off but may occur at any time during the flight. If filters are clogged, flight crew will be alerted with a bypass indication and aircraft may need to be diverted or return to the airport.

The Fuel Quantity Indicator System (FQIS—also known as Fuel Quantity Indicating System) on an aircraft consists of probes located around the fuel tanks which provide readings to the cockpit via the Electronic Centralized Aircraft Monitor (ECAM) system. Microbial biofilm can foul the probes and interfere with measurements leading to incorrect readings being shown instead of the correct fuel quantity.

Most microorganisms produce organic acids when they degrade fuel. The acids are relatively weak, but they can accumulate within biofilms that form on surfaces. The acids can affect the permeability of coatings and if acid comes into contact with the aluminum alloy surface, this can result in etching and pitting of the airframe (Fig. 11). It is documented that microbial growth can impact the properties of composite materials, although there is less documented information on the specific impacts on the composites, which are increasingly used in aircraft construction.

1.3 Brief History

In 1939, uncontrolled microbial contamination in aircraft fuel tanks was first reported to be a contributing cause of sludge accumulation and consequent fuel filter plugging. During the Korean War (1950–1953), fuel filter biofouling was a major cause of fixed and rotary wing aircraft engine failures. By the mid-1960s, the U.S. Airforce (USAF) had accumulated records of incidents of aircraft fuel tank corrosion, as well

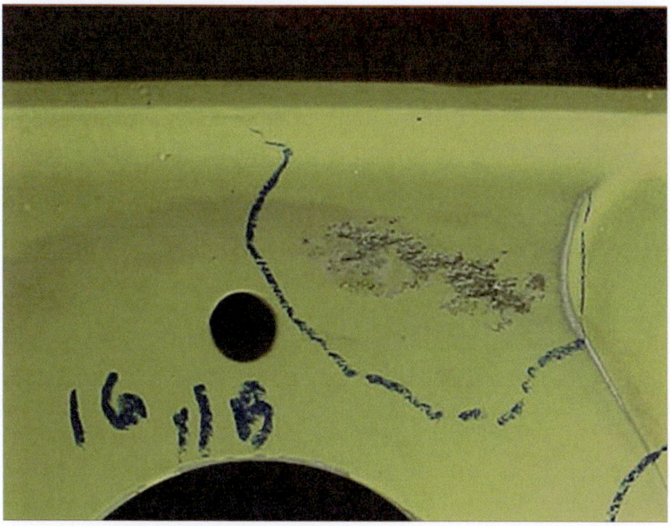

Fig. 11 Etching and pitting of aluminum alloy in an aircraft fuel tank caused by exposure to organic acids generated through microbial activity. Image courtesy of ECHA Microbiology Ltd., Cardiff, UK

as sealant and top-coating polymer deterioration in both commercial and military aircraft. Although microbes were commonly recovered from fuel tanks in which deterioration had occurred, an unequivocal link between microbial contamination and biodeterioration had yet to be established. Biodegradation of hydrocarbon fuels was first reported in the late nineteenth century. Consequently, it was not surprising to recover microbes from fuel systems. Moreover, it seemed plausible that microbes recovered from aircraft fuel systems could have contributed to the observed damage. Before 1963, microbiologically influenced corrosion of aluminum had not been demonstrated. However, research performed at Boeing demonstrated that the Gram-negative bacterium *Pseudomonas aeruginosa* could accelerate the corrosion of 7075 aluminum, whether the aluminum was untreated, anodized, or chromate-treated. Thus, by 1965, the biodeteriogenic role of aircraft fuel system microbial contamination was generally accepted. Surveys conducted by the U.S. Air Force during this period indicated that although the probability of biodeterioration being present increased when microbial contamination was present, it was not inevitable. In some cases, deterioration was observed when no microbes were detected. In other cases, there was no evidence of biodeterioration even though microbial population densities of $>10^4$ CFU mL^{-1} in fuel-associated waters were recovered. At this point, there was little consideration of the role of population characteristics in determining biodeterioration risk.

The late 1950s saw a significant shift from the use of aviation gasoline to aviation turbine kerosene as the predominant fuel used in commercial aviation. Kerosene is inherently more biodegradable than gasoline and because most aviation gasoline at the time contained tetraethyl lead (TEL), making it relatively resistant to microbial growth. Consequently, the more widespread use of aviation kerosene resulted in a significant increase through the 1960s of reports of aircraft operational issues related to microbial growth, notably in aircraft operating in hot, humid environments.

Having introduced the use of ethylene glycol monomethyl ether (EGME) in 1962 as a fuel system icing inhibitor (FSII), the U.S. Air Force (USAF) recognized that the product performed well as a microbial contamination control agent. In fact, EGME's manufacturer obtained a U.S. EPA pesticide registration for the product for use as a fuel treatment antimicrobial pesticide. After EGME was introduced, the USAF discontinued their jet fuel microbiology research. The use of EGME appeared to control microbial contamination and substantially reduce the incidence of the various deterioration phenomena that had been reported in aircraft using untreated fuel.

In the early 1980s, the USAF began replacing JP-4 with JP-8 fuel (Chap. 1). Additionally, due to toxicity concerns, in 1984 EGME was replaced by the less toxic, diethylene glycol monomethyl ether (DiEGME). Not long after these two changes occurred, symptoms of fuel and fuel system biodeterioration were again reported in military aircraft.

In the 1980s, jet fuel was stored in unlined rock caverns in western Europe. These caverns were below the groundwater table and tended to have cave streams running through them. By the mid-80s, it had become apparent that the stored jet fuel was being degraded by aerobic and anaerobic microbes that were present and metabolically active in these natural environments. To reduce the risk of jet fuel becoming corrosive during storage, tanks were installed in the caverns—isolating fuel from the underground streams. Additionally, jet fuel inventories were rotated so that approximately one-third of the fuel was replaced each month. These engineering and inventory management measures were augmented with an aggressive condition monitoring program that included testing for physical and chemical properties of fuel and fuel-associated water samples, and microbial contamination in both fuel and fuel-associated water samples.

During the build up to Operation Desert Storm in 1990, NATO forces stored jet fuel in neoprene bladders. By mid-1991 bladders were beginning to fail. Subsequent investigations attributed the failures to the activity of sulfate reducing bacteria (SRB) growing in traces of water that accumulated in bladder low points.

Before 2005, the aviation fuel literature reported only culturable microbe recoveries (see Sect. 2.7 for an explanation of culturability). In 2005, a chemical microbiology survey was conducted to evaluate microbial contamination in aircraft, JP-8 storage tanks, and fuel delivery trucks at 11 USAF bases around the continental United States. The investigators surmised that several factors contributed to the differences in the types of microbes that they detected and those that had been reported previously. These factors included the following.

- Fuel chemistry—JP-4, Jet A, and JP-8 are slightly different distillation products.
- Fuel additive changes—the introduction of antioxidants, metal deactivators, static dissipator agents, and the switch from EGME to DiEGME since the mid-1960s in military fuels.

- Location—the profiles of microbes recovered from fuel tanks were quite similar to those detected in air samples from nearby. These similarities suggested that airborne, dust particle associated microbes were an important source of fuel tank contamination.

Regarding the switch from EGME to DiEGME, the latter product was never registered as an antimicrobial pesticide, nor is it widely used in commercial jet fuels. Reports of the chemical's toxicity against fuel-contaminating microbes were equivocal. A fuel-soluble dioxaborinane blend (2,2′-((1-methyl-1,3-propanediyl)bis(oxy))bis(4-methyl-1,3,2-dioxaborinane) (67.6% active ingredient—a.i.) + 2,2′oxybis(4,4,6-trimethyl-1,3,2-dioxaborinane) (27.4% a.i.) was first registered for use in 1972. A universally-soluble, isothiazolone biocide (5-chloro-2-methyl-3(2H)-isothiazolone (1.0% a.i.) + 2-methyl-3(2H)-isothiazolone (0.5% a.i.)) was approved for use in jet fuel in 1995.

Despite their availability, the two jet fuel biocides have been used primarily to treat contaminated aircraft fuel tanks. Unlike EGME for which biocidal activity was a secondary function, the fuel-treatment microbicides are not used as fuel additives to prevent growth from occurring (see Sect. 4.1).

In 2018, the manufacturer of the isothiazolinone product removed jet fuel treatment from the product's label. Reportedly, there were a number of instances in which aircraft jet fuel tanks were treated and fuel system problems subsequently developed. The manufacturer opted to withdraw from the market. As of the time of writing, the dioxaborinane-based product is the only fuel-treatment microbicide (see Sect. 4.1).

2 Monitoring Fuel Systems for Microbial Contamination

2.1 *Sampling*

Sampling is discussed in greater detail in Chap. 11. Additionally, several guidance documents [3, 4, 7, 10, 11] provide extensive discussions regarding the collection of samples intended for microbiological testing. In this chapter only the primary considerations will be addressed.

2.2 *Specific Considerations for Microbiological Samples*

- Cleanliness—microbes are ubiquitous in the air, skin, and system surfaces. Care must be taken to ensure that samples contain only the microbes that were in the sampled system. ASTM D7464 [11] provides detailed instructions on how to collect samples for microbiological testing without contaminating the sample with dirt, water, debris, or exogenous microbes.

- Sample perishability—when used in the context of samples to be tested for microbial contamination, the term *perishability* means that as samples age, their microbial populations differ increasingly from those present at the time the sample was collected. Changes are both quantitative and qualitative. Quantitatively, the total bioburden (i.e., population density) can increase, decrease, or remain relatively stable. These phenomena are illustrated in Fig. 12. Qualitatively, the types of microbes present in a sample tend to change over time. When multiple taxa are present, they compete for available resources. Those resources include the various factors reviewed in Sect. 1.1.2. Because the environment within a sample container differs from that in the tank from which the sample was collected, microbial taxa that were originally dominant in the fresh sample can become non-competitive. Conversely, taxa that represented a negligible proportion of the total microbiome in the freshly collected sample can become dominant as the sample ages.

Most commonly, the microbial population's diversity tends to decrease as samples age. However, on occasion, apparent taxonomic diversity increases during sample storage. The term *apparent* is used here because microbes are not created spontaneously in samples during storage. This can happen when taxa originally present at undetectably low numbers find a competitive advantage in the sample container environment, enabling them to proliferate to detectable levels.

When possible, microbiological analysis should be started within 4h after sample collection. When this is not practical, samples can be stored either refrigerated or on ice for at least one day. Although consensus standards [11] prescribe testing within 24 h after collection, there is some evidence that the types of microbiological population changes discussed in this section might not become substantial until samples have been stored for longer than one week.

Analyte heterogeneity—microorganisms are not distributed uniformly in fuel systems. Microbes are most abundant on system surfaces and at fuel-water interfaces,

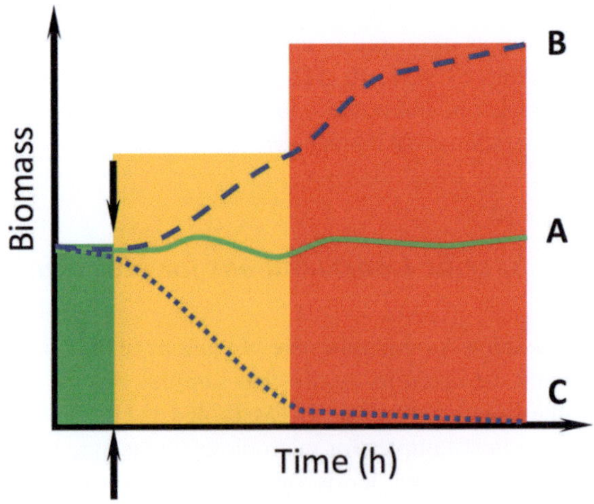

Fig. 12 Fuel sample perishability during storage. The green line indicates that the detectable bioburden has remained stable. The red line illustrates bioburden die-off, and the blue line indicates that the microbial population has increased during storage in the sample container

moderately abundant in continuous phase, fuel-associated waters (i.e., bottoms-water), and least abundant in fuels. The fuel tank zones to which maintenance personnel have easiest access are not necessarily those most likely to have microbial contamination present. For example, if a tank contains pockets of water that are not captured in samples, microbes present in the water will not be detected. Consequently, it is common for bioburdens to be below detection levels (BDL) in specimens collected from tanks that have visible signs of microbial contamination.

2.3 Identifying Sample Collection Points

Optimally, fluid samples are collected from each fuel system's lowest point. This improves the probability of the sample containing multiple phases—including fuel, fuel-associated water, invert emulsion layer, and bottom sludge or sediment. From a practical perspective, it is not always possible to obtain a true, lowest point sample. This section reviews commonly used sample collection points.

- Aircraft—most aircraft wing and fuselage fuel tanks have drain ports designed to facilitate sample collection. Samples should be collected after aircraft have been parked for at least 1 h (preferably, at least 1 h per 30 cm of tank height—particles dispersed in jet fuel settle at approximately 30 cm h^{-1}) to allow any free water and particulate material in the tank to settle.
- Fuel carts and other ground equipment—cart tanks and filter housings often have bottom drains from which samples can be collected. When they do not, bottom samples can be obtained using a fuel-thief sampler or peristaltic pump (Chap. 11).
- Bulk storage tanks—as discussed above and in Chap. 11, bulk storage tank water drains typically have inlets that open $\geq$10 cm above the tank floor. If drain samples are collected, best practice is to collect the initial flow and a post-flush sample (here, *flush* indicates that one volume of the drain line's capacity has been allowed to flow). The initial flow captures microbial contamination within the drain line. Post flush captures fluid from within the tank. The volume flushed should be at least equal to the drain line's volume.

2.4 Sampling and Sample Handling

Traces of dirt and debris from the area around sampling points can contaminate samples and result in microbial loads that exceed the upper control limit. Consequently, particular care must be taken to minimize the risk of sample contamination. Before samples are collected, the exterior surface of drain lines and valves, and sampling devices should be disinfected. The interior (fuel contact) surfaces should also be disinfected. This can be done using a rag that has been saturated with either ethanol, methanol, or isopropanol ($\geq$70% v/v) to wipe down all exposed surfaces. Allow the alcohol to evaporate to dryness before collecting a sample. Chapter 11 and ASTM D7464 [11] provide details regarding sample container selection and labelling. Microbiological parameters are perishable. As explained in Sect. 2.2, during delays between sample collection and testing, microbial populations can change substantially and unpredictably. If testing cannot be started within 4 h after collection, samples intended for microbiological testing should be kept cool (<7 °C) either by placing them in a container with cold packs or ice, or refrigerating them. If placed with cold packs or ice, samples might need to be insulated from direct contact with ice or cold packs to avoid freezing, depending on the type of analysis planned. As discussed in Chap. 11, chain of custody records should include:

- Sample collection date and time
- Sample collector identification
- Sample source—tank identification and, if tank has multiple sample access points, sample point
- Sample disposition—stored in refrigerator, on cold packs or ice, etc.; date and time shipped to testing lab.
- Tests requested
- Date and time received at lab
- Date and time testing initiated—including list of tests performed
- Sample disposition—archived, discarded, etc.

2.5 *Sampling and Testing Frequency*

Sampling intervals should be dictated by the risk that one or more test results is greater than its upper control limit. Optimally, the sampling interval is one-third the expected period for a parameter to exceed its control limit. For example, if the fuel-phase upper control limit of 4000 CFU L^{-1} is typically exceeded at three-year intervals, annual sampling is sufficient. However, if a parameter tends to exceed its control limit within three months, monthly sampling and testing is appropriate. The underlying principle is that more frequent sampling incurs costs but little benefit, but less frequent sampling reduces the cost of sampling and testing but increases the risk of failing to detect out of control conditions in a timely manner.

When data are not available, the primary factor to consider when determining sample collection frequency is the rate at which water accumulates in the tank. Generally speaking, water accumulates most frequently in bulk storage tanks with high turnover rates (fuel volume replaced at least once per week) and aircraft tanks with low turnover rates (aircraft refueled less than once per month). Water accumulation rates in idle aircraft are affected by the frequency with which the dew point is reached where the aircraft are parked. The dew point is the temperature air which the air becomes saturated with water (i.e., when the relative humidity = 100%) and free water condenses out of the air. Invariably, the dew point is reached more frequently in warm, humid climates than in cooler, more arid ones. The longer accumulated water remains in a fuel tank, the greater the opportunity for a robust microbial population to develop. Consequently, increasing the frequency of water draining decreases the microbial contamination risk. When starting up a testing program, or if no testing program exists, the presence of free water in a tank sample should trigger microbiology testing. The data from those tests can then be used to determine the rate at which microbiological contamination trends from negligible to moderate. Although this section offers some guidelines about sampling intervals, see Sect. 2.6 for a fuller discussion.

Aircraft—the IATA guidance document [3] recommends draining aircraft fuel tanks periodically. The interval between draining depends on each aircraft's operating schedule (e.g., continuous operations versus long periods parked between flights) and climate conditions where aircraft are parked. Manufacturers' aircraft maintenance manuals (AMMs) typically specify water drain intervals based on these two factors. As indicated in the preceding paragraph, if there is no history of microbiology testing, and free water is present, it should be tested. If a microbiology testing schedule has been developed, unscheduled testing is needed if the water phase is discolored (i.e., darker than a pale straw color), turbid, or if more than a trace of sediment is visible on the bottom of the sample jar (note: this is the advantage of using clear glass jars for bottom-drain samples).

Fuel carts, and other airport facility fuel handling equipment—depending on a facility's tempo of operations, airport tank truck, fuel cart, hydrant, and filtration unit fuel throughput can vary substantially. Under normal operating conditions, if water is not accumulating in a component, semi-annual testing for microbiological contamination is sufficient (see previous discussion about the impact of positive test results on sampling frequency). For components in which water accumulates sufficiently to require weekly or more frequent draining, quarterly microbiological testing is recommended. Water drained from aircraft or fueling equipment tanks should be observed. If the water is discolored, turbid, or if more than a trace of sediment is visible on the bottom of the sample jar, the water and bottom fuel should be tested for microbiological contamination.

Bulk storage tanks—water drain or bottom samples from bulk storage tanks should be tested semi-annually under normal operating conditions. If the facility does not have a coalescing and particulate filtration unit installed upstream of its bulk storage tanks; quarterly testing is recommended. See the two previous bullets regarding bottoms-water appearance and the need for unscheduled microbiological testing.

2.6 Basis for Determining Sampling and Testing Frequency

As a rule, the optimal return on investment and effort are realized when the interval between monitoring events is approximately one-third of the period during which the monitored parameter drifts from *negligible* to *moderate* (i.e., the threshold at which some maintenance action is required). More frequent monitoring tends to increase condition monitoring costs without providing commensurate value. Less frequent monitoring increases the risk of damage due to untimely maintenance actions. Figure 13 illustrates two condition monitoring frequencies. For the parameter that tends to cross the action level three months after a maintenance action has been performed, monthly monitoring is recommended. For the parameters that tend to cross the action threshold after three-years, annual testing is recommended. When instigating monitoring programs, it is thus useful to establish initial baselines of contamination and assess risk of microbiological growth, taking into account the environmental factors that influence proliferation rates. Further information specific to airports, aviation fuel supply facilities and aircraft fuel tanks is provided in the guidance documents referenced above.

During the period following a maintenance action for microbial control, monitoring should be performed at shorter intervals. Initial testing approximately one week after the maintenance action has been completed is used to confirm that the maintenance action was effective. Additional follow-up monitoring at three times the routine frequency (i.e., if a parameter is monitored routinely annually, decrease the interval to every four months for one year). If the rate at which the parameter drifts to the action level has returned to normal, the routine interval can be restored. However, if the interval between test results yielding actionable bioburdens has decreased, the condition monitoring frequency should be modified as needed. Factors that can contribute to changes in monitoring frequency include:

- Fuel specifications—periodically specifications for aviation gasoline or jet fuels change. For example, military and commercial airlines are now using conventional

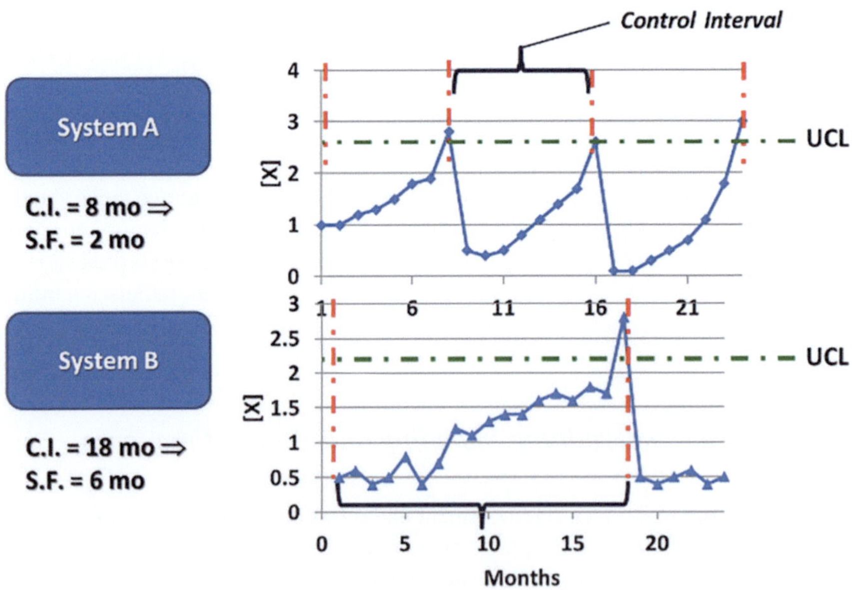

Fig. 13 Condition monitoring intervals based on data trends. C.I.—control interval—period of time between instances of parameter exceeding upper control limit (UCL). S.F.—sampling frequency

jet fuels are blended with a synthetic blending component (SBC). The finished fuel is commonly referred to as semi-synthetic jet fuel. As this chapter is being written, SBCs produced through eight pathways have been qualified for use at blend ratios of 10–50%. Sustainable Aviation Fuel (SAF) is a term applied to describe the final blending of a sustainable SBC with conventional jet fuel. The use of SAF is considered critical for the aviation industry to achieve its decarbonisation goals, and a 100% SAF fuel specification is now under consideration. To date, very little is known about the susceptibility of these fuels to microbial growth. Because semi-synthetic jet fuel and conventional jet fuels share similarities in their properties and chemistries, it could be envisaged that the susceptibility of semi-synthetic jet fuel to microbial growth would also be similar. However, the relative proportions of chemical constituents in SBCs are not the same, or as wide ranging, as those found in conventional fuels, and this could impact overall fuel properties, which in turn could influence the susceptibility to microbial growth. Traditionally, aviation gasoline has contained tetraethyl lead (TEL). Although TEL was used as an anti-knock agent, it was also toxic to microbes. Consequently, microbial contamination was not an issue in aviation gasoline. With the removal of TEL from aviation gasoline, microbial contamination monitoring should be incorporated into AMMs.

- Operations—microbial contamination is more likely to develop in aircraft that spend prolonged periods out of service than active aircraft that are flying daily. Condition monitoring intervals should be adopted to operations. In their AMMs, airframe manufacturers customarily make specific recommendations regarding monitoring frequency and free water mitigation controls depending on the operational status of the aircraft (e.g., long-term parked, active parked or fully operational). However, users are advised to consider the risk of microbial growth based on the environmental factors previously discussed.
- Fuel sourcing—while the majority of aviation fuel uplifted by aircraft will contain only negligible levels of microbial contamination, occasional failures can occur—resulting in transfer of moderate to heavy contamination to aircraft fuel tanks. If aircraft have received fuel from a facility where significant microbial contamination has been detected in storage tanks, hydrants, and other fueling equipment, monitoring frequency should be increased as though contamination had been detected in samples from the aircraft (i.e., at one-third the normal interval, as explained above). Even when levels of contamination in fuel supplied are negligible, the microbes present in fuel delivered can proliferate on-board if environmental conditions on the aircraft are optimal and water is allowed to accumulate.
- Microbiological testing should not be performed immediately after microbicide treatment. The IATA guidance document [3] recommends testing ten days after treatment or after at least five flight cycles, whichever period is longer.

2.7 *Test Methods*

2.7.1 Overview

The perfect test method has yet to be developed. Each of the methods discussed in this section has advantages and disadvantages relative to the others. Note that test method advantages and disadvantages are situational, depending on the testing objective. For example, the time lapse between test initiation and data availability can be several days, but culture testing is the only means for obtaining microbial isolates for additional testing (taxonomic identification, response to biocide treatment, etc.). Generally speaking, there is good agreement among the methods regarding thresholds defining negligible, moderate, and heavy microbial loads. However, as will be discussed in this section, each class of microbiological test methods measures a different parameter. Consequently, there are technical reasons why in some cases results from different methods will not agree. There are three broad categories of microbiological test methods:

- Culture testing—specimens are placed into or onto a nutrient growth medium. As cells divide (proliferate), they become visible as turbidity, masses (colonies), growth medium color change, or some combination thereof.
- Direct counting—individual cells are counted using either a microscope or flow cytometer. Currently, direct count methods are not broadly used to test aviation fuels for microbial contamination.
- Molecular microbiology methods (MMM)—these methods are used to quantify the concentration of biomolecule such as ATP, antigens (Ag), DNA, or others.

As will be explained below, methods vary in terms of the time lag between test initiation and data availability, specificity, user training requirements, and analytes detected (see Table 1). All test methods only detect analytes present in test specimens. Consequently, appropriate sampling, as discussed in Sect. 2.2, is essential for reliable condition monitoring.

2.7.2 Adenosine Triphosphate Testing

Adenosine triphosphate is an energy molecule present in all living cells from bacteria to humans. Although the ATP concentration ([cATP]) per cell varies among organisms, decades of environmental microbiology testing has demonstrated that under many conditions the average [cATP] for bacteria and fungi, respectively, is approximately 1 fg bacterial cell^{-1} and 100 fg fungal cell^{-1} (1 fg $= 10^{-15}$ g). There are two consensus ATP test methods:

- ASTM D7463 [12] Test Method for Adenosine Triphosphate (ATP) Content of Microorganisms in Fuel, Fuel/Water Mixtures, and Fuel-Associated Water, and

Table 1 Test Method Comparison

Test method	Sample volume (mL)	Lag time[a]	Analyte	Specificity
Antigens (ASTM D8070)	150 (F)[b] 15 (W)[c]	15 min	Antigens[d]	Bacteria, Fungi, *H. resinae*
ATP (ASTM D7463)	1000 (F) 1 (W)	10–15 min	Total ATP[e]	Metabolically active cells and cell debris
ATP (ASTM D7687)	20 (F) 5 (W)	10–15 min	Cellular ATP[f]	Metabolically active cells
Culture (ASTM D6974/IP 385)	1–20 (F) 1 (W)	3–5 d	CFU[g, h]	Culturable bacteria and fungi
Culture (ASTM D7978/IP 613)	0.5 (F) 1 (W)	1–4 d	CFU	Culturable bacteria and fungi
Microscopy	0.1–20[i]	1 h	Cells	Varies[j]
qPCR	20 (F) 5 (W)	3 h	DNA	Varies[k]
16S rRNA	20 (F) 5 (W)	1 d	DNA	Bacteria[l]

Notes:
a. Time interval between test initiation and data availability.
b. F—fuel
c. W—water
d. Antigens—three pairs of LFD—one for bacteria, one for fungi, and one specifically for the fungus *Hormoconis resinae.*
e. Total ATP—ATP extracted from whole cells, attached to cell fragments, and in solution.
f. Cellular ATP—ATP extracted from whole cells only.
g. CFU—colony forming units—masses that develop as microbes reproduce on nutrient media.
h. IP 385 provides a specific protocol for assessment of water phase, whereas ASTM D6974 does not.
i. To facilitate detection, specimens can be concentrated by centrifugation or filtration.
j. Various staining methods can be used to identify specific microbes of interest.
k. As discussed in Sect. 2.7.6 primers are used to detect and quantify microbes. Specificity varies based on probe or probes used.
l. An analogous protocol can be used to detect fungi.

- ASTM D7687 [13] Test Method for Measurement of Cellular Adenosine Triphosphate in Fuel and Fuel-associated Water with Sample Concentration by Filtration

Both methods quantify the intensity of light produced when ATP interacts with an enzyme–substrate complex—luciferase enzyme and luciferin substrate). For each molecule of ATP in a specimen, one photon of light is emitted (Fig. 14). The rate of photon emission is detected using a luminometer—a device that has a dark chamber, photo-detector, and associated electronics. Thus, ATP testing is often referred to as ATP luminometry.

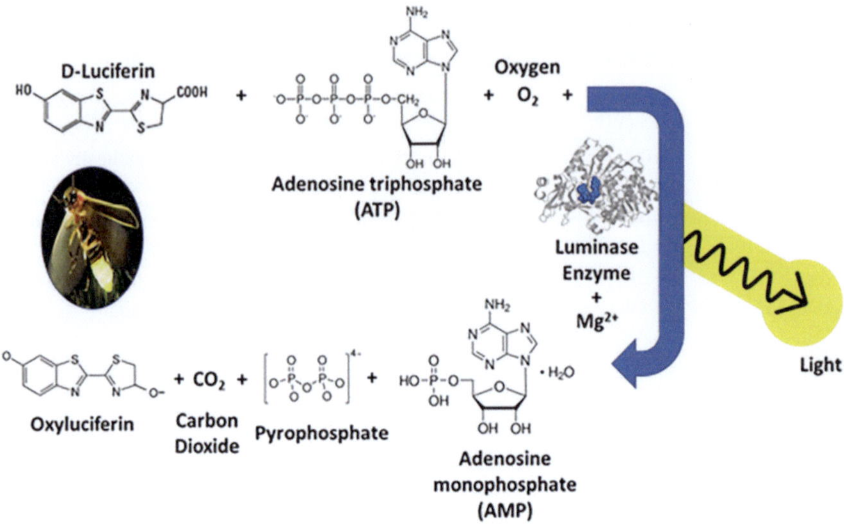

Fig. 14 Luciferin-Luciferase reaction

ASTM D7463 uses an aqueous reagent to extract ATP from a 1 L fuel sample. The method reports total ATP ([tATP]) in relative light units (RLU). For samples comprising of a fuel / water mixture or water only, subsamples obtained from the bulk water layer in the sample container are analysed. An optional add-on test can be used to determine dissolved ATP ([dATP] —all ATP that is not contained within intact cells) so that [cATP] can be estimated by subtracting [dATP] from [tATP]. Using the luminometer specified by Method D7463, the method's limit of detection (LOD) is 1 RLU 500 mL^{-1} when testing fuel phase and 1 RLU mL^{-1} when testing fuel-associated water. The manufacturer's threshold limits are: negligible < 1000 RLU L^{-1} and < 1000 RLU mL^{-1} in fuel and water, respectively; moderate 1000 RLU L^{-1} to 5000 RLU L^{-1} and 1000 RLU mL^{-1} to 5000 RLU ml^{-1} in fuel and water, respectively; heavy > 5000 RLU L^{-1} and > 5000 RLU mL^{-1} in fuel and water, respectively.

ASTM D7687 relies on filtration to separate intact cells from fuel or fuel-associated water specimens. The typical specimen volumes are 20 mL for fuel and 5 mL for fuel-associated water. The method includes a wash step that removes all interferences that have been shown historically to interfere with ATP assays. The method reports [cATP] in pg mL^{-1}, with an LOD = 5 pg mL^{-1} for a 20 mL fuel sample and 20 pg mL^{-1} for a 5 mL fuel-associated water sample (1 pg = 10^{-12} g). The manufacturer's threshold limits are: negligible <10 pg mL^{-1} and <100 pg mL^{-1} in fuel and water, respectively; moderate 10 pg mL^{-1} to <100 pg mL^{-1} and 100 pg mL^{-1} to <1000 pg mL^{-1} in fuel and water, respectively; and heavy ≥100 pg mL^{-1} and ≥1000 pg mL^{-1} in fuel and water, respectively.

Each of the ASTM test methods for ATP can be completed in less than 15 min.

Dormant cells (see Sect. 1.1.2) contain negligible [cATP]. Therefore, ATP luminometry quantifies the abundance of metabolically active microorganisms. Detecting all living cells, ATP test results provide no indication of what types of microbes are present. A modified version of D7687 has been used to differentiate between bacterial and fungal contamination, but the protocol has not been used widely to test jet fuels.

After microbicide treatment, most cells break apart (lyse), but some killed cells remain intact. Consequently, when samples are tested before and after treatment, invariably some detectable ATP remains. With ASTM D7687, a two order of magnitude decrease in [cATP] indicates effective disinfection. When the [cATP] decreases less than two orders of magnitude, a substantial number of cells are injured but not killed, and, typically, the population will recover within five to seven days.

The primary advantages of ATP testing include speed and detection of an indicator molecule that is present in all metabolically active cells. Microbes do not need to proliferate to be detected. However, the basic ATP tests described in D7463 and D7687 do not detect dormant microbes. Nor does ATP testing differentiate between bacterial and fungal bioburdens. A description of D7687 supplemental protocols that can be used to detect dormant microbes and differentiate between bacterial and fungal bioburdens has been published [14].

2.7.3 Culture Testing

Culture tests methods that determine the number of Colony Forming Units (CFU) involve placing a sample on or in a nutrient media and rely on proliferation of each viable (live) microbial particle (one single microbial cell, fungal fragment, spore, or an aggregate of microbial cells) to produce a visible spot referred to as a colony. After several days' incubation, the number of colonies are counted and expressed as CFUs.

The microorganisms detected by culture methods will be dependent on the formulation of the culture medium. In some environments, culture methods may recover only a small percentage of the true species diversity. Studies in fuel systems have shown that industry standard (ASTM or IP) culture methods have a good ability to detect important microbial contaminants [15–18] and show reasonable concordance with both MMM and ATP assays.

CFU tests detect live culturable microorganisms and their dormant spores. In fuel tanks, active microbial growth only occurs in pockets of water or in condensate water films adhered to tank surfaces. Some actively growing molds generate and release spores which are hydrophobic and readily disperse in fuel and thus enable the spread of contamination. It may not always be possible to recover free water in fuel tank drain samples and thus the presence of spores in fuel samples may be the only indicator of microbes being present in a tank.

The culture test methods used for monitoring aerobic microbiological contamination in samples drawn from aircraft fuel tanks include IP 385 method [15], or the technically similar ASTM Test Method D6974 [16], IP 613 [17] or equivalent ASTM

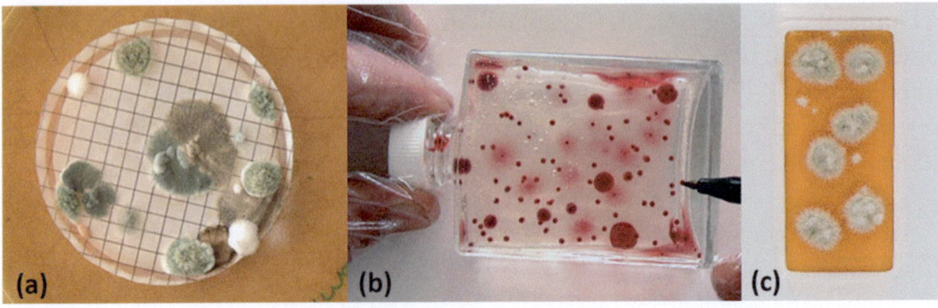

Fig. 15 Culture test methods for monitoring aerobic microbiological contamination in fuel or fuel-associated water—**a** Fungal colonies on an IP 385/ASTM D6974 test filter; **b** Microbial colonies in an IP 613/ASTM D7978 test; and **c** Mold colonies on a dipslide paddle. Images curtesy of ECHA Microbiology Ltd., Cardiff, UK

Test Method D7978 [18]. The time lag between test initiation and data availability depends on the generation times of the microbes in test specimens. Typically, test protocols recommend observing inoculated media for three to five days for bacterial or fungal growth. Using the manufacturers' specified incubation periods is optimal for routine condition monitoring testing. However, for diagnostic testing efforts longer incubation periods can be useful (see Sect. 1.1.2).

The IP 385 [15] or the technically equivalent ASTM D6974 [16] methods are considered by IATA to be the reference microbiological methods. It is a laboratory method requiring microbiological skills. Known volume(s) of fuel are passed through membrane filters which are then washed and transferred to two types of agar plates, one for detection of fungi (yeast and molds) and the other for bacteria. The agar plates are incubated, and then the number of colonies of bacteria, yeasts and mold are counted and expressed independently as number of each type per liter of fuel (Fig. 15a). Procedure B of the IP 385 method enables enumeration of viable microorganisms in fuel-associated water.

In addition to its use for testing samples from aircraft fuel tanks, IP 613/ASTM D7978 is also endorsed by JIG for testing samples from the aviation fuel supply chain [7]. The IP 613/ASTM D7978 test utilizes the same technology as the reference laboratory method, but it is in a format that means it can be used in the laboratory or in the field. No specialist microbiological training is required to perform the test. The test consists of a screw-capped glass bottle containing a thixotropic nutrient gel. The bottle is shaken after addition of a sample of fuel or fuel-associated water, the gel liquefies, and the sample (and any microorganisms in it) are dispersed. The bottle is then tapped so that gel forms a flat layer, and it is then ready for incubation at 25 °C (±3 °C) for up to 4 days. A reactive compound in the gel changes the color of the colonies that develop to a red or purple color so that they can be easily seen and counted (Fig. 15b). The test provides a total count of bacteria, yeasts, and molds rather than individual counts of each type provided by IP 385/ASTM D6974. The number of colonies counted is expressed as CFU L^{-1} of fuel or CFU mL^{-1} of water associated with fuel.

Dipslide tests are not industry standard (IP or ASTM) methods but can be used for testing water phase associated with fuel samples; they are not recommended for testing fuel phase. Dipslide tests are semi-quantitative tests where the number of CFU mL^{-1} of water is estimated by comparison to a chart (Fig. 15c).

IATA provides guidance on threshold contamination limits when using the reference culture methods (IP 385 or ASTM D6974) to assess microbial contamination in sump-drain samples from aircraft fuel tanks. Threshold contamination limits when assessing fuel phase of samples are negligible <4000 CFU L^{-1}; moderate 4000–20,000 CFU L^{-1}; heavy >20,000 CFU L^{-1}. Threshold limits when assessing water phase are negligible <1000 CFU mL^{-1}; moderate 1000–10,000 CFU mL^{-1}; heavy >10,000 CFU mL^{-1}. Levels of contamination of water phase will usually be much higher than in fuel phase, which is why separate guidance is given for fuel phase and water phase.

Other industry documents provide guidance on threshold contamination limits when using IP 385/ASTM D6974 or IP 613/ASTM D7978 to assess bottom or drain samples from tanks and filter vessels in the aviation fuel supply chain [3, 7]. The contamination limits are less stringent than the IATA contamination limits for aircraft tanks and use terminology Normal, Warning and Action to define contamination categories. This reflects the fact that fuel in airport or supply and distribution facilities will pass through several stages of filtration before it is uplifted onto aircraft. Also, any water in supply and distribution facilities could be derived from environmental sources and thus have an inherently higher microbial bioburden.

Sulfate Reducing Bacteria require a separate test because they have special growth requirements; they are anaerobic microorganisms. Test kits are available for field or laboratory assessment of SRB in different types of samples, for example water associated with fuels, sludge or biofilm or corrosion deposits. The test kits are semi-quantitative where the presence of SRB is determined by the rate and extent of the development of a black color, as shown in Fig. 16. The SRB test kits rely on selective media that have been developed to facilitate the detection of culturable SRB. The medium supports SRB proliferation and undergoes a color change as a result of sulfate reduction. Other diagnostic test kits use selective media with specific dyes to facilitate detection of other microbes associated with MIC. Such kits include those for detecting acid producing bacteria, denitrifying bacteria, iron related bacteria, nitrifying bacteria, and others.

The primary advantage of culture testing is that organisms recovered can be used for further testing. Also, in addition to detecting metabolically active, culturable microorganisms, culture tests can detect dormant microorganisms. No single growth medium or set of growth conditions is universally suitable for detecting all microbes present in a sample. Genomic testing has identified types of microbes that had not been previously identified based on culture testing. However, it is rare to find contaminated fuel system samples for which elevated non-culture test results do not correspond with elevated culture test results. As noted above, the delay between test initiation and data availability can be several days to a week, depending on the method applied.

Fig. 16 Example of a test kit for detection of sulfate reducing bacteria (SRB). Presence of SRB is determined by the rate and extent of the development of a black color. Image curtesy of ECHA Microbiology Ltd., Cardiff, UK

2.7.4 Immunoassays

All living cells have antigens (substances that induce an immune system response—i.e., the production of antibodies). Lateral flow devices (LFDs) are a form of immunoassay based on antibody-antigen interactions—using the same principle as early pregnancy tests. LFDs are field versions of the laboratory enzyme-linked immunoassays (ELISAs). The panel of devices used in ASTM D8070 [19] includes three pairs of LFDs. As for ASTM D7463 [12], the test requires an extraction step to partition microbial antigens from fuel-phase samples into an aqueous reagent. The aqueous extract is applied to each of the LFD. One pair of ASTM D8070 LFD has antibodies that bind with antigens generated by a range of bacteria. A second pair has antibodies specific to fungal (both yeast and mold) antigens. A third pair has antibodies unique to the fungus *Hormoconis resinae,* historically reported as a common contaminant of aviation fuel. The test manufacturers suggest that testing fuel-associated water phase samples where available will provide the most representative determination of microbial contamination within a fuel system. For each pair, one LFD detects low levels and the other detects high levels of the target antigens. Interpretation of results is visual (by eye) or using a digital application.

After microbicide treatment, not all antigens deteriorate at the same rate. Consequently, immunoassay testing can detect antigens associated with intact metabolically active cells, cells reactivating from a state of dormancy, intact moribund cells, cell fragments, or those present as extracellular molecules. However, immunoassays will not detect dead or dormant cells. Consequently, immunoassay testing to assess biocide treatment efficacy is not recommended.

The primary advantages of immunoassay testing are speed and that no microbiological expertise or apparatus other than the test kit components are needed. The determination of test line disappearance when results are read by eye can be subjective and consequently contribute to test result variability.

2.7.5 Microscopy

Particulates and free water observed in a sample can be examined using phase contrast light microscopy to assess whether they contain bacteria, yeasts, or molds. This can be complemented with the use of staining techniques and application of specific fluorescent dyes which can inform on the proportion of live and dead microorganisms. The microscopy method IP 472 [20] enables a quantitative assessment of the number of fragments of filamentous fungal material in fuel samples. Figure 17 illustrates how fungal filaments appear when viewed through a microscope at ×400 magnification in accordance with IP 472.

The primary advantage of microscopic examination is that all microbes present in the specimen can be seen. However, the small (at ×1200, the observed field of vision diameter is approximately 100 μm) field of vision means that the limit of detection is approximately 10^7 cells mL^{-1}. Concentration by either filtration or centrifugation can be used to increase the number of cells mL^{-1}. Although specimens can be prepared for microscopic observation within minutes, specialized staining protocols can require 30 min to 1 h to complete. The total time required to complete a direct count test

Fig. 17 IP 472—fungal fragment as seen through a microscope at ×400 magnification. Image curtesy of ECHA Microbiology Ltd., Cardiff, UK

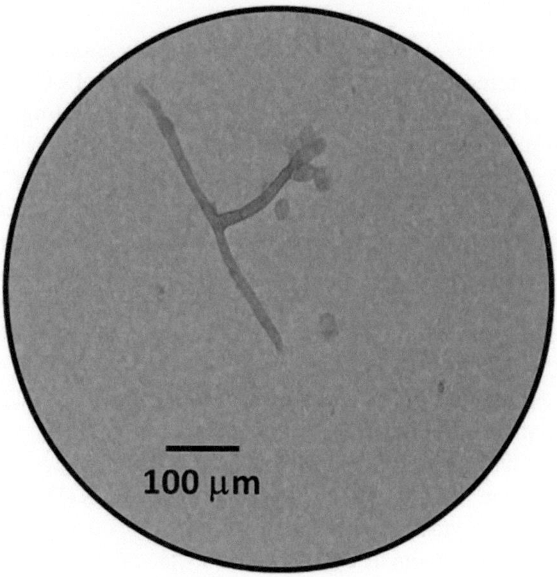

varies substantially depending on the specimen preparation techniques used. It can range from 15 min to 1.5 h. In samples that contain small particles along with microbes, considerable training is required to differentiate between microorganisms and inanimate particles.

2.7.6 Molecular Microbiological Methods

The term "Molecular Microbiological Methods" refers to a growing number of techniques that do not depend on cultivation or microscopy. Although strictly speaking, MMM includes chemical tests such as ATP, carbohydrate, and protein concentration assays, it more commonly refers to genomic, transcriptomic, and proteomic test methods. Genomic tests qualitatively or quantitatively detect DNA. Transcriptomic methods quantify or characterize messenger RNA (mRNA)—the molecule which serves as a transcript of the genetic information found on the DNA translated into a protein through a process mediated by transfer RNA (tRNA) and ribosomal RNA (rRNA) (i.e., the sequence of amino acids in protein molecules is defined by the mRNA's nucleotide sequence). Proteomic methods detect all of the extracted proteins present in a specimen and are used to investigate gene expression. Historically, MMM have been used by highly trained analysists, primarily as research tools. However, in recent years, commercial test kits have been developed—making the technology more generally available. The first steps for all MMM test methods are to concentrate and separate microbes from the specimen matrix (jet fuel or fuel-associated water). The most common techniques for accomplishing these are filtration and centrifugation. Once the microbial cells have been concentrated and isolated, they are lysed to release their contents into a buffer solution. This extract includes DNA, RNA, and proteins.

Genomics—the two most commonly used genomic tests are quantitative polymerase chain reaction (qPCR) and DNA sequencing. Both types of tests rely on the use of a DNA primer molecule. For bacterial qPCR testing, 16S rRNA is used. For fungal testing 18S rRNA or internal transcribed spacer (ITS) is used. For both bacterial and fungal qPCR testing, the primer includes a conserved region (a sequence of nucleotides that appears to be universally present in all cells) and a hypervariable variable region (a sequence of nucleotides that is unique to a single microbe or a group of microbes—such as sulfate reducing bacteria—that share an identical gene). Primer selection is based on the degree of specificity desired (e.g., all microbes with a common function—such as sulfate reduction—or a specific species). ASTM Guide D8412 [21] provides a qPCR protocol and includes a table listing primers appropriate for detecting specific target microbes. DNA sequencing uses primers that have only the conserved region. The next step is amplification. The specimen is added to a mixture that contains the four deoxynucleotide triphosphates (dNTP—adenine (dATP), cytosine (dCTP), guanine (dGTP), and thymine (dTTP)), and Taq DNA polymerase (an enzyme originally isolated from the thermophilic bacterium *Thermus aquaticus* and which catalyzes the formation of DNA double helices from denatured DNA). The dNTPs are labeled with a fluorescent dye so that as the number of DNA copies increases so too does the intensity of fluorescent light emitted. The specimen is then subjected to repeated heating and cooling cycles so that the DNA concentration doubles after each cycle. Theoretically, after 30 cycles, there are 2^{30} (1 billion) copies of each molecule DNA originally in the sample. This process is called amplification (Fig. 18). With qPCR, only DNA that has the gene that matches the primer's hypervariable region is amplified. With DNA sequencing (commonly referred to as next generation sequencing, or NGS), numerous segments (amplicons) or all of the DNA (for total DNA *shotgun* sequencing) is amplified. With qPCR testing, the number of annealing cycles required to create a critical number of gene copies (i.e., the cycle threshold—point at which the specimen's fluorescence is significantly greater than that of a blank, control specimen) is recorded.

Data from specimens are compared against a standard curve so that the number of gene copies per mL (GC mL^{-1}) in the original specimen can be computed. Some protocols use the typical number of copies of the target gene per cell to transform GC mL^{-1} results to cells mL^{-1}. The number of different target microbes for which commercially available qPCR kits are available is growing rapidly [21].

For DNA sequencing, after the amplification process, the different DNA molecules are sequenced (dNTP order is determined) and the sequences are compared both against the other DNA molecules in the sample and against sequences stored in meta databases (DNA sequence libraries). The results are typically reported as relative abundance of each Operational Taxonomic Unit (OTU) present (Table 2), as dendograms (illustrating genetic diversity—Fig. 19), or as Krona plots (illustrating genetic relationships—Fig. 20).

To date, limited genomic testing has been reported for jet fuel. However, those reports suggest that dust particles that enter tanks through vents are a major source of microbes in jet fuel systems. Although the theoretical sensitivity of genomic test methods is one cell in the specimen, the reality is quite different. Currently available methods have detection limits in the range of 1000 GC mL^{-1}. Cell lysis is a significant factor affecting genomic test sensitivity. Chemicals used to lyse cells can also denature DNA. Consequently, reagents used to extract DNA must be sufficiently strong to lyse 100% of the cells but not strong enough to denature the DNA in a specimen. This continues to be an area of research and development.

Transcriptomics—similar to genomics but focused on RNA. Genes are regulated by physicochemical conditions inside and surrounding the cell. Relatively few genes of a given genome are expressed (turned on) at any given time. Genomic testing provides information about what genes are present but provides no insight as to which genes are expressed. Transcriptomics fills this information gap. Knowing whether genes associated with fuel or fuel system biodeterioration are expressed is expected to be a helpful means of better understanding the conditions under which microbial contamination is likely to cause operational problems. Jet fuel transcriptomic testing is still in its infancy but is expected to be a commercially available tool within the next decade.

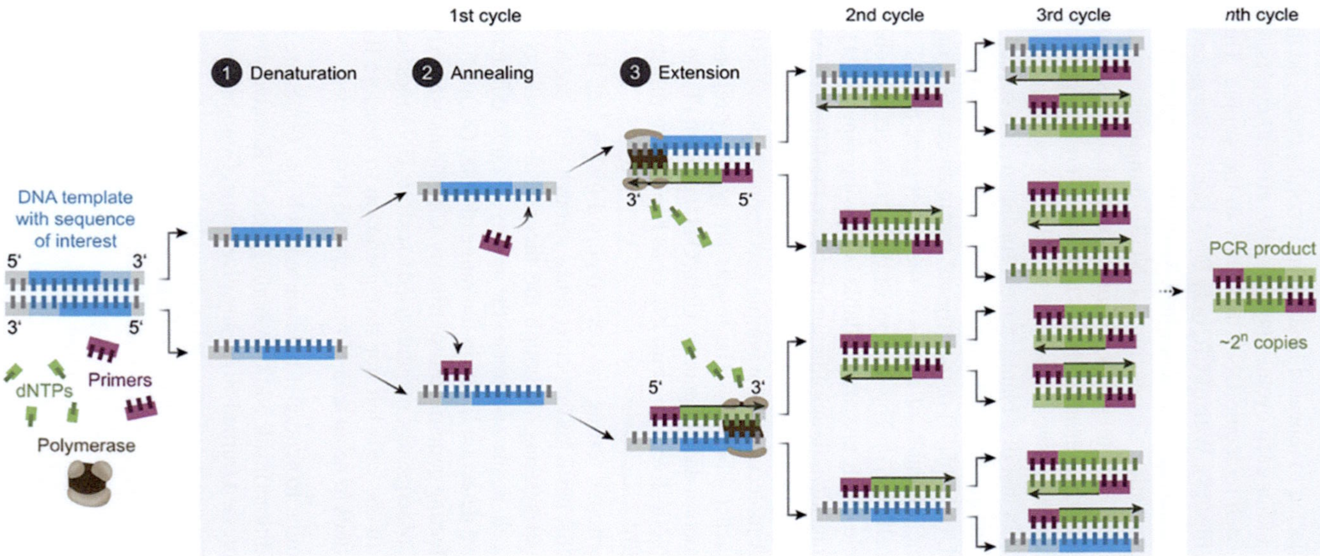

Fig. 18 DNA-amplification during qPCR's denaturing and annealing cycles (*Source* Enzoklop, CC BY-SA 4.0 https://creativecommons.org/licenses/by-sa/4.0, via Wikimedia Commons)

Table 2 16S rRNA-based taxonomic profile of microbiome in sample of bottoms-water under ultra-low sulfur diesel fuel.

OTU	% Abundance (%)
Lactobacillus sp.	66.86
Enhydrobacter sp.	12.18
Methylobacterium sp.	8.22
Unidentified Genus	3.97
Bosea sp.	3.12
Sphingomonas sp.	2.55
Pseudomonas sp.	1.42
Anaerococcus sp.	0.85
Acinetobacter sp.	0.57
Methylotenera sp.	0.57
Streptomyces sp.	0.57
Alishewanella sp.	0.28
Brevundimonas sp.	0.28
Chryseobacterium sp.	0.28
Dethiobacter sp.	0.28
Hydrogenophaga sp.	0.28
Janibacter sp.	0.28
Klebsiella sp.	0.28
Kocuria sp.	0.28
Leptolyngbya sp.	0.28
Stakelama sp.	0.28
Streptococcus sp.	0.28

Proteomics—cells contain structural (e.g., cell wall components) and functional proteins (i.e., enzymes). Proteomics is an evolving field of study focused on understanding how environmental conditions affect what proteins are produced, their individual and relative concentrations, and their configuration (proteins' configurations affect cell structure and enzyme function). As with transcriptomics, little proteomic work has been done with microbial contaminants in jet fuel. However, methods currently under development and refinement are likely to make field test kits available by 2030 or sooner.

2.7.7 Visual Inspection

The "Clear and Bright" test (ASTM Test Method D4176 [22]) is a visual assessment commonly applied to estimate the presence of free water and particulate in fuel samples. Fuel samples that are hazy and contain free water and particulate, will be more likely to contain microbial contamination than samples that are observed to be "Clear and Bright."

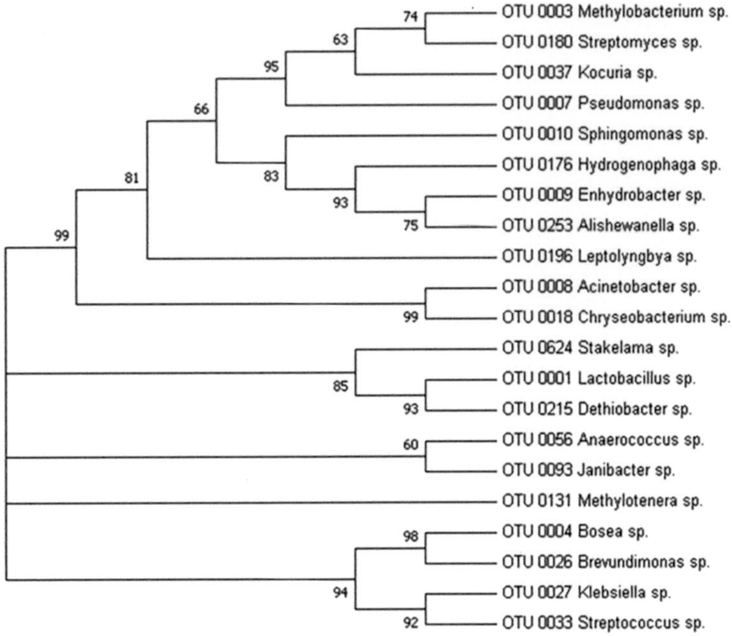

Fig. 19 Genomic profile dendrogram of microbiome in sample of bottoms-water under ultra-low sulfur diesel fuel (from Table 2)

Closer inspection of the characteristics of the free water and particulate present in a sample can inform on whether microbial contamination may be present. If microbial contamination is present, visual examination of fuel tank sump samples may show the presence of translucent, polymeric material (EPS) or soft gray or brown "mat" of material, particularly at the fuel: water interface as shown in Fig. 21. The presence of cloudy water or soft particulate or sludge settled at the bottom of samples can also be indicators of microbial contamination.

It should be noted that significant microbiological contamination may not always be visually evident and samples that are Clear and Bright can still be contaminated with significant numbers of microorganisms. Therefore, visual assessment alone should not be used to determine whether microbiological contamination is present or absent.

2.8 Test Data Interpretation

IATA's Guidance Material on Microbiological Contamination in Aircraft Fuel Tanks [3] lists threshold limits for the reference culture test methods (IP 385 and ASTM D6974). Other culture test methods and test methods based on other technologies are expected to show broad agreement with these contamination limits and have

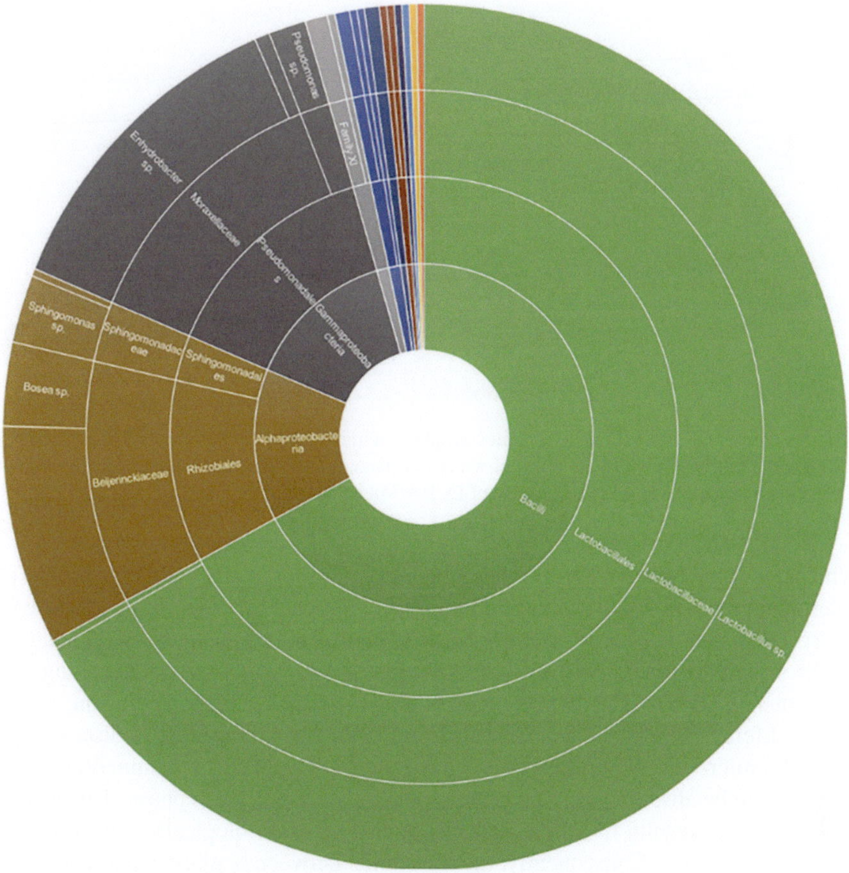

Fig. 20 Genomic profile, Krona plot of microbiome in sample of bottoms-water under ultra-low sulfur diesel fuel (from Table 2)

been published in other documents [3, 7]. Table 3 summarizes categorical ranges for culture and non-culture consensus test methods. For each method three categorical ranges (also referred to as attribute ratings) are defined: negligible, moderate, and heavy contamination. These attribute ratings reflect the collective experience of fuel microbiologists who have compared microbiological test results with other fuel system parameters, including but not limited to: visually observed biofilm on wing tank or sensor surfaces, evidence of tank coating degradation (i.e., discoloration, blistering, or both), and evidence of microbial activity in fuel-associated water [pH < 6, total dissolved solids—TDS –>1 g kg^{-1}, water turbidity and dark color, total suspended solids (TSS) >1 g kg^{-1}, sediment sufficient to cover >10% of the sample bottle's bottom after 1h settling, presence of an invert emulsion (rag) or membranous layer between the fuel and fuel-associated water phases, or any combination of these factors]. The terms "moderate" and "heavy" merely refer to the extent to

Fig. 21 Microbial contamination in jet fuel samples—**a** translucent material comprising of bacterial Extracellular substance (EPS) at the fuel-water interface; and **b** brown floccose fungal growth at the fuel: water interface. Images courtesy of ECHA Microbiology Ltd., Cardiff, UK

which contamination levels are above typical background levels and indicative of microbial proliferation. They do not necessarily indicate that fuel is unfit for use or off specification. However, aircraft OEMs, commercial airlines, and other aviation sector stakeholders use the criteria scores provided in Table 3 to guide testing and maintenance actions. The aim of testing is to detect and highlight significant microbial proliferation before it becomes an operational issue. However, as discussed in Chap. 11, microbial contamination is heterogeneous. Microbial contamination in fuel systems can be absent in samples that do not contain an aqueous phase. This means that negligible results from fuel systems that are microbiologically contaminated are not uncommon. Consequently, the thresholds between negligible and moderate contamination are typically very close to each test method's limit of detection.

The first actions after obtaining any test result that signals moderate or heavy contamination are to compare the microbiological data with data from the other parameters identified in the preceding paragraph. If the microbiological data are not supported by any of the other test results, resample and retest. As discussed in Chap. 11, there is a risk of contaminating samples with microbes coming from other sources (sampling device, rags, technicians' hands or gloves, etc.). Consequently, a moderate or heavy contamination test result should not trigger disinfection or related corrective maintenance actions until after the initial result has been verified.

If retesting is required, best practice is to run two or more different types of microbiological tests, including a repeat of the original test method used. Generally (~70% of the time), different methods yield similar attribute scores. As discussed above, each of the methods covered in this section, each family of microbiological methods measures different attributes of the contaminant population. Test results from different methods can help to diagnose the actual biodeterioration risk better than results from a single method can.

Table 3 Attribute ratings for selected microbiological test methods

Parameter	Sample type	Test method(s)	Units	Contamination assessment		
				Negligible	Moderate	Heavy
ATP	Fuel	ASTM D7463	RLU L^{-1}	<1000	1000–5000	>5000
	Water		RLU mL^{-1}	<1000	1000-5000	>5000
	Fuel	ASTM D7687	pg mL^{-1}	<10	10–100	>100
	Water		pg mL^{-2}	<100	100–<1000	≥ 1000
Culturability	Fuel	IP 385, ASTM D6974, & IP 613,ASTM D7978	CFU L^{-1}	<4000	4000–20,000	>20,000
	Water		CFU mL^{-1}	<1000	1000–10,000	>10,000
Antigen	Fuel	ASTM D8070	μg L^{-1}	<150	150–750	>750
	Water		μg mL^{-1}	<33	33–166	>166

Recognizing the difficulty detecting microbial contamination in aviation fuel systems, best practice is to maintain process control charts that capture microbiological data along with the other parameters discussed in this section. The control charts should be monitored for trends. Microbial contamination is most easily controlled early in the trend toward moderate contamination (see Sect. 4).

3 Microbial Contamination Prevention

3.1 Controlling Water Accumulation

System design—accessibility, minimizing pooling zones, etc.—to minimize the potential for water accumulation is only one of numerous considerations during fuel tank design. Structural integrity during operations is the primary criterion. This means that even in tanks with sampling and drain ports at nominal low points, opportunities are likely to exist for water pooling. During this discussion, it is important to understand that because the specific gravity of water (1.0) is greater than that of jet fuels (~0.82), free water will settle at a rate dictated by Stokes' Law. The settling rate is proportional to water droplets' diameters. Water droplets tend to join together (*coalesce*) as they descend through fuel. Consequently, the settling rate tends to increase with the distance from the fuel's surface. A second fundamental point is that 1 mL of water can support microbial populations >10^6 cells. Thus, traces of water that might be considered negligible from an engineering perspective are relevant regarding microbial contamination control.

Jet fuel bulk storage tank bottom configurations are typically concave to facilitate water drainage to a central low point (Fig. 22a). However, the weight of product over tank bottoms can deform the steel plates that form the tank's floor. The angle between the center of a plate's depression is often greater than the tank floor's designed angle. This creates opportunities for water volumes of 1–10 L to accumulate in each of the depressions (Fig. 22b and c. The total trapped volume can be >1 m^3 free water. Moreover, most tank bottom drains are lines with an inverted elbow—the opening of which is 2–10 mm above the floor—that leads to a horizontal pipe that extends through the tank's shell to a valved discharge fitting (Fig. 23). In theory, the pressure of the full head above the free water phase pushes the water into the drain line. In practice, these drain lines are less than 100% effective. Commonly, jet fuel bulk storage tanks are equipped with recirculation pumps that facilitate bottoms water removal. The fuel passes through a fuel-water separator (Fig. 24) to improve the water-removal process. Additionally, as fuel is transferred from bulk tanks to fueling trucks or fueling hydrants it is passed through filter elements designed to strip out any dispersed water (see Chaps. 9 and 10).

Wing tanks present a unique water collection and removal challenge. Wing tanks generally have a bottom drain close to the aircraft's fuselage, at or near the tank's low point and at various other areas where water is likely to accumulate. However, as illustrated in Fig. 25, wing tanks are ribbed. Although the ribs have holes that allow

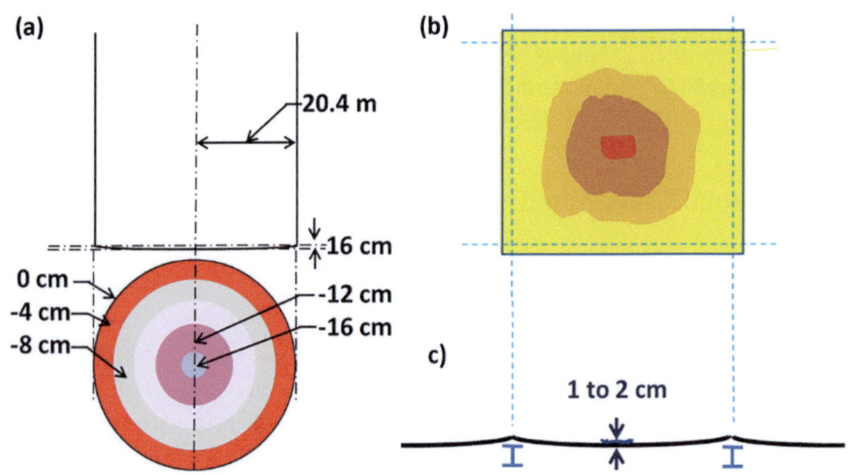

Fig. 22 Jet fuel bulk storage tank water drainage—**a** side and top views of 16000 m^3 (100000 bbl) concave-bottom tank (not to scale); **b** tank deck plate—top view; **c** tank deck plate, side view

Fig. 23 Jet fuel bulk storage tank—water drain

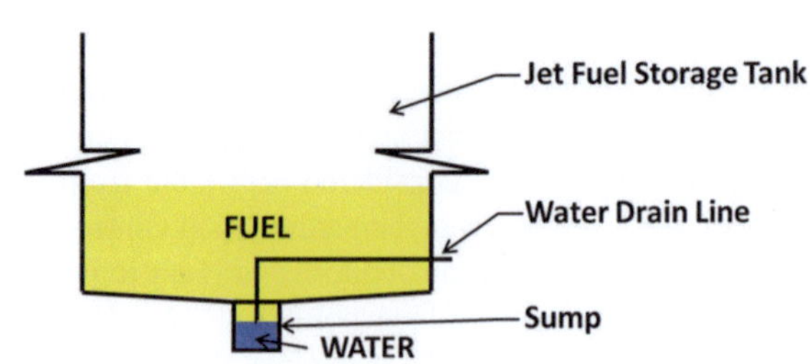

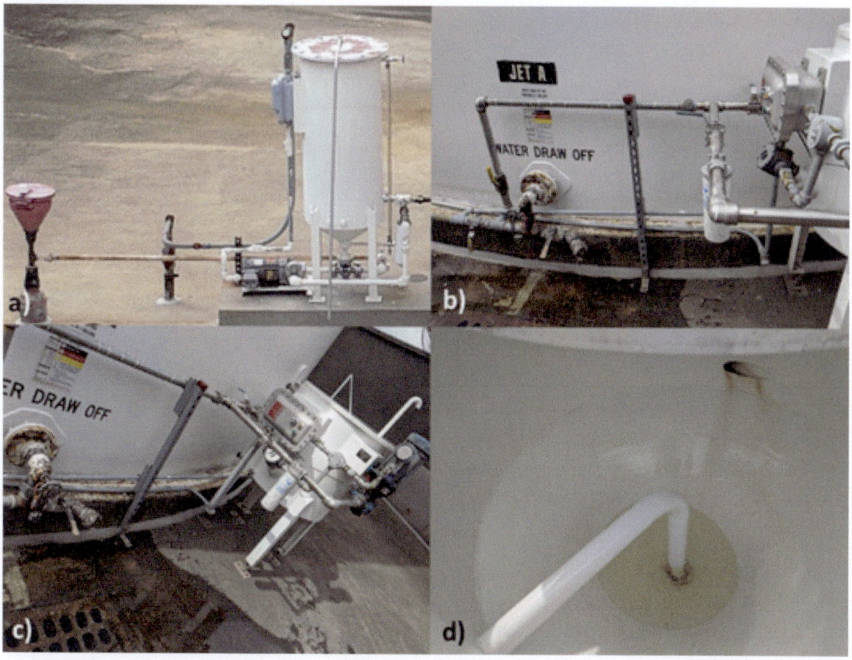

Fig. 24 Jet fuel bulk storage tank—water separator—**a** separator housing; **b** water drain line to separator; **c** water drain line and separator alignment; **d** separator tank interior, showing inlet at top, fuel return line, and water drain (to waste) at bottom. Images courtesy of Biodeterioration Control Associates, Inc., Princeton, NJ, USA

for the free flow of fuel across the tank's space, those holes are typically >1 cm away from the ribs' outer edges.

This is essential for structural integrity purposes but prevents complete water drainage to the drain. Consequently, traces of water are likely to remain trapped along the outboard bottoms of wing tank ribs. It is not uncommon for biofilm accumulation and coating deterioration to be most pronounced between ribs that are furthest from the fuselage.

Water removal—aviation industry stakeholders agree that keeping aviation fuel tanks dry is the most effective strategy for preventing microbial growth. Augmenting fuel storage tank draining practices with recirculation through water coalescing and stripping filtration systems is the best means of capturing free and dispersed water that accumulates in zones from which water cannot be removed by conventional draining practices. See Chaps. 9 and 10 for complete discussions of fuel polishing systems and practices.

Drain intervals—periodic draining is essential to prevent free water accumulation in aviation fuel tanks. Optimally drain intervals are data driven. The rate at which free water will accumulate in a fuel tank depends on the water's state (dissolved,

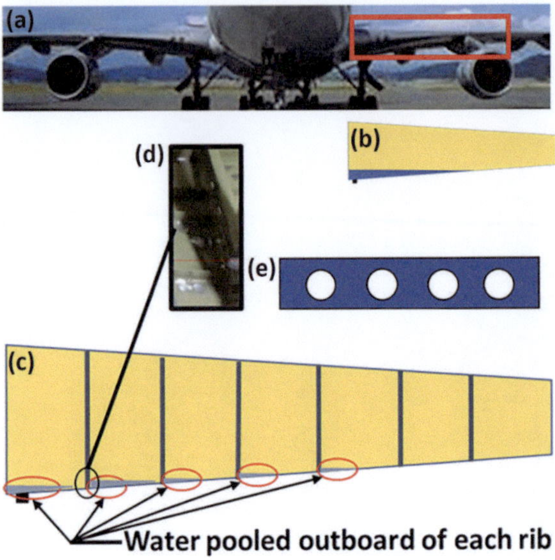

Fig. 25 Water accumulation in wing tanks—**a** Highlighted left wing; **b** schematic of water accumulating in lowest area of wing tank; **c** schematic showing multiple zones where water can accumulate outboard of wing's ribs; **d** photo of bottom of one rib; **e** schematic showing holes through

dispersed, or free), solubility, and concentration. Dissolved water will remain in that state until its concentration exceeds its solubility. At the temperature-dependent saturation point, water droplets will begin to form. Water solubility ranges at different temperatures reflect differences in jet fuel chemistry among refinery runs. At −10 °C water solubility is typically 16 ± 4.5 mg L^{-1} and at 30 °C it is 84 ± 2.8 mg L^{-1}. This means that water can accumulate in its dissolved form while aircraft or on the ground, and separate as dispersed, free, or both forms during flight (for example, at 9.1 km, air temperatures range may be from −57 to −40 °C). Dispersed water appears as fuel haze (cloudiness) due to the dispersion of water droplets that range from <1 to 10 μm diameter.

Over time, water droplets *coalesce*. In static fuel (storage tanks or aircraft sitting on the ground) 10 μm dia water droplets settle at a rate of approximately 1 m h^{-1} based on Stokes' law.

As discussed in Chap. 9, water can become entrained in aviation fuel at each stage of product transport as the fuel moves from refinery tank farms to aircraft. The primary source is tank vents. Tank suction creates a partial vacuum that draws air—including moisture and airborne particles—into the ullage space. Fuel flowing through transport pipelines will also pick up water present in the lines. Because these variables impact the rate of free water accumulation, aircraft maintenance manuals (AMMs) prescribe testing frequencies based on aircraft operations—frequency of flight operations and location of refueling facilities. Water should be drained whenever it is detected.

Remedial Fuel polishing is the process by which water and particulate contamination—including microbial biofilms material and flocs (clumps) of microbial cells—is physically removed from fuel. Typically, polishing operations involve recirculating fuel through a filtration system for at least seven cycles (fuel volumes—e.g., if the tank contains 4 m^3 fuel, ≥28 m^3 would be filtered) until water concentration and particulate loads are below their respective upper control limits. Details of fuel polishing filtration systems are provided in Chaps. 9 and 10. Note that fuel polishing does not eliminate biofilm material from tank surfaces (Sect. 4).

3.2 Condition Monitoring

Effective condition monitoring includes four fundamental elements: collecting the appropriate samples, using proper techniques for collecting and handling samples, sampling at the optimal frequency, and performing appropriate tests.

3.2.1 Using Gross Observations, Physical Test Data, and Chemical Test Data to Augment Microbiological Test Data

The heterogeneous distribution of microbiological contamination has been mentioned repeatedly in the preceding sections of this chapter and in Chap. 11. This heterogeneity creates a risk of biodeteriogenic microbiological contamination remaining undetected although present. This risk can be reduced by using non-microbiological test results to assess whether an active microbial population is present in the tank from which samples have been collected.

- *Gross observations*—hazy fuel or the presence of a free water layer are symptomatic of microbial activity. Microbes produce detergent-like molecules called biosurfactants. Biosurfactants disperse water as droplets in fuel—contributing to fuel haziness. The presence of a membranous zone between fuel and the underlying water provides a stronger indication that the system is home to an active microbial population. Additionally, discoloration (darkening) of a sample's fuel or water phase can reflect microbial activity, as can the presence of suspended solids (i.e., particulates). Each of these symptoms can also be caused by abiotic processes—emphasizing the need for further testing to differentiate between biodeterioration and abiotic deterioration.
- *Physical test data*—Microbial activity tends to contribute to loss of oxidative stability. Thus, reduced induction times can reflect biodeterioration. Similarly, biosurfactant production (described in the previous bullet) can reduce water separability. Microbial metabolites, fragments, and whole cells can contribute to fuel's conductivity. As for gross observations, biodeterioration shares symptoms with abiotic deterioration processes. Consequently, changes in physical test data indicate the need for further testing to identify the cause.
- *Chemical test data*—In-specification jet fuel has an acid number that may be limited from <0.015 mg KOH g^{-1} (Def Stan 91-091 or similar) or <0.1 mg KOH g^{-1} (ASTM D1655 or similar). Consequently, when the acid number is >0.1 mg

KOH g^{-1} it is possible that biogenic, low molecular weight organic acids are contributing to the increase.

4 Microbial Contamination Remediation

4.1 Biocidal Product Use

Technically, biocides can be applied to fuel as a preventative treatment or as a curative treatment. Sometimes the same biocide can be used for both purposes, but dose rates for preventing microbial growth will be lower than the dose rates used for remediation. Microorganisms will be active in water associated with fuel, so the biocide needs to be dispersible in fuel, and the partition coefficient between the fuel and water should be such that the biocide partitions to the water phase. Biocides are only added to aviation fuel at the point of use in aircraft fuel tanks to remediate microbial contamination. This is because of the possibility that multiple additions of biocide could result in overdosing and biocide concentrations exceeding Original Equipment Manufacturers (OEMs) limits. Also, some aviation fuel specifications restrict the use of biocides by fuel suppliers. Only biocides that are approved by airframe, engine, and auxiliary power unit (APU) OEMs can be used for the treatment of aircraft fuel tanks.

At the time of writing, only one biocide, a dioxaborinane-based product, is available to remediate microbial contamination in aviation fuel. From the microbiological point of view, it would be advantageous to have more than one biocide option in the armory. Generally, the overuse or suboptimal use of biocides can result in microorganisms becoming tolerant or resistant to the antimicrobial agent. The industry is eager to find alternative chemistries, but this is not likely to be a quick process. Biocides need to be registered for compliance with the relevant regulations to be legally sold in specific geographical regions, for example compliant with the European Union Biocidal Products Regulation (EU BPR) or registered with the Environmental Protection Agency (EPA) in the US. The costs of regulatory compliance are very high for biocide manufacturers—creating a barrier to the development of new active substances and formulations. Furthermore, the evaluation process for new aviation fuel additives according to ASTM Practice D4054 [23] requires additional investment and time.

Test methods for monitoring microbiological contamination in aircraft fuel tanks work on the principal that they differentiate between negligible, moderate, and heavy levels of contamination (see Sect. 2.6). A biocide treatment of all fuel tanks on an aircraft needs to be performed if moderate microbiological contamination is detected and confirmed. Detection of heavy contamination indicates that there may be some colonization of the fuel tank surfaces by microorganisms, for example as a biofilm (Sect. 1.1.2). In these cases, the treatment procedure requires additional tasks to be completed prior to biocide application; tank entry and inspection of surfaces and components for corrosion and manual cleaning to remove dirt and microbial contamination.

It is important to appreciate that there are several factors that can impact the effectiveness of a biocide treatment. Underdosing of biocide can result in a dramatic reduction of efficacy, whereas overdosing can cause serious hardware compatibility issues. Biocides should always be applied at the correct concentration using a procedure that achieves even dispersion of biocide in the fuel. A dosing procedure of pouring biocides into fuel aircraft tanks via over wing ports and uplift of untreated fuel to aid mixing is no longer advocated. It is recommended that biocide is injected into fuel using metered injection equipment as it is being uplifted onto aircraft. Energy Institute standard EI 1566 [24] provides guidance on suitable injection equipment.

Insufficient biocide contact time will allow some microorganisms to survive the treatment and recolonize the tank. Biocide manufacturers usually recommend a range of contact times; the lower end of the range should be considered the absolute minimum. A factor that is often overlooked is the temperature at the time of biocide application. Biocides become progressively less effective as temperature decreases, meaning that significantly longer contact times may be required to achieve a complete kill. Moreover, flight time should not be included as part of the contact time. This is because the temperature of fuel will drop so low in flight that the biocide will lose its effectiveness and stop killing microorganisms.

A strategy for biocide treatment of storage tanks in the aviation fuel supply chain is the spray application of surfaces with an aqueous biocide solution, for example sodium hypochlorite. This approach avoids concerns over the presence of biocide residues in fuel providing surfaces are thoroughly rinsed after the treatment. Compatibility of biocide with the system and the disposal of biocide treated water need to be considered. Spray and wipe treatment with a 70% isopropanol solution can be a good option for decontamination of small tanks and filter vessels.

4.2 Fuel Polishing and System Cleaning/Disinfection

4.2.1 Disinfection

Commonly, when moderate or heavy microbiological contamination is detected in fuel systems, microbicides (commonly referred to as biocides) are used to disinfect the system. As this chapter goes to press, there is only one biocidal product approved by airframe and gas turbine engine manufacturers for use in jet fuel. The product is a blend of two chemicals: 2,2′-((1-methyl-1,3-propanediyl)bis(oxy))bis(4-methyl-1,3,2-dioxaborinane) (67.6% active ingredient—a.i.; CAS 14697-50-8) and 2,2′oxybis(4,4,6-trimethyl-1,3,2-dioxaborinane (27.4% a.i.; CAS 2665-13-6). The product is registered in the U.S. under U.S. EPA registration number 65217-1 but does not have a Biocidal Products Regulation (BPR) docket number and is therefore not approved for use within the EU. At the time of writing, this product has been given temporary derogation for use in the UK and EU. Before 2019, another

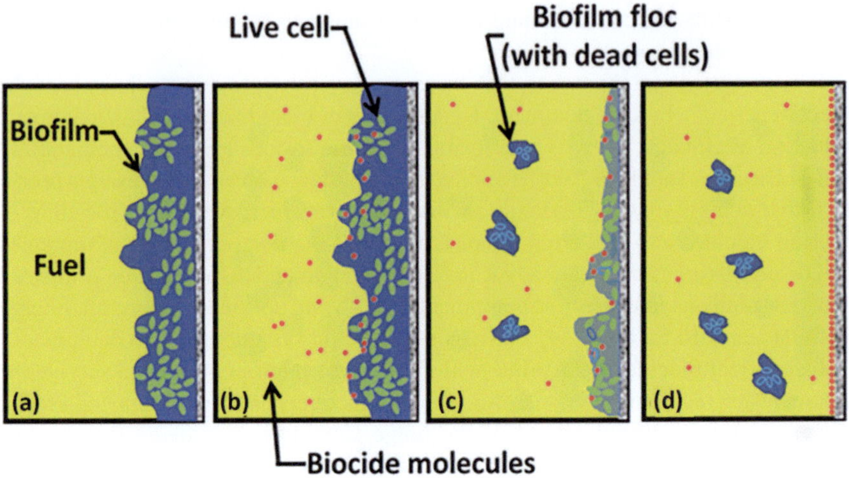

Fig. 26 Effect of biocide treatment on biofilm communities—**a** biofilm on surface before biocide treatment; **b** after initial treatment (red dots are biocide molecules) showing limited biocide penetration into biofilm matrix; **c** after second biocide treatment showing detachment of biofilm flocs; **d** after third biocide treatment surface is disinfected and biocide forms protective film (note—only certain biocides form film)

biocidal product, with both BPR and U.S. EPA registrations was approved for use to treat jet fuel. However, the manufacturer has since removed jet fuel use from their product's label. As an emergency measure, the EU has at least temporarily permitted the dioxaborinane product to be used.

The dioxaborinane product is generally used at 200 mL m^{-3}. Typically, the biocide is proportioned into fuel as described in Chap. 9. Biocide efficacy depends on active ingredient concentration and exposure time. The recommended exposure (soak) for the dioxaborinane product is 36–72 h. Because biocides are consumed as they kill microbes and have difficulty penetrating biofilms, it can be necessary to treat systems more than once, to clean tanks physically, or do both to effectively disinfect tank surfaces. Figure 26 illustrates how multiple biocide treatments disinfect biofilm communities. Aircraft and surface tanks should remain idle during the soak period. Non-oxidizing biocides, such as the dioxaborinane product mentioned above, depend on microbial metabolism for their biocidal activity. Consequently, dormant microbes are likely to survive biocide treatment.

4.2.2 Fuel Polishing

During disinfection, masses (flocs) of biofilm material slough off of tank walls into the fuel and settle to the tank bottom. Consequently, after a fuel system has been disinfected the fuel should be polished to remove the debris and detritus created as microbes were killed. Fuel polishing is a filtration process described in Chap. 9. IATA

[3] recommends that biocide treated fuel be burned through the engines although most often the fuel is offloaded and used for non-aviation purposes.

4.2.3 System Cleaning

- *Bulk storage tanks*—best practice is for fuel to be removed, and the tank is then prepared for safe access. Personnel enter the tank and use a combination of methods to clean all interior surfaces (shell and floor). Cleaning devices include high-pressure hot water or steam. Depending on the amount of residue on surfaces, chemical cleaning agents can be used to facilitate its removal. Typically, shovels, mops, and squeegees are used to remove all traces of deposits and water from the tank floor. Although not required, it is common for tank surfaces and internal fittings to be inspected for damage as part of a cleaning effort. Fuel removed from bulk storage tanks can be polished while the tank is being cleaned.
- *Aircraft tanks*—as with bulk storage tanks, all product is removed, and tanks are rendered gas-free. Depending on tank configuration, high pressure devices can be used to clean internal surfaces, if permitted by the airframe manufacturer. Most aircraft maintenance manuals prescribe having all contaminated internal surfaces wiped clean with lint-free cloths or a stiff brush after they have been washed. A final wipe-down using isopropanol or methanol saturated, lint-free cloths further disinfect internal surfaces and drives off any residual moisture that might be present.

Acknowledgements The authors would like to thank Elizabeth M. Bowman and Myrsini Chronopoulou for their helpful comments and additional insights during the preparation of this chapter.

References

1. ASTM D6469-24 (2024) Standard guide for microbial contamination in fuels and fuel systems. ASTM International, West Conshohocken. https://doi.org/10.1520/D6469-24
2. Passman FJ (2003) Introduction to fuel microbiology. In: Passman FJ (ed) Fuel and fuel system microbiology: fundamentals, diagnosis, and contamination control. ASTM International, West Conshohocken, pp 1–23. https://doi.org/10.1520/MNL47-EB
3. International Aviation Transportation Association (2023) Guidance material on microbiological contamination in aircraft fuel tanks, 6th edn. IATA, Montreal
4. Energy Institute (2019) Guidelines for the investigation of the microbial content of liquid fuels and for the implementation of avoidance and remedial strategies, 3rd edn. Energy Institute, London. ISBN 9781787251151
5. Madigan MT, Martinko JM, Bender KS, Buckley DH, Stahl DA (2015) Brock biology of microorganisms, 14th edn. Pearson, Glenview. ISBN 978-0-321-89739-8
6. Energy Institute (2020) Guidelines on detecting, controlling, and mitigating microbial growth in oils and fuels used at power generation facilities. Energy Institute, London, 127 p. ISBN 9781787251885 (energyinst.org)

7. Joint Inspection Group (JIG) (2023) Technical Information Document #1 (TID#1) Microbial growth in fuel and monitoring strategies (Edition 2). https://www.jig.org/standards-publications/tid-1-microbial-growth-in-fuel-and-monitoring-strategies-edition-2/
8. ASTM D130-10 (2019) Standard test method for corrosiveness to copper from petroleum products by copper strip test. ASTM International, West Conshohocken, PA. https://doi.org/10.1520/D0130-19
9. EI 1581 (2024) Specifications and laboratory qualification procedures for aviation fuel filter/water separators, 7th edn. Energy Institute, London. ISBN 9781787254374
10. Passman FJ (2018) Chapter 8 | Sampling. In: Rand S, Verstuyft A (ed) Significance of tests for petroleum products, 9th edn, pp 119–142. ASTM International, West Conshohocken. https://doi.org/10.1520/MNL120170018
11. ASTM D7464-25 (2025) Standard practice for manual sampling of liquid fuels, associated materials and fuel system components for microbiological testing. ASTM International, West Conshohocken, PA. https://doi.org/10.1520/D7464-25
12. ASTM D7463-25 (2025) Standard test method for adenosine triphosphate (ATP) content of microorganisms in fuel, fuel/water mixtures, and fuel associated water. ASTM International, West Conshohocken, PA. https://doi.org/10.1520/D7463-25
13. ASTM D7687-23 (2023) Standard test method for measurement of cellular adenosine triphosphate in fuel and fuel-associated water with sample concentration by filtration. ASTM International, West Conshohocken, PA. https://doi.org/10.1520/D7687-23
14. Passman FJ, Duguay J, Maradukhel G, Merks M (2014) Adenosine triphosphate testing—recent advances in the differentiation between bacterial and fungal contamination and detection of dormant microbes in fuel and fuel associated water samples. In: Morris RE (ed) Proceedings of the 13th international symposium on the stability and handling of liquid fuels; 06–10 October 2013, Rhodes, Greece. http://iash.omnibooksonline.com/
15. IP 385 (2020) Determination of the viable aerobic microbial content of fuels and fuel components boiling below 390 °C—filtration and culture method. Energy Institute, London. ISBN IP385-7251830
16. ASTM D6974-25 (2025) Standard practice for enumeration of viable bacteria and fungi in liquid fuels—filtration and culture procedures. ASTM International, West Conshohocken, PA. https://doi.org/10.1520/D6974-25
17. IP 613 (2014) Determination of the viable aerobic microbial content of fuels and associated water—thixotropic gel culture method. Energy Institute, London. ISBN IP613-2936900
18. ASTM D7978-25a (2025) Standard test method for determination of the viable aerobic microbial content of fuels and associated water—thixotropic gel culture method. ASTM International, West Conshohocken, PA. https://doi.org/10.1520/D7978-25A
19. ASTM D8070-21 (2021) Standard test method for screening of fuels and fuel associated aqueous specimens for microbial contamination by lateral flow immunoassay. ASTM International, West Conshohocken, PA. https://doi.org/10.1520/D8070-21
20. IP 472 (2019) Determination of fungal fragment of fuels boiling below 390 °C. Energy Institute, London. ISBN IP472-2934869
21. ASTM D8412-21 (2021) Standard practice for quantification of microbial contamination in liquid fuels and fuel-associated water by quantitative polymerase chain reaction (qPCR). ASTM International, West Conshohocken, PA
22. ASTM 4176-22e1 (2022) Standard test method for free water and particulate contamination in distillate fuels (visual inspection procedures). ASTM International, West Conshohocken, PA. https://doi.org/10.1520/D4176-22E01.
23. ASTM D4054-25 (2025) Standard practice for evaluation of new aviation turbine fuels and fuel additives. ASTM International, West Conshohocken, PA. https://doi.org/10.1520/D4054-25
24. EI 1566 (2021) Injection of biocide into jet fuel during underwing pressure refueling for aircraft maintenance, 1st edn. Energy Institute, London. ISBN 9781787252783

Chapter 13
Corrosion in Aviation Fuel Systems

Alec Groysman

Abstract Previous chapters shed light on crude sources and processing, additives, and microbiological growth in aviation fuels. Logically, they connect with corrosion. This chapter explains general principles and fundamentals of corrosivity of aviation fuels, how to control, monitor, and prevent corrosion of metallic parts (and partly polymers) in contact with aviation fuels.

1 Corrosion Definitions

It is worth starting our journey into the world of corrosion and aviation fuels with the definitions of *corrosion*, *corrosivity*, and *aviation fuels*. This is a philosophical approach to any subject under consideration.

Corrosion is the physicochemical interaction between a metallic material and its environment that results in changes in the properties of both the material and the environment [1, 2]. I intentionally emphasized the word *changes* because in most previous definitions the word *corrosion* implies *deterioration* instead of the broader concept, *changes* [3]. This revised corrosion definition reflects current understanding that corrosion includes *constructive* phenomena, such as use of corrosion of a metal in batteries and in cathodic protection, and others [4].

Corrosivity is the ability of an environment to cause corrosion of a metal in a given corrosion system. The latter is a system consisting of one or more metals and those parts of the environment that influence corrosion [1].

Contamination from trace amounts of sodium, potassium, and other alkali metals in aviation fuels can cause high temperature or hot corrosion in the turbine section of the engine [5]. Corrosion caused by exhaust gases after combustion of the fuel is out the scope of this chapter.

A. Groysman (✉)
Haifa, Israel
e-mail: alecgroysman@gmail.com

M. Rabaev et al. (eds.), *Aviation Fuel Problems and Solutions: From Refinery to Wingtip*, https://doi.org/10.1007/978-3-032-05514-9_13

Material includes construction and any equipment contacting aviation fuels. Relevant materials include tankage, pipelines, hoses and pumps, engines, valves, and various fuel-contacting components in aircraft and aviation facilities. The following materials contact aviation fuels: metals, alloys, polymers, and composites. All can be impacted negatively by aviation fuels, additives, corrosive contaminants, and corrosion products.

Tanks and pipelines are made from carbon steel; airplane wing tanks—from aluminium alloys; hoses, gaskets, O-rings, and sealants—from polymeric materials; bushings and bearings—from composites.

Thus, we should know the properties of aviation fuels and materials, interaction of additives and contaminants with fuels and their role to understand potential corrosion mechanisms and how to control deterioration of both.

Aviation fuels can degrade and deteriorate by corrosion products so that the fuels no longer meet product specifications (Chap. 1). This chapter will focus on corrosive contaminants such as water, oxygen, and microorganisms that influence corrosion in jet fuels significantly.

1.1 General Theory of Corrosion

Corrosion is the reaction between a metal and the environment. There are two corrosion mechanisms: by non-electrolytes and in the presence of electrolytes [4].

Non-electrolytes are the substances that do not dissociate into ions and do not conduct electric current when dissolved in water. Electrolytes are the substances whose water solutions conduct electric current on account of free ions (H^+, Na^+, Ca^{2+}, OH^-, Cl^-, $SO_4{}^{2-}$, etc.). Pure water is a weak electrolyte. Sea water is an aqueous solution of strong electrolytes (mostly NaCl and $MgSO_4$). Non-electrolytes include gaseous oxygen (O_2) and liquid sulfur (S_8). When iron meets gaseous O_2 or liquid S_8, it reacts according to reactions:

$$2Fe_{(s)} + 3/2\ O_{2(g)} \rightarrow Fe_2O_{3(s)} \quad (1)$$

$$8Fe_{(s)} + S_{8(l)} \rightarrow 8FeS_{(s)} \quad (2)$$

Thus, if carbon steel contacts hot atmosphere in the furnace or liquid sulfur in petroleum products, it can corrode according to reactions (1) and (2), respectively. These corrosion reactions occur under 'dry' conditions: without water, or more precisely, without electrolyte. Sometimes this mechanism is called 'dry' or 'chemical' corrosion, and there is no electric potential and current on metal contacting non-electrolytes.

The second corrosion mechanism in the presence of electrolytes is more common because water is present in the atmosphere, in soil and can be dissolved in fuels. When carbon steel construction is immersed in water containing dissolved oxygen, the iron corrodes in an anodic reaction (3), and liberated electrons are taken away by dissolved oxygen in cathodic reaction (4). Anodic, or oxidizing, processes occur, namely—passing of positive metallic ions into solution and releasing of electrons (e^-) on the metal surface:

$$Fe_{(s)} \rightarrow Fe^{2+}_{(aq)} + 2e^- \quad (3)$$

Cathodic reaction or reduction—any process of obtaining of electrons:

$$O_{2(g)} + 2H_2O_{(I)} + 4e^- \rightarrow 4OH^-_{(aq)} \quad (4)$$

If iron contacts acidic solution, as well as this (4) reaction, another cathodic process mostly occurs:

$$2H^+_{(aq)} + 2e^- \rightarrow H_{2(g)} \quad (5)$$

Strictly, this reaction can also occur in neutral aqueous solutions.

The solubility of oxygen in water is approximately 8 mg/kg, but this amount is enough for corrosion to occur. Two electrochemical reactions, anodic (3) and cathodic ((4) or (5)), occur simultaneously on the metal surface. Elimination of one of these reactions will decrease and even stop corrosion of a metal. Thus, by removing dissolved oxygen from water, we can significantly decrease corrosion. This corrosion mechanism in the presence of electrolytes is called *electrochemical corrosion*. Now we can give a definition to *electrochemical corrosion*–corrosion that occurs on a metal surface or inside involving two simultaneous reactions, anodic and cathodic, and its rate depends on electric potential. Anode is oxidized and corrodes. Cathode does not corrode.

Usually, in fuel systems, fuel-associated water contains dissolved salts. Therefore, it is important to emphasize that the rate of metal corrosion in such solutions depends on salt content. Corrosion rate will increase with increase of salt concentration in water according to the Eq. (6) [4]:

$$I_{corr} = E/R \quad (6)$$

where:

I_{corr}—an electric current (corrosion rate) on the metal surface.
E—an electric potential difference between cathode and anode areas.
R—an electrical resistance of an electrolyte.

When salt (or any other electrolyte) is added to water, the electrical resistance of the electrolyte solution (R) diminishes, and the corrosion current I_{corr} (corrosion rate) increases accordingly (6). That is why corrosion of carbon steel equipment in sea water is larger than that in tap or river water.

However, the corrosion rate reaches a maximum value with increase of salt concentration and then diminishes because salting out of dissolved oxygen [4]. A cathodic process on the metal surface in neutral aqueous solutions is a reduction of oxygen according to reaction (4). If salt is added to water, the concentration of dissolved oxygen decreases in aqueous solution. This phenomenon is called salting out [4]. If the oxygen concentration in aqueous solution diminishes, there is no consumer for the electrons released in anodic reaction of metal dissolution, and corrosion decreases. To sum up, addition of salts to water influences corrosion of carbon steel by means of two factors: decreases electrical resistance of the solution (and thus increases the corrosion current), and cathodic participant oxygen concentration in water. Up to some salt concentration value in water, the first factor prevails over the second. Then, the more prevailing factor is the concentration of dissolved oxygen. Therefore, the dependence "corrosion rate—salt concentration" is described by the curve with a maximum [4]. We can even add such amount of salt that nearly all dissolved oxygen will be salted out. If there is no cathodic reaction in neutral solution, corrosion may be decreased nearly to zero. Therefore, there is almost no corrosion and … life in the Dead Sea (Israel).

When metallic equipment contacts water, soil or atmosphere of high humidity, electrochemical corrosion occurs with the participation of water and dissolved oxygen. The result is a brick-reddish-brown rust $FeO \cdot Fe_2O_3 \cdot nH_2O$ which is not dissolved in water (Fig. 1).

Fig. 1 The inner corroded surface of carbon steel AISI 1025 pipe contacting kerosene. Diameter—305 mm. Wall thickness—9.2 mm. Shallow pits are revealed on the inner surface. Service period—65 years. Photographer: the author. AISI—American Iron and Steel Institute

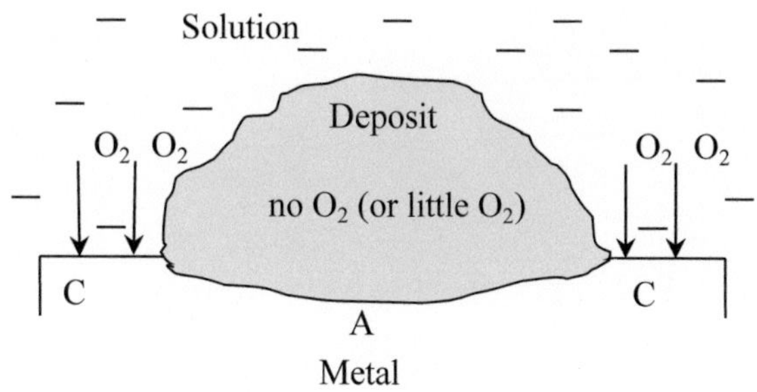

Fig. 2 Differential aeration cells are formed in the presence of any deposits. C—cathode; A—anode

1.2 Differential Aeration Cell

We often observe the uneven spreading of rust inside and outside of tanks and pipelines containing fuels, sometimes shallow pits under deposits on carbon steel surface in fuels, atmosphere, soil or in water, and severe corrosion of various structures and equipment on the interface air/fuel, water/fuel, air/soil, or air/water. If two sites on a carbon steel surface differ in dissolved oxygen concentrations, these sites acquire different electric potentials, and a *differential aeration cell* appears (Fig. 2) [4].

A carbon steel surface with a larger dissolved oxygen concentration on it will be a cathode (oxygen participates in reduction process) and will not corrode. A carbon steel surface with a smaller dissolved oxygen concentration on it will be an anode (electric potential will be lower than that of the area with a higher oxygen concentration) and will corrode. Thus, we may call a *differential aeration cell* by *oxygen concentration cell.*

Differential aeration cells are responsible for the pitting corrosion, crevice corrosion, corrosion of structures and equipment at the interface air/water, air/soil, corrosion in fuels, water, atmosphere, and corrosion in soils. Diverse corrosion phenomena exist. We will describe some of them in the next section.

2 Corrosion Phenomena

Corrosion processes occur primarily on metal surfaces. However, pits and cracks can also occur inside the metal. The corrosion form is a result of a corrosion process. It is possible to define the corrosion phenomenon according to the corrosion form. On the other hand, if we know the corrosion phenomenon type, we may expect what corrosion form, or deterioration type, will happen with a metal: uniform thickness loss, pits, cracks, blistering, delamination, exfoliation, etc.

There are both many metals (85 pure metals and thousands of alloys) and a lot of environmental types, and we might expect many different corrosion phenomena. Despite this, all corrosion phenomena are divided into two types: uniform (general) and non-uniform (localized) corrosion (Fig. 3).

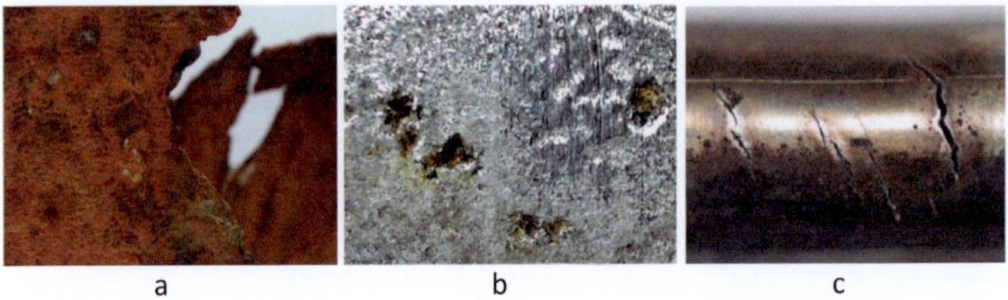

Fig. 3 **a**—Uniform (general) corrosion of carbon steel of the boat, Hof Dor, Israel, Photographer: Inna Kaprovsky. **b**—Pitting corrosion of aluminum alloy Al 5052. **c**—Stress corrosion cracking of stainless steel 316

Uniform corrosion is a process when the entire metal surface corrodes evenly. When we put copper or silver strip in jet fuel to examine the presence of hydrogen sulfide or other sulfur-containing compounds, and the surface of these metals blackens, general corrosion occurs (Eqs. (7) and (8)).

$$2Ag_{(s)} + H_2S_{(g)} + 1/2\ O_{2(g)} \rightarrow Ag_2S_{(s)} + H_2O_{(l)} \quad (7)$$

$$Cu_{(s)} + H_2S_{(g)} + 1/2\ O_{2(g)} \rightarrow CuS_{(s)} + H_2O_{(l)} \quad (8)$$

Localized corrosion is a process when only some definite parts of a metal surface corrode. This corrosion type occurs more commonly than uniform corrosion. Localized corrosion can occur in fuel systems containing water and electrolytes as contaminants. Among them are pitting and crevice corrosion. These corrosion phenomena can occur under deposits on metallic surfaces. Therefore, some authors emphasize *under deposit corrosion* as a separate corrosion phenomenon. Pits of various sizes can be a result of these corrosion phenomena.

We may mention galvanic corrosion, stress corrosion cracking (SCC), exfoliation (type of intergranular corrosion), corrosion fatigue, tribocorrosion (fretting corrosion), erosion (abrasion) and cavitation, filiform corrosion, corrosion of aluminum parts by liquid mercury, and corrosion by exhaust gases and jet efflux. Certainly, galvanic corrosion can occur only in the presence of electrolytes. Despite water and

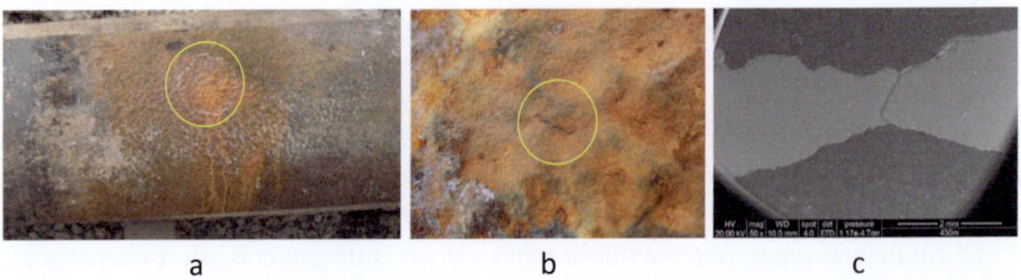

Fig. 4 **a**—General view of the outer surface on the kerosene pipe. **b**—External corroded surface with the crack. **c**—Metallographic cross-section through the pipe, crack area (SEM photo – Scanning Electron Microscope). Photographer: the author

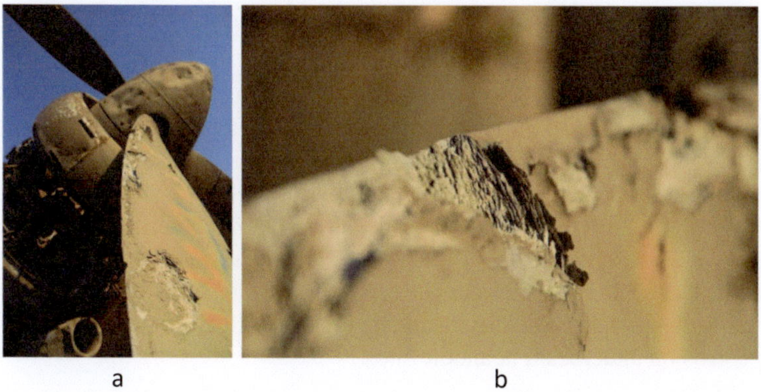

Fig. 5 **a**—Exfoliation of the propeller made from aluminum alloy, atmosphere of the Mediterranean Sea, Israel. **b**—A close-up of part in the picture a. (The Pilots' Memorial, Israel. Photographer Inna Kaprovsky)

salts sometimes being present in aviation fuels, we have not found galvanic corrosion in systems containing aviation fuels. Thus, galvanic corrosion, SCC, exfoliation, and other phenomena occur outside of these systems in aircraft [6, 7]. Some of them are shown in Figs. 4 and 5.

All these phenomena which do not relate to aviation fuels are outside of the scope of this chapter. We will describe pitting, crevice and microbiologically influenced corrosion (MIC) occurring in systems in contact with aviation fuels.

2.1 *Pitting Corrosion*

Pits may appear as a result of presence of chloride (Cl^-) anions in the environment, the existence of differential aeration cells, crevice corrosion, MIC, galvanic corrosion, erosion, and cavitation. First, we shall explain how chlorides cause pitting corrosion.

Passive films are formed on some metals and alloys. For example, a Cr_2O_3 layer can passivate stainless steel, mixed oxides (FeO, Fe_3O_4, and Fe_2O_3) can passivate iron and carbon steels, Al_2O_3 forms a passivating film on aluminum, and a TiO_2 film can passivate titanium. These passive films are susceptible to chloride anion (Cl^-) attack because of defects in these films and the heterogeneity of surface under the films. Defects in the film can be pores, cracks, any imperfections, and inclusions (chemical compounds).

Chloride anions can penetrate through these films and destroy them because of more positive electric potentials at some heterogeneous inclusions (for example, manganese sulfide—MnS), than pure alloy. If chlorides penetrate through passive film, they can attract iron or chromium cations from the alloy lattice. This process is provided by hydrolysis with the formation of acidic medium in the localized point at the metal surface (Eq. (9)):

$$M^{2+}_{(aq)} + 2H_2O_{(l)} \rightarrow H_3O^+_{(aq)} + M(OH)^+_{(aq)} \quad (9)$$

where M^{2+} can be either Cr^{2+} or Fe^{2+}. Ions H_3O^+ and Cl^- make for presence hydrochloric acid in the solution and can reduce the pH to <1 at the imperfections of the passive film. This acid is formed in pits on stainless steel surface. The solution around these pits is neutral. Because of the heterogeneity of stainless steel surface under passive film, pits of various shape, density and size are formed. Depending on the nature of metals and alloys, different low pH values may be obtained in pits on metallic surface. Not only chlorides ions can cause pitting corrosion. Any inorganic and organic deposits can result in formation of differential aeration cells and then to pitting corrosion underneath at anodic sites.

2.2 Crevice Corrosion

Crevice corrosion is a localized corrosion of a metal surface at an area that is shielded from full exposure to the liquid electrolyte because of proximity between the metal and the surface of another material (metallic or non-metallic) and stagnant conditions of the liquid in this area [4]. Crevice means narrow crack or opening, a fissure or cleft. The existence of differential aeration cells can explain this phenomenon. Let us imagine the gap (narrow space, crack, groove, or slot) between two surfaces (metal-to-metal or metal-to-non-metal), wide enough to permit liquid electrolyte entry, but sufficiently narrow to maintain a stagnant zone. Dissolved oxygen in liquid electrolyte inside the gap (crevice) will be used up in a cathodic reaction (4) and its concentration will decrease to zero, as oxygen diffusion into the crevice is restricted. However, the concentration of dissolved oxygen in bulk electrolyte solution near the crevice will remain the same (~8 mg/kg in neutral aqueous solutions at ~20 °C). Thus, a *differential aeration cell* (oxygen concentration cell) is formed. The metallic surface in bulk solution near the crevice will be cathodic (high oxygen concentration) and the surface inside crevice will be anodic (low, nearly zero, oxygen concentration) with appropriate electrochemical reactions (4) and (3). As a result, deep pits may be formed at a carbon steel surface over several months (Fig. 6). Therefore, crevice

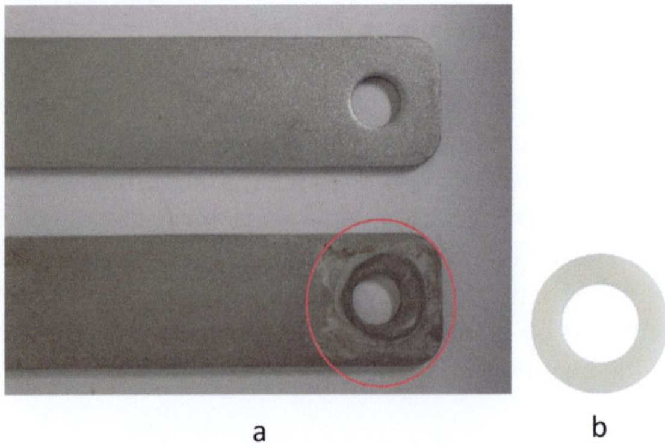

Fig. 6 **a**—Crevice corrosion between a metal and non-metal: the upper—original carbon steel coupon; the lower—carbon steel coupon after the contact with polypropylene washer (**b**) in water for 120 days

corrosion sometimes is considered as a particular form of pitting corrosion which occurs between adjacent or faying surfaces.

Crevice corrosion is initiated by changes in local chemistry within the crevice: decrease of oxygen concentration, decrease of pH, decrease of inhibitor concentration, or increase of Cl^- content. Therefore, crevice corrosion does not always occur due to differential aeration cell. This explains why crevice corrosion is also observed on alloys like stainless steels in chloride solutions where oxygen is absent, and no oxygen concentration cell exists.

Crevices may be of two types: man-made artificial and natural. The former may be unavoidable and may serve a particular design purpose of construction, such as fasteners, gaskets, lap joints, rivets. Other man-made crevices may result during fabrication and assembly. Some of them may be avoidable. Different coatings, sealants, and greases can promote crevice corrosion. Natural crevices may be formed by any deposits: debris, rust, scale, sediments, and biofouling.

The methods of minimizing or prevention crevice corrosion:

- Elimination crevices at the design and fabrication stages, and then during service. For example, design tanks, vessels, pumps, and other equipment for complete drainage; avoid stagnant areas and sharp corners.
- Use welded butt joints instead of bolted or riveted joints. Close crevices in lap joints by continuous welding, caulking, or soldering. Seal lap joints and avoid gaps between pipes and fittings. Minimize use of bolted connections and fasteners.
- Decrease dimensions of the gap: degree of tightness. Tighter crevices can be achieved between nonmetal and metal components than between two metals.
- Inspect equipment.
- Drain equipment completely on shutdowns. Remove deposits regularly.
- Prevent wetting polymeric materials and remove wet packing materials periodically. Use nonabsorbent (non-porous) gaskets and seals (PTFE, etc.).

2.3 Participation of Microorganisms in Corrosion of Metals in Aviation Fuels

Kerosene and jet fuel are sterile when they are first produced because of the high refinery processing temperatures. But they become contaminated with microorganisms during storage and transporting under ambient conditions (see Chap. 12). Once a microbial population becomes established above some value (usually $>10^3$–10^4 microbes in 1 ml of a fuel) and free water is present, it may result in deterioration of fuel quality, haziness, formation of sludge, degradation of fuel additives, filter plugging, appearing of odor, erroneous readings in fuel quantity systems, and *corrosion.*

Microbiologically influenced corrosion (MIC) is corrosion which takes place with the participation of microorganisms on metal surfaces under particular conditions. 'Participation' means the physical presence (see Chap. 12—biofilms discussion), activity, or both of microorganisms in biofilms on the surface of the corroding

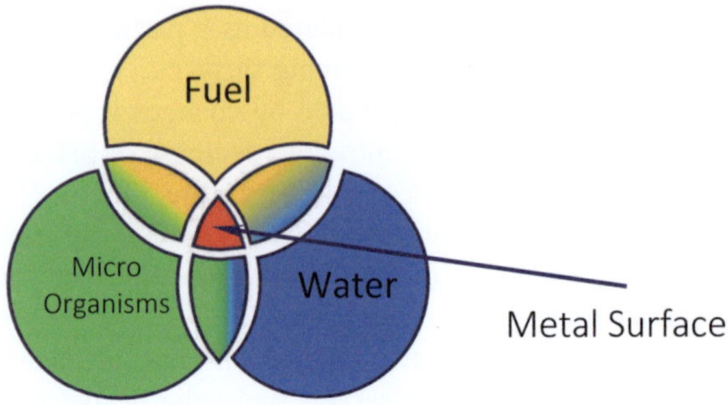

Fig. 7 Necessary and sufficient conditions for MIC on metal surface

material. MIC can occur on interior and exterior surfaces of the bottoms of aboveground and other storage tanks and wing tanks containing kerosene (jet fuel). Three components should be present to occur the MIC: fuel, water, and microorganisms (Fig. 7).

Microorganisms can affect corrosion by the following ways:

- by producing deposits which give rise to differential aeration cells and as a result to crevice and pitting corrosion;
- by creating corrosive metabolites (i.e., organic or inorganic acids, including H_2S);
- by destroying corrosion inhibitors;
- by directly inducing the corrosion reaction (i.e., electron transfer).

It is not easy to prove that corrosion occurs due to microorganisms' activity because the results are pits of various forms associated with chloride or oxygen attack, existence of differential aeration cells (under deposit corrosion), crevice or galvanic corrosion. Microorganisms do not produce a unique corrosion morphology that distinguishes MIC from abiotic corrosion. Prior to diagnosing MIC, it is recommended to examine other corrosion phenomena and mechanisms. Usually, MIC occurs in combination with other types of corrosion (pitting, crevice) which complicate its determination.

How can we prove MIC in tanks containing aviation fuels?

First, the inner surface of bottoms is covered by slippery (sticky, jellylike) sludge (also named slime, biofouling, or biofilm) with specific smell. Second, check for the presence of specific microorganisms (sessile and plankton) which can be responsible for corrosion. Third, to carry out an experiment representing MIC in this specific media. For example, we should take one liter of kerosene deteriorated by microorganisms and to filter 500 ml through biological filter (nitrocellulose membrane filters with a pore diameter of 0.24 μm). Thus, we can obtain a sterile kerosene portion. Then add 100 ml water to both portions of kerosene (sterile and incubated) and put carbon steel coupons in both samples for a minimum 30 days. If microorganisms are responsible for corrosion, severe corrosion will occur in the unfiltered (incubated) sample.

To create biofilms, microbes excrete *extracellular polymeric substances* (EPS) (see Chap. 12 for an explanation of biofilms and their ecology), that form the complex matrix in which biofilm microbial communities exist. EPS deposits can contribute to corrosion processes by creating physicochemical (electropotential and chemical) gradients. Additionally, microbial activity within the EPS matrix includes acid production and extracellular electron transport—both of which can accelerate corrosion rates underneath biofilms.

Biofilm can be present on bottoms of tanks containing crude oil and fuel oil. Grey, black, and greenish layers can be present on bottoms of tanks containing kerosene (jet fuel) and gas oil (diesel fuel). You can detect such slime on the bottom of a barrel (open to the atmosphere) containing kerosene and gas oil after half a year of storage.

The list of microorganisms involved in MIC and the mechanisms by which they influence corrosion is continuously growing. See Chap. 12 for a discussion of the metabolic characteristics of microbes associated with MIC.

Chapter 12 discusses the three general groupings of microbes in terms of their relationship to oxygen. Aerobic, anaerobic, and facultative microbes can contribute to MIC.

MIC, which was linked to heterotrophic bacteria and fungi growth, was first reported in aircraft in the early 1950s. Severe pitting corrosion was detected in jet aircraft fuel tanks (the wings and central torsion boxes) made of aluminum alloys. Biological filaments were found on the aluminium surfaces inside the fuel systems (Fig. 8). Pits were revealed underneath. MIC is observed inboard of the wings and is frequently observed on airplanes operated in warmer and humid climates [8].

During the following 70 years (1950s–2020s), numerous investigations were carried out to understand the causes and mechanism of MIC in aviation fuels and to develop preventive measures [9–22]. Here is the summation of our knowledge for now about MIC in aviation fuels' systems.

Growth of microorganisms in fuel storage tanks occurs at the water/hydrocarbon boundary, and biomass is formed (Fig. 9).

Fig. 8 Microbial filaments in the aircraft fuel aluminum tank. Image courtesy of M. Rabaev

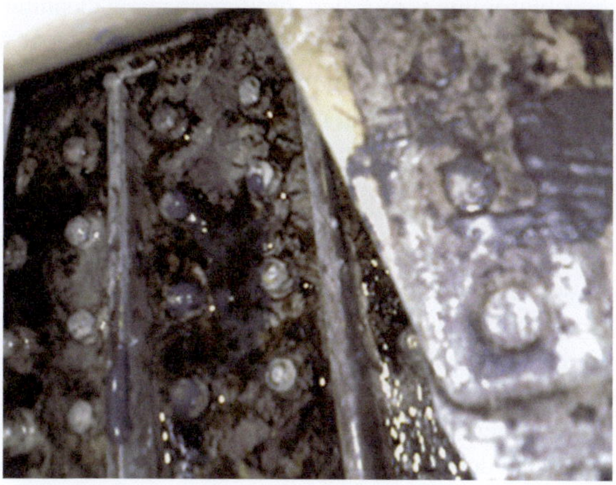

Fig. 9 Microorganisms forming biomass at the water/kerosene interface (this mixture was taken from a kerosene storage tank). Image courtesy of M. Saban

To date, more than 250 kinds of microorganisms have been recovered from jet fuel samples. Traditional culture test methods (Chap. 12) identified fewer than a dozen microbes associated with MIC. The advent of genetics-based test methods has changed this understanding considerably. As discussed in Chap. 12, MIC invariably occurs under biofilms—complex masses of microbes embedded in an EPS. Biofilm communities are complex. The different types of microbes within a given biofilm community interact—contributing to MIC directly and indirectly.

Slime-forming microorganisms influence corrosion in two ways: forming differential aeration cells (and as a result, causing localized corrosion) and excreting organic acids. Microbial production of acids directly etches metal surfaces. When thick biofouling is revealed, only mechanical cleaning may help in the prevention of MIC. Aviation fuels can be corrosive out of contact with microorganisms. We will devote the next section to this.

3 Corrosivity of Aviation Fuels

No doubt that the problem of corrosion of metals in aviation fuels arises mainly from their chemical composition and the environmental conditions during transporting, storage, distributing, and use. The chemical species responsible for corrosion either originate from the crude source and the subsequent distillation and treatment process at oil refineries or arise through water, oxygen and microbiological contamination of the fuel depending on conditions of transporting, storage, and handling facilities.

In order to understand and analyze the corrosivity of kerosene (jet fuel) and aviation gasoline, we should begin with the corrosivity of crude oils. Corrosion characteristics of crude oils are defined by total sulfur (S) content, total acid number (TAN), salt and water content, and microorganisms. These factors influence corrosion by different processes and at different stages of preparation, transporting, storage, and distillation of crudes. Most sulfur compounds are removed by desulfurization processes at oil refineries. However, low concentrations of sulfur can remain in aviation fuels: up to 0.30% (max) total sulfur according to the IP336 and up to 0.0030% (max) mercaptans according to the IP342/ASTM D3227.

Salts and water also are absent in kerosene fraction after special treatment and purification at oil refineries. Total acidity of kerosene is very low (~0.015–0.1 mg KOH/g) and should not influence its corrosivity (see Sect. 4.1). Dissolved oxygen is absent in crude oils but it can appear in kerosene and aviation fuels during transporting, storage, refueling, and use. Despite absence of water in aviation kerosene, water can appear in it like dissolved molecular oxygen from air. As a result, microorganisms can proliferate in kerosene and jet fuel in the presence of water (see Sect. 2.3).

We will analyze corrosive species and their influence on corrosivity of aviation fuels. Kerosene and aviation fuels consist of nonpolar hydrocarbons that are not corrosive to metals and alloys. However, detection of red-brown rust on the carbon steel surface of pipes and tanks contacted kerosene during long time unequivocally shows that dissolved water and oxygen could take part in corrosion (Fig. 1). This corrosivity may be further complicated and compounded by the proliferation of microorganisms in the presence of water. Their activity can influence corrosion by different manner (see Sect. 2.3).

Specification criteria for selected physicochemical properties (related to corrosivity) of jet fuels according to different standards are shown in Table 1.

Table 1 Physicochemical requirements (concerning corrosivity) to jet fuels according to internationally accepted standards [23–26]

Property	Unit	Range of specifications limits
Total acid number[a]	mg KOH/g	Max 0.015–0.10
Total sulfur[b]	% mass	Max 0.25–0.40
Mercaptan sulfur	% mass	Max 0.003–0.005
Corrosion (Copper strip)	Visual rating	Max 1
Water separation	Rating	Min 70 (rating)
Electrical conductivity	pS/m	Between 50 and 600
Particulates	mg/L	Max 1
Total aromatics	% vol	Max 25–26.5

[a]TAN in Def Stan 91-091 is limited to 0.015 mg KOH/g and is usually at least half this level, usually 0.001–0.002 for hydroprocessed fuels

[b]Total Sulfur in most specs is max. 0.3% mass, with relatively little above 0.1% mass and much at or below 500 ppm

We will analyze components and factors causing corrosion of metals in kerosene, jet fuel, and aviation gasoline. Usually, authors comment on the influence of hydrogen sulfide (H_2S), sulfur, mercaptans, and certain organic sulfur- and oxygen-containing compounds on corrosivity of aviation fuels [27, 28]. Among sulfur containing constituents are monosulfides, disulfides, organic polysulfides, thiophenes, and tertiary mercaptans (thiols). Requirements to all aviation fuels is the absence of hydrogen sulfide, sulfur, and other sulfur containing compounds. Examination of corrosivity of aviation fuels according to copper or silver strips confirms that only hydrogen sulfide and elemental sulfur are corrosive to copper and silver at very low concentrations [27]. Some naturally occurring sulfur compounds and organic peroxides were observed to inhibit the corrosive action of elemental sulfur to some extent [27, 28]. It was observed that aviation gasoline became corrosive on prolonged storage because of formation of elemental sulfur. This corrosivity disappeared on exposure to sunlight [28].

Dissolved oxygen and oxygen-containing compounds can influence corrosivity of aviation fuels. In contrast to sulfur and nitrogen-containing compounds whose amounts in aviation fuels depend on their quantities in original crudes, and on the processing used, oxygen-containing compounds not only pass from crudes into aviation fuels but also are formed because of oxidation by dissolved oxygen of unstable (predominantly unsaturated) hydrocarbons in aviation fuels during their storage, transporting, and use (See more about fuel stability in Chap. 7). For instance, many unstable hydrocarbons (mainly alkenes) appear in kerosene due to thermal cracking, and, because of oxidation, oxygen-containing compounds may be formed. The number of oxygen-containing compounds in aviation fuels depends on their composition and oxidation conditions. This process occurs between −50 and +50 °C and depends on the diffusion rate of the molecular oxygen in the organic phase. The oxidation rate of aviation fuels increases at the beginning of their storage. After their saturation by formed oxygen compounds, some oxygen-containing compounds inhibit further oxidation, which decreases to zero.

Corrosion products (rust) on the inner surfaces of carbon steel tanks and pipelines can work as hydrocarbon oxidation catalysts. Storage and transporting of aviation fuels in an inert atmosphere (nitrogen) or isolated from atmospheric oxygen would restrict or prevent their oxidation. Ensuring such conditions, however, is difficult, so oxidation of unstable hydrocarbons in aviation fuels is inevitable. Remember that it is possible to organize the inert nitrogen atmosphere above fuels in storage tanks.

Water and oxygen (as dissolved) appear in kerosene after distillation and treatment in oil refinery units during production, transporting, storage, distribution, and further use. Naval vessels (tankers) use compensated fuel ballast systems. These systems help maintain the vessel's stability by replacing consumed fuel with seawater. This method exposes the fuel tank to marine microorganisms and salts (mainly chlorides), leading to accelerated corrosion and fuel degradation.

Low molecular weight organic acids (C_1 to C_6 mono and dicarboxylic acids) and chlorides dissolved in free water contribute to its corrosivity.

Alkenes (unsaturated hydrocarbons) can be present in aviation fuel in very low amounts. They are easily oxidized by dissolved oxygen to hydroperoxides and peroxides which increase corrosivity of aviation fuels to metals and polymeric materials. This oxidation is accelerated by certain dissolved metals, especially copper. Hydroperoxides and peroxides can turn into organic acids. Saturated hydrocarbons (alkanes and cycloalkanes) can be oxidized slowly by microorganisms, with the formation of hydroperoxides which are corrosive by themselves as well as can they turn into alcohols (ROH), aldehydes (RCHO), ketones ($RCOR_1$), organic acids (RCOOH and HOOC-R-COOH), and then into esters ($RCOOR_1$) and ethers (ROR_1). Some of them can be aggressive to polymers and organic coatings. Corrosion of bronze in jet fuel under water condensation can be enhanced by mercaptans and by products of the oxidation of the fuel.

Dissolution of atmospheric oxygen in aviation fuels is the first stage of further corrosion and oxidation of hydrocarbon components. Thus, the two main "actors", water, and oxygen, are dissolving in enough quantities in aviation fuels and their components and are responsible for electrochemical corrosion to occur. Therefore, solubility of water and oxygen in aviation fuels is critical for occurring corrosion and will be described separately.

3.1 Solubility of Water in Aviation Fuels

The quantity of water dissolved in aircraft fuels is determined by the partial pressure of water in the vapor space above the fuel, temperature, hydrocarbon composition of the fuel, and may be affected using additives such as fuel system icing inhibitors [29, 30].

In spite of requirements of all standards for aviation fuels that they must be free of water, the water can occur in three different forms in fuels:

- dissolved in the fuel;
- a fuel-water emulsion or fine drop dispersion;
- a separate phase (free or undissolved water). This aqueous phase is the most corrosive one.

Polar water molecules are slightly soluble in aviation fuels (composed primarily of nonpolar hydrocarbon molecules) varying from about 30 to 400 mg/kg. Usually, trace amounts (~25–80 mg/kg at 20 °C) of dissolved water are present in kerosene. Certainly, this concentration of water in kerosene is not enough for corrosion to occur. Dissolved water cannot be detected by eye and cannot be removed by settling or filtration.

If the temperature of fuel saturated with water decreases, some of the water dissolved in the fuel will turn into many very small droplets distributed throughout the fuel. Surfactants, which can be produced in situ as microbial metabolites or introduced as contaminants, can cause water in fuel emulsions to develop. In other cases, the water appears as a dispersion of very fine drops.

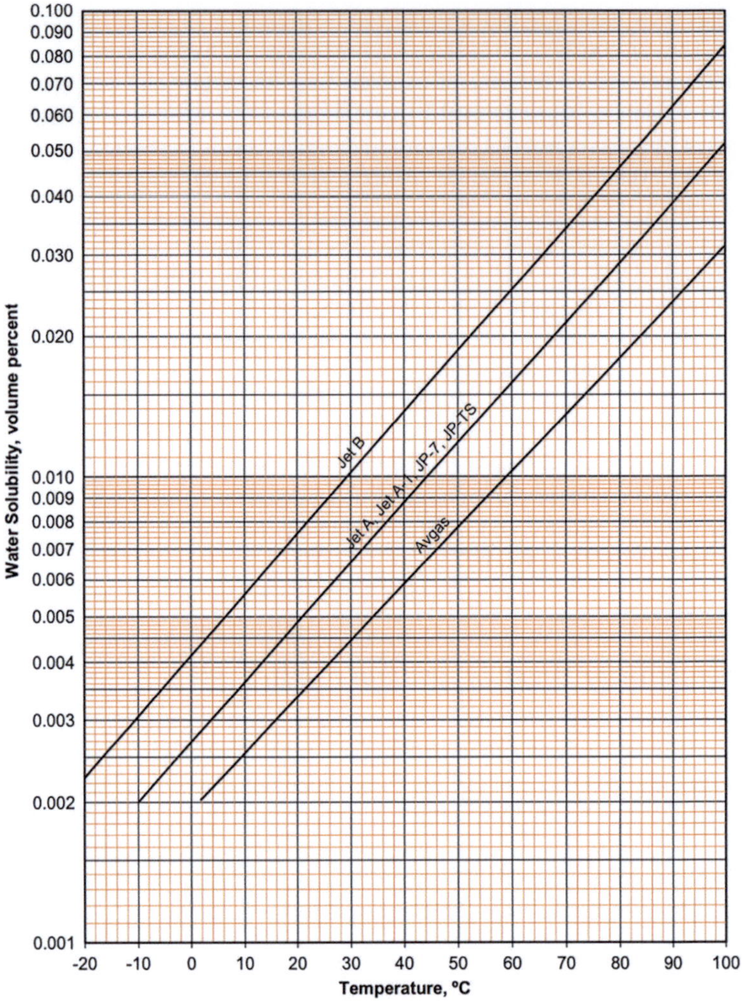

Fig. 10 Water solubility in jet fuels and avgas [30]

Further cooling of aviation fuel gives rise to undissolved or free water. This free water is a critical contaminant in aviation fuels because it plays a major role in corrosion, especially in MIC. In refinery processes, free water is usually removed by a filter separator containing water-absorbing media or by a salt dryer containing sodium chloride [5]. Sometimes kerosene after saturated salt dryers becomes contaminated by chloride ions which result in corrosion of carbon steel tanks and pipelines.

In addition to temperature and relative humidity of air with which aviation fuels contact, solubility of water in fuels depends on fuel composition (Fig. 10).

Water solubility varies as follows among organic chemical species in jet fuel: alkanes (70–100 mg/kg) < cycloalkanes < alkenes (olefins) < aromatics (430–690 mg/kg) [29, 30]. Water solubility varies with temperature. It increases 3–7 times as temperatures increase from 0 to 40 °C [3010).

Water solubility in aviation fuels is greatly influenced by the presence and concentration of aromatic and olefin compounds. Typical jet fuel is a blend of alkanes (75–80%), aromatics (19–24%) and alkenes (0.5–1.5%) [30]. Aromatic polar molecules in aviation fuels contribute to water solubility. Aromatics dissolve five to ten times more water than alkanes containing the same number of carbon atoms (compare curves for ethylbenzene – jet fuels and n-Decane –avgas in (Fig. 10). The maximum amounts of aromatics and alkenes permitted in kerosene (jet fuels) are 25%vol. and 5%vol., respectively.

To sum up, solubility of water in aviation fuels changes with the variation day–night, period of a year, climatic zone, relative humidity, and altitude of flight.

Another important participant in corrosion of metals in aviation fuels is dissolved oxygen.

3.2 *Solubility of Oxygen in Aviation Fuels*

When fuels contact air, its gaseous components (oxygen, nitrogen, carbon dioxide, inert gases, and water vapor) partly dissolve in fuels. Nonpolar oxygen molecules dissolve well in nonpolar hydrocarbons and are transported over large distances by diffusion or convection during various processes of aeration: transporting, pumping over, pouring out, filling, emptying, filtration, and storage of fuels.

Experimental data related to the solubility of oxygen in liquid fuels, alcohols, biofuels, their components, and, for comparison, in water are summarized in [29].

Oxygen dissolved in fuels is spent on two main processes occurring inside of liquid media: oxidation of unstable hydrocarbon components and corrosion of metals. Both processes are limited by diffusion of molecular oxygen in fuels. This means that the rate of both oxidation processes is greater than the rate of diffusion of oxygen, and as a result the concentration of dissolved oxygen in fuels will increase and reach its maximum value called *solubility*. Solubility is a limited, maximum, equilibrium concentration of dissolved oxygen in a liquid which contacts molecular oxygen in gaseous phase:

$$O_{2(\text{gas})} \leftrightarrow O_{2(\text{in liquid phase})} \tag{10}$$

We will describe some main points of solubility oxygen data in pure hydrocarbons and their mixtures (fuels). The main gases of the air, nitrogen (78.09%vol.), and oxygen (20.95%vol.) dissolve in fuels to saturated conditions, namely, to equilibrium concentration described by (10).

- Solubility of oxygen in fuels depends on temperature, pressure, and the fuel type.
- Usually, solubility of oxygen decreases with increase of density, molecular weight, and boiling point of hydrocarbons. Solubility of oxygen is 10–100 times greater in

liquid hydrocarbons and fuels than in water at the same temperature. The solubility of oxygen may be ranged by following order: gasoline (naphtha) > kerosene (jet fuel) > gas oil (diesel fuel) > fuel oil; and hydrocarbons > alcohols > biofuels ≫ water.

- Solubility of oxygen increases with increase of its partial pressure and decreases with increase of temperature.
- Solubility of oxygen in fuels is significantly greater (60–70%) than that of nitrogen. Therefore, if volume ratio of nitrogen and oxygen in the air equals to 3.73: 1, that in fuels equals to 2.07: 1.
- If the fuel is saturated by one gas, another gas cannot dissolve in it. This is a very important point because if once can saturate the fuel with an inert gas (nitrogen), it is possible to prevent dissolution of oxygen in the fuel and thus prevent corrosion of metallic construction and auto-oxidation of the fuel.

In some scenarios, such as during filtration and pumping into a fuel tank, the contact area of liquid fuels with air increases significantly. As the presence of dissolved oxygen in fuels is critical for corrosion, like in water, various methods of removal of oxygen from fuels have been developed.

- Treating of fuels by solid sulfite salts. They are oxidized by dissolved oxygen forming sulfate salts (reaction (11)).

$$Na_2SO_{3(s)} + 1/2\ O_{2(g)} \rightarrow Na_2SO_{4(aq)} \tag{11}$$

- Purging by inert gas (dry nitrogen).
- Vacuum degassing.

However, these methods are not used because of large amounts of fuels needed for treating.

The fuel/water/(oxygen + nitrogen) system exists in a continuously changing variable dynamic equilibrium which is difficult to regulate until the fuel is isolated from ambient air. It is nearly impossible to isolate fuels fully from the environment during production, transporting, storage and use of huge amounts of fuels. Therefore, corrosion with the participation of dissolved oxygen is inevitable, and we should use methods of monitoring and control of corrosion of different metallic constructions contacting fuels.

Oxygen dissolved in aviation fuels can oxidize certain organic compounds contained in them and increase their aggressiveness to metals and polymeric materials.

3.3 Formation of Compounds Aggressive to Metals and Polymers as a Result of Oxidation of Hydrocarbons Contained in Aviation Fuels

The concentrations of alkenes, alkyl aromatic hydrocarbons, alkadienes, and alkene aromatic hydrocarbons in some fuels are not large but these molecules are easily oxidized by dissolved molecular oxygen (O_2) at ambient temperatures. This process is called *auto-oxidation* and can occur during fuel transport and storage. *Auto-oxidation* of these classes of organic compounds forms hydroperoxides (ROOH). Hydroperoxides are intermediate compounds and can exist for some short period. However, hydroperoxides are significantly more reactive than oxygen because the energy bond O–O in ROOH is three times weaker than that O=O in O_2. Hydroperoxides break-up into radicals (ROO•) which are very active chemical particles. They by themselves are aggressive to metals and polymers.

The end products of hydrocarbon oxidation by hydroperoxides are organic oxygen-containing compounds, namely, alcohols, ketones, aldehydes, and organic carboxylic acids. Temperature increases hydroperoxide degradation. Water is a hydroperoxide degradation product. Therefore, water in fuels can appear not only from water vapor contained in the atmosphere but also as a hydroperoxide degradation product. Low molecular weight organic acids dissolved in fuel-associated water, thereby contributing to fuel corrosivity. Alcohols can react with organic acids to form ethers. In addition to organic acids, oxyacids (X–O–H) and hydrogen peroxide (H_2O_2) can form as reaction byproducts. These molecules are also aggressive to metals and polymers. Hydroperoxides degradation is accelerated by increase of temperature and the presence of certain ions of metals, such as copper, iron (mainly Fe^{2+} and iron oxides), cobalt, manganese, tin, and by acidic compounds accumulating in fuels.

$$ROOH + Fe^{2+} \rightarrow Fe^{3+} + RO\cdot + OH^- \tag{12}$$

Thus, hydroperoxide degradation by reaction with ferrous iron ions (Fe^{2+}) results in formation of ferric ions Fe^{3+} and radicals RO• which induce further oxidation of hydrocarbons and formation of organic corrosive compounds containing oxygen atoms. Ferric ions Fe^{3+} play the role of cathodic depolarizer (receiver of electrons) which takes part during oxidation of pure iron according to (13),

$$2Fe^{3+}_{(aq)} + Fe_{(s)} \rightarrow 3Fe^{2+}_{(aq)} \tag{13}$$

and, thus, accelerate corrosion. Ferrous ions Fe^{2+} appearing in process (13) take part in new break-up of hydroperoxides (12). Thus, corrosion products of carbon steel, ions Fe^{2+}, take part in autocatalytic process.

When unsaturated hydrocarbons are depleted in auto-oxidation during storage of fuels, saturated hydrocarbons can be involved in auto-oxidation. Alcohols, sulfur-, nitrogen-, and oxygen-containing organic compounds can inhibit hydrocarbon oxidation. Many hydroprocessed and all synthetic kerosene blending components contain antioxidants to minimise such oxidation.

Hydroperoxides are so good oxidizers that they can oxidize not only hydrocarbons but also organosulfur- and certain nitrogen-containing compounds. For instance, organic sulfur-containing compounds can be oxidized to sulfonic acids which also are corrosive to metals.

It is wrong to think that only aggressive compounds are formed as a result of oxidation of hydrocarbons. Neutral compounds and even corrosion inhibitors can also be formed.

4 Definition of Corrosivity of Aviation Fuels

According to definitions of *corrosion* and *corrosivity* (see Sect. 1), we can use the properties of the environment, the metal, or both for the assessment of corrosivity of aviation fuels. Thus, depending on the physicochemical content of aqueous solution it is possible to give a qualitative assessment of its corrosivity. Quantitative assessment can be given only based on the definition of the corrosion rate of a metal in this environment. Usually, corrosiveness of media is defined by immersion of a metal sample in it during for a reasonable period and calculation of the difference in weight of the sample before and after immersion [4]. Dividing this value by the area of the sample and duration in the media, corrosion loss (g/cm^2) and rate (mm/year) are obtained. If it is possible to define corrosivity of an aqueous solution in a reasonable period (several days), quantitative assessment of the corrosion rate of a metal in aviation fuels (hydrocarbons) is difficult because we would need to keep a metal specimen a considerable time (years) to obtain reasonable corrosion loss.

Owing to the complicated content of aviation fuels where main corrosive factors are the solubility of water and atmospheric oxygen, and the presence of the elemental sulfur (S_8), hydrogen sulfide (H_2S) and other sulfur- and oxygen-containing compounds, it is not easy to define corrosiveness of fuels.

ASTM D7548 test method [31] can be used for determination of accelerated iron corrosion in aviation fuels. This test assumes the most aggressive conditions of two-phase system (aviation fuel +10%vol. water). It is convenient for assessing the efficiency of different corrosion inhibitors in two-phase systems. Instrumentation has made qualitative evaluation of per cent surface area stained feasible and reproducible.

Copper and silver are the most susceptible to corrosion by sulfur compounds which can be present in aviation fuels (see reactions (7) and (8)). Hydrogen sulfide can be found in jet fuel as a result of insufficient stripping after hydrogenation or due to liberation by sulfate-reducing bacteria in stagnant water bottoms of storage tanks. Therefore, the copper and silver strip corrosion tests give an indication of the presence of certain corrosive substances such as sulfur, hydrogen sulfide and other acidic compounds that may corrode metallic equipment. These tests provide qualitative criterion and assesses the relative degree of corrosivity of fuels [32–34].

The tests consist of placing a clean, polished copper or silver strip into 100 ml of the fuel for some period (usually 1, 2 or 3 h) at a certain temperature (100 °C). The strip is removed and compared against a color chart standard that has four (for copper) and five (for silver) degrees of color. If the strip shows no indication of corrosion (the color of strip does not change) the test result is number 1 (for copper) or 0 (for silver). Definite concentrations of corrosive compounds cause tarnishing and blackening of strips' surface as a result generate number 4 or 5. Those tests detects the compounds that could corrode copper and silver systems contacting jet fuels.

Silver is more susceptible to corrosion by hydrogen sulfide than copper [27, 35]. Silver strip corrosion test is sensitive to 0.030 mg H_2S/kg fuel. The copper strip corrosion test reacts positively when H_2S concentration exceeds 0.130 mg H_2S/kg fuel. It is recommended to take off silver jewelry when taking water baths containing hydrogen sulfide.

The silver strip corrosion test IP 227 [36] is more sensitive to the presence of hydrogen sulfide, while the copper strip corrosion test as per ASTM D130 [32] is more sensitive to the presence of elemental sulfur in jet fuels [35]. The standard specifications for aviation fuels ASTM D1655 and MIL-DTL-83133 stipulate compliance with the copper strip test only.

In the past, silver strip test (IP 227 [36]) was used in the British specification DERD 2494 [37] for determination of corrosiveness of jet fuel because some details in the Concord and several military aircraft were made of silver and were equipped with pumps fitted with silver plated slipper pads.

Nowadays there are no silver materials in aircraft systems though silver alloys are used in fuel gauges in tank sender units and in automotive spark-ignition engines. Therefore, a silver strip test is used for the determination of the corrosiveness of gasoline [38, 39].

Jet fuel, meeting specification requirements, may become corrosive to copper strip during transfer through a pipeline or from one storage tank to another due to oxidation of hydrogen sulfide to elemental sulfur [35].

We can conclude that fuel corrosivity's definition is test method dependent.

Due to the complication of definition of aviation fuels' corrosiveness, some countries reinstated silver strip corrosion test in their specifications. For instance, such test was incorporated in Israeli jet fuel specification No. 5563 as a reliable tool for possible detection of free sulfur and hydrogen sulfide presence in jet fuel [35].

4.1 Influence of Acidity of Kerosene on its Corrosivity

Total acid number (TAN—also called *neutralization number*) is a measure of the numbers of milligrams of potassium hydroxide (KOH) needed to neutralize 1 g of crude oil or its distillate fraction [40]. TAN values are used to characterize corrosiveness of crudes and their distillate fractions because of the presence of organic acids (including also naphthenic acids). They are mostly corrosive during distillation (boiling/condensation) and further transporting of some petroleum distillates (kerosene, gas oil, and fuel oil) at temperatures 190–360 °C in refineries.

TAN = 220 to 320 mg KOH/g for most pure naphthenic acids. Some crudes, for instance, from Azerbaijan, contain light naphthenic acids concentrating in kerosene fraction (190–210 °C) [4, 41]. Naphthenic acids that pass from crudes into different fractions are not corrosive at ambient temperatures (~20 °C) of their storage and transporting.

As a rule, there is no correlation between TAN of crude oils/distillates and their corrosiveness. For the same TAN value, molecular size and structure of the acid have an important influence. Some naphthenic acids are very corrosive, others are relatively inert, and high molecular weight naphthenic acids can inhibit corrosion of carbon steel under certain conditions [42, 439.3]. Despite this, some rules of thumb exist for definition of crude oil or distillate corrosion threat. It was suggested using a threshold of 0.5 mg KOH/g of crude oil and of 1.5 mg KOH/g of distillate [41].

TAN values of kerosene are very low, 0.001 mg KOH/g to 0.1 mg KOH/g of distillate [23, 24]. Therefore, fuel acidity does not contribute to aviation fuel corrosivity.

5 Corrosion Mechanism in Kerosene and Aviation Fuels

If rust was detected in tank or pipe containing kerosene/aviation fuels (see Fig. 1), this means that water and dissolved oxygen took part in corrosion of carbon steel structures according to the electrochemical mechanism (see reactions (3) and (4)). The water content in fuels produced at oil refineries is usually low (30–80 mg/kg) and is not sufficient to make them corrosive.

If kerosene contacts carbon steel equipment in these relatively "dry" conditions, the equipment does not corrode. Water vapors may ingress from the atmosphere into fuels during their storage and transporting (which is part of the reason for the strict housekeeping rules employed to drain off water from fuel infrastructures and systems).

Then water vapor may condense as a result of temperature decrease, and the water content in fuels may reach concentrations more than 80 mg/kg [293.1]. Therefore, ASTM D7548 test method provides an addition of 10%vol. water to aviation fuel.

An increase of water concentration in fuels results in a drastic increase in the corrosion rate of carbon steel. For example, increasing the water concentration in gasoline from 80 to 200 mg/kg increases the fuel's corrosivity from 0.001 to 0.4 mm/year (Fig. 11) [29]. The impact of water concentration on corrosion is expected to be similar in other fuels.

The concentration of water in fuel when corrosion rate increases drastically, we call the *critical value*. The critical value of water in gasoline is 200 mg/kg. When 200 mg/kg of water is added to gasoline, all this water is soluble (in the solution gasoline/water). These results suggest that the corrosion process in gasoline/water mixtures

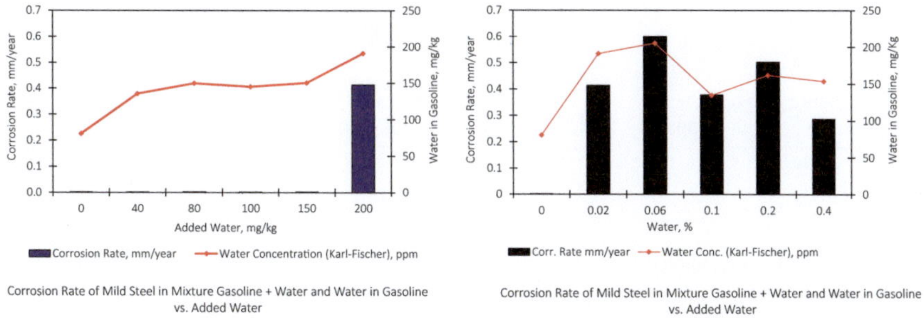

Fig. 11 Corrosion rate of carbon steel and water content in gasoline vs added water to gasoline [29] Soluble water content in gasoline was determined by the Karl Fischer method [44]

is caused by the soluble water and dissolved oxygen in gasoline (one phase). The critical value of water content in kerosene was defined as 1000 mg/kg. The presence of aromatics and oxygenates in fuels and increase of temperature cause the solubility of water in fuels to rise. When temperature decreases water can separate and appear as a free water phase on metal surface.

Another "actor" participating in electrochemical corrosion, *oxygen*, is dissolved significantly better in hydrocarbons than in water (see Sect. 3.2) [29]. The oxygen solubility in hydrocarbons (60–70 mg/kg) is 10 times higher than in the aqueous phase (8 mg/kg). This situation results in a corrosion mechanism similar to atmospheric corrosion in a thin layer of electrolyte [4, 29]. Like the thin water layer formed by oxygen (8 mg/kg) when atmosphere containing ~21% vol. O_2, fuel (containing ~70 mg/kg) form water layer on carbon steel surface by oxygen needed for electrochemical corrosion to occur (Fig. 11a and b).

Another assumption is that the corrosion mechanism in the two-phase system fuel/electrolyte is similar to the corrosion at the splash zone above high tide. In both cases, the organic phase plays the role of the atmosphere that supplies water, oxygen, and other corrosives (acids, chlorides, etc.).

The corrosion process proceeds at the interface between the two phases, metal/fuel/water (like the interface metal/air/water). In a fuel/water environment, a metal is partially wetted by the water, which creates a thin water layer between the metal and organic phase of the fuel. The average thickness of the water layer on the metal surface is 2–10 μm. One can imagine this two-phase system being a differential aeration cell (see Sect. 3.2). These cells mostly are responsible for corrosion in water, in atmosphere, in soil, and in splash zones. An anodic area is formed in the water phase (low oxygen concentration, ~8 mg/kg) and a cathodic area (high oxygen concentration, ~70 mg/kg) in the fuel phase (Fig. 12).

Parameters that affect the corrosion process in the metal/fuel/water system are water concentration in the mixture, the appearance of free water, electrical conductivity of the solution, type and concentrations of electrolytes in water, temperature, and fluid velocity. Different salts and organic acids present in fuels can undergo

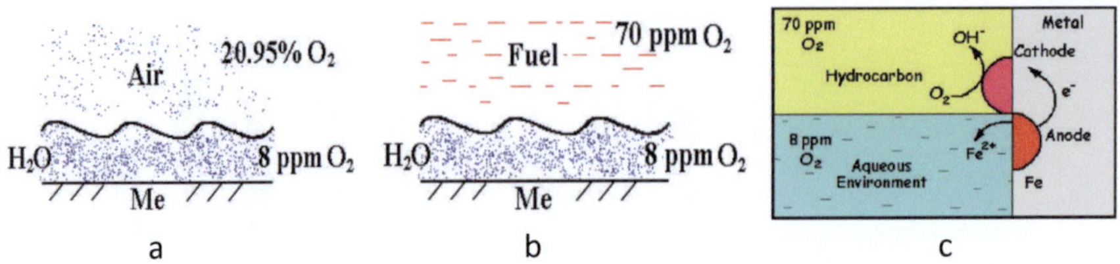

Fig. 12 Corrosion mechanism: **a** – in atmosphere; **b** – in fuel; **c** – on metal/fuel/water environment

extraction from them into the aqueous phase and cause an increase in the corrosion rate of carbon steel (Table 2).

These data show a drastic increase in water conductivity after contact with fuels. This means that fuels are the source of anions (Cl^-, SO_4^{2-}, NO_3^{2-}, $SO_4{}^{2-}$, and $NO_3{}^{2-}$) and low molecular weight organic acids, and which can contribute to increased fuel-associated water corrosivity. Decrease of water pH (5.8 for the blank) after contact with fuels (4.7–5.2) also confirms the presence of acidic salts and organic acids. Organic acids can appear in fuels as a result of oxidation of hydrocarbons during their treatment or storage, as well as wastes of the microorganisms' growth. Some additional water also can appear in fuels as a result of proliferation of microorganisms (see Sect. 2.3).

SEM (scanning electron microscopy) and EDS (energy dispersive spectroscopy) results showed that the corrosion products formed on the carbon steel coupons from the kerosene-water mixture consisted of iron and oxygen. The morphology of the corrosion products (rust) formed on these coupons is shown in Fig. 13a. This is the place to show the beauty *of rust*. Our imagination can compare the rust after magnification by 350 with the Moon surface (Fig. 13b).

Table 2 Chemical composition of the aqueous phase after contact of deionized water with fuels (7 days, T = 25 °C) [29]

Parameter of aqueous phase	Unit	Deionized water (blank)	Aqueous phase after contact with			
			Gasoline	**Naphtha**	***Kerosene***	**Gas oil**
pH	–	5.8	*5.0*	5.2	*4.8*	4.7
Conductivity	μS/cm	0.8	*74–205*	51–57	*38*	47–122
Cl^-	mg/kg	0.04	*57*	2	*4*	9–13
$SO_4{}^{2-}$	mg/kg	0	*2*	2	*1*	0.2
$NO_3{}^-$	mg/kg	0	*1–3*	2–4	*0.04*	0.01
HCOOH	mg/kg	0	*3–29*	0.6–2.6		
CH_3COOH	mg/kg	0	*28–42*	7–20		

Note: The chemical composition of the aqueous phase was determined after 7 days of contact with fuels while experiencing intensive agitation

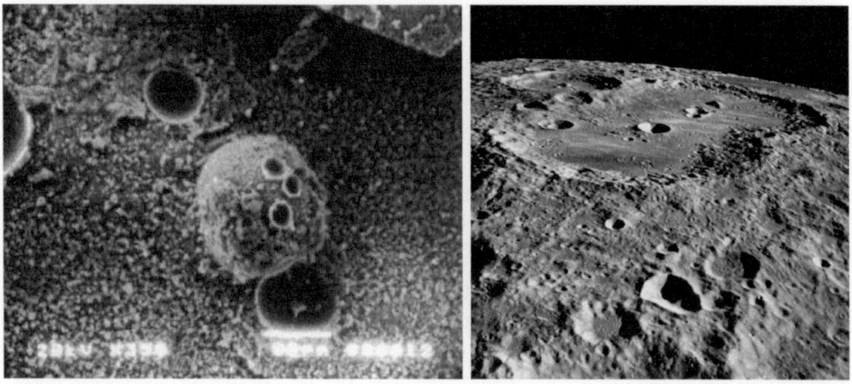

a b

Fig. 13 **a**—SEM photo of rust formed on the carbon steel coupons in kerosene-water mixture after 7 days immersion (×350) (Image courtesy of A. Groysman). **b**—the Moon's surface (Image courtesy of NASA)

Thus, the main cause of corrosion in the fuel-water mixtures is the presence of water and dissolved oxygen. These data support the electrochemical mechanism of the corrosion of carbon steel in fuels containing small quantities of water.

Pitting corrosion is dominant when the water concentration in the fuel-water mixture is <0.1%. Above this value, uniform corrosion is more dominant. The addition of small concentrations of salt (~10 mg NaCl/kg H_2O) to the aqueous phase of the fuel-electrolyte mixture increases the corrosion rate of carbon steel by 23–73% [29]. This fact also proves the electrochemical mechanism occurring in the two-phase fuel-electrolyte systems.

The main recommendation to diminish or prevent corrosion in aviation fuels is to "dry" them down to values that do not exceed "critical" water concentrations.

6 Electrical Conductivity of Aviation Fuels and Their Corrosivity

The electrical conductivity of a liquid solution is an ability to conduct electric current by means of ion migration. Electrical conductivity of aviation fuels is a very important property because of the possibility of generating static electricity and influence corrosion of metals. In this section, electrical conductivity of kerosene and its influence on metallic corrosion will be discussed.

6.1 Character of Electrical Conductivity of Aviation Fuels

Kerosene and aviation fuels consist of nonpolar hydrocarbons which do not dissociate into ions under usual environmental conditions. Therefore, formation and accumulation of electrical charges in liquid fuels usually occurs while the fuel is flowing or filtering. Polar organic compounds and inorganic impurities contained in fuels increase their electrical conductivities. Levels of oxidized products (hydroperoxides, peroxides, alcohols, aldehydes, ketones, and organic acids) may increase during storage of fuels because of interaction of the hydrocarbons with dissolved oxygen. The presence of these substances increases electrical conductivity of fuels and can intensify electrochemical corrosion. For instance, purification of kerosene from impurities decreases its electrical conductivity 10–100 times. *Antistatic (static dissipator) additives* are used to dissipate static electricity and prevent sparking (see Chap. 5). They are fuel-soluble chromium substances, polymeric sulfur- and nitrogen-containing compounds, and quaternary ammonium salts. All these chemicals may possess corrosion inhibitive properties to metals (see Sect. 9.3).

Electrical conductivity of fuels increases with rising of boiling range because the amount of non-hydrocarbon compounds (sulfur-, oxygen-, nitrogen-containing compounds, and compounds with metallic ions) also increases.

The greater the electrical conductivity of fuels, the greater the possibility of electrochemical corrosion. Generally electrical conductivities of gasoline and kerosene are close. For aviation fuels which are transported to airports, this value may raise tenfold during transporting. Increase of temperature of fuel on 20 °C causes two-fold raise of electrical conductivity of fuels.

6.2 Influence of Electrical Conductivity on Corrosivity of Aviation Fuels

Organic acids, sulfur-containing compounds and different ions are dissolved in small amounts of water contained in aviation fuels. These compounds, in addition to dissolved H_2O and O_2, may be the cause of corrosion of carbon steel pipes, pumps, filters, and tanks (during filling-emptying operations) when fuels move.

It is obvious that corrosion depends on electrical conductivity of substances which move (fuels–water with other contaminants). The more electrical conductivity (the less electrical resistance) the liquid has, the higher the ability to carry the electric current on the metal surface between anode and cathode sites, and the corrosion current is consequently greater (see Eq. (6)).

For example, the electrical conductivity of liquid pure fuels is very low, less than 1 pS/m, and their corrosivity is consequently very low, close to zero. The electrical conductivity of pure water is more, 10^6 pS/m, and its corrosivity is more. Ions H^+ and OH^- are responsible for the electrical conductivity of pure water. Various cations and anions which are present in aqueous solution are responsible for its electrical

Table 3 Electrical conductivity of various liquid media and corrosion rate of carbon steel in them [29]

Liquid media	Specific electric conductivity at 20 °C, S/m	Corrosion rate of carbon steel[a], mm/year
Cooling water (industrial)[b]	0.2	0.6–1
Potable water[c]	0.05	0.2–0.3
Demineralised water	10^{-4}	0.1
Gasoline	10^{-10}	0.001
Kerosene	10^{-12}	0.0005
Gas oil	10^{-12}	0.0001
Gasoline +0.02% wt. water	Two-phase system	0.4
Kerosene +0.02% wt. water	Two-phase system	0.5

[a]Corrosion rate of carbon steel was determined by the weight loss method with intensive agitation during a week at 25 °C
[b]Cooling water in the oil refining plant (without any treatment by inhibitors, biocides, and anti-scaling agents)
[c]Potable water in Israel

conductivity. The electrical conductivity of aqueous electrolyte solutions is high, 10^{-3} to 40 S/m, and they are aggressive towards metals (Table 3).

Corrosion rate of carbon steel in gasoline is 100 lower than in demineralized water, but two-fold more than in kerosene and ten-fold more than in gas oil. Thus, various fuels behave differently regarding carbon steel. We can compare the electrical conductivity of these liquid media. The electrical conductivity of gasoline one million times lower than that of demineralized water and 100 times more than that of kerosene and gas oil. Small concentrations of water (0.02% wt.) in gasoline and kerosene result in increase 1000 times of corrosion rate of carbon steel. Thus, we should prevent any ingress of water into aviation fuels.

7 Difference in Corrosivity of Conventional and Alternative Aviation Fuels

Conventional aviation fuels (jet and avgas) are produced from fossil sources in the refining processes. Their corrosivity is determined by acidity, water, oxygen, sulfur compounds and microorganisms' content, and was analyzed in previous chapters. During the last 15 years (2010–2025), alternative aviation fuels derived from renewable resources have been investigated [45–50]. These nominally synthetic fuels contain paraffinic and other hydrocarbon compounds.Synthetic blending components that meet aviation fuel specifications can be blended with conventional aviation fuels. It was shown that despite various production methods and some difference in chemistry of these fuels, their corrosivity does not differ from corrosivity of conventional aviation fuels [45].

8 Corrosion Case Studies Related to Corrosiveness of Aviation Fuels

Data about aircraft failures induced by aviation fuels is scarce and dispersed in literature (see Chap. 15). Some cases are described and analyzed below. Corrosion cases not concerning aviation fuel contact, such as galvanic [51] or fatigue corrosion [52], or general, pitting, crevice, exfoliation, SCC, and hydrogen embrittlement [53] are out of our concern in this chapter.

Case 1. Plane destruction because of the presence of water in the fuel system (Fig. 14).

Two pilots were seriously injured and were hospitalized for a long period, for treatment and then for rehabilitation.

Turbid liquid and significant amount of water were found in the right-wing fuel tank.

The cause of failure—water intrusion into the aircraft's fuel system because of distortion or gap that prevented the sealing of the refuelling opening using the lid.

Case 2. Emergency landing following engine trouble during the flight. The destruction of the aircraft and injuries of the pilot, Japan.

The cause of failure—stoppage of the engine because of the restriction of the fuel flow by the blockage of the carburetor inlet screen by rust (Fig. 15). Automotive gasoline instead of avgas was filled in four tanks made from aluminum. Corrosion was observed in three tanks. Infrared absorption spectrometry analysis of deposits formed in the tanks showed the presence of water. The aircraft was parked in the open air for about five months in South Korea. It is considered possible that the aircraft was also parked in the open air for a long period previously. As a result, it is estimated that gasoline in the fuel tanks evaporated, creating gum, peroxides, and organic acids, that the gum accumulated in large quantities, and that the organic acids and moisture that had collected in the fuel tanks caused the aluminum tanks to corrode.

a b c

Fig. 14 **a**—Heavy landing in a wheat field accompanying by passenger compartments and engine detaching and plane destruction, Israel, 2016. **b**—Rust marks on the right-wing fuel filler cap. **c**—No similar marks on the left-wing fuel opening cap [54]

a b c

Fig. 15 **a**—Corrosion and rust inside of strainer and filter. **b**—Part of carburetor and inlet screen filled with grease-like (slime) foreign matter. The latter consisted mainly of aluminum. Microbes were not detected. **c**—Aircraft emergency landing [55]

9 Corrosion Prevention and Control in Systems Containing Aviation Fuels

Anti-corrosion preventive measures of systems for transportation and storage aviation fuels must be started at the stage of engineering design and correct use of standards. Design of compatible materials, hydrophobic basement (foundation) for aboveground storage tanks (AST), constructions (storage tanks, transfer lines, and other fuel-handling equipment and hardware) with as few crevices as possible, suitable protective coatings, systems for injection of corrosion inhibitors, scavengers of corrosive substances (H_2S, H_2O, O_2), biocides, and cathodic protection systems are viable. It is important also to design and to plan corrosion monitoring methods which will be able to follow up the state of materials, the corrosiveness of the environment, and the efficiency of anti-corrosion control measures. All these measures must be carried out at the stage of design, erection of tanks and pipelines, assembly, correct monitoring and inspection, cleaning, and maintenance of equipment intended for contact with fuels. Therefore, corrosion because of pure design, inadequate surface preparation and protection, surface contaminations, galvanic couples, and flux inclusions is avoidable.

Three main factors influence corrosion: a metal type, the environment, and conditions at the interface between a metal and the environment. Therefore, we can use the properties of a metal, of an environment, and of a border metal–environment for corrosion control measures in systems containing aviation fuels. In the light of this, preventive methods can be divided into three groups: measures dealing with metals (selection of materials), measures dealing with the environment (treatment of fuels), and measures dealing with the interface metal–environment (use of coatings and cathodic protection).

Selection of materials means design and use of metals/alloys, polymeric and composite materials compatible with aviation fuels and other environments. Anti-corrosion preventive measures dealing with the environment are based on treatment of aviation fuels, atmosphere, and water, namely on fuel composition and impurities in it, as well as on the composition of vapor and aqueous phases which can be present in tanks and pipelines containing aviation fuels. These measures include removal of the corrosive components by a suitable procedure, for instance, drying, coalescing, deaeration, use of corrosion inhibitors, and anti-bacterial treatment. Anti-corrosion preventive measures of systems in contact with aviation fuels are described below.

9.1 Choice of Materials

Liquid aviation fuels are stored in tanks and are transported through pipelines, ships, ocean tankers, barges, railroad tank cars, and tanker trucks. Above storage tanks (AST) sometimes are equipped with floating roofs and pontoons. Sea water can be used for filling tanks in ship tankers for ballast. In these cases, residual sea water may be mixed with fuels. Physicochemical resistance of all constructive materials is vital in preserving environment and aviation fuels from deterioration. We will describe shortly materials which are used in contact with aviation fuels.

9.1.1 Metals and Alloys

The most widely used materials in airframe structures and components are aluminum, steel, titanium, and magnesium alloy systems [56].

Carbon steel, stainless steel and aluminum alloys are used as materials of AST, UST (underground storage tanks), pipes, truck tanks, car tanks, ship tanks, tankers, railroad tanks, floating roofs and pontoons, pumps, and their components (rotors, etc.), filters, hydrants, dispensers, etc. Materials used in the construction of tanks should comply with API 620 [57], API 650 [58], and they may be carbon steel, austenitic stainless steels 304, 304L, 316, 316L, 317, and 317L, duplex stainless steel, and aluminum. The duplex stainless steel may be SAF 2205 (UNS S31803), 2003 (UNS S32003), 2101 (UNS S32101), SAF 2205 (UNS S32205), 2304 (UNS S32304), Ferralium alloy 255 (UNS S32550), 255+ (UNS S32520), SAF 2507 (UNS S32750), and Zeron 100 (UNS S32760). UNS—Unified Numbering System, a designation system relating to a specific metal or alloy [59].

Carbon steel and stainless steel are ferrous alloys because iron is the main component. Other alloys (based on aluminum, titanium, copper, etc.) are called nonferrous alloys.

Carbon steel is an alloy containing iron (Fe) and carbon (C) at concentrations from 0.008 to 2 wt%, and small amounts of other elements (Mn, Cr, Ni, Mo, Cu, Si, S, P). Generally, tanks and pipelines are made from low-carbon (mild) steel (Fe+ 0.1 to 0.3 wt% C). Nowadays low-carbon steel ASTM A516 Grade 70 (UNS K02700) is widely used as a material of AST and UST. Carbon steel has the advantage of lower capital cost. Carbon and low-alloy steels (Ni–Cr–Mo) are prone to general, pitting, SCC, and hydrogen embrittlement. The disadvantage of carbon steel constructions is higher life cycle cost due to increased maintenance and costs associated with corrosion protection. Carbon and low-alloy steels require the application of inorganic coatings (e.g., ion vapor deposited aluminum) and/or organic coatings (e.g., epoxy or polyurethane) to resist corrosion in the environments associated with aviation.

Stainless steel is used for manufacturing new small tanks, car tanks, and as details of floating roofs in AST. Stainless steel is an alloy of iron with chromium content above 10.5 wt%. A tenacious passive film of chromium oxide (Cr_2O_3) is formed on stainless steel surface and is responsible for protective properties in pure atmosphere, water, and fuels. Stainless steels UNS S30400 (Fe+ 18 to 20 wt% Cr+ 8 to 12 wt% Ni) and UNS S31600 (Fe+ 16 to 18 wt% Cr+ 10 to 14 wt% Ni+ 2 to 3 wt% Mo) usually are used in fuel systems. Stainless steel is prone to localized corrosion: pitting, crevice, SCC, and MIC (see Sect. 2).

Aluminum alloys are used for manufacturing floating roofs, pontoons, and fixed roofs in AST (geodesic dome). Aircraft store fuel in their wings made from aluminum alloys. Each wing tank has a pump that supplies fuel to a manifold that feeds engines. Filter/separators are made from aluminum or carbon steel (coated by epoxy paint).

Aluminum and *zinc* are used as metalizing protective coatings on inner surfaces of truck tanks and stationary tanks (capacity 5–50 m^3) made of carbon steel. Zinc coatings also are used for protection of outer surfaces of truck tanks [29].

Aluminum is resistant to water, hydrocarbons, hydrogen sulfide, acetic acid, and disodium silicate solutions because it is protected by aluminum oxide films that rapidly form in atmosphere [41]. The corrosion resistance of aluminum alloys is strongly influenced by their chemical composition as a consequence of alloying elements and impurities. Principal alloying elements include zinc, magnesium, silicon, copper, and manganese [56].

Aluminum and its alloys are susceptible to general, pitting, crevice, galvanic corrosion, SCC, exfoliation, and MIC.

Special requirements exist for materials in contact with jet fuel. For instance, galvanized steel, zinc, copper, and their alloys are not recommended for use in contact with jet fuel because copper and zinc may catalyse oxidation of jet fuel and thus lead to a deterioration of its quality.

Titanium and titanium alloys provide low density, high strength-to-weight ratios, and high corrosion resistance. The excellent corrosion resistance of titanium is due to the formation of a protective titanium oxide film during atmospheric exposure [56]. Titanium alloys have high resistance to pitting, galvanic corrosion, crevice corrosion, erosion-corrosion, and corrosion fatigue in aviation environments. Aqueous chloride solutions, methanol, ethanol, and ethylene glycol can cause SCC of titanium alloys.

Magnesium and its alloys are the most corrosion-prone materials used in aircraft structural components [56, 60]. Magnesium is alloyed with aluminum, zinc, lithium, rhenium, thorium, and silver. Nickel, iron, copper, and cobalt have deleterious effects on the corrosion resistance of magnesium. Critical concentrations are 5, 170, and 1300 mg/kg for nickel, iron, and copper, respectively [56]. Aluminum and zinc are the most widely used alloying elements in magnesium because they have high solubility and give rise to some of the highest strengths. Magnesium alloy AZ91 (Mg–9Al–1Zn) is a commonly used alloy that exhibits good weldability and low porosity. Magnesium alloys are subject to general, pitting, crevice, galvanic, filiform, SCC and fatigue corrosion and must be surface treated and protected. Breakdowns of passive films composed from magnesium oxide and hydroxide result in rapid corrosion of the magnesium alloys, especially where galvanic couples are present, that is, steel inserts, fasteners, and mounting areas. This is especially prevalent in naval aviation. Corrosion prevention for magnesium components should be carried out at the design stage including minimizing heavy metal contaminants, flux inclusions, contacts with other metals, and inadequate surface preparation.

9.1.2 Polymeric Materials

The choice of polymeric materials depends on the purpose of their use. For additional information on how fuel interacts with non-metals, please refer to Chap. 6.

Seals (O-rings, gaskets, packers, plugs, repair clamps), washers and nuts, tubing, fuel returned lines, valve sleeves, flexible joints, diaphragms, pulsation damper bladders and bellows, protective coatings, adhesives, foams, films, thermal insulators, and insulations of electrical wiring using in tanks, pipelines, pumps, and filters contacting fuels and their components are made from polymeric materials. Mostly NBR (nitrile butadiene rubber), Teflon (polytetrafluoroethylene), and Viton (fluorocarbon-based fluoroelastomer) are used for these articles.

High density polyethylene (HDPE) is used as membrane material for secondary containment in AST. However, polyethylene and polypropylene are not recommended as construction materials for long contact with petroleum products.

Hydroperoxides in jet fuel have a deleterious effect on elastomers in aircraft fuel systems (see Sect. 3). Phenolic antioxidant additives are injected into aviation fuels to prevent formation of hydroperoxides (see Chap. 5). Removal of dissolved oxygen leads to a decrease in the peroxidation rate.

Flexible pipes (hoses) used for the drainage of rainwater from the roofs of AST are also made from polymeric materials. Outer surface of these pipes is in contact with fuels. Inner surface of these pipes is in contact with rainwater. Usually Buna-N (NBR) or Viton are used for flexible hoses. Polymers' compatibility with aviation fuels are described in [29] and in Chap. 6.

9.1.3 Composite Materials

A composite is a multiphase material, and its properties are the function of the properties of the constituent phases (metal, polymer, and ceramic), their relative amounts, and the geometry of the dispersed phase. The properties of composite are improved relative to properties of constituent phases. Composites are used in aircraft applications such as radomes, engine blades, propellers, rotors, single aisle wings, wide body wings, fin, tailplane, and fuselage sections.

Fiberglass (glass-reinforced plastic—GRP or glass-fiber reinforced plastic—GFRP) is used for rigid pipes for aviation fuels transportation, protective coatings inside AST, and UST manufacture. It is made of a polymer matrix reinforced by fine fibers of glass. A polymer matrix may be polyester, vinyl ester, epoxy, or polypropylene. Fiberglass is resistant to aviation fuels. The modulus of glass, however, is low compared with that of metals and so it was not until the introduction of carbon reinforcements (carbon fiber reinforced plastic—CFRP) that primary composite structures were developed. Nowadays there are three main types in aircraft use: carbon fiber-, glass-, and aramid-reinforced epoxy. There are others, such as boron-reinforced (a composite formed on a tungsten core). Although the volume of composites used for aircraft applications is a relatively small percentage of total use, the key aspects are light-weight, high-strength, high-stiffness, good fatigue resistance, and corrosion resistance. The main disadvantage in using CFRP/GFRP is delamination of layers, so weakening the structure and its reliability. Water trapping inside a panel through pin holes and small cracks speed up delamination if the panel is not replaced at proper time. Another disadvantage of using composites is that they are not biodegradable. However, they do have some positive impact on the environment because of their longevity and fuel-savings.

9.2 Coatings

Organic and metallic coatings are used as protective coatings for tanks and pipelines containing aviation fuels. Organic coatings found the widest use among all protective methods of inner and outer surfaces of AST and UST containing aviation fuels, as well outer and inner surfaces of underground and submerged pipelines for aviation fuels transportation. The use of protective coatings allows not only to prevent corrosion in tanks, but to maintain the quality of fuels, to reduce losses of volatile organic components of fuels (thus protect the environment and keep fuel quality), and to reduce wear of pontoons and seals in floating roofs. The requirements to coatings using in contact with aviation fuels are: resistance to all fuel components, abrasion resistance (because of movement of floating roofs with pontoons), resistance to cold and hot water (90 °C) and surfactants (during cleaning of AST), and coatings should not affect the fuels physicochemical properties (quality) during long storage (above 5 years).

The API RP 652 standard [61] recommends two types of coating systems for the protection inner surfaces of new and old bottoms in AST: thin (<500 μm) and thick (>500 μm). Unfortunately, this standard does not include new generic types of coatings which were developed and examined for aviation fuels during the last 20 years (2005-2025). You may find the list of such coatings in [29

They are epoxies, epoxy phenolic, epoxy novolac, polyurethane, polysiloxane, polyvinylchloride, and vinyl ester (Table 4).

Examples of practical use of coatings of tanks for storage of fuels are given in Fig. 16.

Table 4 Coating systems for anticorrosion protection of inner surfaces of tanks containing kerosene and aviation fuels[1] [29]

No	Generic type	Thickness, μm
1	Inorganic zinc silicate	150
2	Epoxies of various cross-linkers	200–1500
3	Polyvinylchloride	200
4	Silicone-epoxy	250
5	Epoxy phenolic	300
6	Epoxy novolac	200–400
7	Polysiloxane	300
8	Polyurethane	500
9	Epoxy reinforced with glass and mineral flakes	500
10	Glass filled epoxy with rust convertor, corrosion inhibitor and passivator	600
11	Vinyl ester with acrylic copolymer	1250
12	Epoxy vinyl ester	1500
13	Vinyl ester	1500

a b c

Fig. 16 **a**—Inner surface of the bottom coated by epoxy novolac. Average thickness is 570 μm (minimum 550 μm). Welds and patches are well coated (have additional thickness). AST is intended for storage of kerosene. **b**—Outer surface of the floating roof with coating system: surface tolerant aluminum mastic epoxy (125 μm) + surface tolerant mastic epoxy (125 μm) + polyurethane (50 μm). Minimum thickness is 300 μm. AST is intended for the storage of gasoline. Outer surface contacts the atmosphere at the oil refinery and sometimes spillage of gasoline. **c**—The 1st layer—epoxy primer (100 μm); the 2nd layer—epoxy high build (100 μm); the 3rd layer—polyurethane (50 μm). Total—250 μm. Photographer—the author

Antimicrobial coatings on aluminum surface based on poly (hexamethylene guanidine) hydrochloride-sodium stearate (PHMG-SS) have been developed recently [62].

[1] Compatible also to gasoline containing oxygenates (MTBE to 15% vol.) and aromatics (BTX to 30% vol.). MTBE—Methyl tert-butyl ether. BTX—benzene, toluene, xylene.

9.3 Corrosion Inhibitors

Corrosion inhibitors are chemical compounds which, when present in a system at a sufficient concentration, reduce the corrosion rate of a metal without significantly changing the concentration of the corrosive agent [1].

As discussed in Chap. 5, corrosion inhibitors are a class of fuel additives.

Microorganisms also find favorable food and conditions for their proliferation at the interface water/fuel. Even if elemental sulfur, H_2S and mercaptans are absent in fuels, they may appear as a result of break-up of complicated organic sulfur-containing compounds.

High molecular weight carboxylic (fatty) acids, aliphatic amines with long chains, the amine salts of carboxylic acids, aliphatic polyamines and polyamides are used as corrosion inhibitors of carbon steel in contact with aviation fuels. Examples are the DCI-4A and DCI-6A. The following corrosion inhibitors were used in the past and some are still used in aviation fuels [63]: Nalco 5403, Nalco 5405, Hitec 580, Hitec 515, Tolad 245, Tolad 249, Tolad 351, Tolad 4410, Unicor J, Mobilad F-800, Spec-Aid 8Q22, RPS-613, Apollo PRI-19, PWA-536. Certainly, corrosion inhibitors for aviation fuels are not limited by only these proprietary types.

Commercially available corrosion inhibitors are typically combinations of a fatty acid dimer (dicarboxylic acids—the active ingredients) and ethylbenzene or naphthalenes (serving as a solvent). A typical dimer used is dilinoleic acid [64].

These inhibitors are surface-active polar molecules whose polar ends adhere to the metal surface and form the protective film on it. The molecule's oil-soluble tail forms a water repellent (hydrophobic) layer over the metal. As most corrosion inhibitors are surfactants, they can also play the role of lubricity improvers (lubricity improver additives—see Chap. 5).

Corrosion inhibitors were introduced in 1940s and can be added to aviation fuels in concentrations between 5 and 24 mg/kg. As fuels pass through transfer lines or are stored, the concentration of corrosion inhibitor decreases and a corresponding decrease in both the corrosion protection and the lubricity of the fuel occurs. This loss of corrosion inhibitor can cause fuel pump failure. Quantitative gel permeation chromatography is used for determining 1–35 mg/kg of these corrosion inhibitors in aviation fuels [64]. Seawater (not fresh water) effectively removes the corrosion inhibitor from aviation fuel.

Corrosion inhibitors play polyfunctional role as they prevent corrosion of metals and thus avoid participation of metal ions (e.g., Fe^{2+}) in oxidation of hydrocarbons in fuels.

9.3.1 Organic Acids as Corrosion Inhibitors

Carbon steel and aluminum alloys corrode in acidic solutions at $pH < 4$. However, the type of acid influences significantly the corrosion resistance of these alloys. Carbon steel corrodes in diluted H_2SO_4 and HNO_3 but possesses high resistance in concentrated H_2SO_4 and HNO_3. This phenomenon is called passivity [4]. Aluminum is also passivated in concentrated HNO_3. There are other examples when addition of inorganic acids H_3PO_4 and H_2S in small concentrations (100–200 mg/kg) can inhibit corrosion of carbon steel in neutral aqueous solutions because of the formation passive thin layers of iron phosphates and sulfides on the surface of carbon steel [4, 41]. In some units at oil refineries, this property of iron sulfide passive films is used for corrosion control.

In addition, carboxylic acids can react with iron and form insoluble salts on the carbon steel surface which also prevent the penetration of corrosive species and their further contact with the steel surface. Thus, mechanism of protection may be physical adsorption or chemical reaction of inhibitors with a metal surface and passivating it. Passivation is a metastable condition that can be destroyed by any outer intervention: temperature, stresses, high velocity flow, appearing of aggressive components (i.e., chlorides), and surface roughness quality.

It was found that some organic carboxylic acids (R–COOH) can function as corrosion inhibitors of iron and its alloys, aluminum, copper and its alloys, and zinc [65–70].

Phosphonic acids and their derivatives also possess inhibitive properties [65].

The ability of carboxylic acids to inhibit corrosion in automotive cooling systems was first reported at the beginning of 1950s [71]. Aromatic carboxylic acids, such as benzoic acid and its derivatives, have been extensively studied and are used in Europe as corrosion inhibitors in commercial coolants.

Organic acids can react with metals forming passive protective layers on metallic surfaces. Carboxylic acids form hydrophobic layers on the surface of many metals by adsorption or reaction of carboxyl groups (–COOH) on a positively charged metal surface. Salts of carboxylic, phosphonic and dialkylphosphoric acids possess high inhibitor efficiency by ability of chemisorption and self-assembly on the surfaces of Fe, Cu, and Zn [65].

The adsorption of the studied compounds via the donated electrons from the $-COO^-$, and the $C = C$ bonds could accept the feedback electrons from steel. The adsorption capacity of inhibitor increases with the increasing distance between the C=C bond and the COO^-. The carboxylate inhibitors adsorb on steel surface by forming Fe–OOC–C_x compounds and the C=C bonds could enhance the adsorption process. The alkylene tail in the inhibitors could form a hydrophobic film to protect the steel [72]. Alkylcarboxylic acids (stearic, myristic) and alkylphosphonic acids are used for hydrophobization and superhydrophobization of the metals surface in aqueous-organic solutions [65, 70]. Noteworthy is the use of octadecylphosphonic acid for imparting anticorrosion, wear-resistant, hydrophobic thin coatings [65].

It is not recommended to inject corrosion inhibitors based on carboxylic acids into jet fuel pipelines because these acids remove deposits and clean inner surface of pipelines. As a result, filters are blocked, and the periodicity of their replacement can be increased drastically, up to once per few days.

Inhibition efficiency of these carboxylic acids depends on the number and position of carboxyl groups (COOH). Increase of the chain length increases the inhibiting power of aliphatic acids. For instance, usually naphthenic acids are very corrosive to carbon steel at high temperatures (T > 200 °C). However, high molecular weight naphthenic acids are corrosion inhibitors [69].

Carboxylic acids can inhibit pitting corrosion of carbon steel in alkaline medium in the presence of chlorides [73].

We reveal on these examples the main principle of philosophical dialectics in science. Metals can be corroded by acids. But—metals can be protected by acids. Humans can die without water—Humans can be drowned in water.

Carboxylic acids used as lubricity improvers are the same substances that are used as corrosion inhibitors. Both corrosion and lubricity are surface phenomena. Therefore, it is not too surprising that corrosion inhibitors also improve lubricity and thus serve a dual function: improve the fuel's lubricity as well as protecting the fuel system against corrosion.

9.4 Cathodic Protection (CP)

The electrochemical mechanism of corrosion in electrolytes allows the use of electric current and electric potential to protect metals from corrosion. Therefore, electrochemical methods work only in solutions of electrolytes and cannot work in fuels and other non-conductive media. These methods are inapplicable in aviation fuels, even in the presence of small amounts of water of about several hundreds mg/kg. However, when the outer surface of pipelines and tanks contact soil, the latter is an electrolyte and electrochemical methods are applicable in this case. We describe shortly the principles of cathodic protection.

Carbon steel pipe in contact with electrolyte is an anode that corrodes according to the reaction (14):

$$Fe_{(s)} \leftrightarrow Fe^{2+}_{(aq)} + 2e^{-} \tag{14}$$

Cathode does not corrode. Thus, if we turn this pipe from an anode to cathode it will not corrode. We can reach this if we connect iron to a metal possessing a lower electric potential, for example, zinc, aluminum, magnesium, or their alloys [4]. The metal which has a lower electric potential will be an anode, will corrode and will protect iron in a solution of electrolyte (wet soil or seawater).

Another way to suppress the anodic dissolution (14), namely, to change the direction of this reaction from right to left, is to connect the iron to the negative pole of a direct current power supply. The electrons will move to the iron, and reaction (14) in right direction would slow down to a negligible value.

Thus, iron will serve as cathode and will not corrode. This is an example of 'beneficial' galvanic corrosion and the principle of cathodic protection (CP). We meet in this case the constructive role of corrosion. Anode (zinc, aluminum, or magnesium) corrodes and protects a cathode (iron) from corrosion. Zinc, aluminum, and magnesium are called sacrificial anodes, sometimes anodes of galvanic type.

In CP, metallic equipment is connected to a metal with a lower electrical potential or to negative pole of power supply and turns completely into a cathode, which does not corrode. This method is realized for protection of outer surfaces of underground and submerged pipelines for transportation aviation fuels, outer surfaces of bottoms of AST in contact with soil or sand, and outer surfaces of UST shell containing aviation fuels. When we are talking about CP of outer surfaces of metallic tanks and pipelines in contact with soil there is no matter what kind of fuel and the media in general is inside.

Numerous standards exist for implementation of CP of the outer surfaces of pipelines and AST bottoms, testing methods and monitoring its effectiveness [74–79].

10 Summary

We made a long journey into the world of corrosion and corrosivity of aviation fuels, and learned that literature has little data except in the case of MIC. Here are important findings in this journey.

- Aviation fuels are not corrosive to metals if they meet the requirements of standards in Table 14.1. They become corrosive when water and molecular oxygen appear in them. There are no criteria how much these contaminants should be present in aviation fuels to corrosion occur. Maximum concentration of dissolved oxygen in aviation fuels occurs after their puping/filtration. Dissolved water in fuels as a rule does not cause a corrosion problem. Water-fuel emulsion and free water are potentially corrosive in fuel systems. Dissolved water, however, may also be a problem in the sense because it can become free water as water-saturated fuel is cooled.
- The causes of corrosion of different parts of AST containing kerosene/jet fuel and preventive anti-corrosion measures are summarized in Table 5.

Table 5 Corrosion causes and corrosion prevention in kerosene/jet fuel AST

Part position of AST-subjected corrosion	Corrosion cause[a]	Corrosion prevention method
Bottom	MIC	Drainage and cleaning Coatings (see Table 4) Biocides
Roof	H_2O vapor; H_2S; O_2	Coatings (organic—see Table 4) Inert atmosphere (nitrogen) VCI[b] Dehumidification Scavengers

[a]These causes result in general and pitting corrosion in AST
[b]VCI—Vapor Corrosion Inhibitor

- The corrosion mechanism is electrochemical, similar to atmospheric corrosion of metals under thin layers of electrolytes. Usually, carbon steel pipes and tanks suffer from such corrosion with formation of rust. This rust can serve as a passive protective layer during decades.
- Unfortunately, the same rust can work as a hydrocarbon oxidation catalyst with formation of hydroperoxides. Water in fuels can appear not only from water vapor contained in the atmosphere but also as a result of the break-up of hydroperoxides. Therefore, it is important to clean periodically rust from, the system and to use anticorrosion coatings (see Table 4). Storage and transporting of aviation fuels in an inert atmosphere (nitrogen) or isolated from atmospheric oxygen would restrict or prevent their oxidation.
- Aviation fuel cleanliness means the absence of solid particulates (mostly, rust and dirt), free water, and microorganisms. Mercaptans and other sulfur-containing compounds in very low concentrations do not influence corrosivity of aviation fuels. Total acidity with values shown in Table 1, Chap. 14 also does not influence their corrosivity.

Aviation fuels may contain different additives (up to 0.1% mass.) for improving or preservation of their properties. Most additives not influence corrosivity of aviation fuels. Avgas contains tetraethyl lead. The latter attacks the hot section components in engines.

- As moisture is a key element to the corrosion process, aircraft located in coastal areas and areas with humid climates are more susceptible than aircraft located in arid regions. The fact is that there are very few areas of the world where the air is so dry that you do not need to worry about corrosion. Even in desert areas, daily changes of temperature and humidity can significantly increase the water levels in aviation fuels. Everyone needs to consider corrosion regardless of their geographic location.
- MIC is a common factor in fuel system corrosion. To diminish or prevent MIC, it is strictly recommended to follow free water presence and microbiological contamination of both drain water and fuels with periodicity at least once a month. "The rule of thumb" is 103 (3 superscript or 1000) (Colony Forming Units—CFU) microorganisms counting in ml of fuel or drain water. It is important to follow the tendency of these values with time. Sudden increase of CFU shows possible MIC development and necessity of immediate filtration or cleaning from the sludge.
- Two authoritative standards, ASTM and EN (European standard), give contradictory recommendations regarding the use of biocides. Therefore, we do not recommend using biocides without careful analysis and experience.
- Of particular concern are multi product pipelines that transport both aviation fuels and biofuels containing alcohols and FAME (fatty acid methyl esters). We should be aware that use of biofuels can clean inner surface of old pipes from deposits that can contaminate aviation fuels and impact their quality.

Corrosion prevention and control in aviation systems can be achieved with careful material selection, enhanced, strategically placed and regular drainage, systematic and adequate use of sealants to eliminate the presence of moisture entry into structural joints, use of modern primers and coatings, and careful corrosion monitoring and inspection.

Acknowledgements The author would like to thank Francois Ropital, Jason Lee and Brenda Little for their helpful comments and additional insights during the preparation of this chapter.

References

1. ISO 8044:2020 (2020) Corrosion of metals and alloys—vocabulary, 29 p
2. Groysman A (2020) Terminology confusion in corrosion science and engineering. Israel Chem Eng 6:18–30
3. ASTM G193—22 (2022) Standard terminology and acronyms relating to corrosion. ASTM International, West Conshohocken, PA, USA
4. Groysman A (2010) Corrosion for everybody. Springer, Dordrecht, p 368
5. Aviation fuels technical review. Chevron Products Company, USA (2007), 94 p
6. CAP 1570 (2017) Corrosion and inspection of general aviation aircraft. Civil Aviation Authority, UK, 59 p
7. Aviation maintenance technician handbook—general: FAA-H-8083-30A (ASA FAA Handbook), 696 p (2018)
8. Ajorlolo A (2006) Corrosion in commercial aviation. In: Cramer SD, Covino BS Jr (eds) ASM handbook, Volume 13C: Corrosion: environments and industries. ASM International, pp 598–612
9. Dexter SC (2003) Microbiologically influenced corrosion, corrosion: fundamentals, testing, and protection, vol 13A, ASM handbook. ASM International, USA, pp 398–416
10. Digman WJ (1962) Effects of fuel contamination on corrosion of aircraft fuel systems. SAE Technical Paper 620357. https://doi.org/10.4271/620357
11. Corrosion, Vol. 13, Ninth edition of metals handbook. ASM International, USA, pp 114–122, 314–315 (1987)
12. Ward CB (1963) Corrosion resulting from microbial fuel tank contamination. Mater Perform 2(6):10–17
13. Churchill AV (1963) Microbial fuel tank corrosion: mechanism and contributory factors. Mater Perform 2(6):18–23
14. McGregor JM (1963) Microorganisms fail to grow in furane-lined jet fuel storage tanks. Mater Perform 2(6):24–29
15. Miller RN, Herron WC, Krigrens AG, Cameron JL, Terry BM (1964) Research program shows micro-organisms cause corrosion in aircraft fuel tanks. Mater Perform 3(9):60–67
16. Sheridan JE, Nelson J, Tan YL (1971) Studies on the 'Kerosene Fungus' Cladosporium Resinae (Lindau) De Vries—Part I. The problem of microbial contamination of aviation fuels, tuatara. J Biol Soc 19(1):21–40
17. Ayllon ES, Rosales BM (1988) Corrosion of AA 7075 aluminium alloy in media contaminated with Cladosporium Resinae. Corrosion 44(9):638–643
18. de Meybaum BR, de Schiapparelli ER (1980) A corrosion test for determining the quality of maintenance in jet fuel storage. Mater Perform 19(8):41–44
19. de Schiapparelli ER, de Meybaum BR (1980) Microbiological corrosion in terminal storage tanks for aircraft fuel. Mater Perform 19(10):47–50
20. Iannuzzi M, Rosales BM (2003) Microbiologic contamination and corrosion of aircraft integral fuel tanks—evaluation and risk control. Mater Perform 42(6):62–66
21. Maruthamuthu S et al (2011) Microbial corrosion in aviation turbine fuel (Jet A 1) transporting pipeline. J ASTM Int 8(6):1–15

22. McNamara CJ et al (2005) Corrosion of aluminum alloy 2024 by microorganisms isolated from aircraft fuel tanks. Biofouling 21(5–6):257–265
23. Defence Standard 91-091. Turbine fuel, aviation kerosine type, Jet A-1 NATO Code: F-35. Joint Service Designation: AVTUR, Issued by the United Kingdom (U.K.) Ministry of Defence. Glasglow, UK. https://www.dstan.mod.uk/
24. ASTM D1655-21a. Standard specification for aviation turbine fuels. ASTM International, USA, 20 p (2021)
25. CAN/CGSB 3.22-2012(R2017) (2012) Wide-cut type aviation turbine fuel (GRADE JET B). Canadian General Standards Board, 13 p
26. ASTM D4054-22 (2021) Standard practice for evaluation of new aviation turbine fuels and fuel additives. ASTM International, West Conshohocken, PA, USA
27. Tripathi RP, Gulati IB, Pandey SS, Rohatgi HS (1973) Copper and silver corrosion by aviation turbine fuels. Ind Eng Chem Prod Res Dev 12(3):227–231
28. Bentur S, Babitz M, Stern AB (1962) Effect of ultraviolet irradiation on the corrosiveness of aviation gasoline. J Chem Eng Data 7(1):117–119
29. Groysman A (2014) Corrosion in systems for storage and transportation of petroleum products and biofuels. Springer, Dordrecht, p 297
30. Handbook of aviation fuel properties, 4th edn. CRC Report No. 663. Coordinating Research Council, 5755 North Point Parkway, Suite 265, Alpharetta, GA 30022, USA
31. ASTM D7548-16a (2022) Standard test method for determination of accelerated iron corrosion in petroleum products. ASTM International, USA
32. ASTM D130—19 (2019) Standard test method for corrosiveness to copper from petroleum products by copper strip test. ASTM International, USA
33. ASTM D7095-17 (2023) Standard test method for rapid determination of corrosiveness to copper from petroleum products using a disposable copper foil strip. ASTM International, USA
34. ASTM D849—15 (2019) Standard test method for copper strip corrosion by industrial aromatic hydrocarbons. ASTM International, USA
35. Geva J et al (2011) Hydrogen sulfide diffusion in jet fuels and its fate during fuel transportation and storage. In: IASH 2011, the 12th international conference on stability, handling and use of liquid fuels, Sarasota, Florida USA, 16–20 October 2011
36. IP227 (1999) Determination of corrosiveness to silver of aviation turbine fuels silver strip method (withdrawn without replacement in 2001)
37. DERD 2494 (1990) Aviation turbine fuel kerosene type AVTUR F-35, Superseded by DEF Standard 91-091 [23]
38. ASTM D7671-21 (2021) Standard test method for corrosiveness to silver by automotive spark-ignition engine fuel—silver strip method. ASTM International, USA
39. ASTM D7667-21 (2021) Standard test method for determination of corrosiveness to silver by automotive spark-ignition engine fuel—thin silver strip method. ASTM International, USA
40. ASTM D3242 (2023) Standard test method for acidity in aviation turbine fuel. ASTM International, USA
41. Groysman A (2017) Corrosion problems and solutions in oil refining and petroleum industry. Springer, Dordrecht, p 356
42. Messer B, Tarleton B, Beaton M, Phillips T (2004) New theory for naphthenic acid corrosivity of athabasca oilsands crudes. Paper 04634, CORROSION 2004, NACE International, Houston, TX, USA
43. Groysman A, Penner J (2008) Process for inhibiting naphthenic acid corrosion, WO/2009/053971, USA
44. ASTM D4928—12 (2018) Standard test method for water in crude oils by Coulometric Karl Fischer Titration.ASTM International, USA
45. Pires APP et al (2018) Chemical composition and fuel properties of alternative jet fuels. BioResources 13(2):2632–2657
46. Popp J, Lakner Z, Harangi-Rákos M, Fári M (2014) The effect of bioenergy expansion: food, energy, and environment. Renew Sustain Energy Rev 32:559–578

47. Maurice LQ, Lander H, Edwards T, Harrison WE (2001) Advanced aviation fuels: a look ahead via a historical perspective. Fuel 80(5):747–756
48. Doliente SS et al. (2020) Bio-aviation fuel: a comprehensive review and analysis of the supply chain components. Front Energy Res 8(Article 110):38
49. ASTM D7566-21 (2021) Standard specification for aviation turbine fuel containing synthesized hydrocarbons. ASTM International, West Conshohocken, PA, USA
50. Rumizen MA (2021) Qualification of alternative jet fuels. Front Energy Res 9(Article 760713)
51. Anes V et al (2016) Galvanic corrosion of aircraft bonded joints as a result of adhesive microcracks. Proc Struct Integr 1:218–225
52. Sujata M et al (2019) Unraveling the cause of an aircraft accident. Eng Fail Anal 97:740–758
53. Findlay SJ, Harrison ND (2002) Why aircraft fail? Mater Today 18–25
54. Final safety investigation report. Accident file No. 99-16. Israeli Ministry of Transportation and Road Safety, 22 p. In Hebrew (2016)
55. Aircraft accident investigation report AA2006-4, Privatery owned piper J3C-65, G-kirk, Kanazawa city, Ishikawa prefecture, October 21, 2005. September 2006, Aircraft and railway accidents investigation commission ministry of land, infrastructure and transport, Japan (2006)
56. Benfer JE (2006) U.S. navy aircraft corrosion. In: Cramer SD, Covino BS (eds) ASM handbook, Volume 13C: Corrosion: environments and industries. ASM International, pp184–194
57. API 620 (2013) Design and construction of large, welded, low-pressure storage tanks, 12th edn. API, USA
58. API 650 (2020) Welded tanks for oil storage, 13th edn. API, USA
59. ASTM E527-16 (2023) Standard practice for numbering metals and alloys in the unified numbering systems (UNS). ASTM International, West Conshohocken, PA, USA
60. Ghali E (2006) Magnesium and magnesium alloys. In: Winston Revie R (ed) Uhlig's corrosion handbook, 2nd edn. Wiley &Sons, Inc., pp 793–830
61. API RP 652 (2020) Lining of aboveground petroleum storage tank bottoms, 5th edn. API, USA
62. Zhang J et al (2021) Surface coating on aluminum substrate with polymeric guanidine derivative to protect jet fuel tanks from microbial contamination. Surf Coat Technol 422, Paper number 127521
63. Baumi J, Bertosse CM, Guedes CLB (2020) Aviation fuels and biofuels. IntechOpen, 21 p
64. Hardy DR, Black BH, Wechter MA (1986) Quantitative determination of corrosion inhibitors in middle distillate jet fuels by gel permeation chromatography. J Chromatogr A 366:351–361
65. Kuznetsov YI, Redkina GV (2022) Thin protective coatings on metals formed by organic corrosion inhibitors in neutral media. Coatings 12:149
66. Wysocka J, Cieslik M, Krakowiak S, Ryl J (2018) Carboxylic acids as efficient corrosion inhibitors of aluminum alloys in alkaline media. Electrochim Acta 289(1):175–192
67. Suriya Prabha A, et al (2022) 12—Corrosion inhibition by carboxylic acids—an overview. In: Environmentally sustainable corrosion inhibitors, fundamentals and industrial applications, pp 245–271
68. Moussa MN et al (1990) Carboxylic acids as corrosion inhibitors for aluminum in acidic and alkaline solutions. Anti-Corros Method Mater 37(3):4–8
69. Groysman A et al (2005) Corrosiveness of acidic crude oil and its fractions. Mater Perform 44(4):34–39
70. Žerjav G, Milošev I (2014) Carboxylic acids as corrosion inhibitors for Cu, Zn and brasses in simulated urban rain. Int J Electrochem Soc 9:2696–2715
71. Mercer WC (1993) An investigation of carboxylic acids as corrosion inhibitors in engine coolants, Symposia paper STP 251156S, In: Beal RE (ed) Engine coolant testing, vol 3, STP 1192-RB. ASTM, pp 44–62
72. Lin B, Zuo Y (2019) Corrosion inhibition of carboxylate inhibitors with Different alkylene chain length on carbon steel in an alkaline solution. R Soc Chem 9:7065–7077
73. Almobarak NA, El-Naggar MM, Al-Mufraj RS, Al-Zoghbi OA (2014) Carboxylic acids: pitting corrosion inhibitors for carbon steel in alkaline medium and in the presence of chlorides. Chem Technol Fuels Oils 50:170–178

74. U.S. Environmental Protection Agency, Part II, 40 CFR, Code of Federal Regulations, Parts 280 and 281, Underground storage tanks containing petroleum (2015)
75. AMPP Standard SP0285-2021 (2021) Corrosion control of underground storage tank systems by cathodic protection. AMPP, USA
76. API RP 1632 (R2010) (1996) Cathodic protection of underground petroleum storage tanks and piping systems, 3rd edn. American Petroleum Institute, USA
77. AMPP Standard SP0193-2016 (2016) External cathodic protection of on-grade carbon steel storage tank bottoms, USA
78. AMPP Standard SP0169-2013 (2013) Control of external corrosion on underground or submerged metallic piping systems, USA
79. UL1746 (2007) UL standard for safety, external corrosion protection system for steel underground storage tanks, Northbrook, IL, Underwriters Laboratory, USA

Chapter 14
Test Methods and Their Significance for Aviation Fuel Quality Testing

Konstantin Tartakovsky, Marlin D. Vangsness, and Moshe Rabaev

Abstract The remarkable safety of air travel compared to land transport is achieved through systematic risk management, failure investigation, and continuous improvement. A crucial area where these principles are applied is in preventing fuel system issues caused by aviation fuel contamination, which can pose acute and long-term risks during storage, transport, or use in aircraft fuel systems. Fuel quality assurance (QA) laboratories employ standardized protocols to ensure that aviation fuels comply with the specifications and limits defined in standard test methods and to detect many contaminants that can cause considerable short-term damage to the fuel system. However, routine analysis cannot identify numerous other contaminants that might severely damage the ground and aircraft fuel systems. In addition, to prevent re-contamination, the nature and source of the contamination must be determined. Where contamination is suspected, analytical chemistry laboratories have methods for deep fuel analyses that can meet these challenges. This chapter provides a general overview of some contaminants often found in fuels and suggests ways to detect them.

K. Tartakovsky (✉)
Independent Consultant, Winnipeg, MB, Canada
e-mail: kosta002@gmail.com

M. D. Vangsness
Independent Consultant, Yellow Springs, OH, USA
e-mail: marlin.vangsness@sbcglobal.net

M. Rabaev
Independent Consultant, Moshe Rabaev Consulting, Beer Sheva, Israel

Blechner Center for Industrial Catalysis and Process Development, Department of Chemical Engineering, Ben-Gurion University of the Negev, Beer Sheva, Israel

M. Rabaev
e-mail: rabaevm@gmail.com

M. Rabaev et al. (eds.), *Aviation Fuel Problems and Solutions: From Refinery to Wingtip*, https://doi.org/10.1007/978-3-032-05514-9_14

1 Introduction

This chapter provides a general overview of the contaminants commonly found in fuel and practical methods for investigating them. Understanding how the fuel system works, what materials are used in the fuel system, how maintenance is performed, and other factors are necessary to resolve problems and prevent recurrences. A fuel contamination study's success depends on cutting-edge analytical instruments and collaboration between QA fuel laboratory staff, fuel engineers, and an analytical chemistry laboratory. Equipping the analytical laboratory with diverse instruments specifically tailored to address the complexities of aviation fuel contamination facilitates a more holistic approach. Such an approach is vital for conducting in-depth contamination analysis and achieving effective problem resolution. A thorough understanding of the standard tests performed in the QA laboratory and familiarity with the operation of standard testing equipment can provide the analytical researcher with the critical information required for an effective investigation.

Fuel Contamination

Any organic or inorganic contamination that may impair the operation of a fuel system or engine system is called fuel contamination. Not all fuel contaminants, however, require extensive study. Some contaminants are typical, and remedies are available. For example, water is a typical fuel contaminant entering fuel systems through tank ventilation, rain, groundwater penetration, condensation, or coming in transport. Regular sump draining and filter water separators will reduce the water content to manageable levels. The presence of some particular contaminants (particles caught in filters) is not usually a cause for worry. However, system components that fail before their average lifetime are prime indicators of contamination that may require a dedicated investigation. Contamination that cannot be removed by settling, draining, filtration, or common additization may cause the aircraft's fuel system to malfunction and threaten flight safety. Even very low levels of some contaminants may have serious consequences after long-term exposure.

Mechanical contamination is dangerous when the particles overwhelm filters or when many particles with sizes below the pore size of the filter pass through a filter system and build up in the fuel system. The Standard Test Method for Particulate Contaminants in Aviation Fuel by Line Sampling (ASTM D2276) tells us enough about the fuel quality in this case. Annex 14.1: Selected Standard Test Methods for the Analysis of Contamination in Aviation Fuels, provides a list of selected standard test methods for analysis of contamination in aviation fuels. However, there are times when we need to know the source of the unusual contamination.

Figure 1 shows turbine blades corroded after burning fuel probably contaminated by an atypical contaminant phosphorus-based hydraulic liquid.

Figure 2 depicts a pattern associated with pollen atypical contamination that should be avoided. A pollen pattern could have entered the fuel tanks through the vent inlets. This could be a common type of external contamination during a blooming period.

Fig. 1 Advanced hot corrosion damaged turbine blade (Image courtesy of M. Rabaev)

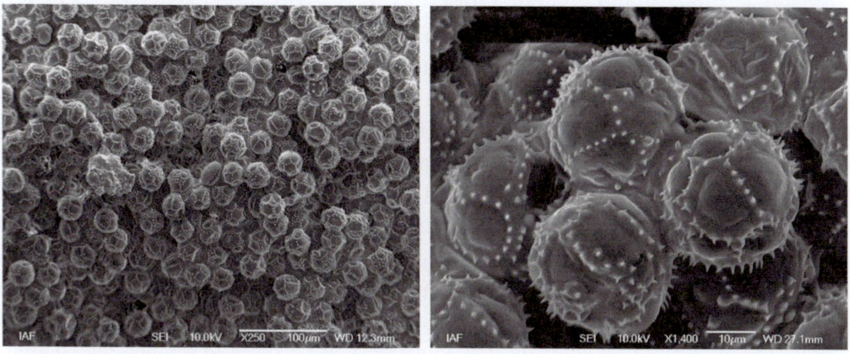

Fig. 2 SEM image of plant pollen contaminants on an aircraft fuel strainer (Image courtesy of M. Rabaev)

Impurities in fuel can arise from the solubilization of components in elastomers that are not fuel-resistant. This can result from incorrect formulation of elastomer causing abnormal swell or installation of inferior components. Sometimes incorrect fuel or penetration of some aggressive contaminants may damage sealants, coating, fuel bladders, or O-rings. This can lead to migration and subsequent release of additives, such as plasticizers or stabilizers from the elastomer. Sometimes, the damage is discovered long after it has been caused and the cause has already passed.

External contaminants of non-human origin may include natural elements such as sand, dirt, microbial growth, and pollen. These materials, which are not part of the fuel system, can inadvertently contaminate the fuel. On the other hand, human-related external contaminants can include excessive amounts or incorrect types of additives, as well as external artificial components like biodiesel, which are clearly defined as materials external to the fuel system. Vangsness (2015) thoroughly lists different contaminants and their origins within the aircraft [1]. A third source of contamination is internal contamination caused by changes in chemical composition during storage, exposure to temperature, or interaction with surrounding structural materials—metals and non-metals. Most contaminants may be discovered using analytical techniques explained in this chapter.

2 Analytical Chemistry Laboratory

Analytical Chemistry Laboratory for fuel analysis is a research facility encompassing various analytical fields, including at least chromatography, spectroscopy, thermal analysis, elemental analysis, and visual and SEM microscopy. A broader range of analytical disciplines within the laboratory provides more precise outcomes in fuel research. Analysts must be flexible and develop different methods to address research questions. Maintaining scientific integrity through transparent methods and protocols is suggested to comply with ISO 17025 certification standards, which is recommended but not mandatory for such laboratories.

In many cases, analyst's decisions are not based on limits and guidelines for standard fuel analysis results. The decision is based on the laboratory analyst's professional experience, the results obtained and data processed. Furthermore, these decisions and recommendations must be clear and comprehensible to fuel engineers, who rely on this information to make informed choices about fuel or fuel systems usage.

Collaboration between the analytical chemistry laboratory, the fuel engineer, and the fuel quality assurance (QA) laboratory representative is essential to the success of any fuel research. Each party brings its perspectives, experiences, and data into the discussion. The fuel laboratory begins by receiving suspected fuel samples and conducting an initial fuel analysis. The laboratory analyst could request repeat analysis and view the results independently. This enables them to verify essential data and generate ideas for future research based on these standardized tests. The fuel engineer should describe what transpired, explain the system, and provide a history of similar problems that have occurred in the past.

Updating team members on intermediate findings during the analytical investigation is beneficial. Addressing questions and uncertainties from various disciplines through brief team meetings can streamline decision-making for fuel engineers using solid data. A fuel lab representative's involvement in research enhances their expertise, enabling them to tackle future challenges and pay attention to issues during routine fuel analysis. The success of an analytical study hinges on the close collaboration and trust between the fuel laboratory and fuel engineers.

3 Reference Samples

Reference samples are essential for pinpointing the source of contamination by comparing chemical, physical, visual, and tactile data (e.g., from handling with fingers or tweezers). In many cases, finding a match can conclude the investigation; in others, it may lead to further research. However, matching a sample to a reference should always prompt questions about specificity. For instance, if FTIR analysis indicates "nylon," confirming the specific type may require melting point tests or additional Raman analysis. Similarly, detecting a generic "hydrocarbon" can lead to numerous questions, necessitating deeper analysis to identify the exact compound. Even chromatographic results may be inconclusive without isomer-specific techniques, such as Cold EI GC-MS, which improve molecular-ion visibility and reduce fragmentation [2]. In contrast, discovering phthalates in a residue might be sufficiently definitive, requiring no further analysis. Conversely, there are situations where clear "fingerprints" can identify a contaminant specifically, for example as synthetic engine oil, differentiating it from other oils—even among different lots of the same product.

Reference samples are needed for reliable chemical analysis. This may be a valid reference sample of the suspected contaminant and data related to the material, such as chemical spectra, composition, and other chemical and physical properties. If the suspected contaminant is a commercial product, getting all the information the manufacturer provides is paramount. The availability of both actual control and data samples simplifies, accelerates, and ensures the accuracy of fuel contamination analysis, allowing fuel contamination to be chemically searched for at the desired low detection level.

Commercial products, such as engine oil, grease, aviation soap, hydraulic or cooling fluid, additives, and solvents, can contaminate aviation fuel. These products are typically mixtures of various chemical components formulated during production. In most cases, their specific composition is proprietary to the manufacturer, and the formulation may vary from batch to batch, even for the same product. For example, engine oil MIL-PRF-23699[3] may contain more than 30 different compounds.

While the reader might infer that every analytical lab has an extensive repository for actual reference samples, this is generally false. Analytical laboratories do not usually keep samples from previous work; a laboratory should be clean and free of materials, to avoid cross-contamination between different cases and contamination of sensitive analytical instruments. Information about materials is stored in software libraries that contain all available information and reports from researchers collected over years of work in the laboratory. All this information, especially previous reports, allows researchers to find similar past cases and get advice from colleagues who wrote the reports, even if they are no longer in the lab. These data are the intellectual property of the laboratory. In addition, commercial software libraries and access to different materials databases make it possible to gather information about many materials that have never been in the lab.

Fig. 3 Fuel samples (some of them were collected incorrectly) from the investigation of light aircraft accident (Image courtesy of K. Tartakovsky)

Preparing for fuel research should always be preceded by creating laboratory databases, purchasing all available software libraries, and gaining access to online material databases. Before conducting a thorough investigation, it is desirable to conduct a preliminary study on the case, particularly what is expected to be found in the fuel, and, if possible, to obtain actual reference samples.

4 Sampling

A critical step in any fuel contamination investigation is the collection of representative or appropriate, diagnostic fuel samples. If contamination in a specific part of the fuel system is suspected, trained personnel should collect samples taken from both upstream and downstream of that location. Samples should be collected correctly and in sufficient quantities, as described by sampling procedures, to allow for standard fuel and analytical lab tests. Figure 3 shows different samples collected during an investigation of fuel quality in light aircraft accident. It is easy to distinguish that some of them were collected incorrectly. Chapter 11 deals with sampling issues, while Fig. 1 provides a comprehensive guide on fuel sampling techniques.

A description of what was seen and smelled when the field samples were collected, such as how the samples looked, separated into phases, and changed colour, can aid in determining the source of contamination.

5 Preliminary Inspection

When looking at fuel contamination, it is essential to start with a thorough evaluation, including visual and tactile checks. It is critical to document these findings clearly and concisely. The written report should include well-lit, high-resolution images accompanied by a scale to convey the investigator's viewpoint accurately.

Acknowledging that certain aspects, such as tactile sensations, cannot be fully represented in pictures or by analytical devices is crucial. Characteristics like softness, hardness, flexibility, crumbliness, peel resistance, and additional substances can be identified through touch and sight within minutes, offering valuable insights into the contamination. Tactile examinations can be carried out using a single-ended dental probe.

Rubber (elastomer) particles are usually flexible and can be stretched, but plastic particles are mostly flexible, but not stretched. Methyl methacrylate-based materials are crumbly, epoxy glues are much harder to crush, and phenolic-based materials are more difficult to crush than epoxy. Many metallic particles can be differentiated by their luster, while a magnet will attract ferromagnetic particles. Biocontamination might appear as dark sludge or slime and have a sour or rotten odour. Sand and soil contamination are often more easily detected visually than through analytical instruments. Changes in fuel colour can also provide hints about potential contamination and warrant further standard testing.

For instance, a written report that describes a contaminant resembling blue jelly can be more advantageous than relying solely on photos or chemical analyses. It is prudent, however, to refrain from using senses such as smell and taste during these inspections due to potential safety hazards.

For example, improper preliminary investigation led to a long and expensive inquiry that could have been resolved in the early stages. This case involved contamination in a helicopter filter, prompting an extensive investigation through multiple chemical analyses. Despite the considerable time and effort spent on these analyses, they failed to identify a clear cause for the consistent filter clogging, as reported by the client. It was only through visual inspection that a manufacturing defect was discovered: most filter folds were closed, leaving only a few open for the fuel to flow through. These open folds quickly became clogged. In this instance, the frequent filter clogging did not result from fuel contamination. This case underscores the importance of prioritizing visual preliminary inspection in fuel contamination investigations, as it can uncover defects that might be overlooked in chemical analysis.

6 Source of Contamination Materials in Fuel

Recognizing the origins of a specific contamination allows for adopting effective prevention and removal strategies. Non-metallic particles from internal aircraft systems and external sources typically include paint, coatings, sealants, adhesives, fibres, rubber, and plastic polymers. Metallic contamination can originate from both internal and external sources. Contaminants originating in the fuel supply system may include polyethylene particles, plasticizers, fuel additives, other fuels and oils, and metal chips from pipeline maintenance or structure works. Environmental contaminants often involve soil materials, sand, and biologically derived materials, while biological contamination includes such as fungi and other microorganisms. Malfunction or sabotage-related contamination varies depending on the specific case.

7 Mitigating False Positives and Negatives in Analysis

False positive refers to the misidentification of compounds such that a harmless material or even the absence of contaminants is identified as a harmful contamination issue. This error wastes time and effort and may neglect further analysis of harmful materials. The cure for this error is further analysis, either by more sensitive testing or a more specific method, to verify that a suspected compound is present. If the analysis indicates a high concentration of material, verification is usually not problematic; however, when a harmful material is found near the limit of quantification (LOQ), obtaining a definitive verification of its presence may be difficult. It is then advisable to assign a probability for the identification and report its probability for inclusion in the overall analysis. False-negative results represent a more severe risk, potentially overlooking contaminants that could adversely impact the aircraft's fuel system over short and long durations.

Errors originating from secondary contamination occurring during sample acquisition, transport, or laboratory handling may be reduced or eliminated by comparing multiple and upstream samples. Laboratory contamination will typically fall into recognizable categories, including but not limited to:

- Instrumental contamination, such as using silicone oil for various laboratory purposes.
- The presence of KBr powder utilized in Fourier transform infrared (FTIR) analysis.
- Phthalates can originate from plastic pipes and containers.
- Siloxanes, a common contaminant in gas chromatography mass spectrometer (GC-MS) analysis.
- Organophosphates, another frequent intruder in GC-MS analysis.
- Squalene, a contaminant often transferred from human skin.
- Aliphatic hydrocarbon oils.

Aliphatic hydrocarbon oils present a significant challenge in the laboratory setting, as they can permeate from numerous sources, including technical dry air used in Gum (ASTM D381) analysis. It is virtually impossible to anticipate every potential source of interference that might lead to inaccurate results.

It is essential to note that the compounds above are potential contaminants and categorizing them as interferences would be a misrepresentation.

7.1 Sampling Protocol

Sampling Procedure Adherence to standardized fuel sampling procedures is paramount. Refer to Fig. 9.1, Chap. 9 for a comprehensive guide on fuel sampling techniques that minimize the risk of contamination.

Secure Sample Storage Store samples under prescribed conditions to maintain integrity, preventing potential degradation or contamination over time.

Analytical Solvents Employ high-purity HPLC or GC-MS-grade solvents for analysis to prevent the introduction of contaminants during the analytical process.

Reference Fuel and Suspect Components If it is possible to obtain a local sample that is confirmed to be free of contamination, this sample should be used to establish a baseline for analysis. Comparing this baseline with the sample under investigation allows for identifying the presence or absence of harmful compounds based on observed differences in the analyses.

7.2 *Analytical Rigor*

Repeat Analysis Conduct at least two analyses for each fuel sample to ascertain the consistency of results. Document any exclusions or deviations meticulously.

Calibration Standards Standard solutions of single suspect components are frequently needed to verify tentative identifications. They should be prepared in concentrations near the unknown material to compare instrument response properly.

7.3 *Blank or Background Sample Analysis*

In aviation fuel analysis, blank sample analysis is necessary to establish the baseline level of potential impurities in reagents, instruments, or the ambient environment. Particularly during gum analysis, delineated in the ASTM D381 guidelines, false readings often emerge from residues or contaminants introduced during preceding laboratory analysis or technical air/steam contaminations.

7.4 *Cross-Contamination*

Cross-contamination can occur when residues from a previously analyzed sample inadvertently affect subsequent samples. For example, even if low contamination levels are suspected, analysing blank samples immediately before and after the probable contaminated case is prudent in GC analysis. This practice ensures that any potential cross-contamination is identified and avoided.

7.5 *Instrumental and Methodological Verification*

In cases of questionable results, consider using alternative analytical instruments or techniques to verify the findings, thereby eliminating potential sources of error.

7.6 *Contingency Measures*

External Contamination Checks In the event of suspected external contamination, initiate duplicate analysis and reference checks to ascertain the source of contamination and rectify it promptly.

Collaborative Analysis Engage with external laboratories or agencies for collaborative analysis to validate findings and ensure the utmost accuracy in results.

8 Laboratory Equipment Requirements for the Study of Fuel Contamination

For practical fuel contamination analysis, it is essential to have both high-tech and simple tools in the laboratory. These can be put into several groups, focusing on sample preparation, microscopic examination, chemical identification, and visual inspection.

(1) Sample Preparation Equipment

   (a) Evaporator or oven for removing fuel and releasing higher-boiling liquids.
   (b) Centrifuge for separating liquids based on density and particle sedimentation.
   (c) Vacuum filtration kit with a vacuum pump for filtering solid particles
   (d) A magnet for detecting and separating ferromagnetic particles.

(2) Microscopes

   (a) Stereomicroscope for initial particle examination
   (b) Scanning electron microscope with energy dispersive X-ray spectroscopy (SEM–EDS) for enhanced magnification and elemental analysis, as shown in Fig. 4.
   (c) Optical microscope for inspecting biological contaminants, as shown in Fig. 5.

(3) Chemical Identification Instruments

   (a) Fourier transform infrared (FTIR) spectrometer for initial chemical identification.

Fig. 4 SEM image of fibers and particles found in a fuel filter (×2000) (Image courtesy of M. Rabaev)

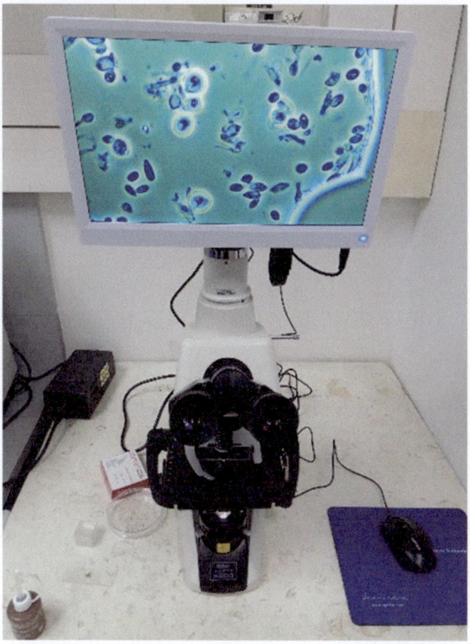

Fig. 5 Microscopic view of biocontamination (Image courtesy of K. Tartakovsky)

(b) Raman spectrometer for cases where FTIR is not applicable or for additional spectral information.
(c) X-ray fluorescence (XRF), Rotrode, or inductively coupled plasma (ICP) spectrometers for elemental analysis.
(d) Gas chromatography (GC) coupled with a mass spectrometer (MS) detector, or other GC detectors, is widely used for fuel investigation and compositional analysis.

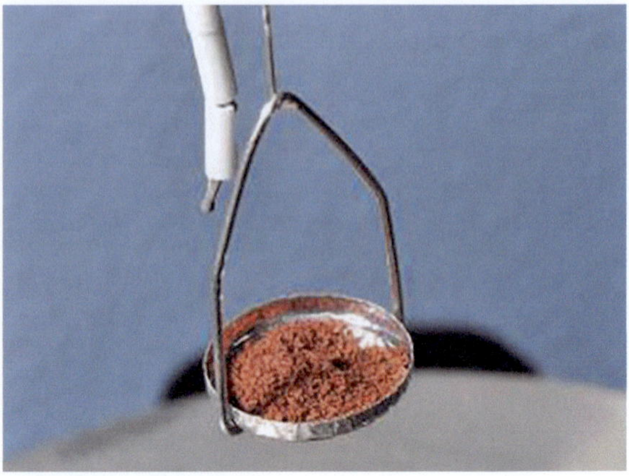

Fig. 6 The combustion remains from TGA analysis. The red color testifies to the presence of iron oxide, Q500 TA (The diameter of the platinum dish is 10 mm) (Image courtesy of K. Tartakovsky)

    (e) Thermal gravimetric analyzer (TGA) and differential scanning calorimeter (DSC) for various investigations, see Fig. 6.

(4) Visual inspection tools

    (a) Clear sample containers for easy observation of contaminants.
    (b) Refractometer for measuring the refractive index.
    (c) Hydrometer for measuring specific gravity.
    (d) pH test strips.
    (e) Flashlight (torch).
    (f) Thin-layer chromatography (TLC) kit.
    (g) UV light for illuminating contaminants and enhancing visibility.

In addition to possessing laboratory equipment, it is also beneficial to have field analysis tools. The accurate collection of samples in the field by trained professionals using appropriate sampling instruments and adhering to correct procedures is crucial, underscoring the importance of acquiring proper field sampling gear. Furthermore, it is advised to maintain a diverse range of instruments for executing field analyses. Standard fuel testing equipment includes a filtration kit. For analytical field research, the primary instruments are portable FTIR spectrometers with Attenuated Total Reflectance (ATR) and portable biological microscopes, which are essential for investigating biological contaminants. Photography equipment is also necessary.

All field equipment and tools should adhere to Foreign Object Debris (FOD) requirements for the specific location, as the field laboratory kit might be used near or inside the aircraft. This consideration helps to prevent damage to sensitive equipment and maintain operational safety.

A well-equipped lab for analysing fuel contamination should have advanced analytical tools and simple tools, such as tools for preparing samples, microscopes, chemical identification tools, and tools for visual inspection. These tools and instruments make it easy to find and analyse contaminants in different fuel types.

9 Extraction Techniques for Sample Preparation in Filter Analysis

Filter analysis typically involves extracting and analysing material present in the filter. Technicians can use tweezers to remove particles or fibres. Alternatively, an ultrasonic bath can extract fine particles and dirt. In the ultrasonic bath method, the contaminated filter is placed in a container with a cleaner, such as SBP 60/95 LNH (Solvent Base Petroleum, boiling range 60–95 °C, Low-Naphthenic Hydrocarbon type) or another suitable light hydrocarbon cleaner. The container is then sealed with aluminum foil and placed in a water-filled ultrasonic bath. This container is agitated for 15–30 min without heat. After the bath, the filter is removed, allowing the liquid to evaporate. The resulting residue is dried in an oven and then transferred to a desiccator for cooling.

To prevent false positive results, it is important to verify that the solvent used is pure and leaves no residue on the glass walls. This can be confirmed by conducting a control experiment using only the pure solvent to check for residue.

10 Limit Detection and Quantification for Contamination Detection in Aviation Fuel

The detection threshold (often called limit of detection—LOD) signifies the smallest observable signal or its corresponding measurable quantity that can be confidently identified during an analysis. On the other hand, the quantification threshold (often called LOQ) represents the minimum concentration of a substance that can be accurately and precisely measured. Various factors influence these thresholds, including the instrument's responsiveness to the specific substance, the makeup of the sample, and the practicality of obtaining a pure sample to set an appropriate baseline. Use limits are established to provide a safe operational level, and while exceeding these limits may initiate a request for investigation, they are unrelated to instrumentation limits. The laboratory analyst must ensure that the method's LOQ is equal to or lower than the concentration limits required by the fuel system specification to ensure reliable detection and quantification of potential contaminants.

11 Analytical Instrument Methods

This summary presents an overview of various, but not all, analytical methods and their corresponding instruments, which include molecular analysis spectrometers (such as FTIR or Raman), elemental analysis tools (ICP, Rotrode spectrometer, and SEM–EDS), thermal analysis devices (DSC and TGA), gas chromatography systems (e.g., GC-MS, GC-FID), and microscopes. A concise explanation of each instrument's operation will be provided to aid in contamination analysis, with more in-depth information available in the referenced materials. The discussion will also cover optimal accessories for each method and showcase real-life examples demonstrating these techniques' critical role in detecting fuel contamination. To ensure precision and reliability in the analytical results, the analyst must cross-validate findings from each instrument with results from other devices.

11.1 Fourier Transform Infrared Spectrometer

The Fourier transform infrared (FTIR) spectrometer is an analytical instrument used to evaluate the vibrations of chemical functional groups. Frequently used as a primary device for investigating fuel contamination, FTIR offers ease of use, minimal sample preparation, and produces data in molecular absorption spectra. It effectively detects contaminants such as non-volatile fuel residues and particulate contamination from Gum tests and has demonstrated its usefulness in initial assessments for visible microbial contamination. Owing to its efficiency, low sample preparation needs, and accessibility for less experienced operators, the FTIR spectrometer is a valuable technique for both field and laboratory applications. A molecular absorption spectrum occurs when covalent bonds absorb infrared energy at frequencies corresponding to their resonance frequency. These frequencies depend on the atoms within the bond, the electron cloud, neighbouring functional groups, conjugation within the molecule, and low-energy bonding such as hydrogen bonding and solvation effects. However, ionic or metallic bonds fall outside the IR energy range, and symmetrical bonds (such as C–C, O_2, N_2, etc.) and symmetrical vibration modes typically do not display significant spectra due to their unchanging dipolar moment during vibration.

A molecule comprises various functional groups, each possessing one or more vibration modes, such as elongations and deformations. These modes produce unique peak positions, intensities, and widths, which can serve as molecular fingerprints. These fingerprints enable substance identification or classification through spectral library searches. Alternatively, identification can be achieved by examining the presence or absence of specific peaks or peak groups, as a given functional group will always have its absorption bands in the same range. However, some variations may occur due to solvation and other mixing effects.

In real-life situations, spectra are often obtained from mixtures of compounds rather than isolated compounds. Depending on inter- and intra-molecular interactions, the resulting spectra can range from a simple sum of individual spectra to a sum of significantly modified ones. Despite these complexities, automated identification can work exceptionally well when analysing unique compounds.

Several accessories are used to test samples in an FTIR device. The most helpful accessory for fuel pollution testing is an ATR accessory with a diamond crystal. Diamond crystals are strong, suitable for any area of the measured spectrum, and have good transmittance over the full range from 400 to 4000 cm^{-1}. The depth of light penetration into a sample applied to the crystal is relatively large, and the diamond crystal is extremely robust, both chemically and mechanically. For samples containing black charcoal, as in rubber chip analysis, a germanium crystal is preferred to avoid absorption saturation. However, germanium crystals can only be used from 600 cm^{-1} and may break or be scratched when a hard sample is pressed excessively.

Pyrolysis FTIR analysis is another technique that is useful for studying rubber samples. It focuses on the polymer within the rubber, eliminating interference from additives such as black carbon or talc during the identification process. However, this method is destructive to the sample and requires a larger sample size than Germanium ATR analysis, which can be preferred in a small sample. Rubber pyrolysis can be performed in compliance with ISO 4650 [4] and ASTM D3677 [5] standards, with slight modifications to obtain a more useful spectrum with less water content. The condensed pyrolysis products are collected on a glass slide, which is used to close the open side of the tube slightly during the pyrolysis process, as shown in Fig. 7. The pyrolysis product from the glass slide is then transferred to a diamond ATR crystal for analysis using the ATR method, yielding a spectrum that solely contains rubber pyrolysate and excludes several additives.

Additionally, the potassium bromide (KBr)-FTIR analysis technique is also useful for sample preparation. It entails grinding the sample using an agate pestle and mortar to create a fine powder. This powder is then mixed with KBr powder and ground together. The resulting mixture is pressed using a KBr press assembly to form a transparent KBr pellet containing the sample, which can be further analyzed using a transmission accessory in FTIR. While this approach may be more time-consuming, it often yields better spectra than ATR analysis. This method is suitable for all solid materials that can be ground or crushed, except rubber samples, and can also be adapted to analyze liquid samples.

Lastly, the FTIR microscope represents a sophisticated and powerful analytical tool that substantially augments the capabilities of material analysis. This advanced instrument facilitates the investigation of particulate matter down to dimensions as

Fig. 7 Rubber pyrolysis condensate collection on slide glass for further FTIR examination (Image courtesy of K. Tartakovsky)

minute as several microns. Simultaneously generating an FTIR spectrum and a high-resolution magnified image of the particle reveals complex underlying structures and chemical compositions. The microscope is typically furnished with an ATR germanium probe, an essential component that markedly enhances its analytical performance. Used predominantly for opaque particles, the germanium probe is integral to the system, while a pressure diamond cell is used to examine transparent samples. The FTIR microscope's ability to furnish detailed images and spectral data on microscopic particles underscores its scientific appeal.

Some examples of artificial non-metallic species that can be analyzed by FTIR, commonly found in aircraft systems may include components from many sources, including, but not limited to:

- Paint/coating systems: epoxy, polyurethane, alkyd enamel, phenolic resin
- Sealing systems: polysulfide and silicone sealants
- Adhesives: epoxy, polymethacrylate, cyanoacrylate
- Fibers: cellulose (cotton-like), nylon, polyester, fiberglass
- Rubber: NBR (nitrile butadiene rubber), silicone rubber, polyurethane rubber
- Plastic polymers: polyvinyl chloride (PVC), Polyethylene terephthalate (PET)
- Particulate cleaners: polymethacrylate, urea–formaldehyde, polyester, melamine–formaldehyde, phenol–formaldehyde, acrylic

Some examples of non-metallic particles not originating from aircraft systems typically involve (natural or human-related from supply lines) :

- Biological sources: leaves, insect parts, pollen, hairs, wood, etc.
- Soil sources: various soils, different sands, calcium carbonate, and other minerals
- Human-related sources: plastic sheets, fabrics, glove fragments

Moreover, a practical FTIR analysis for Gum residue assessment can be conducted following the ASTM D381 guidelines. This method evaluates the fuel Gum content by evaporating the sample using a 160 °C airflow. It is crucial to avoid flushing or washing the glass containing the suspected fuel samples before the final residue assessment, ensuring that the Gum content can be accurately determined. Figure 8 illustrates the residue obtained from samples of contaminated fuels in the Gum apparatus before flushing.

Typical contaminants identified in Gum testing include phthalate-based fuel tube softeners, engine and lubricant oils, hydraulic fluids, and aging or oxidation products.

11.2 Near-Infrared Spectroscopy

Near-Infra Red (NIR) spectroscopy is one of the established techniques in fuel analysis that uses near-infrared light's interaction with molecular vibrations to generate spectra, which are the unique "fingerprints" of the sample's molecular composition.

In the ASTM guidelines D6122, D8321, and D8340, the value of NIR is emphasized for both qualitative chemical identification and quantitative spectrometric

Fig. 8 Residue obtained from samples of contaminated fuels (Image courtesy of M. Rabaev)

measurements. ASTM D6122 addresses using NIR to identify aromatic hydrocarbon types in middle distillates, D8321 focuses on quantifying water content in petroleum products, and D8340 determines the total sulfur content in bioheat fuels. These guidelines highlight NIR's capacity to provide a rapid, straightforward, and effective method for analysing various facets of petrochemical products.

As a portable instrument, NIR permits prompt sample analysis and results for various fuel parameters. Its effectiveness and adaptability are comparable to other established standard analysis techniques. Quantative measurements yielded by the NIR spectrometer include water content, aromatic content, Research Octane Number (RON), Motor Octane Number (MON), and distillation properties such as API gravity, density, and viscosity. The NIR spectrometer is a suitable alternative to various instruments in a field laboratory due to its small size and extensive range of capabilities, enhancing results accuracy.

11.3 Raman Microscope

Raman spectroscopy is also a spectroscopic technique that can identify vibrational modes of molecules, detect rotational and low-frequency modes of systems, and generate unique structural fingerprints for compound identification. It is particularly useful in identifying fuel contamination, as it can provide detailed information on the molecular structure and composition of various contaminants. Raman microscopy can resolve and analyze particles and regions of interest as small as 1 micron, compared to about the 10-micron detection limit of FTIR microscopes, offering more precise analysis of contaminants at a smaller scale.

Raman and infrared (IR) spectroscopy are complementary techniques, as each method examines different aspects of a given sample. While IR is sensitive to the vibration of end groups and highly polar bonds, Raman is more informative about backbone structures and symmetric bonds. Combining both techniques provides twice as much information about the molecular structure as obtained using either technique alone.

One area where Raman spectroscopy has proven particularly useful in fuel contamination identification is the analysis of iron oxides. Iron has at least seven known types of oxides: Hematite, Lepidocrocite, Akageneite, Goethite, Maghemite, Magnetite, and Wustite [6]. Raman spectroscopy can effectively distinguish most of these oxides, except Magnetite and Wustite, which have nearly identical spectral results. Identifying the type of iron oxides can help establish the oxidation mechanism and parameters, which in turn can be instrumental in identifying the source and cause of contamination within a system.

Another example is the detection of Molybdenum disulfide (MoS_2), a material used in solid and liquid lubrication applications. It can be applied to various mechanical parts to prevent friction, leading to degradation. The Raman microscope can detect MoS_2, especially with the 780 nm laser. MoS_2 often contains Antimony trioxide (Sb_2O_3), and both their spectra can be found in material libraries.

Strontium chromate ($SrCrO_4$) is another compound that can be identified using Raman spectroscopy. In aircraft, the paint system may consist of several layers, including a primer, a base paint, and an outer paint. The base paint in such systems could be made of $SrCrO_4$, which can be detected in Raman spectroscopy using the 532 nm laser.

Raman spectroscopy is also valuable in examining carbon materials such as graphite, amorphous carbon, etc., which can be analyzed to identify burning products in fuel systems. Additionally, Raman spectroscopy can analyse liquid-phase materials, such as dissolved contaminants in fuel or water phases. The technique can even analyse samples through clear glass or water, using the confocal optics of Raman spectrometers. Despite its limitations, such as fluorescence interference, Raman spectroscopy remains valuable for studying fuel contamination, particularly when combined with complementary techniques like infrared spectroscopy.

11.4 Scanning Electron Microscope Coupled with Energy-Dispersive X-ray Spectroscopy

One of the powerful tools in the analytical chemist's arsenal for the elemental analysis of solid contaminants is a scanning electron microscope (SEM) equipped with an energy-dispersive X-ray spectroscopy (EDS) detector. An SEM microscope can provide information regarding the sample surface topography and composition with a magnification of up to six orders of magnitude. Thus, high-resolution images obtained using SEM can reveal phase separations and three-dimensional structures.

A typical SEM instrument is limited to solid samples and requires a dry, solvent-free sample that can withstand the high vacuum inside the microscope chamber.

Another consideration is the sample size. The sample under investigation must be small enough to be mounted on the specimen stage (~10 cm^2).

11.5 Gas Chromatography—Mass Spectrometer

Gas Chromatography-Mass Spectrometry (GC-MS) is a well-established instrument commonly used for identifying contaminants in fuels. It allows for the analysis of compounds across a broad temperature range, making it suitable for various fuel components. GC-MS separates sample components based on boiling point and their affinity to the column surface (maybe uncoated or coated to separate by polar or surface affinity). Components are eluted and ionized, carried through the mass filter, and analyzed with the charge-to-mass ratio, providing the resultant chromatograph.

One significant strength of GC-MS is its ability to process a wide array of volatile and semi-volatile compounds, making it adept at detecting common organic contaminants in contaminated fuels, such as aromatic hydrocarbons and chlorinated compounds. The high-resolution capabilities of GC-MS facilitate the analysis of complex mixtures. Nonetheless, GC-MS has limitations as a fuel contamination analysis tool. The complex mixtures of hydrocarbons in fuels like aviation fuel or diesel can present a significant challenge, often resulting in a 'hedgehog' pattern in chromatograms. This pattern is characterized by numerous overlapping peaks representing myriad compounds, making identifying individual contaminants difficult. While gasoline samples may be less complex, they exhibit numerous chromatographic peaks requiring resolution and identification.

Sophisticated techniques such as GC×GC (two-dimensional gas chromatography, see Sect. 11.7) may be required for a thorough separation and identification process. This approach can offer a more detailed separation by polarity. However, it produces a considerable volume of data points that can be complex to interpret and compare, particularly when examining a questionable sample.

It is also important to note that GC-MS cannot analyse solid materials such as metals and plastics, which might contaminate fuel. Additionally, the analysis of biological contaminants generally falls outside the scope of GC-MS capabilities.

The versatility of GC-MS allows it to handle a broad array of samples and contaminants, providing detailed insights into their composition.

11.6 Injection Using a ChromatoProbe

The ChromatoProbe injection technique is ideal for testing heavy compounds or fluids like grease, as it allows nonvolatile samples to be inserted into the injector while minimizing instrument contamination. Such contamination could cause a loss in sensitivity, potentially leading to a loss of detector signal. Not all components of these substances may vaporize, leaving residues in the micro vial. To avoid detector saturation, use a small amount of substance for testing with the ChromatoProbe. For liquid analysis, use as little substance as possible with a flow split at 100ml/min, while for solids, use more material and adjust the split to 40ml/min. Starting with less material and reducing the split is recommended rather than increasing the material if the signal is insufficient. Always check for potential contaminants and conduct an empty test tube run to test for these. Any remaining residues can be further tested with methods like FTIR or SEM.

11.7 GC×GC-MS—Enhanced Detection and Analysis of Contamination in Aviation Fuel via Comprehensive Two-Dimensional Gas Chromatography

While GC-MS is practical for analysing organic contaminants in complex samples, it can struggle with intricate mixtures like aviation fuel due to peak overlaps. This, coupled with the unpredictable and diverse nature of contaminants, especially at low concentrations, emphasizes the need for a more advanced system. Comprehensive two-dimensional gas chromatography (GC×GC) has emerged as a robust, superior technique. Providing improved chemical separation capabilities, GC×GC offers substantial potential for extensive use in fuel contamination analysis, addressing the limitations posed by traditional GC-MS systems [7].

A typical GC×GC setup involves a GC instrument featuring two different column types interfaced via a modulator. The first column, usually long, is tailored for primary compound separation. A semi-polar is advantageous as the primary column for fuel analysis. It interfaces via a modulator, with a second, shorter column featuring a distinct internal coating for secondary compound separation. A non-polar column is suitable for the second stage.

The modulator collects compounds from the first column at regular intervals, facilitating their transfer to the second column for further separation. Two modulator types exist: flow and cryogenic. Flow modulators incorporate a valve connecting both columns and housing a storage loop. The loop fills up and is subsequently pushed by the carrier gas to the second column. Conversely, cryogenic modulators use liquid nitrogen, temporarily freezing compounds from the first column before releasing them to the second column through heating.

A GC×GC-MS setup produces a 3D chromatogram. Here, the x-axis signifies the retention time of the compounds reaching the modulator, the z-axis denotes the separated compounds, and the y-axis represents the intensity of the separated compounds. The mass spectrum at each point adds another dimension to the output.

To summarize, GC×GC technologies, despite their complexities, represent a state-of-the-art solution in contamination detection and analysis, significantly enhancing both the resolution and sensitivity of contaminant measurements.

11.8 Elemental Analysis

Fuel elemental analysis is an invaluable tool for assessing the health and efficiency of fuel systems. This process provides crucial insights into various aspects of the system, including early-stage corrosion, coating degradation, contamination accumulation, wear of moving parts, and penetration of external contaminants.

Silver and copper strip testing allow easy identification of active/corrosive sulfur in the fuel. In a widely adopted standard method, a polished copper strip is exposed to the fuel at an elevated temperature, and discoloration indicates the presence and levels of corrosive sulfur. Silver, being susceptible to active sulfur compounds [8], is immersed as a strip in the fuel sample, and the tarnishing on the strip indicates the level of active sulfur in the fuel.

It is worth noting that certain microbes can produce sulfur compounds as part of their metabolic processes. Microbes such as *Desulfovibrio* and *Desulfotomaculum* reduce sulphates to hydrogen sulphide, a corrosive gas [9]. Therefore, detecting active sulfur might also indicate microbial activity in the fuel system.

Regular application of fuel elemental analysis allows for early detection and mitigation of these issues, thus preventing costly repairs or system failures. This practice can prolong the fuel system's lifespan, increase its efficiency, and ensure the quality of the fuel it delivers.

Advanced instruments like inductively coupled plasma mass spectrometry (ICP-MS) are needed for measurements at the ppb level. It is important to note that elemental analysis does not examine the main components of fuel, such as hydrogen or carbon, nor oxygen. To effectively use elemental analysis, researchers must understand the characteristic elemental background of reference fuels and identify the source of contaminants by connecting abnormal results to chemical compositions. This process can be challenging since elemental results are not specific to particular chemical compounds, making it difficult to differentiate between different sources of contaminants in some cases.

Various techniques in elemental analysis exhibit differing sensitivity, repeatability, and reproducibility, as outlined in the ASTM standards D6728 and D7111. Two commonly used devices for fuel analysis are the rotating disc electrode (RDE), also known as the rotrode spectrometer, and inductively coupled plasma optic emission spectrometer (ICP-OES). While these devices share similarities in operating principles, they differ in their method of atom excitation and have distinct applications. RDE, for example, is primarily used for jet engine oil analysis to detect wear metals and requires minimal sample preparation. Fuel analysis with RDE is performed according to ASTM D6728, but this method isn't intended for gasoline tests. The lower detection limit for RDE is ppm for most elements, and it can test concentrations higher than 100 ppm without diluting the sample. In contrast, the inductively coupled plasma optical emission spectrometer (ICP-OES) offers significantly lower detection limits, often in the parts-per-billion (ppb) range, making it more suitable for trace-level determinations in fuels and lubricants.

Occasionally, an analysis of particles in a fuel filter becomes necessary. Detecting contaminants within the fuel can be challenging; however, they become more accessible for analysis once removed through filtration. The standard approach for analysing these contaminants entails dissolving particles in various acids and testing the liquid using an ICP spectrometer.

Suitable instrumentation is essential to conduct non-liquid material analysis, including greases, solids, and metals. The substances must first be "dissolved" in blank oil. For example, a small amount of grease is dissolved in a blank oil vial, followed by ultrasonication for approximately 10 min. Although it is impossible to achieve complete grease dissolution, this process generates a uniform suspension suitable for analysis using an RDE device.

While electron microscopes can perform elemental analysis tests, RDE and ICP-OES devices provide cost-effective alternatives. Identifying the elemental composition of the suspected grease aids in determining whether it is the source of contamination in the fuel. However, the grease suspension method is incompatible with ICP devices due to the risk of undissolved grease lumps obstructing the instrument's tubing.

For the analysis of solid samples, such as soil or paint, a small portion is crushed and ground using a mortar and pestle, resulting in a fine powder with particle sizes smaller than 10 microns. This powder is then suspended in blank oil and analyzed exclusively with an RDE device. For ICP instruments to analyse solid materials, the specimen must first be digested into liquid form, enabling the assessment of its metal content. The choice of acid for dissolving the sample should be compatible with the test substance.

RDE and ICP instruments are vital in analysing various materials, including fuels, oils, greases, and solids. Proper sample preparation and understanding of each device's capabilities are crucial for obtaining accurate and reliable results in elemental composition analysis. Furthermore, researchers should be aware of the characteristic elemental background of reference fuels and the potential sources of contaminants to interpret the results effectively and address any issues related to fuel system performance and reliability.

11.9 Analysis Using ICP-OES and ICP-MS

The different viscosities of petroleum and many of its derivatives make it challenging to analyse these samples directly by ICP-OES (Optical Emission Spectroscopy) or ICP-MS. As a consequence, the aspiration of organic liquids into the plasma typically requires additional equipment or sample pretreatment to maintain plasma stability and reduce carbon buildup. Years of ICP-OES used to determine wear metal content in lubricating oils have resulted in standard practices giving quantification in the range of 50–100,000 ppb, depending on the metal. Moving to ICP-MS has resulted in much lower quantification limits but imposed additional requirements for sample handling, sample preparation, and attention to the details of instrumentation operation.

Analysis of combustible organic samples like naphtha, vacuum gas oil (VGO), diesel, or jet fuel using ICP-MS is more challenging compared to conventional aqueous samples due to the higher carbon content. Intermittent extinguishing of plasma and carbon deposition in the interface region can occur due to the incomplete ionization of carbon, which has a high ionization potential of 11.3 eV. Reducing the carbon load on the plasma is accomplished by using a cooled spray chamber for high vapor pressure organics and dilution of low vapor pressure organics (usually xylene).

Current state-of-the-art instrumentation uses two or three multipole mass filters to select a particular ion m/e to be captured by a sensitive detector. The first mass filter allows the addition of reactive or collision gases to eliminate specific interferent species and enforce an absolute mass floor for further m/e separation. Carbon-based interference on key analytes, such as chromium or vanadium, causes additional concerns but can typically be overcome using kinetic energy discrimination (KED) on single quadrupole instruments (instruments with a multipole reaction cell upstream from the final mass filter quadrupole). Reactive gases such as hydrogen have been proven to remove intense argon-based polyatomic interferences that form within the plasma as it cools.

The use of triple quadrupole ICP-MS systems can further improve the detection of some elements challenged by severe interferences, such as sulfur. One should use internal standardization to compensate for potential drift during a longer sequence and improve accuracy. Typically, Yttrium and Indium are spiked at a concentration of 20 $\mu g \cdot L^{-1}$ to each blank, standard, or sample and used as internal standards in the experiment. Although the selection of the internal standard should ideally match the mass range and ionization potential of the analytes in the best possible manner, a suitable correction is nevertheless achieved also for analytes of considerably lower (such as boron, sodium, magnesium, and aluminum) and higher mass (such as mercury and lead).

The following is provided by Thermo Scientific to provide some examples of available detection limits. A Thermo Scientific™ iCAP™ RQ ICP-MS equipped with 250 $mL \cdot min^{-1}$ additional mass flow controller was used in the following study. The system was operated in conjunction with a Teledyne CETAC™ ASX-560 autosampler (Teledyne CETAC Technologies, Omaha, NE, USA). To address the challenge of carbon deposition, researchers introduced a constant flow of pure oxygen into the spray chamber. Pure helium and hydrogen were introduced to the collision/reaction cell (CRC) using dedicated mass flow controllers to enable operation with KED and hydrogen reaction mode in a single method, analysing 26 various elements in a single sample aspiration. While the use of an inert gas such as helium KED is a practical approach to remove polyatomic interferences, the use of a reactive gas such as hydrogen has shown effective removal of highly intense polyatomic interferences, mainly argon and carbon-based polyatomic interferences. Other reactive gases, such as oxygen and ammonia, can also be used for interference removal but require a triple quadrupole ICP-MS instrument to pre-filter the ion beam and avoid the formation of additional interferences with other elements contained in the sample. Table 1 lists the Instrument Detection Limits (IDL) for various elements in organic fluids using optimized methods.

Table 1 Typical instrument detection limits (IDL) obtainable in organic fluids using methods optimized by element [10]

	m/z	CRC gas	IDL (μg/L)
Boron	11	H_2	0.189
Sodium	23	H_2	0.098
Magnesium	24	H_2	0.314
Aluminum	27	He	0.122
Potassium	39	He	0.059
Calcium	40	H_2	0.158
Titanium	48	He	0.019
Vanadium	51	He	0.006
Chromium	52	H_2	0.018
Manganese	55	He	0.01
Iron	56	H_2	0.016
Iron	57	He	0.029
Cobalt	59	He	0.002
Nickel	60	He	0.009
Copper	63	He	0.034
Zinc	66	He	0.265
Arsenic	75	He	0.005
Selenium	80	H_2	0.021
Molybdenum	98	He	0.001
Silver	107	He	0.009
Cadmium	111	He	0.029
Tin	118	He	0.008
Antimony	121	He	0.001
Barium	137	He	0.024
Mercury	202	He	0.006
Lead	208	He	0.001

12 Analysing and Understanding Plasticizer Contamination in the Fuel System

Fuel flow across polymer surfaces can physically dislodge or solvate plasticizer molecules, which may leave the polymer through migration. This can lead to accumulation, forming gums in fuel and on surfaces, and increasing the risk of clogging or malfunctioning fuel components.

Simultaneously, fuel or fuel additives can interact with the plasticizers within a polymer, effectively replacing them and causing their release. For analysis to occur, this material must be collected either from the fuel or the collection surface.

The advanced analytical techniques such as FTIR, TGA, GC-MS, ChromatoProbe-GC-MS, and Pyrolysis Gas Chromatography-Mass Spectrometry (Py-GC-MS) in detecting and quantifying plasticizers in fuel might be successfully utilized.

This procedure first involves extracting plasticizers from the fuel by evaporation, leading to a heavy, non-volatile residue that probably contains the plasticizers. This residue is subsequently placed on an ATR diamond crystal, where an infrared spectrum is recorded.

TGA can enhance compound differentiation; however, it may not be able to differentiate all compounds as comprehensively as chromatographic methods can.

GC-MS may be used for even more detailed and specific analysis. This method effectively analyses a broad spectrum of plasticizer compounds with specific ions, such as the 149 m/z for phthalates, serving as unique identifiers. It is limited by the boiling point of the compound, which must occur within the temperature range of the GC oven and capillary.

The ChromatoProbe-GC-MS method is highly recommended due to its sensitivity to different components and accurate quantitation. Based on thermal extraction, this method allows the analysis of plasticizers in polymers using small quantities of a solid sample that can be placed directly in the gas chromatography (GC) injector. ChromatoProbe-GC-MS provides a valuable point of comparison with the plasticizer contamination analysis performed on fuel samples using GC-MS. Enabling the compounds from the polymer to be analyzed similarly facilitates the comparison of results from contaminated fuel and the polymer, thereby aiding in pinpointing the source of fuel contamination.

In conclusion, a thorough understanding of the presence and behaviour of plasticizers in fuel systems, aided by these sophisticated analytical techniques, is pivotal in maintaining the performance and longevity of fuel systems, particularly those containing polymers with plasticizers.

13 Navigating Challenges in Detecting Hydraulic Fluid Contamination in Jet Fuel: A Multifaceted Analytical Approach

In this section we will describe challenges associated with different techniques employed in the analysis of phosphate ester-based hydraulic fluids, which are widely recognized for their fire resistance and are extensively used in aerospace applications. However, these fluids pose significant risks when contaminating jet fuel, potentially triggering engine damage. The hydraulic fluid can cause polymer seals to swell, especially O-rings used in aircraft fuel systems. In the intensely heated environment of aircraft turbines, it can induce hot corrosion of the turbine blades [11].

Since phosphorus makes up approximately 10% w/w of these hydraulic fluids, it should theoretically be possible to use the phosphorous signal at below 0.1 ppm to detect oil contamination in the fuel. However, the phosphorus background from other sources in clean fuel may be more significantly higher than one ppm, as opposed to the anticipated 0.1 ppm. This unexpected discrepancy renders the ICP-OES approach unsuitable for the proposed analysis without the availability of an uncontaminated sample of the exact fuel to use for baseline comparison.

Signal from tri-butyl phosphate may be used as a marker with GC-MS to try quantifying the main component in hydraulic fluids. The challenge lies in differentiating the base ion at $m/z = 99$ of the hydraulic fluid from the saturated aliphatic ion fraction inherent in uncontaminated jet fuel, which also contains said ion. Given that the objective is to detect this ion as a fluid marker at one ppm level or lower, any overlap with ions from the fuel matrix hinders the concurrent identification and quantification of the compound. Despite attempts at chromatographic separation through extended heating programs and various stationary phases of the column, the complexity of the fuel matrix thwarted all efforts.

The next step involves optimizing the sample preparation using the solid-phase extraction (SPE) technique. This method allows concentrating tributyl phosphate on the solid phase column while flushing out the rest of the sample components, thus eliminating the chemical matrix and reducing background noise. The tributyl phosphate is solvated in methanol and injected into the GC-MS instrument with a Rxi®-5 MS column. A sensitivity level of 0.1 ppm of hydraulic fluid in jet fuel may be achieved.

14 Conclusion

Aviation hydrocarbon fuels, characterized by their complex mixtures, including additives, dissolved water, and various environmental components, face a broad spectrum of contamination. The range encompasses system degradation products, commingled liquids from pipeline carryover, residues from incomplete carrier cleaning, and unexpected compounds entering fuel systems. These contaminants significantly complicate the identification and quantification process, underscoring the need for advanced analytical methods.

The evolution of analytical methodologies, offering increased sensitivity and precision, is pivotal in addressing these contamination challenges. Enhanced methods enable the detection of contaminants at lower thresholds, which is crucial for a comprehensive understanding of fuel composition and ensuring safety and efficacy.

Beyond the capabilities of standard fuel QC laboratories, which may only indicate a problem with the fuel but lack the methodologies and instruments required for complex analyses, specialized analytical laboratories are essential for investigating fuel contamination. These laboratories have the expertise, equipment, and procedures to handle various contamination cases effectively and can provide detailed insights into the nature, origin and extent of contamination.

Moreover, the effective management of fuel contamination requires a synergistic approach, combining the expertise of researchers and field engineers. Collaboration between these groups ensures that laboratory findings are correctly interpreted and practically applied. This partnership is vital for developing scientifically sound and operationally feasible contamination mitigation strategies.

In summary, the complexity of hydrocarbon fuel analysis demands advanced analytical techniques and a collaborative approach between research and engineering professionals. Maintaining high safety and performance standards is essential, especially in aviation industries where fuel quality is paramount. Such a comprehensive approach ensures that fuel contamination challenges are addressed effectively, safeguarding the fuel quality and the systems that depend on it.

Acknowledgements The authors would like to thank David Abdallah and Reuven Falkovich for their helpful comments and additional insights during the preparation of this chapter.

Annex 1: Selected Standard Test Methods for the Analysis of Contamination in Aviation Fuels

Class	Institute	Number	Description
ASTM guide	ASTM	MNL44-3RD-EB	Guide to ASTM test methods for the analysis of petroleum products, liquid fuels, and lubricants
Chemical contamination	ASTM	D381	Standard method for gum content in fuels by jet evaporation
Chemical contamination	ASTM	D7821	Standard test method for field test determination of urea concentration in diesel exhaust fluid (DEF)
Chemical contamination	ASTM	D7959	Standard test method for chloride content determination of aviation turbine fuels using chloride test strip
Chemical contamination	ASTM	D1094	Standard test method for water reaction of aviation fuels
Chemical Contamination	EI	IP 225	Determination of copper in light petroleum distillates—spectrophotometric method
Sulfur content	ASTM	D1266	Standard test method for sulfur in petroleum products (lamp method)
Sulfur content	ASTM	D2622	Standard test method for sulfur in petroleum products by wavelength dispersive x-ray fluorescence spectrometry

(continued)

(continued)

Class	Institute	Number	Description
Sulfur content	ASTM	D3227	Standard test method for (thiol mercaptan) sulfur in gasoline, kerosine, aviation turbine, and distillate fuels (potentiometric method)
Sulfur content	ASTM	D4294	Standard test method for sulfur in petroleum and petroleum products by energy dispersive x-ray fluorescence spectrometry
Sulfur content	ASTM	D4952	Qualitative analysis for active sulfur species in fuels and solvents (doctor test)
Sulfur content	ASTM	D5453	Standard test method for determination of total sulfur in light hydrocarbons, spark ignition engine fuel, diesel engine fuel, and engine oil by ultraviolet fluorescence
Sulfur content	EI	IP 336	Petroleum products—determination of sulfur content—energy-dispersive-X-ray fluorescence method
Sulfur content	EI	IP 342	Petroleum products—determination of thiol (mercaptan) sulfur in light and middle distillate fuels—potentiometric method
Hydrocarbons content	ASTM	D1319	Standard test method for hydrocarbon types in liquid petroleum products by fluorescent indicator adsorption
Hydrocarbons content	ASTM	D1840	Standard test method for naphthalene hydrocarbons in aviation turbine fuels by ultraviolet spectrophotometry
Hydrocarbons content	ASTM	D6379	Standard test method for determination of aromatic hydrocarbon types in aviation fuels and petroleum distillates—high performance liquid chromatography method with refractive index detection
Hydrocarbons content	ASTM	D8267	Standard test method for determination of total aromatic, monoaromatic and diaromatic content of aviation turbine fuels using gas chromatography with vacuum ultraviolet absorption spectroscopy detection (GC-VUV)

(continued)

(continued)

Class	Institute	Number	Description
Hydrocarbons content	ASTM	D8305	Standard test method for the determination of total aromatic hydrocarbons and total polynuclear aromatic hydrocarbons in aviation turbine fuels and other kerosene range fuels by supercritical fluid chromatography
Hydrocarbons content	EI	IP 156	Determination of hydrocarbon types in petroleum products—fluorescent indicator adsorption method
Hydrocarbons content	EI	IP 436	Determination of aromatic hydrocarbon types in aviation fuels and petroleum distillates—high performance liquid chromatography method with refractive index detection
Particulate contamination	ASTM	D2276	Standard test method for particulate contaminant in aviation fuel by line sampling
Particulate contamination	ASTM	D7619	Standard test method for sizing and counting particles in light and middle distillate fuels, by automatic particle counter
Particulate contamination	EI	IP 423	Determination of particulate contaminant in aviation turbine fuels by laboratory filtration
Particulate contamination	EI	IP 564	Determination of the level of cleanliness of aviation turbine fuel — automatic particle condition monitor
Particulate contamination	EI	IP 565	Determination of the level of cleanliness of aviation turbine fuel—portable automatic particle counter method
Particulate Contamination	EI	IP 577	Determination of the level of cleanliness of aviation turbine fuel—automatic particle counter method using light extinction
Particulate contamination	EI	IP 216	Determination of particulate contaminant of aviation turbine fuels by line sampling
Microbial contamination	ASTM	D6469	Standard guide for microbial contamination in fuels and fuel systems
Water contamination	ASTM	D3240	Standard test method for undissolved water in aviation turbine fuels

(continued)

(continued)

Class	Institute	Number	Description
Water/Particulate contamination	ASTM	D8148	Standard test method for spectroscopic determination of haze in fuels
Water/Particulate contamination	ASTM	D4176	Standard test method for free water and particulate contamination in distillate fuels (visual inspection procedures)
Water separation	ASTM	D3948	Standard test method for determining water separation characteristics of aviation turbine fuels by portable separometer
Water separation	ASTM	D7224	Standard test method for determining water separation characteristics of kerosine type aviation turbine fuels containing additives by portable separometer
Water separation	ASTM	D8073	Standard test method for determination of water separation characteristics of aviation turbine fuel by small scale water separation instrument
Additive measurement	ASTM	D5059	Standard test methods for lead and manganese in gasoline by X-ray fluorescence spectroscopy
Additive measurement	ASTM	D3341	Standard test method for lead in gasoline—iodine monochloride method (Withdrawn 2022)
FAME content	EI	IP 583	Determination of the fatty acid methyl esters content of aviation turbine fuel using flow analysis by Fourier transform infrared spectroscopy—rapid screening method
FAME content	EI	IP 585	Determination of fatty acid methyl esters (FAME), derived from bio-diesel fuel, in aviation turbine fuel—GCMS with selective ion monitoring/scan detection method
FAME content	EI	IP 590	Determination of fatty acid methyl esters (FAME) in aviation turbine fuel—HPLC evaporative light scattering detector method
FAME content	EI	IP 599	Determination of fatty acid methyl esters (FAME) in aviation turbine fuel—gas chromatography using heart-cut and refocusing

(continued)

(continued)

Class	Institute	Number	Description
FAME content	ASTM	D7797	Standard test method for determination of the fatty acid methyl esters content of aviation turbine fuel using flow analysis by fourier transform infrared spectroscopy—rapid screening method
Rubber analysis	ASTM	D3677	Standard test methods for rubber—identification by infrared spectrophotometry
Elemental analysis	ASTM	D6595	Standard test method for determination of wear metals and contaminants in used lubricating oils or used hydraulic fluids by rotating disc electrode atomic emission spectrometry
Elemental analysis	ASTM	STP1468-EB	Elemental analysis of fuels and lubricants: recent advances and future prospects
Elemental analysis	ASTM	D7691	Standard test method for multielement analysis of crude oils using inductively coupled plasma atomic emission spectrometry (ICP-AES)

References

1. Vangsness MD (2015) Best practices for fuel system contamination detection and remediation final report. https://apps.dtic.mil/sti/pdfs/AD1031879.pdf. Accessed Oct 5, 2022
2. Amirav A, Neumark B, Elkabets O, Yakovchuk A (2025) Cold EI—the way to improve GC-MS and increase its range of applications. Mass Spectrom Rev
3. MIL-PRF-23699. Performance specification: lubricating oil, aircraft turbine engine, synthetic base, Nato code numbers: O-152, O-154, O-156, and O-167
4. ISO 4650 International Organization for Standardization. Rubber, raw, natural—specification. Geneva, Switzerland
5. ASTM D3677. Standard test methods for rubber—identification by infrared spectrophotometry. ASTM International, 100 Barr Harbor Drive, PO Box C700, West Conshohocken, PA 19428-2959, United States
6. Bellot-Gurlet L, Neff DD, Reguer S, Monnier J, Saheb M, Dillmann P, Neff D, Réguer S (2009) Raman studies of corrosioers formed on archaeological irons in various media. J Nano Res 8:147–156. https://doi.org/10.4028/www.scientific.net/JNanoR.8.147
7. Melder J, Zinsmeister J, Grein T, Jürgens S, Köhler M, Oßwald P (2023) Comprehensive two-dimensional gas chromatography: a universal method for composition-based prediction of emission characteristics of complex fuels. Energy Fuels 37:4580–4595. https://doi.org/10.1021/acs.energyfuels.2c04270

8. Hu Z, Yang G, Lou F, Wang L (2002) Exploration on silver strip corrosion by active sulphide. Petrol Process Petrochem 33
9. Passman FJ (2003) Introduction to fuel microbiology. In: Passman FJ (ed) Fuel and fuel system microbiology: fundamentals, diagnosis, and contamination control. ASTM International, West Conshohocken, pp 1–23. https://doi.org/10.1520/MNL47-EB
10. Surekar B, Kutscher D (2020) Addressing the challenges of routine determination of elemental impurities in refinery products using a robust ICP-MS approach. Thermo Fisher Scientific Application Note 44465
11. Stamker Gertopski D, Tartakovsky K, Rabaev M (2019) Identification and quantification of phosphate ester-based hydraulic fluid in jet fuel. SAE Int J Fuels Lubr 12. https://doi.org/10.4271/04-12-01-0003

Chapter 15
Aircraft Failure Analysis Induced by the Fuel or Fuel System

Albert Schmidt, Moshe Rabaev, Yalfal Siyum, Haim Judah, and Liel Mazar

Abstract This chapter provides a comprehensive analysis of aircraft failures induced by fuel or fuel system issues. The failures are categorized into four main types: fuel starvation, fuel exhaustion, fuel contamination, and wrong fuel usage. The chapter explores how these failures can lead to safety incidents or accidents, supported by real-world case studies. It highlights the importance of proper fuel system maintenance, monitoring, and adherence to safety regulations. The chapter also addresses the role of international regulatory bodies, such as the International Civil Aviation Organization (ICAO), and national agencies, such as the Federal Aviation Administration (FAA) and the European Union Aviation Safety Agency (EASA), in setting certification standards for fuel systems. Despite global standards, compliance varies, and the chapter emphasizes the need for harmonized safety measures across jurisdictions. Additionally, the analysis delves into the risks posed by various

A. Schmidt (✉)
Founding Partner, TASK4S, Tel Aviv-Yafo, Israel
e-mail: albertronit@gmail.com
URL: https://task4success.com/

M. Rabaev
Independent Consultant, Moshe Rabaev Consulting, Beer-Sheva, Israel

Blechner Center for Industrial Catalysis and Process Development, Department of Chemical Engineering, Ben-Gurion University of the Negev, Beer-Sheva, Israel

M. Rabaev
e-mail: rabaevm@gmail.com

L. Mazar
Independent Consultant, Jerusalem, Israel
e-mail: liel.mazar@cservices.co.il

Y. Siyum · H. Judah
Israeli Air Force, Tel Aviv-Yafo, Israel
e-mail: Yalfalsi@idf.il

H. Judah
e-mail: Haim.judah@idf.il

M. Rabaev et al. (eds.), *Aviation Fuel Problems and Solutions: From Refinery to Wingtip*, https://doi.org/10.1007/978-3-032-05514-9_15

contaminants—such as water, solids, and microbes—within fuel systems, and examines the dangers of using incorrect fuel types. The chapter concludes by stressing the crucial role of technical investigators and fuel engineers in conducting thorough failure analyses and implementing preventive measures to enhance aircraft safety.

1 Chapter Goals and Structure

This chapter discusses potential failures in aviation fuel systems that may lead to accidents and safety incidents. Problems with fuel and fuel systems that result in aircraft safety incidents and accidents are mainly classified into four groups: starvation, exhaustion, contamination, and wrong fuel. The chapter will focus on the different problem categories from various perspectives, using real-life scenarios to emphasize the roles of technicians and pilots, as well as the importance of understanding regulations, original equipment manufacturer (OEM) instructions, and the intricacies of the aircraft. Additionally, it provides insights into each component of the fuel system and presents examples of accidents from different investigative bodies, such as the National Transportation Safety Board (NTSB, USA), the Air Accidents Investigation Branch (AAIB, UK), the Aviation Incidents & Accidents Investigation (AIAI, Israel), etc.

2 The Airworthiness System, Requirements, and Standards

In 1944, a civil aviation conference was held in Chicago, inviting representatives from 55 countries. The conference concluded with 52 countries signing the Convention on International Aviation, also known as the Chicago Convention. The document, known as Doc 7300 [1], established by the International Civil Aviation Organization (ICAO) consists of 96 articles that set the legal framework for international civil aviation, including principles such as state sovereignty over airspace, aircraft registration, air traffic regulations, and safety measures. The document also includes technical annexes that cover operational standards for aircraft, airports, and safety protocols. Amendments are made through the ICAO Assembly to keep the convention updated with advancements in aviation. The convention plays a key role in facilitating international cooperation for safe and efficient civil aviation. This pivotal agreement established the standards and procedures for global air navigation with the primary goal of fostering the development of international civil aviation "in a safe and orderly manner" ensuring that air transport services would be established "on the basis of equality of opportunity and operated soundly and economically". ICAO was subsequently established under the agreements reached at the conference. ICAO's core mission is to assist its 193 member states (as for 2023) in achieving uniformity in civil aviation regulations, standards, procedures, and organization. Since the convention, ICAO has issued over 12,000 international Standards and Recommended

Practices (SARPs), all of which have been agreed upon by consensus among ICAO's Member States. However, it is essential to recognize that despite the existence of ICAO standards, it is not a global regulator; rather, national regulations enforced by sovereign states must be adhered to by air operators using applicable airspace and airports.

Article 31 of the Chicago Convention states that "every aircraft engaged in international navigation shall be provided with a certificate of airworthiness issued or rendered valid by the state in which it is registered". Article 33 of the Convention states that "Certifications of airworthiness and certificates of competency and licenses issued or rendered valid by the contracting states in which the aircraft is registered, shall be recognized as valid by the other contracting states, provided that the requirements under which such certificates or licenses were issued or rendered valid are equal to or above the minimum standards which may be established from time to time pursuant to this convention".

Article 37 states, "Each Contracting State[1] undertakes to collaborate in securing the highest practicable degree of uniformity in regulation, standards, procedures, and organization in relation to aircraft, personnel, airways and auxiliary services in all matters in which such uniformity will facilitate and improve air navigation". The article also states that ICAO shall adopt and amend from time to time, as may be necessary, international standards and recommended practices and procedures dealing with aircraft airworthiness (among other matters). These three articles establish the basis for the international airworthiness certification and safety oversight, primary implemented by national aviation authorities in alignment with ICAO standards. As a result of these articles and Annex 8 [2] to the convention (titled "Airworthiness of Aircraft"), ICAO published Doc 9760 (Airworthiness Manual) [3] as the minimum airworthiness standards mentioned in Article 33 and Annex 8. In these documents, ICAO defines several essential terms and concepts for understanding how global aviation ensures aircraft safety:

- Airworthy: The status of an aircraft, engine, propeller, or part when it conforms to its approved design and is in a condition for safe operation.
- Type Certificate: A document issued by a Contracting State to define the design of an aircraft, engine, or propeller type and to certify that this design meets the appropriate airworthiness requirements of that state.[2]
- Appropriate airworthiness requirements: The comprehensive and detailed airworthiness codes established, adopted, or accepted by a contracting state for the class of aircraft, engine, or propeller under consideration.

---

[1] States members of ICAO are also named Contracting States (Article 1 of the Chicago Convention) or Members States, i.e., states that ratified or adhered to the Chicago Convention.

[2] The contracting states issue type certificates through governmental organizations that set the standards for the manufacture, performance, and required test results per ICAO standards for any entity that desires to become an OEM, supplying the aero industry within the "Contracting State" jurisdiction. Examples are the Federal Aviation Administration (FAA) for the United States, the European Aviation Safety Agency (EASA) for most of Europe, and the Civil Aviation Authority (CAA) for the United Kingdom.

- Anticipated operating conditions: Conditions known from experience, or which can be reasonably anticipated to occur during the aircraft's operational life. They consider operations for which the aircraft is made eligible, and the conditions so considered being relative to the meteorological state of the atmosphere, the configuration of the terrain, the functioning of the aircraft, the efficiency of the personnel, and the factors affecting safety in flight. Anticipated operating conditions do not include the following extremes:
  - Instances that can be effectively avoided using operating procedures.
  - Instances that occur so infrequently that requiring the standards to be met under circumstances that would set a level of airworthiness higher than what practical experience has shown to be necessary.
- Power plant: The system comprises all the engines, drive system components (if applicable), propellers (if installed), accessories, ancillary parts, and fuel and oil systems installed on an aircraft excluding the rotors for the helicopter.

From the perspectives of fuel and fuel systems, numerous relevant requirements and information can be found in Annex 8 and Doc 9760 [3] concerning the certification of fuel and fuel systems. For example, Annex 8 states that "The fuel, oil, air induction and other systems associated with the powerplant shall be capable of supplying each engine it's established requirements, under all conditions affecting the functioning of the systems (e.g., engine power, aircraft attitudes, accelerations, atmospheric conditions, fluid temperatures) within the anticipated operating conditions". Another example from Doc 9760 states that when approval for Extended Diversion Time Operation (EDTO) is required, "The aeroplane fuel system should maintain the engine inlet fuel pressure and flow to all operable engines throughout any diversion. Fuel necessary to complete the EDTO should be available to the operating engine(s) after an engine failure and/or other system failures unless the combination is shown to be extremely improbable. Alerts should be displayed to the flight crew when the fuel available to the engines falls below the required level to complete the operation. These alerts should include provisions for abnormal fuel management or transfer between tanks and possible loss of fuel". Doc 9760 also addresses requirements for fuel venting and emission, fuel balance concerning flight stability, and environmental conditions such as the effects of a lightning strike on electronic fuel control.

These provisions lack specificity and theoretically allow for compliance with these requirements to vary among aircraft designers or certifying agencies. As previously mentioned, ICAO does not function as a regulatory body.

When a new aircraft is designed, a Contracting State shall issue a type certificate for the airframe and engine based on the appropriate airworthiness requirements of the Contracting State. The FAA of the USA and the Europe's EASA are the two prominent central certifying agencies. In the regulations of FAA and EASA, there are more detailed requirements regarding the airworthiness of fuel and fuel systems. FAA and EASA are divided into four primary categories: large aircraft, small aircraft, large rotorcraft, and small rotorcraft. The exact definitions of those four primary categories are provided in Table 1. There are also certification requirements for other aircraft types, such as gliders, air balloons, and Unmanned Air Vehicles (UAV).

The certification specification regulations described in Table 1 are subdivided into various sections, including general requirements, structural design requirements, and flight requirements. Each certification specification includes a dedicated section for powerplant requirements, which also encompasses regulatory requirements regarding the air vehicle fuel system. The fuel system certification specification regulatory requirements of the FAA and EASA are highly harmonized, allowing for the worldwide operability of aircraft. Additionally, the requirements of airplanes and rotorcraft are very similar, and in general, large aircraft and rotorcraft have more stringent regulatory requirements.

Table 2 summarizes the fuel system certification specification regulatory requirements of normal-category airplanes based on 14 CFR Part 23 (Airworthiness Standards: Normal Category Airplanes). As previously mentioned, the regulatory requirements for small airplanes and rotorcraft under the FAA and EASA are very similar.

Table 1 Aircraft type definitions as defined in FAA's and EASA's corresponding regulation titles

Regulation title under FAA	Regulation title under EASA	Definition of aircraft type applicability
14 CFR[a] Part 23 Airworthiness standards: normal category airplanes	CS[b]-23 Normal category airplanes	Certification in the normal category applies to airplanes with a passenger seating configuration of 19 or less and a maximum certified take-off mass of 8618 kg (19,000 pounds) or less
14 CFR Part 25 Airworthiness standards: transport category airplanes	CS-25 Large airplanes	Certification in the transport/large category applies to airplanes not eligible for certification under normal airplane category or under other 'special' certification rules
14 CFR Part 27 Airworthiness standards: normal category rotorcraft	CS-27 Small rotorcraft	These certification specifications apply to small rotorcraft with maximum weights of 3175 kg (7000 lbs) or less and nine or fewer passenger seats
14 CFR Part 29 Airworthiness standards: transport category rotorcraft	CS-29 Large rotorcraft	Certification in the transport/large rotorcraft category applies to rotorcrafts not eligible for certification under normal rotorcrafts category or under other 'special' certification rules
14 CFR Part 33 Airworthiness standards: aircraft engines	CS-E Engines	Turbine and reciprocating engines

[a]CFR—Code of Federal Regulations
[b]CS—Certification Specifications

Table 2 Summary of fuel systems certification specification regulatory requirements in 14 CFR Part 23

Topics addressed in regulation	Summary based on 14 CFR Part 23 (Normal category airplanes)
System independence	Each fuel system must: (1) Be designed and arranged to ensure independence between multiple fuel storage and supply systems, ensuring that failure of any one component in a system will not result in the loss of fuel storage or supply of another system (2) Be isolated from personnel compartments and protected from hazards resulting from unintended temperature influences (3) Be designed to prevent significant loss of stored fuel from any vent system due to fuel transfer between fuel storage or supply systems, or under likely operating conditions (4) Provide fuel for at least thirty minutes of operation at maximum continuous power or thrust (5) Be capable of safely jettisoning fuel if necessary for landing
Lightening protection	Be structured to prevent ignition of the fuel within the system caused by direct lightning strikes, swept lightning strokes to areas prone to such incidents, or by corona or streaming at fuel vent outlets
General	(1) Supply necessary fuel to ensure the proper functioning of each powerplant and auxiliary power unit under all foreseeable operating conditions (2) Provide the flight crew with a method to determine the total usable fuel available and maintain an uninterrupted fuel supply when the system is operating correctly considering potential fuel fluctuations (3) Provide a mechanism to safely remove or isolate stored fuel from the aircraft and prevent improper refilling (4) Ensure fuel retention under all likely operating conditions and minimize hazards to occupants during any survivable emergency landing. For level 4 airplanes, account for the failures of overload of the landing system (5) Implement measures to prevent hazardous contamination of the fuel stored and supplied to each powerplant and auxiliary power unit

Table 3 summarizes the fuel system certification specification regulatory requirements of transport category airplanes based on current FAA 14 CFR Part 25 (Airworthiness Standards: Transport Category Airplanes—up to date as of 9/12/2023). As mentioned earlier, the regulatory requirements for large airplanes and rotorcraft under the FAA and EASA are very similar.

The certification specification regulations of fuel systems are performance-based, meaning they outline the desired system performance without specifying mandatory design patterns or requirements to achieve it. Additionally, the regulation does not prescribe a specific method to demonstrate compliance. This is a significant consideration because while the FAA and EASA offer guidance on acceptable means of compliance, the actual methods are not publicly accessible and are confined to discussion between the regulator and the airframe OEM. Several standards and industry guides exist regarding fuel systems, design, and components, such as ASTM F3063

Table 3 Summary of fuel systems certification specification regulatory requirements based on 14 CFR Part 25

Topics addressed in regulation	Summary based on 14 CFR Part 25 (Transport category airplanes)
General	(a) Each fuel system should ensure proper fuel flow and pressure under all likely operating conditions for which the engine and aircraft are certified (b) Each fuel system must be arranged to prevent flameout in the event of air introduction into the system (c) The fuel system will operate correctly with fuel saturated with water at 26.7 °C (80 °F) and having 0.20 cm^3 of free water per liter (0.75 cm^3 per US gallon), cooled to the most critical condition for icing likely to be encountered in operation
Fuel system analysis and test	(a) Proper functioning fuel systems under all probable operating conditions must be demonstrated through analysis and any necessary tests required by the Administrator. If tests are required, they must be conducted using the airplane fuel system or a test article that mimics the operating characteristics of the portion of the fuel system being tested (b) The potential failure of any heat exchanger using fuel as one of its fluids may not lead to a hazardous condition Note: Topic not covered in transport/large rotorcraft regulations
Fuel system independence	Each fuel system must allow for engine isolation (failure of one engine will not affect the other engines or will allow continued safe operation) by supplying fuel to each engine in an independent system or by any other acceptable method
Lightning protection	Defines types of lightning and their effects, as well as the requirement to design the fuel system to prevent catastrophic fuel vapor ignition
Fuel flow	Each fuel system must provide at least 100% of the fuel flow required under every intended operating condition and maneuver. Compliance must demonstrate that fuel pressure remains within engine limits under adverse usable fuel quantity, emergency pumps deliver that proper fuel flow, and that the automatic system can divert fuel to the engine between tanks, without the attention of the flight crew's intervention in case of useable fuel depletion Note: In transport/large rotorcraft regulations, there is also a requirement to apply critical fuel flow properties and filter blockage
Flow between interconnected tanks	If fuel can be transferred from one tank to another in flight, the fuel tank vents, and the fuel transfer system must be designed to prevent structural damage to the tanks due to overfilling
Unusable fuel supply	The unusable fuel quantity for each fuel tank and its components must be determined to be no less than the quantity at which the first evidence of engine malfunction occurs under the most adverse fuel feed condition for all intended operations and flight maneuvers involving fuel feeding from that tank. Fuel system component failures need not be considered

(continued)

Table 3 (continued)

Topics addressed in regulation	Summary based on 14 CFR Part 25 (Transport category airplanes)
Fuel system hot weather operation	The fuel system must perform effectively in hot weather conditions. Compliance may be shown by pressurizing the fuel to prevent vapor formation or by conducting flight test (or ground test with simulating flight conditions) with a fuel temperature at least 43.3 °C (110 °F) and additional required test parameters
Fuel tanks general	Provisions for performance-based design, tests, and installation performance design are described
Fuel tank expansion space	Each fuel tank must include an expansion space of at least 2 percent of the tank capacity. It should not be possible to fill the expansion space with the aircraft in its typical ground attitude. Additionally, pressure fueling systems should be addressed
Fuel tank sump	Each fuel tank must be equipped with a sump having an effective capacity, in the typical ground attitude, of not less than the greater of 0.10 percent of the tank capacity or one-sixteenth of a gallon, unless operating limitations are established to prevent the accumulation of water exceeding the sump capacity. Each fuel tank must allow drainage of any hazardous quantity of water when the airplane is in the ground. Each fuel tank sump must have an accessible drain allowing complete drainage, discharge clear of each part of the airplane, and with manual or automatic means for secure locking in the closed position
Fuel tank filler connections	Fuel tank filler connection prevents fuel from entering any part of the airplane other than the tank itself. Additionally, any recessed filler connection capable of retaining fuel must be equipped with a drain that discharges clear of each part of the airplane. Each filler cap must provide a fuel-tight seal, and each fuel filling point must have a provision for electrically bonding the airplane to ground fueling equipment
Fuel tank vents	Fuel Tank vents should maintain an acceptable pressure differential between the interior of the fuel tank and the exterior at refueling, flight, or any other normal condition. They should be designed to prevent moisture accumulation on the ground, blockage from ice or dirt, and to prevent explosion and other hazards
Fuel tank outlet	Each fuel tank outlet or booster pump must be equipped with a fuel strainer. For reciprocating engine-powered airplanes, this strainer must have 8 to 16 meshes per inch, while for turbine engine powered airplanes, it must prevent the passage of any object that could restrict fuel flow or damage any fuel system component. The clear area of each fuel tank outlet strainer must be at least five times the area of the outlet line. Each finger strainer must be accessible for inspection and cleaning

(continued)

Table 3 (continued)

Topics addressed in regulation	Summary based on 14 CFR Part 25 (Transport category airplanes)
Pressure fueling system	(a) Each connection in the pressure fueling system fuel manifold must have means to prevent the escape of hazardous quantities of fuel from the system if the fuel entry valve fails (b) An automatic shutoff means must be provided to prevent the quantity of fuel in each tank from exceeding the maximum approved quantity. This means must: (1) Allow checking for proper shutoff operation before each tank fueling (2) Indicate failure of the shutoff means to stop the fuel flow at the maximum approved quantity for each tank at each fueling station (c) Means must be provided to prevent damage to the fuel system in the event of failure of the automatic shutoff means prescribed in paragraph (b) of this section (d) The airplane pressure fueling system (excluding fuel tanks and fuel tank vents) must withstand an ultimate load that is two times the maximum pressures likely to occur during fueling, including surge. The maximum surge pressure must be determined with any combination of tank valves intentionally or inadvertently closed (e) The airplane defueling system (excluding fuel tanks and fuel tank vents) must withstand an ultimate load that is 2.0 times the load arising from the maximum permissible defueling pressure (positive or negative) at the airplane fueling connection
Fuel tank explosion prevention	Detailed requirements for preventing fuel tank overfill are provided in the regulations
Fuel pumps	(a) Main pumps: Each fuel pump required for proper engine operation or required to meet the fuel system requirements (excluding those in paragraph (b) of this section) is considered a main pump. Provision must be made for bypassing each positive displacement fuel pump, other than a fuel injection pump approved as part of the engine, for each main pump (b) Emergency pumps: Emergency pumps or another main pump must supply each engine immediately after the failure of any main pump (excluding a fuel injection pump approved as part of the engine)

(continued)

Table 3 (continued)

Topics addressed in regulation	Summary based on 14 CFR Part 25 (Transport category airplanes)
Fuel system lines and fittings	(a) Each fuel line must be installed and supported to prevent excessive vibration and withstand loads due to fuel pressure and accelerated flight conditions (b) Each fuel line connected to components of the airplane between which relative motion could occur must have flexibility provisions (c) Each flexible connection in fuel lines subjected to pressure and axial loading must use flexible hose assemblies (d) Flexible hose must be approved or suitable for the specific application (e) Flexible hose affected by high temperatures must not be used where excessive temperatures occur during operations or after engine shutdown (f) Each fuel line within the fuselage must allow a reasonable deformation and stretching without leakage
Fuel system components	Fuel system components in an engine nacelle or in the fuselage must be protected from damage that could result in spillage sufficient to pose a fire hazard during a wheels-up landing on a paved runway under each of the conditions prescribed in § 25.721(b)
Fuel valves	In addition to the requirements of § 25.1189 for shutoff means, each fuel valve must be supported so that no loads resulting from their operation or accelerated flight conditions are transmitted to the attached lines
Fuel strainer or filter	There must be a fuel strainer or filter between the fuel tank outlet and the inlet of either the fuel metering device or an engine-driven positive displacement pump, whichever is nearer the fuel tank outlet. This fuel strainer or filter must: (a) Be accessible for draining and cleaning and must incorporate a screen or element that is easily removable (b) Have a sediment trap and drain, except it need not have a drain if the strainer or filter is easily removable for drainage purposes (c) Be mounted so that its weight is not supported by the connecting lines or by the inlet or outlet connections of the strainer or filter itself unless adequate strength margins under all loading conditions are provided in the lines and connections (d) Have the capacity (with respect to operating limitations established for the engine) to ensure that engine fuel system functioning is not impaired when fuel is contaminated to a degree (with respect to particle size and density) greater than that established for the engine in Part 33 of this chapter

(continued)

Table 3 (continued)

Topics addressed in regulation	Summary based on 14 CFR Part 25 (Transport category airplanes)
Fuel system drains	(a) Drainage of the fuel system must be accomplished using fuel strainer and fuel tank sump drains (b) Each drain required by paragraph (a) of this section must: (1) Discharge clear of all parts of the airplane (2) Have manual or automatic means for positive locking in the closed position (3) Have a drain valve: (i) That is readily accessible and can be easily opened and closed (ii) That is either located or protected to prevent fuel spillage in the event of a landing with the landing gear retracted
Fuel jettisoning system	Provisions for installing a fuel jettisoning system and their requirements are detailed in the regulation
Crash resistance	This topic is not discussed in 14 CFR 25, but it is discussed on 14 CFR 27, airworthiness standards: transport category rotorcraft

(Standard Specification for Aircraft Fuel Storage and Delivery [4]), SAE AIR7975 (Aircraft Fuel System Design Guidelines [5]), SAE MAP749 (Aircraft Turbine Engine Fuel System Component Endurance Test Procedure [6], and SAE AIR1408B (Fuel Systems Spec and Standards [7]).

3 Statistics of Fuel-Related Accidents

ICAO's integrated Safety Trend Analysis and Reporting System (iSTARS) is a key repository of aviation safety occurrences data, encompassing a dataset of safety-related incidents worldwide. According to the authors' research, the dataset comprises more than 18,000 records of safety occurrences spanning from 2008 to 2020. Among these records, about 7000 contain information about the occurrences categories, as defined by the European Coordination Centre for Accident and Incident Reporting Systems (ECCAIRS) Aviation 1.3.0.12 Data Definition Standard. Within this dataset, about 140 occurrences are categorized as fuel-related, according to ECCAIRS, which defines such occurrences as incidents where "One or more power plants experienced reduced or no power output due to fuel exhaustion, fuel starvation/mismanagement, fuel contamination/wrong fuel, or carburetor and/or induction icing". Another significant source of aviation safety occurrence data are the NTSB databases. One of them, The Case Analysis and Reporting Online (CAROL) database, containing information on aviation investigations since 1983, holds a total of about 22,000 records of aviation safety occurrences between 2008 and 2020. Among these, about 1400 has fuel event classifier attached . The event categorization used in CAROL is based on definitions provided by the Common Taxonomy

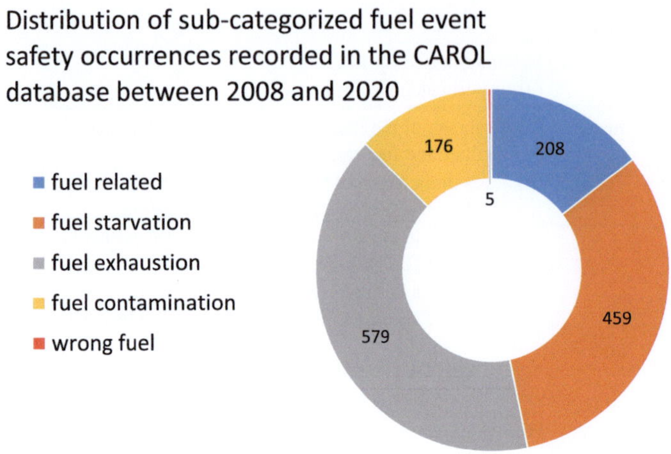

Fig. 1 Safety occurrences categorized as fuel events between 2008 and 2020

Team, which closely aligns with ECCAIRS definitions. The fuel event category is further subdivided into distinct subcategories such as fuel exhaustion, fuel starvation, fuel contamination, fuel related, and wrong fuel. While the subcategories are self-explanatory, the "fuel-related" category encompasses a mix of fuel occurrences including contamination, starvation, and mismanagement, with approximately 40% of incidents attributed to carburetor icing (as for 2023, records from this time can be added and/or classifications can be updated in the future). Figure 1 illustrates the distribution of sub-categorized fuel event safety occurrences recorded in the CAROL database between 2008 and 2020.

Table 4 shows how the safety occurrences are distributed per operation classification of the FAA regulations.

Table 4 The distribution of safety occurrences per operation classification of the FAA regulation

Operation classification of the FAA	Exhaustion	Starvation	Related	Contamination	Wrong fuel	Total
General aviation (FAR[a] 091)	529	432	192	155	4	1312
Agricultural aircraft operations (FAR 137)	27	9	1	16	0	53
Air taxi commuter (FAR 135)	8	11	2	3	0	24
Public aircraft	5	2	1	0	0	8
Rotorcraft, external load operations (FAR 133)	3	2	0	1	1	7

(continued)

Table 4 (continued)

Operation classification of the FAA	Exhaustion	Starvation	Related	Contamination	Wrong fuel	Total
Non-US commercial	3	1	2	0	0	6
Unknown	1	2	3	0	0	6
Not reported	2	0	4	0	0	6
Air carrier (FAR 121)	0	0	2	0	0	2
Total	579	459	208	176	5	1427

[a]FAR—Federal Aviation Regulations. These are a set of rules issued by the FAA

From the table, it is evident that most safety occurrences happen to general aviation aircraft operated by small organizations or individuals (1312 out of 1427 occurrences fall within the general aviation category). This observation could also explain why 'fuel exhaustion' is the most common failure mode, as detailed in Sect. 7.3.

4 The Role of the Technical Investigator, Fuel System Engineer, and Fuel Engineer in Failure Analysis

In failure analysis, it is insufficient to merely state that the fuel expert should provide facts and understanding of the fuel and fuel systems, along with any occurring failures. The primary responsibility of the technical investigator is to effectively apply their engineering skills and knowledge within a complex and multidisciplinary process. Systems and their realities are not conveniently categorized. The truth remains intricate at every level of examination. Therefore, fuel experts should extend their interests and efforts into related fields, as well as their own knowledge and scientific expertise. This approach aims to provide understanding and measures to prevent future failures.

5 Fuel-Induced Aircraft Failure

This section aims to explore the valuable insights gained from accidents and incidents, specifically those related to fuel quality issues. We will focus on understanding the different phenomena and failures associated with the fuel itself, as well as the contaminants that may be present in it. Through examining these aspects, we can uncover their complex relationship with various scenarios and risks encountered in aircraft operations.

5.1 Water

One of the most common contaminants of aviation fuel is water. The Energy Institute publishes EI 1550 "Handbook on equipment used for the maintenance and delivery of clean aviation fuel" [8] which provides actionable guidance on water contamination and prevention methods. One of the primary quality requirements for aviation fuel is to be visually free from water. Water can infiltrate fuel through various means, including environmental precipitation (via loose fuel tank entries), in-tank condensation due to temperature and humidity fluctuations, and faults in operational and infrastructure systems [9]. Water contamination in aviation fuels can have adverse effects on aircraft performance and safety. It can promote microbial growth, leading to biofilm formation (Chap. 12), filter blockages (Chap. 10), and corrosion of fuel system components (Chap. 13). Water in the fuel can also cause icing, resulting in blockages and engine issues. Accumulation and migration of water through the fuel system from the refinery to the wingtip are described in Chaps. 9 and 11.

Described below are some Accident Reports (ARs) describing real safety occurrences in which water was a contributing factor:

AR Index	001	**Date**	04/22/1992	**Location**	USA, California
Investigation Agency	NTSB	**Accident Number**	LAX92MA183 [10]	**Aircraft**	de Havilland DHC-6-200
Event Type	Accident	**Injuries**	16 Fatal, 6 Serious	**Aircraft Damage**	Destroyed
Fuel Narrative	Fuel starvation ***due to*** water fuel contamination ***caused by*** inadequate measures taken against water contamination, both at the airport and within the aircraft				
During takeoff, the right engine lost power, causing the right wing to lower to approximately 90°, and the airplane subsequently impacted the ground adjacent to the runway. Upon investigation, it was discovered that the forward fuel tank, which supplies fuel to the right engine, contained approximately 8 gallons of a heavily contaminated mixture consisting of water, an emulsifying agent, and bacterial growth. The aircraft was refueled by a refueling truck. The refueler personnel stated that the aircraft did not show any signs of draining after refueling. The refueling truck was filled from the airport's underground tank the evening before the accident, which was found to contain the same contaminated mixture as the truck and the airplane.					

AR Index	002	**Date**	23/12/2016	**Location**	Israel
Investigation Agency	IAIA	**Accident File No**	99–16 [11]	**Aircraft**	M20J Mooney
Event Type	Accident	**Injuries**	2 Serious	**Aircraft Damage**	Destroyed
Fuel Narrative	Fuel starvation ***due to*** water fuel contamination was ***caused by*** a defect in the filler cap housing, allowing rain to penetrate the aircraft fuel tank, and the failure of flight crew to detect water contamination on sump samples				

(continued)

(continued)

Seconds after take-off, the engine lost power, and the plane turned sharply to the right, making a heavy landing in a low wheat farm field. Post-accident examination revealed water in the right-wing tank, center tank, fuel pump, and air fuel unit. Further investigation concluded that the main reason for water penetrating the right-wing tank was a defect in the filler cap housing. The flight crew sampled the aircraft's fuel and found evidence for water on the left-wing tank and none in the right-wing tank. The investigator believed that under certain lighting conditions, clear water from the fuel tank could be interpreted as 100LL gasoline (Fig. 2). It is important to mention that a similar accident occurred to the same specific aircraft 4 years earlier (IAIA 107-12), and it was concluded that the filler cap was defective.

It is evident from the examples how crucial it is to conduct corrective and proper maintenance, as well as adopting water contamination mitigation measures for fuel, both in the aircraft and on the ground.

5.2 Solids

Ice, debris, dirt, microbes, etc. in aircraft fuel tanks can clog fuel supply lines and filters, ultimately depriving the engines of the necessary fuel. Fuels can hold water, which may freeze in cold weather or during high-altitude flights, as well as particulate matter in suspension. When fuels are introduced into airport storage systems, they are typically left in a supply tank for a settling period to allow water and solid debris to settle out. If left unmonitored or unfiltered, fuel contaminants can reach hazardous levels.

Despite protective coatings often applied to fuel tanks and pipes used in ground systems, many are made of steel or its alloys, which can lead to particulate contamination from rust and scale. The presence of even small quantities of water ensures

Fig. 2 Sample drained from the right-wing tank in different lighting conditions (actually 100% water) [11]

that most distribution processes will contribute to rust contamination. Other sources of particulates include airborne solids entering through damaged hoses and filters (especially rubber particles and fibers), and solids from microbial infestation. Additionally, construction materials such as chips, sealing material particles, residues from construction materials, drilling products, abrasives, rags, gloves, and more can induce particulate contamination.

Below are more AR examples of safety occurrences regarding solid contamination:

AR Index	003	**Date**	03/11/2015	**Location**	USA, Washington
Investigation Agency	NTSB	**Accident Number**	WPR16LA021 [12]	**Aircraft**	MD Helicopters 369D
Event Type	Accident	**Injuries**	None	**Aircraft Damage**	Substantial
Fuel Narrative	Fuel starvation ***due to*** naphthenate fuel contamination, ***caused by*** contamination from unknown source blocking fuel flow through fuel pump				
While conducting longline operations, the helicopter experienced an engine power loss. The pilot initiated an autorotation and attempted to land on flat terrain at the bottom of the hill. However, the helicopter touched down on a slope, causing the tail boom followed by the skids to impact the side of a hill. Subsequently, the helicopter came to rest on its side. Examination following the accident revealed debris throughout the fuel system. The fuel bypass valve inlet port screen was found with a brown debris (spongelike/crystalline debris, see Fig. 3). This debris, located on the fuel pump bypass valve inlet port, inside the pump, and in the centrifugal pump, obstructed fuel flow sufficiently to result in fuel starvation and engine power loss. Chemical analysis of the debris consistently shows the presence of naphthenates, similar to those removed during the refining process. It was reported that a month prior to the accident, the company performed biocide treatment. However, no evidence was found that the microbicide used by the operator contributed to the dissolution of the naphthenates, and the reason for the separation of naphthenates from the fuel could not be determined.					

AR Index	004	**Date**	13/04/2010	**Location**	Hong Kong
Investigation Agency	Accident Investigation Division, Civil Aviation Department, Hong Kong	**Accident Report**	Aircraft Accident Report 2/2013 [13]	**Aircraft**	Airbus A330-342
Event Type	Accident	**Injuries**	Minor	**Aircraft Damage**	Substantial
Fuel Narrative	Loss of engine thrust control ***due to*** Superabsorbent Polymer (SAP) fuel contamination ***caused by*** stiction of fuel metering units by SAP uplifted to the fuel				

(continued)

(continued)

The plane approached Hong Kong International Airport with a control problem on both engines. Upon landing heavily, fire and smoke emerged from the wheels, resulting in injuries to a total of 57 passengers and six cabin crew during the evacuation.
The investigation revealed that fuel contaminated with SAP particles, uplifted to the plane, had caused stiction in the fuel metering units of both engines, leading to the total seizure of these components. This ultimately resulted in the loss of thrust control for both engines during landing. A series of causal factors had led to the uplift of contaminated fuel to the aircraft, including the presence of saltwater in the hydrant system and low flow rates.
SAP contamination is a specific case of solid contamination—see Sect. 5.2 for more details.

Solid impurities can enter fuel through various ways. Having a suitable filtration system upstream of refueling the plane is crucial for preventing contamination. Additionally, maintaining a robust quality surveillance program to ensure the functionality of the filtration system is necessary. This includes measuring the filter pressure difference during operation to prevent overpressure. Furthermore, performing a visual inspection of the fuel in accordance with ASTM Test Method D6986 [14] during preflight can be a valuable tool for detecting fuel contamination. This topic is also discussed in Chap. 9.

5.3 Microbial Contamination

Microbial contamination refers to the presence of microorganisms, such as bacteria, fungi, etc., in aviation jet fuels. These microorganisms can proliferate and form solid debris (i.e. biomass), leading to the clogging of fuel filters. Microbial growth is typically found in environments where water is present in a fuel system (without the use of a biocide) and can result in issues such as malfunctioning fuel level gauges, fuel system blockage, and fuel tank corrosion. Water plays a crucial role in microbiological growth. Regular draining of any accumulated water helps keep fuel systems as dry as possible, aiding in the control of microbial growth. The primary method for assessing the presence of microbiological growth in storage tanks and filters is the daily clear and bright test of sump samples. Indicators such as discolored water, a lacy interface between the fuel and water layers, or organic debris in the fuel or water layer are all signs of potential microbiological activity, requiring immediate investigation and expert advice. Aviation fuel harbors a diverse range of microbial contamination. Microbial contamination in aviation fuel and systems is covered in Chap. 12.

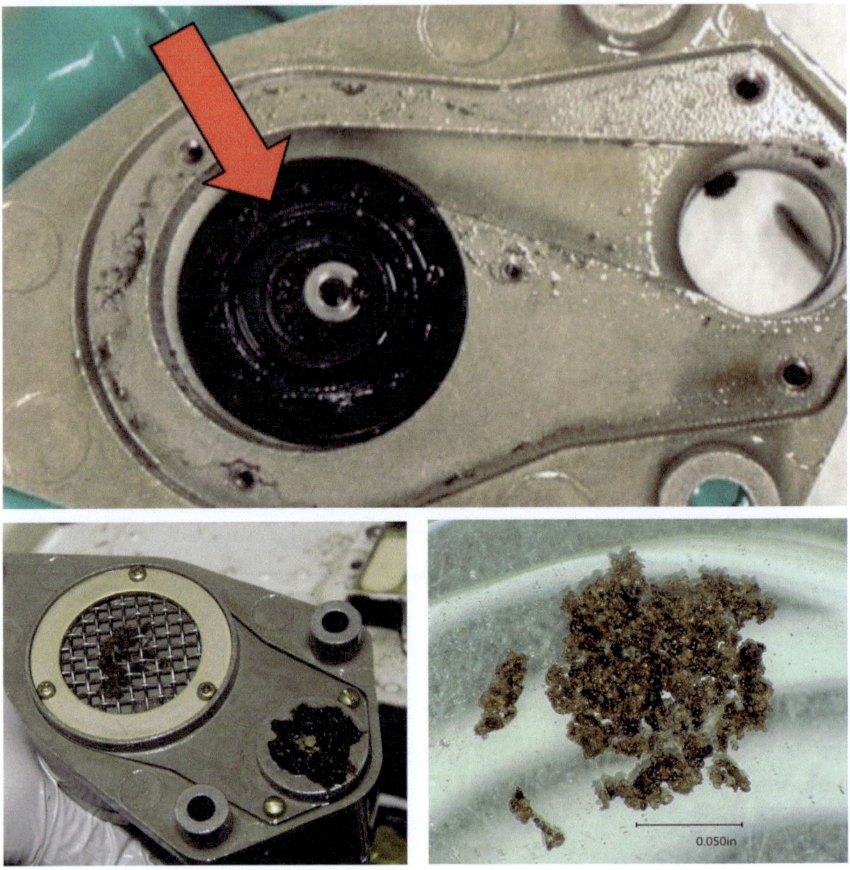

Fig. 3 Debris presence on fuel pump screen and bypass valve inlet

There are many examples of aviation failures attributed to microbiological contamination. Some of them are given in the AR below:

AR Index	005	**Date**	04/01/2002	**Location**	Australia
Investigation Agency	ATSB	**Aviation Safety Investigation Report**	200,200,007 [15]	**Aircraft**	GAF Nomad N22C
Event Type	Incident	**Injuries**	None	**Aircraft Damage**	None
Fuel Narrative	Fuel exhaustion ***due to*** microbial contamination that disrupted the fuel quantity transmitter units ***caused by*** an unknown reason, as the operator did not use adequate measures against microbial contamination				

(continued)

(continued)

During the flight, the pilot noticed a slight yaw to the left and felt the aircraft decelerate. He then saw that the left engine's low oil pressure and generator warning lights had illuminated. While responding to the left engine failure, the pilot noted that both low fuel pressure warning lights had illuminated and that the right engine had failed. The pilot then subsequently executed a successful forced landing. Post-incident inspections of the fuel tanks revealed contamination with microbiological material. Each of the four usable fuel tanks contained one fuel quantity transmitter unit. All four transmitter units were found to be contaminated with microbiological material. These transmitter units consisted of a float housed inside an aluminum tube. Fuel entered the tube through small holes in the bottom, and the level of the float was electrically sensed and transmitted to the fuel gauges in the cockpit. The presence of microbiological material in the fuel quantity transmitter units obstructed the flow of fuel into the tube and affected the electrical circuits at the bottom of the tube. As a result, the contaminated fuel quantity transmitters indicated that there was still usable fuel in the tank when there was none, resulting in fuel exhaustion. It is important to note that the transmitter units had been cleaned less than 1000 h prior to the occurrence. Additionally, the operator's fuel supplier added Fuel System Icing Inhibitor (FSII) to the bulk fuel supply (FSII acts as a biostat when used continuously), and the operator also added a biocide to the aircraft fuel tanks each month. A similar incident was also reported in a Hercules plane of the Israeli Air Force in the early 1990s.

Microbial contamination of jet fuel can present challenges such as plugged filters, leading to fuel flow issues, injector malfunctions, problems with the fuel gauge indicator, and microbiologically influenced corrosion (MIC). The easiest way to limit microbial growth is to minimize the presence of water in the fuel system by regularly draining water from the bottom of the fuel tanks (called "draining"), as well as from low points in pipelines, filtration equipment, and other fuel system equipment. Microbial growth can be mitigated by draining water, applying chemical treatments to eliminate the growth, and then manually cleaning the tank (see Chap. 12).

5.4 *Super Absorbent Polymer Contamination*

The aviation industry has been dealing with the issue of particle migration in the water-absorbing media used in filter monitors for years. The superabsorbent polymer (SAP) media in the filter monitors of the dispenser, when exposed to water, transforms into a gel, restricting the flow of fuel to the aircraft. Failure of the monitor cartridge can result in the uptake of fuel contaminated with SAP into the aircraft fuel system, leading to significant operational issues in the engines. The refueling process is a critical juncture to prevent contaminations. If the pressure drop (DP) across the equipment filtration system exceeds the designed value or range, the equipment operator should immediately halt refueling and replace the monitor cartridge. For more information and details, see JIG Bulletin No 132 "Phase out of Filter Monitors from the JIG Standards [9] and IATA Super-absorbent Polymer (SAP) Special Interest Group—Data summary and proposed roadmap" [16]. See the AR below:

AR Index	006	**Date**	2008 – 2009	**Location**	Unknown
Investigation Agency	USA Military	**Technical Interim Report**	INTERIM REPORT TFLRF No. 402 [17]	**Aircraft**	USA Military Helicopters
Event Type	Incident	**Injuries**	None	**Aircraft Damage**	None
Fuel Narrative	Operational/flight safety issues ***due to*** SAP fuel contamination				
In January 2010, the US Army published a report investigating operational field issues encountered by the US Army Aviation Engineering Directorate involving helicopters. All these field issues involved contaminated fuel problems that posed operational and flight safety concerns. Two of the incidents were associated with plugged fuel filters directly caused by SAP migration from water absorbent monitors. The third incident also resulted in a plugged fuel filter, but this debris originated from an unknown, external source.					

5.5 *Diesel Exhaust Fluid*

Diesel Exhaust Fluid (DEF) is a NOx emissions reduction agent in ground-based diesel engines. Comprising 32.5 wt% urea and 67.5 wt% water [18], DEF is utilized in vehicles featuring the Selective Catalytic Reduction (SCR) system. It is administered into the diesel fuel tank of a refueler vehicle by means of systems that are separate from the jet fuel delivery systems. Should DEF be inadvertently mixed with jet fuel, it may undergo reactions with certain chemical components of jet fuel, resulting in the formation of crystalline deposits within the fuel tank system. These deposits have the potential to traverse the aircraft fuel system to the engines, where they can accumulate on filters, fuel metering components, other fuel system components, or engine fuel nozzles. Consequently, jet fuel contaminated with DEF fails to meet standard jet fuel specifications and cannot be used for aviation propulsion. An example of DEF contamination is detailed in the AR provided below:

AR Index	007	**Date**	2017–2019	**Location**	Different locations
Investigation Agency	FAA	**Safety Risk Assessment Report**	Safety Assessment for Jet Fuel Contamination with DEF [19]	**Aircraft**	Different aircraft
Event Type	Incidents	**Injuries**	None	**Aircraft Damage**	None
Fuel Narrative	Operational/flight safety issues ***due to*** DEF fuel contamination				

(continued)

(continued)

In all incidents, DEF was mistakenly added to the Fuel System Icing Inhibitor (FSII) tank on the fuel truck by airport maintenance personnel. DEF reacts with the jet fuel, forming crystals that clog the fuel filters and nozzles, resulting in engine shutdowns. In these three incidents, a total of 15 civilian and three military aircraft were refueled with contaminated jet fuel, leading to emergency landings for six civilian and three military aircraft. Figure 4 shows an auxiliary fuel tank with DEF deposits.

Eppley Field Airport in Omaha, Nebraska (OMA), November 18–21, 2017

During this event, the FSII reservoir on a refueling truck was mistakenly filled with DEF instead of FSII. Consequently, seven civilian aircraft, along with several military aircraft, were fueled with the jet fuel contaminated with DEF, leading to emergency landings for three civilian and two military aircraft. Additionally, six more aircraft were serviced with equipment exposed to DEF. In response to this incident, SAIB HQ-18-08R1 was issued on December 26, 2017.

Miami-Opa Locka Executive Airport in Miami, Florida (OPF), August 12–16, 2018

During this event, DEF was mistakenly mixed into the FSII reservoir of a refueling truck during an off-truck leak repair. As a result, jet fuel containing DEF was directly injected into the fuel tanks of five aircraft, causing two emergency landings. Nine more aircraft were refueled with equipment exposed to DEF. In response, the FAA issued SAIB HQ-18-28 on September 13, 2018, Safety Alert for Operators (SAFO) 18015 on November 13, 2018, and the Office of Airports (ARP) Awareness and Guidance Letter on October 29, 2018. Additionally, this event prompted the initiation of the Industry Working Group, and Airlines for America (A4A)/ National Air Transportation Association (NATA) Bulletin 2018.4 was released in December 2018.

Punta Gorda Airport in Punta Gorda, Florida (PGD), May 9, 2019

During this event, DEF was mistakenly mixed into a 5-gallon FSII transfer container by an operator. As a result, three aircraft were serviced from this contaminated container, leading to significant consequences: Two Citation 550 aircraft experienced engine flameouts during flight, one at 35,000 feet and the other at 36,000 feet. Both aircraft managed to land safely, albeit with one engine inoperative. Additionally, an Eclipse 500 aircraft unknowingly had DEF directly injected into its fuel system, resulting in post-flight inspection revealing contamination and rendering the aircraft un-airworthy. Discussions with airport personnel at Punta Gorda revealed that they, including the operator, were aware of the risks associated with DEF and had implemented measures, including training and procedures, to avoid inadvertently adding DEF to FSII tanks. Subsequently, SAIB HQ-18-08R2 was issued on June 12, 2019, as a response to this incident.

UDR-TR-2019-80—Chemical Analysis Investigation of Jet Fuel Contamination Event: Punta Gorda Airport, FL, May 9th, 2019

After the jet fuel contamination event at Punta Gorda Airport, The University of Dayton Research Institute fuel science group conducted a chemical analysis study. The investigation revealed that the "drum" sample of FSII was pure and unadulterated. However, samples taken from the trucks showed heavy contaminations with water and urea. The presence of water and urea was likely due to the addition of DEF. Chemical analysis suggests that the FSII pail of the KPGD refueling truck was the source of urea, which contaminated the fuel uploads being treated with FSII.

Refueling systems that require DEF and FSII or any other additives, in Jet fuel may encounter confusion between DEF and these additives. Because DEF and FSII are both clear liquids, this similarity can lead to further confusion among maintenance personnel. To prevent confusion between DEF and FSII, it is crucial to carefully inspect the labels on the containers of the materials. Additionally, it is advisable to avoid using common containers for storing these materials and to refrain from using

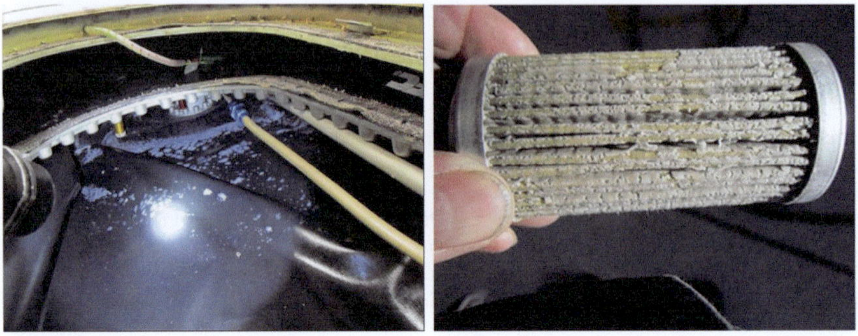

Fig. 4 Auxiliary fuel tank with DEF deposits, left; Engine fuel filter contaminated with DEF, right [19]

non-standard containers. Therefore, the key to avoiding potential contamination problems lies in maintaining vigilance for prevention.

5.6 *Kathon™ FP 1.5 Biocide*

According to the ASTM D1655—22a specification [20], the only approved biocide additive for aviation fuel is Biobor® JF[3] (see Chap. 12). Although Kathon ™ FP 1.5[4] Biocide was listed in earlier versions of Specification D1655, it is no longer included in this specification. Like all additives, biocidal additives should only be used following the manufacturer's recommendations, with specific guidelines provided by aircraft engine and airframe manufacturers. End users must be informed of and agree to the presence of biocide additives in their jet fuel supply, with instructions available in the appropriate documentation, such as technical manuals or guidance materials. It is also essential to refer the most current edition of ASTM D1655, as well as any other relevant specifications such as DEF STAN 91-091, to ensure compliance with industry standards and Aircraft Maintenance Manual (AMM). It is important to note that there are strict limits on the maximum concentration of Biobor® JF, and maintenance doses (half of the maximum) can be used intermittently.

While biocide offers significant benefits when used correctly, improper use can lead to damage or serious incidents. The FAA has issued an airworthiness directive (AD) for certain Boeing Company Model 737-8 and 737-9 (737 MAX) airplanes, requiring the removal of Kathon™ FP 1.5 biocide from use in fuel tanks. The AD also mandates the installation of a fuel limitation placard and revising the existing airplane flight manual (AFM) to prohibit engine operation with Kathon™ FP 1.5 biocide in a fuel tank. This action follows a report indicating that excessive Kathon™

[3] Biobor® JF is a registered trademark of Hammonds Technical Services, Inc. 910 Rankin Rd., Houston, TX 77073.

[4] Kathon™ FP 1.5 Biocide is a trademark of Lanxess GMBH, Cologne, Germany.

FP 1.5 biocide added to fuel and running through the engines can cause significant engine anomalies. The FAA is issuing handbooks to address the unsafe condition [21]. See the following AR example for more details:

AR Index	008	**Date**	26/02/2020	**Location**	UK
Investigation Agency	AAIB	**AAIB Special Bulletin**	S1/2020 [22]	**Aircraft**	Airbus A321-211
Event Type	Incident	**Injuries**	None	**Aircraft Damage**	Minor
Fuel Narrative	Operational/flight safety issues ***due to*** Kathon^{TM} FP 1.5 overdose in fuel ***caused by*** a miscalculation of a maintenance engineer				
At approximately 500 ft above ground level during takeoff, a loud noise was heard, accompanied by flames from the tailpipe of the No. 1 engine as it surged. The crew immediately issued a MAYDAY call and initiated a right turn to return to the airport. Within two minutes, parameters related to the No. 2 engine began fluctuating, and the crew received an indication of engine stall. Despite these challenges, the aircraft managed to land successfully. During the investigation, fuel samples were collected from the drain valves of both right and left-wing fuel tank water and sent to the laboratory for analysis. The samples did not meet the JET A-1 specification requirements in terms of appearance and water separation. They displayed a distinct brown liquid layer beneath the fuel layer, as seen in Fig. 5. Trace element analysis of both the fuel and the bottom brown layer revealed similar spectra to a reference Kathon^{TM} FP 1.5 sample, but with a higher water content. The laboratory's analysis indicated contamination with undissolved Kathon^{TM} FP 1.5. The darker bottom layer primarily comprised Kathon^{TM} FP 1.5, along with some unknown products and water. The operator's work card instructed for biocidal treatment with fuel mixed with Kathon^{TM} FP 1.5 biocide at a concentration of 100 ppm by volume but did not provide instructions on performing the calculation. The maintenance engineer, unfamiliar with the term "ppm," searched the internet for calculator settings and conversions. Due to an incorrect calculation, the engineer administered an amount of 3750 ppm, approximately 37 times the recommended dose.					

Fig. 5 Fuel samples drained from left- and right-wing fuel tanks [22]

5.7 *Vapor Lock*

Vapor lock occurs when the fuel delivery to the engines is disrupted due to the collection or trapping of vapor in the system. Volatile fuels, like gasoline, tend to evaporate as atmospheric pressure decreases with altitude, particularly when the fuel is warmer. Those conditions exacerbate the likelihood of vapor lock [23]. For examples of aviation failures attributed to vapor lock see the ARs below:

AR Index	009	**Date**	04/13/2006	**Location**	USA, Kansas
Investigation Agency	NTSB	**Accident Number**	DEN06LA063 [24]	**Aircraft**	Rumle/ Williams/ Boomhower Starduster II
Event Type	Accident	**Injuries**	None	**Aircraft Damage**	Destroyed
Fuel Narrative	Fuel starvation ***due to*** vapor lock ***caused by*** the installation of the fuel line too close to the hot exhaust pipes				

The pilot reported a routine takeoff and ascent. However, at approximately 200 feet above the ground, the engine unexpectedly lost power. Promptly, the pilot initiated a forced landing. Upon touchdown, the pilot applied "hard brakes" to avoid a ravine at the end of the grass strip. Consequently, the aircraft overturned onto its back, causing significant damage to the upper wing, vertical stabilizer, and rudder. Upon inspection of the engine, it was discovered that the fuel lines had been routed within 1–1/4 inches of the dual exhaust pipes. FAA regulations stipulate a minimum separation of 6 inches for certified airplanes to mitigate the risk of fuel boiling in the lines and subsequently cause vapor lock. No other anomalies were found.

AR Index	010	**Date**	28 May 2017	**Location**	England
Investigation Agency	AAIB	**AAIB Bulletin**	Bulletin 11/ 2017 [25]	**Aircraft**	Europa, G-NDOL
Event Type	Accident	**Injuries**	1 Fatal	**Aircraft Damage**	Destroyed
Fuel Narrative	Fuel starvation ***due to*** vapor lock ***caused by*** the incorrect installation of the fuel vapor return line, which routed vapor back to the system instead of the atmosphere				

The aircraft departed from Coal Aston Airfield but struggled to maintain a normal rate of climb. As it approached the end of the runway, the aircraft turned to the left and then stalled, descending into an adjacent field. Investigation revealed a partial loss of engine power, likely caused by fuel vapor disrupting the fuel supply. The root cause was identified as the misconnection of the fuel vapor return line to the inlet of the fuel selector valve instead of the fuel tank. This misalignment led to vapor being directed back to the engine rather than returning to the fuel tank to re-condense. Using E5 Mogas, a highly volatile fuel, this misconfiguration likely contributed to the accident.

Vapor lock poses a serious risk, potentially blocking fuel flow and causing engine failure. To mitigate this risk, various design considerations are implemented, such as keeping fuel lines away from heat sources, incorporating booster pumps to maintain pressure in the lines, and ensuring proper venting of the fuel system. Additionally, fuel quality is crucial, as higher vapor pressure increases the likelihood of vapor lock. Mogas (short for motor gasoline) describes standard automotive gasoline that is used in cars and other ground vehicles. It is distinct from Avgas (aviation gasoline), which is specifically formulated for use in aircraft. While Mogas can sometimes be used in certain types of small aircraft, it is generally not suitable for high-performance aircraft engines that require Avgas. Avgas is specifically formulated for aviation use, often with higher octane ratings and different additives to meet the needs of aircraft engines (for more details about gasoline for aviation see Chap. 4).

6 Failures in Aircraft Fuel System Units and Components

This section addresses failures observed in fuel system units and components, encompassing tanks, quantity gauges, pumps, valves, filters and strainers, carburetors, fuel lines, the Fuel–Oil Heat Exchanger (FOHE), and the Fuel Control Unit (FCU). Subsequent sections will provide examples of real-world failures.

6.1 Fuel Tanks

Fuel tanks exhibit significant variability in size, location, material, and quantity from one aircraft to another, contingent upon the relative size and complexity of the aircraft in which they are installed. In its simplest form, a fuel tank comprises a suitably sized metal (or rubber) container with an entry port and a discharge port, located in the airframe to gravity-feed an aircraft engine using the associated fuel line. In a modern multi-engine passenger or cargo aircraft, the fuel system is likely to encompass multiple fuel tanks located in the wings and the fuselage. Each tank will be equipped with necessary internal fuel pumps, along with associated valves and plumbing to supply the engines, facilitate refueling and defueling, isolate individual tanks, and in larger applications, enable fuel dumping or transfer between tanks to optimize the aircraft's center of gravity. Several accidents have been attributed to aircraft fuel tank issues, including aircraft imbalance due to improper fuel management, engine failure stemming from contamination entering the tank, or a fuel leak.

AR Index	011	**Date**	21/06/1998	**Location**	USA, Oregon
Investigation Agency	NTSB	**Accident Number**	SEA98FA109 [26]	**Aircraft**	Nolley BD-5T
Event Type	Accident	**Injuries**	None	**Aircraft Damage**	Substantial

(continued)

(continued)

Fuel Narrative	Fuel starvation occurred ***due to*** a clogged fuel filter ***caused by*** the introduction of fuel containing an ethanol additive into fuel tanks sealed with a compound incompatible with alcohols, resulting in debonding of the compound from the fuel tank walls
During the assembly of the home-built aircraft, its wet-wing fuel tanks were coated with a sealer (Randolph #802) that is not compatible with alcohol fuel additives. Jet A fuel containing an FSII additive (Prist, which is essentially diethylene glycol monomethyl ether) was in its fuel tanks during the aircraft's three-week engine test run and flight sequence. This resulted in the sealer debonding from the fuel tank walls and partially blocking an in-line fuel filter. As a consequence of this blockage, approximately 25 min into the fourth test flight, the aircraft experienced a total loss of power. During the subsequent forced landing, the aircraft nosed over when one of its wings became entangled in tall vegetation.	

AR Index	012	**Date**	07/10/2012	**Location**	Antigua
Investigation Agency	ECCAA	**Accident Report**	7AC/1/99 [27]	**Aircraft**	BN2A-26 Islander
Event Type	Accident	**Injuries**	3 fatal, 1 serious	**Aircraft Damage**	Destroyed
Fuel Narrative	Fuel starvation ***due to*** water fuel contamination inside a fuel tank, ***caused by*** anomalies in the aircraft's fuel filler neck and cap				
During a climb, at an altitude of 200–300 feet above ground level, the aircraft experienced a slight descent without yawing or rolling. However, it then yawed to the right, followed by a roll to the right, and pitched nose-down into an incipient spin to the right. According to the investigation, significant rainfall and anomalies in the aircraft's fuel filler neck and cap led to the presence of water in the right-hand fuel tank 2. Shortly after takeoff, the water in the right-hand fuel tank entered the engine fuel system, causing the engine to stop running. It was found that no pre-flight water drain check was carried out. Such a check would have allowed the presence of water in the right-hand fuel tank to be detected, and corrective action could have been taken.					

6.2 Fuel Quantity Gauges

All aircraft must be equipped with some form of fuel quantity indicator in their fuel systems, which can range from simple visual gauges to more complex electrical ones, depending on the aircraft's complexity. Accidents related to fuel quantity gauges typically involve malfunctions or inaccuracies, which can mislead pilots about the amount of fuel remaining in the tanks and potentially lead to fuel exhaustion. The issues may arise due to mechanical or maintenance problems, as well as microbial growth in the fuel tanks. Further examples can be found in the investigations which are described in Chap. 12.

In many small aircraft lacking electrical gauges, it is particularly important to visually inspect fuel levels in the tanks before flight to ensure proper fuel quantity assessment. Below are accident reports concerning fuel quantity gauges. See AR Indexes 013 and 014:

AR Index	013	**Date**	August 2005	**Location**	Italy
Investigation Agency	ANSV	**Final Report**	ATR 72-202 TS-LBB [28]	**Aircraft**	ATR-72-202
Event Type	Accident	**Injuries**	16 Fatal	**Aircraft Damage**	Destroyed
Fuel Narrative	Fuel exhaustion ***due to*** the incorrect procedure used for replacing the fuel quantity indicator				

While cruising approximately 23,000 feet, about 50 min after take-off, the right engine (no.2) shut down, followed by the left engine (no. 1) approximately 100 s later. The crew decided to divert to the nearest airport for a precautionary landing. After gliding for 16 min, the aircraft ultimately ditched in Italian territorial waters, breaking into three sections upon impact. Tragically, 16 of the 39 occupants lost their lives. The investigation revealed that insufficient fuel had been loaded prior to flight. This was because the flight crew relied solely on the fuel quantity gauges, which indicated 1,800 kg of fuel (900 kg in each wing) despite there being no fuel at all. It was determined that the incorrect quantity gauge had been installed the day before the accident by the maintenance personnel.

AR Index	014	**Date**	October 2016	**Location**	USA, Washington
Investigation Agency	NTSB	**Accident Number**	WPR17LA014 [29]	**Aircraft**	Aerospatiale AS350
Event Type	Accident	**Injuries**	None	**Aircraft Damage**	Substantial
Fuel Narrative	Fuel exhaustion ***due to*** a malfunctioning fuel quantity gauge ***caused by*** internal mechanical failure due to improper maintenance				

During the agricultural application flight in the helicopter, the pilot reported that the fuel pressure warning light illuminated. Upon checking the instruments, the pilot observed low fuel pressure and that the fuel quantity indicated between 40 and 50%. Subsequently, the engine began to lose power. The pilot initiated an autorotation and prepared the helicopter for landing in the flattest available area. The helicopter subsequently made a hard landing, resulting in significant damage to the tail boom and main rotor assembly. Post-accident examination of the helicopter revealed that the fuel tank was intact, undamaged, and completely empty. Based on this evidence, it was determined that the engine likely lost power due to fuel exhaustion. Surprisingly, despite the absence of fuel, the fuel quantity gauge indicated that 42% of the fuel remained (approximately 60 gallons). Further inspection revealed that the float guide pin was separated from the float assembly at the tack weld. This separation allowed the float to slide down the center torque shaft, preventing the potentiometer drive plate at the bottom from rotating with changes in fuel quantity. Consequently, the fuel quantity gauge displayed an inaccurate fuel level.

Reviewing the airplane's maintenance records revealed that the fuel transmitter had been overhauled approximately 11 months before the accident. Given this information, it is likely that during the overhaul, maintenance personnel improperly welded the float pin guide to its mount and subsequently failed to adequately inspect it, resulting in its eventual separation from the assembly at the welded area.

The primary issue with fuel quantity gauges occurs when they indicate sufficient fuel when there is insufficient fuel to complete the mission. This discrepancy is a common cause of engine failures, with numerous examples where the pilot noticed that there was no fuel despite the gauges indicating otherwise. The reliability of fuel quantity indication is crucial for flight safety. Accepting inaccuracies and deviations can lead to the normalization of deviance–a phenomenon in which known deviations are assumed to be constant and, therefore, ignored. This behavior is both human and organizational, acting as a driving force behind failures or accidents.

6.3 Fuel Pumps

Fuel pumps play a critical role in the aircraft fuel system. Depending on the aircraft configuration, there can be several fuel pumps serving different functions. Boost pumps may be installed in fuel tanks located below the engines to lift fuel to the engines, while other pumps may transfer fuel between tanks or dump tubes. Fuselage centerline tanks typically have pumps to transfer fuel to wing tanks for balancing purposes. In small aircraft such as General Aviation, Piper Cubs, Cessna's, and Taylor Craft, fuel tanks are located above the engines and rely on gravity to feed fuel. Modern helicopters usually have fuel tanks below the engines, with boost pumps. Large commercial aircraft such as Boeing, and Airbus models typically feature engines mounted beneath the wings and are gravity-fed, but they also incorporate low-pressure fuel pumps at the engine fuel inlet to ensure proper fuel flow.

Fuel pumps are responsible for moving fuel as needed and providing the correct pressure to the engines. Damage or contamination in the fuel pumps can disrupt fuel flow, leading to fuel starvation or flight path deviations.

The following examples illustrate incidents of aircraft failure due to fuel pump malfunctions:

AR Index	015	**Date**	16/06/2002	**Location**	USA, Illinois
Investigation Agency	NTSB	**Accident Number**	CHI02LA166 [30]	**Aircraft**	Jacobs Rutan Vari Viggen
Event Type	Accident	**Injuries**	1 Fatal	**Aircraft Damage**	Destroyed
Fuel Narrative	Partial loss of engine power occurred ***due to*** an excessively rich mixture ***caused by*** a faulty accelerator pump housing seal				

(continued)

(continued)

The airplane was destroyed upon impact with the terrain after losing control following a partial loss of engine power during the initial climb from the runway. A witness reported that the airplane had an unusually long take-off roll, lifting off approximately 300 feet from the end of the 4000-foot runway. Once airborne, the wings were observed to be "rocking and rolling". The pilot transmitted on the radio that he was having trouble climbing. Another witness stated that the airplane entered a slow turn to the right with a high angle of attack. The witness observed the airplane's nose falling to the right, and the aircraft impacted the terrain nose first. A fuel leak was found around the accelerator pump housing on the carburetor. Upon testing the engine on a stand, it was found to be running excessively rich. A contributing factor to the accident was determined to be the partial loss of engine power due to an excessively rich mixture caused by a faulty accelerator pump housing seal.

AR Index	016	**Date**	20/10/2007	**Location**	USA, Alaska
Investigation Agency	NTSB	**Accident Number**	ANC08LA004 [31]	**Aircraft:**	BELL 206B
Event Type	Accident	**Injuries**	1 Minor 1 Serious	**Aircraft Damage**	Substantial
Fuel Narrative	Fuel starvation occurred ***due to*** solid fibrous fuel contamination ***caused by*** an improper refueling procedure that introduced the contamination to the fuel tank and fuel pumps				

While the helicopter was hovering about 60 feet above the ground, the engine lost power, causing the helicopter to descend rapidly and impact the ground, resulting in substantial damage to the fuselage and rotor system. Upon examination, it was discovered that both electric fuel boost pump intakes were clogged with a fibrous material. These pumps serve as conduits for the engine-driven fuel pump, and if they become clogged, no fuel can reach the engine. Similar fibrous material was also found in the airframe fuel filter. Subsequently, during testing at a facility, the engine started and ran without any issues. Upon further inspection of the airframe, the fuel tank was cut open, revealing a triangular shaped 4″ piece of an absorbent pad inside. It was determined that the contract fueler had routinely inserted the fuel nozzle through a hole in the plastic packaging of a bundle of absorbent pads into the edges of the pads to prevent fuel from dripping on the tundra during the refueling process, which occurred every 7–8 min. Samples of the material taken from the fuel pumps, the piece found in the fuel tank, and an exemplar pad provided by the fuel contractor were sent to a laboratory for testing. The tests revealed that all the samples were essentially identical. The contamination of the fuel pumps most likely occurred when the fuel nozzle tip cut or captured portions of the absorbent pads as it was placed in the pad bundles, and these cut portions were then pumped into the fuel tanks.

Fuel pumps are a vital component of the fuel system and ensuring their proper performance is paramount for maintaining the "Safety of Flight". Consequently, the design of the fuel system must have "built-in" protections for these devices. For example, the pump design should prioritize robustness, maintainability, and reliability. Additionally, the fuel system must include screens and filters to eliminate the majority of solid particle fuel contamination (refer to the Filter Section below). Mechanical or electrical failures in a pump can disrupt the fuel flow and engine feed, potentially leading to fuel starvation scenarios. It is important to note that for General Aviation, the operational time for (piston) engine fuel pumps may reach

2000 h before requiring inspection or overhaul. However, accumulating these 2000 h may take three or four years. In commercial aviation, the main fuel pumps for large gas turbines may have operational times ranging from 12,000 to 16,000 h between overhauls and these hours can be accrued within two years of operation. Operational issues with pumps are particularly serious due to their implications for the Safety of Flight.

6.4 Valves

Valves are also integral and important components of the overall fuel system. Malfunctioning fuel valves can lead to a significant accident or incident, as these components directly impact the fuel supply both to and within the engine.

The primary function of fuel valves is to control the flow of fuel. Many valves respond automatically to levels of fuel pressure from the upstream fuel pump. Some valves are electrically operated by control from the cockpit (such as cross-feed valves used for balance). Some valves permit fuel flow in only one direction (no reverse flow permitted); these are called "Check Valves".

Here are some incidents related to fuel valves:

AR Index	017	**Date**	10/02/1999	**Location**	USA, California
Investigation Agency	NTSB	**Accident Number**	LAX00LA002 [32]	**Aircraft**	Zdybel GLASAIR FT
Event Type	Accident	**Injuries**	None	**Aircraft Damage**	Substantial
Fuel Narrative	Fuel starvation occurred ***due to*** the owner/builder's improper installation of the fuel selector valve and his failure to placard the operating positions of the valve handle				

The pilot had recently finished building the airplane, which has accrued a total of 16 h of flight time. The pilot mentioned that he had previously placed placards on the fuel selector positions, but they fell off, and he never bothered to replace them. While on an extended downwind for landing, the engine suddenly stopped running, causing him to land short of the runway and collide with a ditch, resulting in the aircraft nosing over. The pilot reported that the engine stopped running just after he finished changing fuel tank positions. Subsequent to the accident, the engine was successfully run three times during the investigation. The pilot later admitted that he believed he had not properly indexed the fuel selector during the accident flight. Post-accident examination of the fuel selector revealed that the shaft had an improper alignment, leading to excess play in the handle.

AR Index	018	**Date**	10/12/2000	**Location**	USA, California
Investigation Agency	NTSB	**Accident Number**	LAX01LA012 [33]	**Aircraft**	Tolle VENTURE MODEL 20

(continued)

(continued)

Event Type	Accident	**Injuries**	1 Minor	**Aircraft Damage**	Substantial
Fuel Narrative	Fuel starvation occurred ***due to*** a mechanical failure of a fuel sump tank check valve ***caused by*** unreported reason				
Shortly after the pilot took off in his experimental airplane, the engine experienced a partial loss of power. As a result, the pilot made a forced landing on a rural road and collided with a pole during rollout. An examination of the airplane revealed that a check valve in its fuel sump tank was mostly stuck in the closed position, leading to engine fuel starvation.					

Incidents or accidents involving valve components may occur due to mechanical failure of the valve, a series of fuel management errors, improper installation of the fuel selector valve, or a lack of indication of valve position.

6.5 Fuel Filters

Filters and strainers in aircraft fuel systems protect the engine(s) from solid contaminants. Chaps. 9 and 10 discuss ground systems filters and fuel cleanliness procedures designed to minimize water and solid contaminants that reach the aircraft fuel system. The permitted levels of solid contaminants and free water in into-plane fuel are specified in aviation regulations and are considered very strict.

Even though fuel supplied to the aircraft is very clean, the fuel filters and strainer within the aircraft are meant to protect the engine fuel supply from solid contamination originating from the aircraft itself or brought into the fuel system from outside sources. These contaminants can come from residual sealing materials, organic coatings, from the fuel tank, or maintenance aids such as cloths, gloves, etc. Additionally, some aircraft fuel tanks include anti-sloshing or explosion suppression foam, which can deteriorate over time and produce solid particles.

Moreover, some aircraft have sensitive components, such as fuel controllers, that use fuel for metering. Such components could be disrupted even by a small amount of solid contamination, so they are usually specifically protected by design with filters. In advanced and redundant fuel systems, there are often indications (to pilots/maintenance personnel) of clogged filters. For such scenarios, bypass lines are usually present to prevent immediate fuel starvation and engine failures.

Here is an example of an accident related to a fuel filter:

AR Index	019	**Date**	02/24/2007	**Location**	USA, California
Investigation Agency	NTSB	**Accident Number**	SEA07LA066 [34]	**Aircraft**	Scherer Europa
Event Type	Accident	**Injuries**	None	**Aircraft Damage**	Substantial

(continued)

(continued)

Fuel Narrative	Fuel starvation occurred ***due to*** a blocked fuel filter ***caused by*** a piece of rubber detached from a deteriorated fuel line
The engine lost power approximately 80 feet above the runway. The aircraft impacted a raised tank on the left side of the west end of the runway, bounced, and came to rest about 150 feet from the end of the runway. After the accident, it was discovered that the fuel line leading to the fuel filter had deteriorated. A small piece of rubber from the fuel line was blocking the filter's inlet. This was likely what led to the fuel starvation of the engine.	

Fuel filters are a crucial component of the fuel system, safeguarding engines, and sensitive components from solid contamination. However, they require monitoring and maintenance to prevent them from obstructing fuel flow and normal engine operation.

6.6 Carburetor

A carburetor is a device that mixes fuel and air for a spark-ignition engine. It typically consists of a fuel chamber, choke or throttle plate, fuel jets, a venturi-shaped airflow passage, and an accelerator pump. A float-actuated valve controls the amount of fuel. The choke adjusts airflow for starting a cold engine, while the throttle plate modulates airflow as the engine warms. A common issue is carburetor icing, where water vapor freezes inside the device. Solutions include carburetor heat or non-stick coatings to prevent icing. Here are some ARs involving engines equipped with carburettors.

AR Index	020	**Date**	October 21, 2005	**Location**	Japan
Investigation Agency	ARAIC	**Accident Number**	AA2006-4 [35]	**Aircraft**	PIPER J3C-65
Event Type	Accident	**Injuries**	1 minor	**Aircraft Damage**	Destroyed
Fuel Narrative	Fuel starvation occurred ***due to*** fuel contamination blocking the carburetor's inlet screen ***caused by*** prolonged parking in open air, resulting in an increased gum content of the gasoline				

(continued)

(continued)

While flying along a straight-line route, the engine suddenly misfired and stopped. The aircraft was forced to make an emergency landing on a road that was under construction. During the landing the light aircraft collided with a mini-truck, turned rightward, hit a normal-size cargo truck, and stopped. Although the aircraft was destroyed, fortunately, there was no post-crash fire. Post-accident examination revealed that a large amount of foreign matter had accumulated in the vicinity of the carburetor's inlet screen, blocking most of it (see Fig. 6). It is estimated that the fuel flow was restricted by this foreign matter, leading to the engine trouble.
The aircraft was parked in the open air for approximately five months in South Korea. It is possible that the aircraft had also been parked in the open air for an extended period previously. As a result, it is estimated that gasoline in the fuel tanks evaporates, resulting in the formation of gum, peroxide, and organic acid. The gum accumulated in large quantities, and the organic acid and moisture that had collected in the fuel tanks caused corrosion of the aluminum tanks, thereby forming the foreign matter discovered.

AR Index	021	**Date**	September 14, 2008	**Location**	USA, Texas
Investigation Agency	NTSB	**Accident Number**	DFW08FA228 [36]	**Aircraft**	American champion (ACAC) 7GCBC
Event Type	Accident	**Injuries**	1 Fatal	**Aircraft Damage**	Destroyed

(continued)

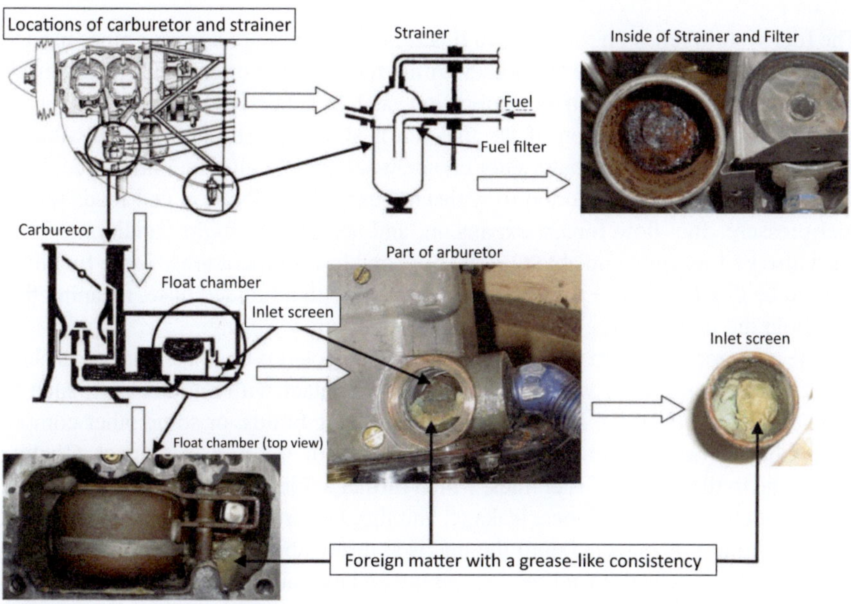

Fig. 6 A large amount of foreign matter accumulated around the carburetor's inlet screen, blocking most of it

(continued)

Fuel Narrative	Fuel starvation occurred ***due to*** carburetor icing
The private pilot took off on a flight with the intention of remaining in the traffic pattern. Winds were from 010 degrees at 16 knots, gusting to 20 knots. A carburetor icing chart indicated the potential for "moderate icing at cruise power or serious icing at descent power." Through a GPS and engine monitoring system, the accident flight was partially reconstructed. Upon takeoff, the airplane turned left and entered the downwind leg. As the airplane was rolling out on the downwind, a reduction in throttle was recorded. The airplane continued on the south-southwest on the downwind leg for at least 14 s. As the airplane approached the base turn, the first recorded attempt to actuate the throttle resulted in an increase in manifold pressure and spike in fuel flow. Manifold pressure fluctuates as the airplane made the base turn to final. Approximately 3/4 through the base turn, the airplane rolled out and flew a straight approach to the open field. Several witnesses observed the airplane approaching the open field without engine power. After "barely clearing" a transmission wire, the airplane was observed to roll quickly right and impact the ground "nose first." These statements are consistent with the airplane entering a stall/spin. An examination of the airframe and engine revealed no anomalies that would have precluded the production of the engine power or the safe operation of the airplane.	

Carburetors require regular maintenance and attention, especially in conditions where ice or foreign substances may form in the system. Applying carburetor heat or using anti-stick coatings, such as Teflon, can help prevent these issues.

6.7 Fuel Lines

The fuel system relies on fuel lines to function. Fuel lines are hoses or pipes used to transfer fuel from one place to another within the airframe or engine. The materials used for fuel lines depend on the application, such as general aviation or commercial aircraft. For general aviation, if the line run is short and straight, it is possible to use a non-metal fuel lines, but a safer choice would be metallic fuel lines. Fuel lines must be installed and supported to withstand excessive vibration and loads due to fuel pressure, fuel flow forces, expansion, and accelerated flight conditions. They must also be fuel-tight and able to resist high and low temperatures. Some fuel lines need to be flexible. Those types of lines are built with a braided hose, meaning that they hold the fuel pressure using a braided metal wire covering.

The common hazard to which fuel lines are subjected is chafing. Chafing is abrasive damage that occurs when a fuel line is in contact with another part, such as another line, a part of the structure, electrical wiring braids, or some other component. Chafing occurs during the normal vibration of the part in contact. Chafing could lead to the failure of fuel lines, which further subjects the airframe to different hazards such as fuel imbalance, leakage, and fire. Unfortunately, in many instances, chafing will not be detected until the line is severely abraded and the unmistakable odor of fuel is detected. Installing fuel piping and hydraulic fluids or oil lines across electrical wiring, engine controls, or close to exhaust pipes is highly discouraged. Nevertheless, those combinations are necessary for some military aircraft. Where hoses and aluminum lines pass through a bulkhead, protect them with rubber grommets. When two aluminum lines are in contact with each other, chafing can take place on both if the lines are not properly separated, secured, or supported. During maintenance, it is recommended that the fuel lines between connectors be carefully checked for sufficient clearance. Chafing is a known phenomenon in aircraft and on engines; therefore, maintenance should address it. There are additional possible failure mechanisms of fuel lines, such as fatigue of seam welding in flanges and wear of joints and fittings. Those mechanisms, like chafing, are also driven by vibrations.

Here are some examples of fuel lines failures:

AR Index	022	**Date**	September 20, 2020	**Location**	Israel
Investigation Agency	IAF	**Accident Number**	Internal	**Aircraft**	F16-I
Event Type	Accident	**Injuries**	None	**Aircraft Damage**	A crack in the Y fuel line
During inspection, a technician noticed a crack in a fuel line. This was a Y-shaped pipe, installed in the engine feed line, that transfers fuel from rear fuel tanks to the Fuel Flow Proportioner (FFP). A failure of this pipe can cause an imbalance and, in extreme scenarios, exhaustion of engine feed tank.					

AR Index	023	**Date**	October, 2011	**Location**	Israel
Investigation Agency	IAF	**Accident Number**	Internal	**Aircraft**	F16-I
Event Type	Accident	**Injuries**	None	**Aircraft Damage**	A fraction/ crack of fuel line
During flight, a fuel pipeline failure occurred. The pipe most likely failed as a result of crack progression in a high-cycle, low-stress fatigue mechanism. The cause is apparently an increased mechanical strain resulting from wrong production dimensions of the pipe, deviating from the original drawings. Two such cases were reported during that period of time.					

The fuel lines allow fuel to flow around the aircraft as needed and are the means to deliver fuel to the engine(s). The fuel line is a static component in the fuel distribution system and is normally sized to allow fuel, liquid contamination such as water, and many forms of solids contamination to flow freely and unobstructed from tank to tank, and from tank to engine.

For these lines to function satisfactorily, they need to be safely routed within the airframe and within the engine, supported and clamped to withstand vibration, and kept away from surrounding structures and any sources that could cause chafing on their surfaces.

The safety risk arises when any line cracks, frets on its surface, or experiences fitting failures at line connectors, which can lead to significant fuel leakage and the possibility of fire or explosions in the aircraft or the engine, potentially resulting in the loss of the aircraft, passengers, and crew. Frequent maintenance attention is a key factor for the "Safety of Flight".

6.8 Fuel Oil Heat Exchanger

The Fuel Oil Heat Exchanger (FOHE; also known as fuel cooled oil cooler or FCOC) cools the engine oil by using the fuel as a heat sump. The FOHE is a hybrid crossflow/counter-flow design, including a matrix of fine tubes. Fuel enters the top of the FOHE and passes through the tubes, while hot oil enters the main body of the FOHE and circulates around the fuel tubes. The temperature of the fuel after passing through the FOHE is higher then its entry temperature. The FOHE fuel tube matrix can consist of over 1000 small tubes that are crimped at various locations along their length to increase thermal transfer efficiency. The crimps at the inlet of the tubes are slightly smaller in diameter than the remainder to prevent small debris from becoming lodged in the matrix. The tubes protrude by approximately 4 mm from the matrix top plate, separating the fuel from the oil and extending into the fuel in the inlet chamber (see Fig. 7).

The icing of the FOHE is a problem that could cause a blockage in the fuel system. Since water is always present in jet fuel, ice will form as the fuel temperature cools below zero degrees Celsius (32 °F). Laboratory testing, as further described in Aircraft Accident Report 1/2010 by AAIB [22], has shown that ice can accumulate within the tank and fuel system tubing at cruise power settings. Over many hours at low power settings, the ice accumulation can reach an amount where if there is

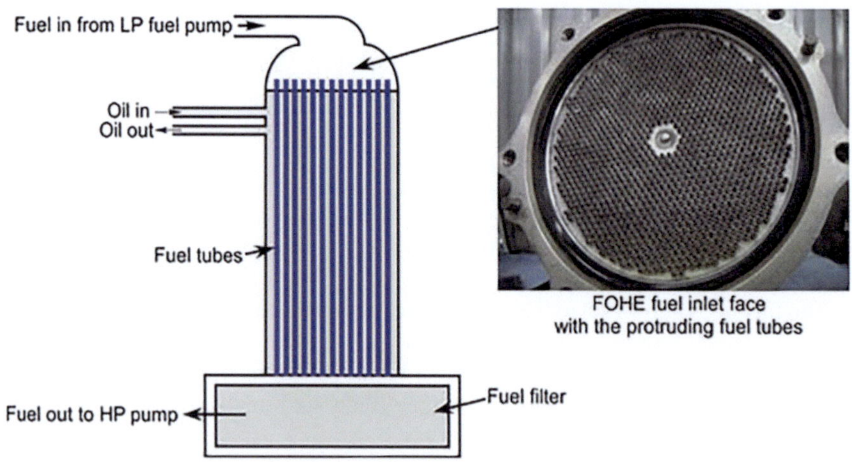

Fig. 7 Schematic description and FOHE fuel inlet face [22]. The schema illustrates the flow of fuel and oil through the system. The arrows indicate the following directions: Fuel enters from the low-pressure fuel pump (LP fuel pump) and passes through the fuel tubes, shown as blue lines. Oil flows in and out of the heat exchanger (Oil in/Oil out) parallel to the fuel tubes. After the fuel travels through the heat exchanger tubes, it exits towards the high-pressure fuel pump (HP fuel pump). In the top right corner is a close-up image showing the FOHE fuel inlet face with protruding fuel tubes extending from the mesh screen

a sudden increase in fuel flow, the ice surges out of the lines onto the face of the FOHE, causing a partial blockage.

Here are two accidents that are related to this phenomenon (the rollback events on the Boeing 777 aircraft powered by Rolls-Royce engines). The first example is broadly analyzed in the systematic analysis section.

AR Index	024	**Date**	17 January 2008	**Aircraft**	Boeing 777–236 ER
Investigation Agency	AAIB	**Aircraft Accident Report**	Aircraft Accident Report 1/2010 [37]	**Location**	England, London Heathrow Airport
Event Type	Accident	**Injuries**	46 minor	**Aircraft Damage**	Aircraft damaged beyond economical repair
Fuel Narrative	Fuel starvation occurred ***due to*** icing of FOHE ***caused by*** cruising in extremely cold weather without adjusting fuel flow rates				

The flight from Beijing to London (Heathrow) was uneventful, and the engine operation was normal until the final approach. The auto-throttles commanded an increase in thrust from both engines, and the engines initially responded. However, at a height of about 720 feet, the thrust of the right engine reduced, and seven seconds later, the thrust on the left engine reduced too. The investigation identified that the reduction in thrust was due to restricted fuel flow to both engines. It was determined that accreted ice released from within the fuel system, causing a restriction to the engine fuel flow at the face of the FOHE, on both engines. The restriction likely occurred just before the final acceleration cycle. Ice, formed from water naturally present in the fuel, accumulated at the face of the FOHE. The FOHE is the only component in the fuel system that could be demonstrated to collect sufficient ice crystals to cause the fuel restrictions observed during the accident flight (see Fig. 8). The fuel samples from the plane complied fully with the specifications for Jet A-1. The sampled fuel had a fuel freeze temperature of –57 °C (–70.6 °F) and a water content of between 35 and 40 parts per million (ppm). The samples were clear of any microbial growth and did not contain any anti-icing additives such as FSII, nor was there any evidence of contamination of "bio-fuel".

AR Index	025	**Date**	26 November 2008	**Location**	USA, Montana
Investigation Agency	NTSB	**Accident Number**	DCA09IA014 [38]	**Aircraft**	Boeing 777-200ER
Event Type	Incident	**Injuries**	None	**Aircraft Damage**	None
Fuel Narrative	Fuel starvation occurred ***due to*** icing of FOHE ***caused by*** cruising in an extremely cold weather without adjusting fuel flow rates				

(continued)

(continued)

While in cruise flight at flight level at about 39,000 feet, the plane experienced a single engine rollback. No injuries were reported. The flight crew reported that the airplane's autopilot and auto-throttle systems were engaged when, about an hour after two routine step climbs were performed to maximum climb thrust, they detected that the No. 2 engine was operating below the commanded thrust level. The flight crew descended the airplane to FL310 and performed the engine response non-normal checklist procedure for clearing ice from the engine FOHE by temporarily commanding idle thrust on both engines (in accordance with FAA AD 2008-19-04). The engine recovered and responded normally, and the flight continued to Atlanta, where it landed without further incident. See AD 2009-05-11 that followed this case.

Operation of the FOHE must be analyzed during different airplane flight modes to determine if it is correctly sized for the engine application. Due to the various liquid temperature ranges that may occur, the FOHE operation also needs to be analyzed for both the oil and the fuel system at steady-state and transient operating modes.

The FOHE plays a critical role in the aircraft engine operation by maintaining required lubricating oil temperatures. Because of the importance of the FOHE and the hazards related to the icing phenomenon (two rollback events occurring within a year), the investigation determined that the only acceptable solution to this safety vulnerability was a redesigned FOHE that would eliminate the potential of ice blockage. The company responsible for the engines ("Rolls-Royce") provided a redesigned FOHE unit to prevent ice accumulation conditions 12 months later for retrofit on their engines.

6.9 Fuel Control Unit (FCU)

Fuel control units in modern commercial aircraft gas turbine engines consist of an electronic unit containing all the fuel schedules (pounds mass of fuel versus power lever angle) with ambient temperature modifiers and any other electronic modifiers about the engine's performance expectation and/or emergency conditions. Additionally, a mechanical unit physically handles and meters the fuel that flows through the engine. Many military aircraft also have mechanical or electro-mechanical fuel controllers to provide redundancy.

The primary task of the engine FCU is to meter the fuel flow demanded by the power lever (throttle) setting to obtain thrust. The FCU helps reduce the pilot workload for an aircraft engine by providing compensating changes as the aircraft encounters variations in the atmosphere that can affect airspeed or altitude. It acts as an intermediary between the operator's controls and the engine's performance to address these issues.

The control senses engine rotor speeds and adds fuel until the actual rotor speeds match the electronic values set in the control by the power lever position. To prevent exceeding operating limits, the FCU has functions to ensure the operation is maintained within limits for engine temperature, pressure, and shaft speeds.

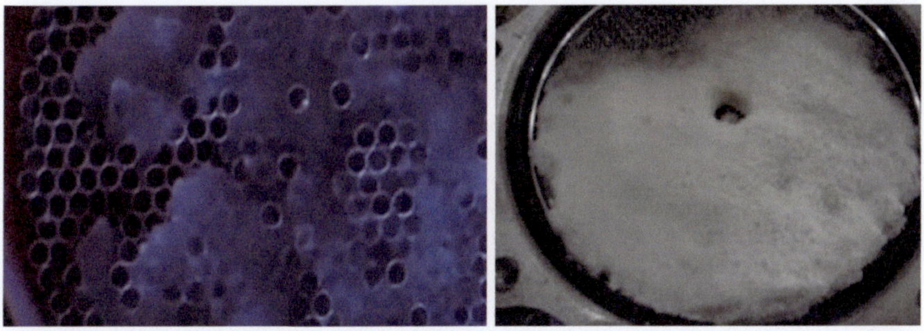

Fig. 8 Ice accumulated on the face of a FOHE as shown in an experiment conducted during the investigation [37]

An example of a mechanical FCU is a small turbine engine FCU that uses a diaphragm-and-spring setup. Incoming fuel pushes the diaphragm one way, while the spring pushes against it to maintain the exact fuel flow to the engine, balancing incoming fuel pressure with incoming bleed air pressure. In the case of an "unregulated" fuel supply, the fuel pump always delivers maximum performance from the fuel tank, with excess fuel being routed back to the fuel tank [39].

Large engine FCUs use a physical metering hole with a sliding door that is positioned to provide the necessary opening for the required fuel to flow to the engine combustor.

The advantages of electronic FCUs are that they can be programmed for infinite flight situations, can account for fuel specification differences, and supply only the required amount of fuel.

Presented are some accident reports related to the fuel control unit:

AR Index	026	**Date**	March 27, 2003	**Location**	USA, Texas
Investigation Agency	NTSB	**Accident Number**	FTW03FA118 [40]	**Aircraft**	Bell 407
Event Type	Accident	**Injuries**	2 Fatal, 3 Serious	**Aircraft Damage**	Destroyed
Fuel Narrative	Partial loss of engine power ***due to*** erratic fuel flow metering to the engine ***caused by*** the single point failure of the PLA potentiometer in the hydro-mechanical fuel control unit				

(continued)

(continued)

While establishing a 125-foot out of ground effect hover during a search and recovery mission of Columbia Space Shuttle debris, the helicopter lost power without warning, descended rapidly into a forest with 80-foot-tall trees, and impacted the ground. During post-accident component examinations, anomalies were noted while performing the rigging procedure for the Hydro-Mechanical Unit (HMU) during an engine test cell run. These anomalies involved the actual position of the throttle Position Lever Angle (PLA) and the readings obtained from the Electronic Control Unit (ECU). During the test cell run, the Full Authority Digital Electronic Control System (FADEC) controlled engine operated erratically in the Auto mode when rigged with its original Electronic Control Unit (ECU) and HMU. The engine control was then changed from auto to manual mode, and the engine responded to throttle input as required; however, the ECU readings were erratic. After a slave ECU was installed, the engine operated with similar erratic readings. A slave HMU was installed, and the engine operated normally without erratic readings. The ECU's non-volatile memory was downloaded and revealed no faults on the accident flight. Extensive electrical and mechanical testing of the HMU revealed severe signal fallout on the PLA signal, which was found to be random in location, bi-directional, and present during both rotational and stationary operation of the PLA input of the HMU. The probable cause of this accident is related to partial loss of engine power due to erratic fuel flow metering to the engine resulting from the single point failure of the PLA potentiometer in the hydro-mechanical fuel control unit.

AR Index	027	**Date**	16 December 2002	**Location**	Canada, British Columbia
Investigation Agency	TSB	**Aviation investigation report**	A02P0320 [41]	**Aircraft**	Sikorsky S-61N
Event Type	Accident	**Injuries**	3 Minor	**Aircraft Damage**	Destroyed
Fuel Narrative	Operational and flight safety issues ***due to*** faulty fuel control unit ***caused by*** a sticking pressure-regulating valve component due to fuel contamination				

(continued)

(continued)

While the helicopter was climbing through about 1,000 feet above sea level (asl) at about 65 knots, the crew became aware of an intensifying whining sound, followed by a single, loud bang. Immediately, the number 1 engine lost power, and the number 2 engine did not automatically compensate for the power loss. The pilot-in-command (PIC) lowered the collective lever to enter autorotation and pushed the cyclic stick forward. Acrid smoke filled the cockpit, and flames appeared from the lower left section of the main rotor gearbox in the cabin. During the roll-out, the helicopter struck other power lines across the road, and the main rotor blades severed a large tree on the left side of the road. The helicopter veered right, and the tail rotor and tail pylon struck the same tree and broke away from the fuselage. The helicopter then started to vibrate severely, with large airframe oscillations, but it remained upright and stopped at the right-hand edge of the road. The three occupants received minor injuries, and the helicopter was substantially damaged. Post-accident examination revealed that the right engine FCU fuel filter contained a significant number of mixed contaminants; however, it could not be determined if the filter had gone into bypass. Further disassembly revealed that the pressure regulating valve (PRV) in the FCU was jammed with contaminant significantly different from that found in the filter. Earlier bench tests had showed that the PRV was sticking; a sticking or immoveable PRV would cause unstable stator vane actuator (SVA) operation, engine starting difficulties, and inconsistent topping settings. Collective experience from US operators of this FCU shows that sticking or jammed PRVs also leads to unpredictable and degraded engine performance. The microscopic and infrared analysis of the debris found in the PRV determined that it comprised particulates of chipboard, bleached cellulose, paint, and metal; the FCU filter contaminant comprised cellulose, paint, human hair, and unidentifiable fibers. No contamination was found from the fuel samples taken. This finding was not the major cause of this accident but is another major finding that can contribute because movement of the pressure-regulating valve in the right engine fuel control unit was restricted by contamination, thereby causing unstable stator vane actuator operation, engine starting difficulties, inconsistent topping settings, and unpredictable and degraded engine performance.

7 Aircraft Accidents and Incidents Systemic Analysis

This section provides an overview of aircraft accidents and incidents involving complex scenarios with fuel and fuel systems. It also describes investigation practices and techniques. Full scenarios of failure and safety incidents involving the fuel and fuel systems are presented as they unfolded in real events. This provides an opportunity to understand fuel system phenomena in the context of actual reality.

7.1 The Concept of System Complexity in Air Accident Investigation and Failure Analysis

An air accident investigation aims to find the root cause of the accident. However, fundamentally, accident investigation and failure analysis are about developing an understanding of a complex reality. When dealing with air accident investigations and failure analysis, there is a need to conceptualize what the system under investigation is. Of course, when we investigate the fuel system, it is simple–the investigation is within the borders of the fuel system. But we have to take a broader look to explain a complete aircraft failure. The fuel system is embedded in the airframe and operates in conjunction with other systems, for example:

- Fuel system components are fixed to the structure.
- The fuel is supplied to the engine in different regimes and is governed by a controlling system or the pilot (or both).
- The fuel system may depend on electrical power.
- The fuel system operates in different atmospheric conditions.
- The fuel is managed by ground personnel and by the pilot throughout the flight.
- The pilot's performance is subject to various physical and social (including organizational) factors.
- The fuel is supplied from various ground systems.
- The entire aircraft is licensed and maintained under certain regulations.

These are just a handful of examples, each of which can be further expanded. But they are aimed to illustrate the concept of complex interaction. It is fair to say that in every fuel system component (or any other system), numerous systemic interactions are embedded. These interactions materialized in the design of the component, and its integration into the system, and therefore, define its "sensitivity" to variations in different contextual factors (other aircraft systems, aerodynamic stability, human decision-making and performance, environmental factors, operational procedures, etc.).

The scope of the investigation or breadth of a definition of a system being investigated is a practical matter. For example:

- In the case of a small, privately owned aircraft accident, the definition of the system would be in the spectrum of the aircraft, the pilot, ground infrastructure, and regulation.
- In the case of an accident involving a large airliner, the definition of the system would encompass the global aviation ecosystem.

This applies to the generally accepted air accident investigation framework. But regardless of the complexity of the system, the investigating principle stays the same: determining the root cause of an accident is the functional goal of the system being investigated. The investigator must map and analyze the relevant component interactions to find it. These interactions can be purely technological, techno-human, or socio-technical. Understanding their interconnecting influences and patterns is understanding complexity. Because the aviation industry is generally organized, root causes of failures are phenomena to be found in system complexity.

7.2 Fuel Starvation

This is a highly common and acute accident scenario attributed to fuel-related issues. This scenario usually involves a complex blend of technical and human factors. Firstly, it is implied that there is fuel onboard the aircraft, but it is not feeding the engine, thus causing an engine flameout and subsequent loss of pilot control over the aircraft, leading to impact. Each stage in this sequence has its mechanism, both dependent and independent of others.

Let's first address the fuel itself. As was shown in previous paragraphs, there are various properties and anomalies of fuel as a substance that can prevent it from effectively feeding the engine, such as water and microbial contamination, or other solid or particle contamination. Contamination usually blocks the proper fuel flow to the engine, either blocking filters, injectors, or other sensitive components. The unfolding of events following this blockage depends on the engine fuel system design, the actions of the pilot, and the mission circumstances. In small aircraft, such fuel blockage may result in acute engine cuts. In these cases, whether the incident develops into an accident largely depends on the pilot's ability to effectively perform emergency procedures, such as maintaining airspeed, restarting the engine, or performing an emergency landing. In some cases, an emergency landing may be impossible due to terrain and topographic circumstances. Large commercial or military aircraft incorporate different system redundancies, such as multiple engines, independent fuel systems in the aircraft, filter bypasses, and so on, which allow fuel flow to continue, usually providing some indications to the pilot.

Interestingly, when investigating very redundant systems, there is still one single critical point of failure–the fuel itself. A vivid illustration of such a case is the BA Boeing 777 landing on January 17, 2008, in Heathrow after a long flight from Beijing (mentioned in Sect. 6.8). In this accident, the crew experienced a dual engine disruption during the final approach, probably caused by ice accumulation in the fuel lines. The ice source was "normal" condensation due to the low fuel and ambient temperature at cruising altitude. As the investigation showed, the steady flight regime at cruising altitude without much throttle manipulation (and therefore fuel flow) caused the ice to accumulate in the fuel lines. Upon descent towards the final approach, the ice became dislodged and blocked the fuel flow in the engine heat exchanger. Notably, this occurred almost simultaneously in both engines, which are separate independent systems. This issue resulted in a hard landing, landing gear collapse, and injuries with no fatalities (see Fig. 9). Interestingly, because the engines are operated at minimal power during the final approach, the crew's detection of the problem may not have been timely for effective emergency decision-making. However, the Engine Pressure Ratio (EPR) indicator began to show subnormal readings (below 1), indicating that both engines were not producing enough thrust. This, along with close monitoring of the approach, allowed the crew to maintain the speed above stall speed, despite losing altitude. Later, the crew adjusted the landing configuration (reduced drag by clicking up the flaps a notch) to gain enough ground distance to land within the airport, by undershooting the runway.

Fig. 9 The BA Boeing 777 at Runway 27L undershoot shortly after the accident [37]

Interestingly, initially, the investigators could not find any malfunction or failure in any of the aircraft's systems. Additionally, the quality of the fuel and ground fuel supply chain systems were thoroughly inspected, but no clues were found. The investigators had an important lead: signs of cavitation erosion in the outlet port of the high-pressure fuel pumps of both engines, Fig. 10. This evidence suggested that the outlet fuel pressure was, in fact, lower than the inlet, indicating that something was restricting the fuel flow.

Because the fuel was the only link between two independent engines and their respective fuel systems, the investigators constructed a large experimental facility to recreate the conditions of the accident. Eventually, they turned to the fuel icing scenario. It was found that Ice was accumulating inside the fuel pipes in a specific fuel temperature condition: between –8 °C (17.6 °F) and –20 °C (–4 °F). They termed this range "the sticky range", indicating that at this temperature, small particles of ice in the fuel begin to accumulate on the inner side of fuel pipes. Experiments conducted on a special test rig showed that accumulated ice, when detached in the course of fuel flow manipulations, could block the FOHE, Fig. 8.

In an airliner such as the Boeing 777, some indications alert the crew before the actual engine flameout occurs, but it can be very different in other scenarios.

Fuel starvation can be caused by various failures in the aircraft's fuel system or the engine fuel system, as broadly illustrated in the above examples. These failures include different fuel contaminations that block filters, fuel lines, etc., and are linked to fuel quality management, both inside the aircraft and in ground systems.

Another aspect of a fuel starvation scenario is the failure of critical components. A vivid example is an Israeli Air Force (IAF) A4 accident in 1998. During a regular training mission, the pilot experienced an abrupt engine shutdown. After a few restart

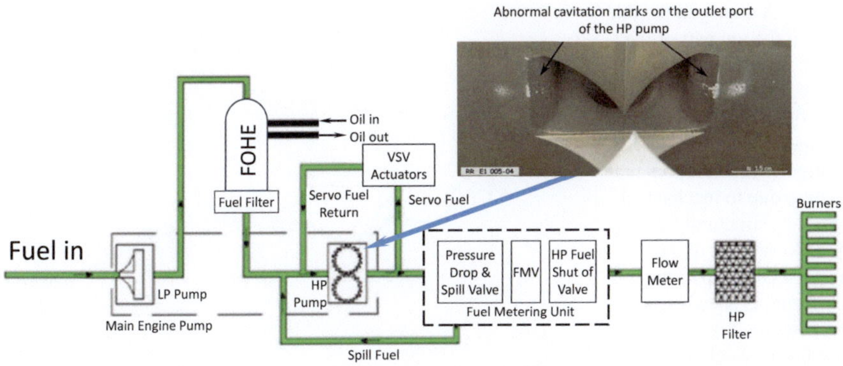

Fig. 10 Schematic figure (Left) [37]: a flow system of fuel. The system starts with the main engine pump, feeding fuel into the low-pressure (LP) pump. The fuel then flows through the FOHE. After passing through a fuel filter, the fuel is directed to a high-pressure (HP) pump. The HP pump has an outlet for spill fuel, which is returned to the system, and a servo fuel return line. Fuel continues to a Fuel Metering Unit (FMU) that includes a shut-off valve and a flow meter, and then on to the burners after passing through an HP filter. Cavitation Image (Right): The top right part of the image shows a close-up photograph of cavitation damage on the outlet port of the high-pressure (HP) pump

attempts, the pilot safely ejected. The crash site investigation revealed two things. First, their aircraft carried enough fuel, as evidenced by the massive spillage and a severe ground fire. Second, the damage to the engine by the impact was typical of a non-operating engine. Further investigation revealed a failure of the engine feed fuel line as a direct cause of the engine failure. The fuel line was of a flexible type and served as a connecting component between the engine and the aircraft fuel system. This flexible line was a braided hose, meaning that it held the fuel pressure using a braided metal wire covering. It was found that the failure of the hose was due to fatigue of that wire cover (see Fig. 11).

The fatigue gradually caused the braided layer to deteriorate until it could not hold the working pressure of the fuel. The cause of the fatigue, in turn, was a preload, an unwanted torquing of the hose caused by improper installation. Flexible hoses do not permit any torsion in installation and provide swivels as an interface between the hose and the fittings to relieve the torsion. Proper installation requires knowledge of that and practicing a specific installation technique. So, in this case, improper maintenance practices caused an aircraft loss and endangered the pilot. Fortunately, the hose failed during high altitude flight, making it easier for the pilot to perform necessary emergency procedures. Failure of this kind during takeoff, for example, would have been much more stressful, challenging, and dangerous.

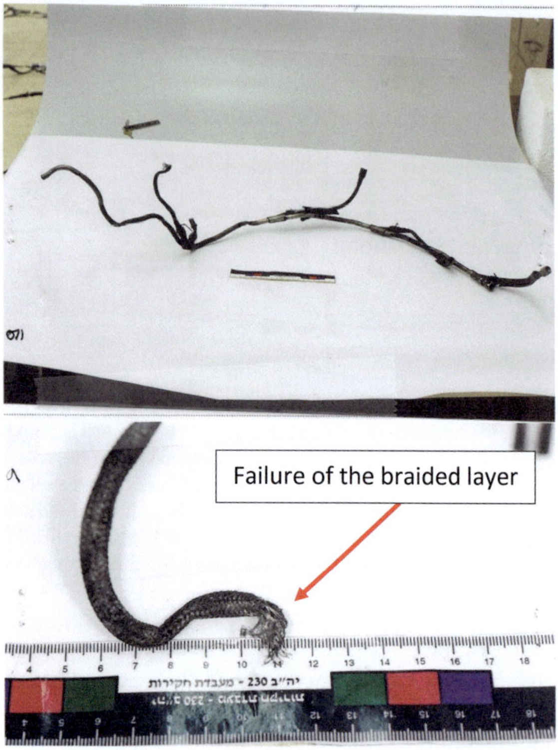

Fig. 11 A damaged flexible hose. In the top part of the image, the entire hose is visible, twisted and showing signs of significant wear and damage. The hose has undergone deformation and twisting due to mechanical impact or structural failure. In the lower part of the image, a close-up of the hose's end is shown, revealing the braided metal wire layer, which has deteriorated and been damaged. The braided layer appears to have lost its ability to contain the fuel pressure

7.3 Fuel Exhaustion-Accidents in Which Aircraft have Become Devoid of Usable Fuel

Fuel exhaustion (having no usable fuel remaining on the aircraft) failure scenarios are highly dependent on the human element, often involving a fuel quantity miscalculation. Instances of fuel exhaustion leading to air crashes are rare in large airliners but more common in small aircraft. This discrepancy is not due to the size of the aircraft but rather attributable to standard operating procedures (SOPs). It is a standard practice for airlines to uplift fuel sufficient for the planned flight leg, plus an airline-specified amount or percentage of fuel in excess to account for possible weather problems en route. Additional fuel may also be added to account for engine-out emergency diversion to alternative airports. While airliners are not immune to fuel exhaustion, these SOPs help reduce the possibility of it happening.

Most fuel exhaustion accidents occur during less structured missions in environments where a single pilot does "ad hoc" planning and implementation. In these cases, the pilot may uplift only enough fuel to fly point-to-point, or decisions made during the flight may push the system beyond its limits.

In other cases, technology—or the lack thereof—can contribute to fuel exhaustion. In 2020, an aircraft—a self-built kit double-decker—flamed out during flight within the CTR (controlled traffic region) of one of Israel's airports [42]. Fortunately, the pilot managed to perform an emergency landing, damaging the aircraft without sustaining any injuries. The fuel gauges in the aircraft were essentially clear tubes in which the fuel level could be visible, as shown in Fig. 12.

Because the aircraft had a tail wheel configuration, the actual fuel level indication was only possible when lifting the tail and leveling the aircraft. This process requires two people–one to lift the tail and one to check the fuel level, as shown in Fig. 13.

The investigation revealed that the flight was performed without sufficient fuel quantity for the planned flight duration. One might ask, how did it come to this? At first glance, it does not seem very logical for a pilot to take off without fuel. However, complex system analysis is based on the premise that there is an inner rationale in every system's behavior. The aircraft was owned by a few partners who flew recreational flights now and then, and not very often. The flights were relatively short, as revealed by the investigation.

The last time the aircraft was refueled was approximately a year before the accident. So, probably, the pilot had a concept formed that there should be enough fuel because refueling is a rare event. This, in combination with a very crude fuel gauge and the lack of standard procedure for fuel consumption calculation, led to the pilot's unawareness of the actual fuel quantity on board. This gradual acceptance of partial or insufficient means is known as the "normalization of deviance".

Another example of a fuel quantity management mistake is the Israeli Robinson helicopter incident, which occurred in May 2015, as shown in Fig. 14.

The helicopter commercially transported people to and from a religious event in northern Israel. During the last planned leg, a "fuel low" light illuminated the cockpit. The helicopter was just minutes from its destination. Still, the pilot decided to land at

Fig. 12 The plane fuel quantity gauge [42]

Fig. 13 A self-built kit double-decker aircraft. The aircraft is not leveled when standing on the ground [42], (Image courtesy of Micha Sender)

Fig. 14 A Robinson helicopter similar to the one from the incident that occurred in May 2015 [43], (Image courtesy of GB Grunfeld)

a nearby international airport because, upon the illumination of the “fuel low” light, there were approximately 3 min of flight available.

The pilot had initially calculated the fuel amount required for the mission but did not include sufficient spare fuel. The onboard quantity indicator, as shown in Fig. 15, is an analog gauge, making it challenging to accurately monitor fuel quantity, particularly when tanks are nearly empty. An important safety feature of helicopters

Fig. 15 The Robinson helicopter fuel quantity gauge. An analog gauge, making it challenging to monitor fuel quantity accurately [43]

is their ability to land immediately, although this may not always be feasible, such as when flying above mountains, forests, or dense urban areas.

The combination of events is as follows. A private commercial flying, lacking clear SOPs to ensure schedule flexibility, coupled with air vehicle instrumentation that does not allow for close monitoring of fuel quantity, can create a critical situation for the safety of flight. Therefore, implementing a robust plan to carry spare fuel is a safe strategy.

7.4 Fuel as Weight and Balance Factor

Accidents where fuel affected aircraft aerodynamic stability in flight or plays a role during takeoff or landing are significant. Fuel is a key factor in determining any aircraft's takeoff weight and aerodynamical stability. The weight of the fuel can impact the aircraft's aerodynamic stability during flight or become a critical factor during takeoff or landing. However, aircraft designers typically consider these factors to ensure safety and performance, as described in the aircraft weight and balance handbook by the FAA [44], see the cover on Fig. 16.

The FAA's Aircraft Weight and Balance Handbook states: "Weight is a major factor in airplane construction and operation, and it demands respect from all pilots and particular diligence by all maintenance personnel. Excessive weight reduces the efficiency of an aircraft and the available safety margin if an emergency condition should arise" [44].

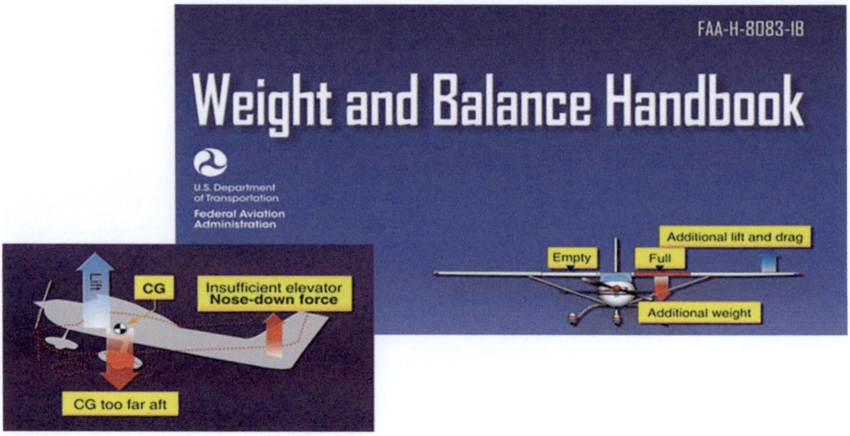

Fig. 16 The aircraft weight and balance handbook by the FAA

All aircraft have defined maximum takeoff weights; fuel calculations must adhere to these limits. Attempting takeoff with an overloaded aircraft can lead to failure to reach takeoff speed, resulting in a runaway overrun or a stall and crash. Such incidents often involve pilot miscalculations of either the total takeoff weight or runway length, influenced by human and organizational factors.

7.4.1 Piper PA-46-350P Accident: Crash After Takeoff, Chofu City, Tokyo Metropolitan, Japan at Around 10:58 JST, July 26, 2015

The aircraft crashed into a private house shortly after takeoff from Runway 17 of Chofu Airport, see Fig. 17. The crash resulted in the death of the captain and one passenger, with three passengers sustaining serious injuries. Additionally, one resident died, and two others suffered minor injuries. The aircraft was destroyed, and a fire broke out, engulfing the house and causing damage to neighboring properties.

The investigation revealed that the aircraft exceeded the maximum take-off weight by approximately 58 kg (fully fueled and overloaded with passengers and baggage). This caused the aircraft's center of gravity to shift aft in its range of movement, resulting in a high nose attitude and difficulty in control. The takeoff was performed without reaching the required speed (approximately 5 knots) because of runway length. Interestingly, a failure scenario of this kind is almost entirely predetermined before the pilot pushes the throttle forward. It all relies on the pilot's ability to gather data, perform accurate calculations, and exercise proper decision-making. If take-off parameters are not met, the take-off should not be attempted. Fuel can, for that matter, always be traded off with other payloads, but that results in range reduction. An aircraft overweight and with poor CG location, if the takeoff is accomplished, makes subsequent flight choices challenging. The pilot must recognize the insufficient speed and rate of climb promptly, push the nose down to level flight, and build up airspeed. If

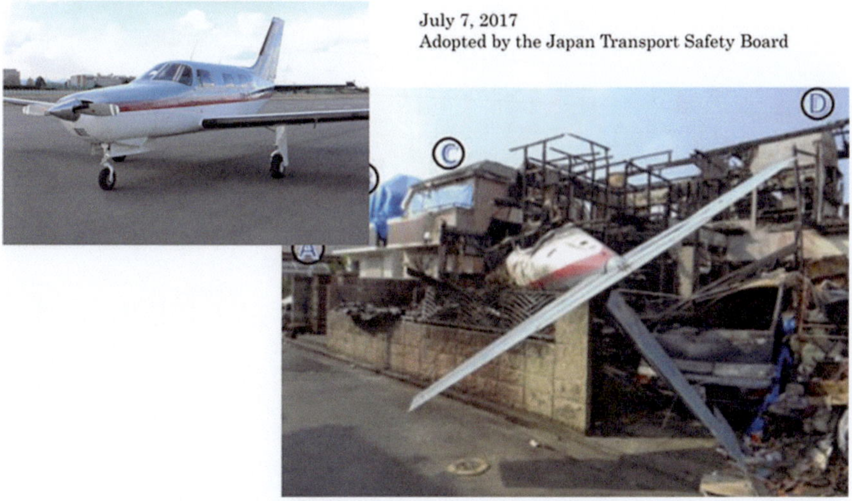

Fig. 17 The Piper PA-46-350P accident crash site Top Left Image: the aircraft in good condition, on a runway. Bottom Right Image: the aircraft after a crash. The aircraft appears severely damaged and partially burned, resting amidst the wreckage of a building [45]

this does not work, the only option is to perform an emergency landing straight ahead at whatever location is available. This is to avoid a stall because, with insufficient airspeed, an attempt to turn will make the aircraft uncontrollable. In this accident, the investigation concluded that the pilot "prioritized climb over speed". This suggests that the pilot either did not understand the situation or could not plan and implement emergency measures.

Building and maintaining proficiency by private pilots is a worldwide challenge. Without supporting requirements and organizational infrastructure, it is up to every individual pilot to maintain proficiency and demonstrate adequate decision making, even under mission pressure.

7.4.2 C-21A Accident: Crash During Emergency Landing, Alexander City, Alabama, US, April 17, 1995 [46]

On April 17, 1995, another example of an accident caused by fuel impacting weight and balance occurred, involving a C-21A (military version of the Learjet 35A) (see Fig. 18). The aircraft was operated by the US Air Force. The crew declared an emergency because of a fuel problem and diverted the flight to Thomas C. Russell Field in Alexander City, Alabama, US. The crew was maneuvering the aircraft for a visual approach to the airport when the aircraft struck terrain. The two flight crew members and six passengers were killed.

According to the US Air Force Investigation, an electric fuel pump malfunction resulted in fuel being pumped into the left wing but prevented fuel from being

Fig. 18 C-21 Learjet, Photographer: Senior Master Sgt. David Lipp [47]

transferred to the right wing during normal transfer procedures. In fact, due to this malfunction, the fuel would also transfer from the right wing to the left wing. The flight crew incorrectly believed that fuel in the left wing had become "trapped" and that both engines were using fuel from the right wing.

This condition is caused by a severe imbalance. When the crew used normal procedures to balance the wing tanks, the fuel imbalance worsened. After considerable, unsuccessful efforts by the crew to balance the aircraft (including dumping fuel), the decision was made to divert to an alternate airfield for landing. During the emergency landing attempt, the aircraft lost its lateral stability and impacted the ground.

7.5 Fire and Explosion

7.5.1 Inflight Fire and Explosion

The flammability of fuel is a constant risk in aviation due to its inherent properties. By design, fuel is isolated from sources of heat and ignition. Failure in that isolation seriously threatens the safety of crew and passengers and the integrity of the aircraft.

The flammability of aircraft fuels is defined by properties such as vapor pressure, flash point, fire point, minimum ignition energy, auto ignition temperature, and more. All these parameters interact in a complex manner; thus, fire is a complex process controlled by both chemical and physical factors. Fires as failures involving air

Fig. 19 The aircraft engulfed in flames. The accident was filmed from the terminal [48]

operations and aviation fuel can occur under certain heat balance and sustainable environmental conditions.

An accident of this kind occurred in Naha airport on August 20, 2007. A China Airlines Boeing 737-800, flight 120, landed and caught fire upon reaching a stop at the terminal. All 165 crew and passengers were safely evacuated, but the aircraft was destroyed (See Fig. 19).

An investigation conducted by the Japan Transport Safety Board revealed that the fire originated from a fuel spillage caused by an improperly installed leading-edge slat mechanism, resulting in a loose bolt rolling around inside the leading-edge compartment in the right wing. This compartment is isolated from the adjacent fuel compartment of the wing. The leading-edge slats are extended before landing to increase wing lift at low speeds and retracted after landing is completed. During this retraction, the leading-edge main track in the right-wing pushed the loose bolt through the wing's structure, breaching the fuel compartment and causing the fuel to flow out of the aircraft. As the aircraft was taxiing, the spilled fuel was washed away from the aircraft, but when the aircraft reached a full stop at the terminal, the fuel splashed onto the right engine's hot exhaust pipe and ignited (see Fig. 20). The crew was informed about the fire and initiated an emergency evacuation. In approximately a minute, the fire caused an explosion in the right wing, eventually resulting in the loss of the aircraft. It is well-known that fire in an aircraft can progress very rapidly, posing significant survivability issues for crew and passengers. A crew's ability to carry out evacuation within seconds could make the difference between simply losing the aircraft and experiencing a humanitarian disaster. This accident serves as an example of the emergent nature of a system's complexity. In regular linear understanding, a faulty maintenance task of installing a leading-edge mechanism should only affect the operation of that mechanism. However, as reality shows, everything in complex systems is interconnected in too many ways to be forecasted.

When uncontrolled fuel ignition occurs in a confined compartment of the aircraft, the pressure wave generated from the combustion can compromise the aircraft's structure–in simple terms, it can cause an explosion. The conditions in aircraft fuel tanks are of concern in such explosion scenarios. Ullage is the gaseous space above

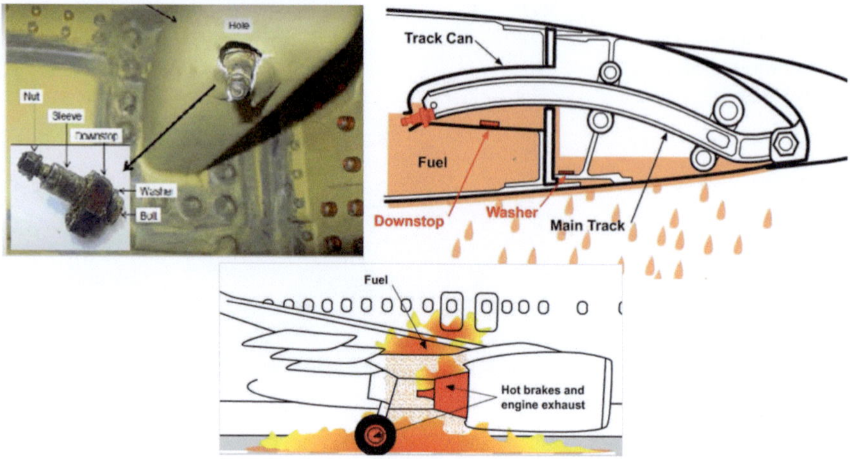

Fig. 20 The image consists of three key illustrations. Top Left Section: A close-up image showing a nut, sleeve, washer, and bolt assembly. Top Right Section: A schematic diagram showing how fuel can leak from a section of the aircraft's wing structure. The diagram shows fuel leaking out of the wing track, potentially due to an improperly secured washer, and dripping out of the assembly. Bottom Section: A drawing showing a scenario where fuel has leaked onto the aircraft's hot brakes and engine exhaust [48]

the liquid fuel that contains fuel vapor and air. The tank is vented to equalize the pressure with the outside. As the tank breathes, the ullage vapor concentration changes. During the climb, the total pressure decreases, and vapor concentration increases. The opposite occurs during decent. Thus, flammability limits vary with altitude. Ullage flammability is mainly governed by fuel temperature and altitude.

An important milestone in understanding the risk of in-flight fuel explosion was the investigation and subsequent studies of the TWA-800 accident. This accident involved a Boeing 747-100 that exploded and crashed into the Atlantic Ocean near East Moriches, New York, off New York's Long Island, on July 17, 1996, [49].

The NTSB investigation determined that the probable cause was: "an explosion of the center wing fuel tank (CWT), see Fig. 21, resulting from ignition of the flammable fuel/air mixture in the tank. The source of ignition energy for the explosion could not be determined with certainty, but of the sources evaluated by the investigation, the most likely was a short circuit outside of the CWT that allowed excessive voltage to enter it through electrical wiring associated with the fuel quantity indication system". Wiring as either a part of the fuel system or just routed through the tank is common on many aircraft. One of the final NTSB report findings was: "Existing standards for wire separation may not provide adequate protection against damage from short circuits".

A study on the flammability of the fuel conditions in the CWT of TWA 800 was conducted back in 1997 [51]. It was noted: "On some flights, this tank is often empty except for whatever residual fuel (50 to 100 gal) remains from the previous use. This was the case at TWA 800. The tank is vented to the atmosphere and the volume of the

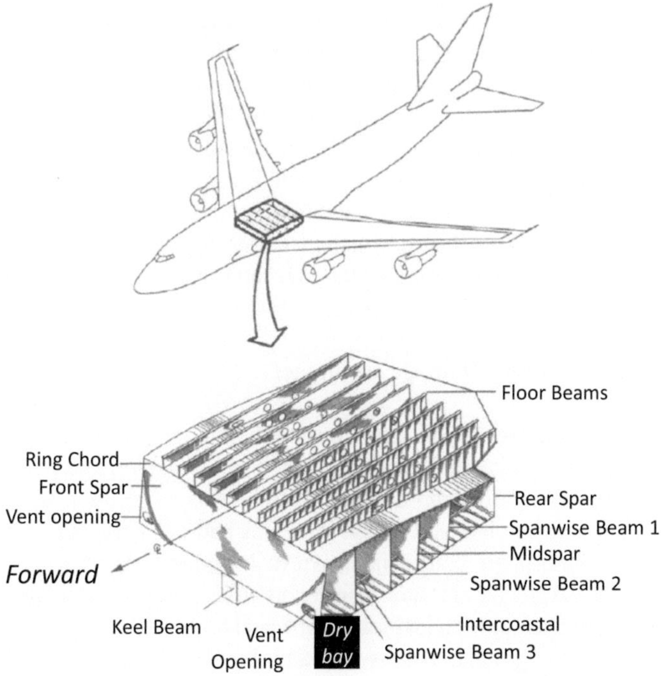

Fig. 21 A cross-section of the Boeing 747-100 wing center section [50]

tank above the fuel (ullage) contains mainly air. The vaporization and mixing of the residual fuel into this air can result in the formation of an explosive mixture within the tank. A main concern is the range of conditions under which Jet A-air mixtures can be ignited, i.e., are flammable. The key question is: What was the composition and concentration of fuel vapor within the tank at the time of the explosion? The answer depends on the entire history of the fuel up to the time of the explosion".

The study traced the history of the fuel since the refueling of the CWT in Greece for the last leg before the accident and conducted a series of fuel property tests in different environmental conditions (such as fuel temperature changes inside the tank, changes in ambient pressure during the climb to 15,000 feet, expected vapor pressure, etc.). The study concluded: "The fuel vapor in the ullage of the TWA 800 CWT was flammable at the time of the explosion and the estimated peak pressures are sufficiently high that structural failure is a credible consequence of flame propagation within the tank ullage".

The NTSB investigation concluded that the fuel tank flammability issue reflected an awareness of the industry's insufficient systemic approach. The investigation stated, "A fuel tank design and certification philosophy that relies solely on the elimination of all ignition sources, while accepting the existence of fuel tank flammability, is fundamentally flawed because experience has demonstrated that all possible ignition sources cannot be predicted and reliably eliminated" [52].

The TWA 800 accident is one in which the crew could not have done anything to avoid the tragic consequence. The entire event transitioned from a routine flight to a structural breakdown and uncontrollable situation in one instance.

Many commercial airliners aircraft, especially Boeing 737 models, now have fuel tank inerting systems. These systems are designed to lower the flammability of the fuel tank ullage volumes by increasing the nitrogen levels in the air within these spaces. OBIGGS (Onboard Inert Gas Generation System) units remove Oxygen from the ullage air, dumping it overboard, which increases the Nitrogen content of that air to well in excess of 90%, thus reducing the fuel vapor flammability in the tank [53].

7.5.2 Fire as Secondary Damage on Impact

Fuel plays a primary role in post-crash fires. When an aircraft disintegrates upon impact, fuel and other flammable materials, such as lubricating and hydraulic oils, are released into the air, forming a white cloud of vaporizing fuel. If not ignited, this cloud is sometimes observed by eyewitnesses and cameras. However, a rapid release into the atmosphere causes intensive fuel mixing with oxygen, forming a highly flammable explosive mixture. During an aircraft impact accident, there are many potential ignition sources, including hot wheel brakes, engine parts, and sparks from friction with the ground. This is why a high potential for post-crash fire risk exists.

A famous attempt to reduce this risk was made by NASA and the FAA in 1984 using a controlled impact demonstration (CID), which involved the deliberate crash of an unmanned Boeing 720 aircraft to demonstrate a reduction of post-crash fires through the use of anti-misting fuel [54]. This was achieved by adding a high molecular weight long-chain polymer to the fuel. The additive, ICI's FM-9 blended with Jet-A fuel, forms an anti-misting kerosene (AMK), which is solid when quiescent and a liquid when agitated. AMK had demonstrated the capability to inhibit ignition and flame propagation of the released fuel in simulated impact tests. However, this had to be compensated for by a special mechanical mechanism—a degrader—that would cause the high molecular-weight long-chain polymer to break down (liquefy) before the fuel was fed into the engines. This was necessary because AMK, as a solid, cannot be introduced directly into a gas turbine engine due to its inability to vaporize and form a combustible fuel–air mixture. The actual experiment was performed on December 1, 1984: (see Fig. 22).

The experiments showed that the actual impact demonstrated that the anti-misting additive tested was not sufficient to prevent a post-crash fire in all circumstances, though the reduced intensity of the initial fire was attributed to the effect of AMK [54]. The development of safer fuels continues to be an important challenge.

Fig. 22 These images are a series of six photographs depicting a controlled crash test of a Boeing 707. Top Left: The aircraft is airborne, moments before touching down. It seems to be in the process of an emergency landing or a controlled crash. Top Right: The aircraft contacts the runway, with the rear section beginning to lower onto the ground. Middle Left: The plane slides across the ground, creating a large cloud of smoke or dust as its crashes at high speed. Middle Right: Fire begins to engulf the plane as it slides along the runway, with flames starting to spread, though the aircraft remains relatively intact. Bottom Left: A massive fireball erupts from the plane's rear, with thick black smoke billowing into the sky. Bottom Right: The aircraft is fully engulfed in flames and heavy smoke, nearly destroyed, with few intact sections remaining [55]

7.5.3 Prevention of Fuel Tank Explosions by Explosion Suppressive Materials

Reticulated polyurethane foams have a unique purpose within aircraft fuel tanks, serving to suppress explosions and mitigate surge and noise in situations where the risk of a fuel tank explosion exists or where surge and noise control is required. Explosion Suppressant Foams (ESF) find practical application in both stationary and portable fuel storage and containers. Military aircraft are highly vulnerable to fires and explosions due to combat threats, particularly gunfire, including High-explosive Incendiary (HEI) rounds. During the late 1960s, the United States Air Force (USAF) began using reticulated polyester polyurethane foam to counteract fires and explosions within fuel tanks. However, it was soon discovered that those materials were prone to hydrolytic degradation. To address this issue, improved materials were developed, resulting in what is commonly known as "blue foam". Although blue foam exhibited better hydrolytic stability, it also possessed significantly higher electrical resistance properties compared to the original foam materials, leading to a capacitance effect and static charge potentials exceeding 10,000 V. Following the integration of the blue foam kits, the USAF encountered fuel tank fires in A-10 and C-130 aircraft. Despite discovering numerous remnants of fuel tank fires in the C-130 fleet, no aircraft losses were directly attributed to fuel tank fires. This static electrical discharge issue prompted the development of conductive foams, which meet the specifications outlined in MIL-PRF-87260. These conductive foams are now being manufactured and installed in numerous USAF and USN aircraft.

Acknowledgements We would like to thank Joel Schmitigal and Stan Seto for the chapter's deep professional review.

References

1. ICAO (1944) Convention on international civil aviation—Doc 7300. ICAO.int
2. ICAO (2018) Annex 8—Airworthiness of aircraft. International Civil Aviation Organization
3. ICAO (2014) Airworthiness manual, 3rd edn, Doc 9760. ICAO.int
4. ASTM F3063 Standard Specification for Aircraft Fuel Storage and Delivery, ASTM International, West Conshohocken, PA, https://doi.org/10.1520/F3063_F3063M-21
5. SAE International (2021) Aircraft fuel system design guidelines
6. SAE International (2018) SAE MAP749: Aircraft turbine engine fuel system component endurance test procedure. SAE International
7. SAE International (2020) Fuel systems spec and standards. SAE International
8. Energy Institute, & Joint Inspection Group, EI 1550: Handbook on equipment used for the maintenance and delivery of clean aviation fuel. Energy Institute (2019)
9. Joint Inspection Group (JIG), Bulletin No. 132 Phase out of filter monitors from the JIG Standards, Standards & Publications (2020)
10. National Transportation Safety Board (NTSB) (1993) Aviation accident final report: LAX92MA183, Perris Valley, California. National Transportation Safety Board, Washington, D.C.
11. Ministry of Transport and Road Safety (2016) Final safety investigation report—engine failure during flight due to water in fuel: Mooney M20J, 4X-ARO, December 23, 2016 (Accident File No. 99-16) (Hebrew). Ministry of Transport, State of Israel
12. National Transportation Safety Board (2017) Aviation accident final report: WPR16LA021, Sedro-Woolley, Washington. NTSB, Washington, D.C.
13. Civil Aviation Department, Hong Kong Special Administrative Region, Report on the accident to Airbus A330-342 B-HLL operated by Cathay Pacific Airways Limited at Hong Kong International Airport, Hong Kong on 13 April 2010, Hong Kong (2013)
14. ASTM International. Standard test method for free water, particulate and other contamination in aviation fuels (visual inspection procedures). ASTM International
15. Australian Transport Safety Bureau (ATSB). Aviation safety investigation report 200200007: government aircraft factories N22C nomad, fuel exhaustion. Australian Transport Safety Bureau, Canberra (2002)
16. IATA SAP Special Interest Group. IATA super-absorbent polymer (SAP) special interest group—data summary and proposed roadmap. International Air Transport Association (IATA) (2017)
17. Bessee GB (2010) Analysis of debris from helicopters from the field: interim report TFLRF No. 402, U.S. Army TARDEC, Southwest Research Institute (SwRI), San Antonio, TX

18. ISO (2019) 22241-1:2019 Diesel engines—NOx reduction agent AUS 32—Part 1: quality requirements
19. Federal Aviation Administration (FAA), Safety Risk Management Team. Safety risk assessment report: jet fuel contaminated with diesel exhaust fluid (DEF). Federal Aviation Administration (FAA), Washington, D.C.
20. ASTM International. Standard specification for aviation turbine fuels
21. Federal Aviation Administration (FAA) (2019) Aviation maintenance technician handbook—airframe, vol 2. FAA
22. Air Accidents Investigation Branch (AAIB). Special bulletin S1/2020: Airbus A321-211, G-POWN, fuel system contamination—Kathon biocide overdose, Crown Copyright
23. Maxwell S (1970) Aviation fuels. G. T. Foulis & Co. Ltd.
24. National Transportation Safety Board (NTSB) (2000) Aviation investigation final report, accident number: DEN06LA063. National Transportation Safety Board, Washington, D.C.
25. A. A. I. B. (AAIB), AAIB bulletin 11/2017. Department for Transport, United Kingdom, Aldershot, Hampshire (2017)
26. National Transportation Safety Board (NTSB) (1998) Aviation investigation final report, accident number: SEA98FA109. National Transportation Safety Board (NTSB), Washington, D.C.
27. E. C. C. A. A. (ECCAA), Final accident report: VP-MON, Britten-Norman BN2A-26 Islander, ECCAA, Antigua
28. Agenzia Nazionale per la Sicurezza del Volo (ANSV) Final Accident Report: ATR 72-202 TS-LBB, ditching off the coast of Capo Gallo (Palermo—Sicily). Agenzia Nazionale per la Sicurezza del Volo, Rome, Italy (2005)
29. National Transportation Safety Board (NTSB) (2017) Aviation investigation final report, accident number: WPR17LA014. National Transportation Safety Board, Washington, D.C.
30. National Transportation Safety Board (NTSB) (2002) Aviation investigation final report, accident number: CHI02LA166. National Transportation Safety Board, Washington, D.C.
31. National Transportation Safety Board (NTSB) (2007) Aviation investigation final report, accident number: ANC08LA004. National Transportation Safety Board (NTSB), Washington, D.C.
32. National Transportation Safety Board (NTSB) (1999) Aviation investigation final report, accident number: LAX00LA002. National Transportation Safety Board (NTSB), Washington, D.C.
33. National Transportation Safety Board (NTSB) (2000) Aviation investigation final report, accident number: LAX01LA012. National Transportation Safety Board (NTSB), Washington, D.C.
34. National Transportation Safety Board (NTSB) (2007) Aviation investigation final report, accident number: SEA07LA066. NTSB, Washington, D.C.
35. Aircraft and Railway Accidents Investigation Commission (ARAIC) (2006) Aircraft accident investigation report: Piper J3C-65 G-KIRK. Ministry of Land, Infrastructure and Transport, Tokyo, Japan
36. National Transportation Safety Board (NTSB) (2008) Aviation investigation final report, accident number. NTSB, Washington, D.C.
37. Air Accidents Investigation Branch (AAIB) (2010) Aircraft accident report 1/2010: report on the accident to Boeing 777-236ER, G-YMMM, at London Heathrow Airport on 17 January 2008. Department for Transport
38. National Transportation Safety Board (NTSB) (2008) Aviation incident final report: DCA09IA014, Boeing 777-200 engine rollback. NTSB, Washington, D.C.
39. Horne TA (2021) System rundown: focus and engine-outs. AOPA
40. National Transportation Safety Board (NTSB) (2003) Aviation accident final report: FTW03FA118, Bell 407, crash during aerial application flight. NTSB, Washington, D.C.
41. Transportation Safety Board of Canada (2002) Loss of engine power/collision with tree—Hayes Helicopter Services Limited Sikorsky S-61N (Shortsky) Helicopter C-FHHD, Lake Errock, British Columbia. TSBC, Gatineau, Quebec

42. The Chief Investigator, Ministry of Transport, Israel (2021) Final safety investigation report on fuel exhaustion, engine failure, and emergency heavy landing (report no. 40-20) (Hebrew). Ministry of Transport and Road Safety, Israel
43. Chief Investigator, Ministry of Transport, Israel (2015) Final safety investigation report on low fuel warning and emergency landing at Ben Gurion Airport (report no. 43-15) (Hebrew). Ministry of Transport and Road Safety, Israel
44. Federal Aviation Administration (2016) Aircraft weight and balance handbook, vols. FAA-H-8083-1B. U.S. Department of Transportation
45. Aircraft and Railway Accidents Investigation Commission (ARAIC), Japan (2017) Final aircraft accident report: Piper PA-46-350P Malibu Mirage JA4060 Crash in Chofu City, Tokyo. Aircraft and Railway Accidents Investigation Commission (ARAIC), Japan
46. National Transportation Safety Board (NTSB) (1995) Aviation accident final report: C-21A Learjet Crash, Alexander City, Alabama, Washington, D.C.
47. Official United States Air Force Website (n.d.). "https://www.af.mil/News/Photos/igphoto/2002864307/mediaid/5476938/
48. Aircraft and Railway Accidents Investigation Commission (ARAIC). https://www.faa.gov/lessons_learned/transport_airplane/accidents/B18616
49. Planefilms, TWA Flight 800 B747: How the accident happened. Planefilms (2021). https://www.planefilms.com/2021/01/twa-flight-800-b747-how-accident.html
50. National Transportation Safety Board (NTSB) (1996) In-flight breakup over the Atlantic Ocean: trans World Airlines Flight 800, Boeing 747-131, Near East Moriches, New York. National Transportation Safety Board, Washington, D.C.
51. Shepherd JE (1977) Jet A explosion experiments: laboratory testing. Graduate Aeronautical Laboratories California Institute of Technology
52. Cramoisi G (2013) Air crash investigations: a disastrous spark—the crash of TWA 800. Lulu.com
53. Reynolds LT, Eklund IT, Haack AG (2001) Onboard inert gas generation system/onboard oxygen gas generation system (OBIGGS/OBOGS) study: Part II: Gas separation technology—state of the art, national aeronautics and space administration. Glenn Research Center
54. Federal Aviation Administration & NASA (1987) Full-scale transport controlled impact demonstration program. U.S. Department of Transportation
55. NASA (2004) Controlled impact demonstration (CID) photo gallery. https://www.dfrc.nasa.gov/Gallery/Photo/CID/index.html.

Index

M. Rabaev et al. (eds.), *Aviation Fuel Problems and Solutions: From Refinery to Wingtip*, https://doi.org/10.1007/978-3-032-05514-9

G

H

I

MIX
Papier aus verantwortungsvollen Quellen
Paper from responsible sources
FSC® C105338
FSC
www.fsc.org

If you have any concerns about our products,
you can contact us on
ProductSafety@springernature.com

In case Publisher is established outside the EU,
the EU authorized representative is:
Springer Nature Customer Service Center GmbH
Europaplatz 3, 69115 Heidelberg, Germany

Printed by Libri Plureos GmbH
in Hamburg, Germany